D1789245

AVIAN GUT FUNCTION IN HEALTH AND DISEASE

Poultry Science Symposium Series
Volumes 1–28

*Out of print
Volumes 1–24 were not published by CABI. Those still in print may
be ordered from:

Carfax Publishing Company
PO Box 25, Abingdon, Oxfordshire OX14 3UE, UK

Avian Gut Function in Health and Disease

Poultry Science Symposium Series
Volume Twenty-eight

Edited by

G.C. Perry

Department of Clinical Veterinary Science, University of Bristol, UK

www.cabi.org

CABI is a trading name of CAB International

CABI Head Office
Nosworthy Way
Wallingford
Oxon OX10 8DE
UK

Tel: +44 (0)1491 832111
Fax: +44 (0)1491 833508
E-mail: cabi@cabi.org
Website: www.cabi.org

CABI North American Office
875 Massachusetts Avenue
7th Floor
Cambridge, MA 02139
USA

Tel: +1 617 395 4056
Fax: +1 617 354 6875
E-mail: cabi-nao@cabi.org

© CAB International 2006. All rights reserved. No part of this publication may
be reproduced in any form or by any means, electronically, mechanically, by
photocopying, recording or otherwise, without the prior permission of the
copyright owners.

A catalogue record for this book is available from the British Library, London, UK.

A catalogue record for this book is available from the Library of Congress,
Washington DC, USA.

ISBN-10:   1-84593-1807
ISBN-13:   978-1-84593-1803

Produced and Typeset in 10/12pt Souvenir Light by Columns Design Ltd, Reading
Printed and bound in the UK by Biddles, Kings Lynn

CONTENTS

CONTRIBUTORS

T. Acamovic, *Avian Science Research Centre, Scottish Agricultural College, West Mains Road, Edinburgh, EH9 3JG, UK; (address for communication) SAC, Ayr Campus, Ayr, KA6 5HW, UK; e-mail: Thomas.Acamovic@sac.ac.uk*

J. Apajalahti, *Alimetrics Ltd, Höyläämötie 1, FIN-00380, Helsinki, Finland; e-mail: juha.apajalahti@alimetrics.com*

S.S. Árnason, *Department of Physiology, University of Iceland, IS-101 Reykjavik, Iceland*

R.K. Beal, *Division of Immunology and Immunopathology, Institute for Animal Health, Compton, Newbury, Berkshire, RG20 7NN, UK*

M.R. Bedford, *Zymetrics, Chestnut House, Beckhampton, Marlborough, Wiltshire, SN8 1QJ, UK; e-mail: Michael.Bedford@onetel.net*

M. Bisgaard, *Department of Veterinary Pathobiology, The Royal Veterinary and Agricultural University Stigbojlen 4, DK-1870, Frederiksberg C., Denmark*

S.C. Bishop, *Roslin Institute, Midlothian, EH25 9PS, UK; tel: +44 (0)131 527 4200; fax 44 (0)131 440 0434 ; e-mail: stephen.bishop@bbsrc.ac.uk*

D.P. Blake, *Institute for Animal Health, Compton, Newbury, Berkshire, RG20 7NN, UK; tel: 44 (0)1635 577293; fax: 44 (0)1635 57726; e-mail: damer.blake@bbsrc.ac.uk*

L. Bohez, *Department of Pathology, Bacteriology and Avian Disease, Research Centre on Veterinary Food Safety and Zoonoses, Faculty of Veterinary Medicine, Ghent University, Salisburylaan, 133, B-9820, Merelbeke, Belgium*

F. Boyen, *Department of Pathology, Bacteriology and Avian Disease, Research Centre on Veterinary Food Safety and Zoonoses, Faculty of Veterinary Medicine, Ghent University, Salisburylaan, 133, B-9820, Merelbeke, Belgium*

M.S. Chadfield, *Department of Veterinary Pathobiology, The Royal Veterinary and Agricultural University, Stigbojlen 4, DK-1870, Frederiksberg C., Denmark*

J.P. Christensen, *Department of Veterinary Pathobiology, The Royal Veterinary and Agricultural University, Stigbojlen 4, DK-1870, Frederiksberg C, Denmark; e-mail: jpc@kvl.dk*

K. Cole, *Department of Poultry Science, University of Arkansas, Fayetteville, Arkansas 72701, USA*

S.R. Collett, *The University of Georgia, College of Veterinary Medicine, Poultry Diagnostic and Research Center, 953 College Station Road, Athens, Georgia, 30602-4875, USA; e-mail: colletts@uga.edu*

I.F. Connerton, *Division of Food Sciences, School of Biosciences, University of Nottingham, Sutton Bonington Campus, Loughborough, LE12 5RD, UK; tel: +44 (0)115 9516119; fax: +44 (0)115 9516162; e-mail: ian.connerton@nottingham.ac.uk*

P.L. Connerton, *Division of Food Sciences, School of Biosciences, University of Nottingham, Sutton Bonington Campus, Loughborough, LE12 5RD, UK; tel: +44 (0)115 9516119; fax: +44 (0)115 9516162*

R.G. Cooper, *Department of Physiology, University of Zimbabwe, Mount Pleasant Drive, Harare, Zimbabwe*

A. Corzo, *Department of Poultry Science, Mississippi State University, Box 9665, Mississippi 39762,USA; fax: 662 325 8292*

T.F. Davison, *Formerly Division of Immunology and Immunopathology, Institute for Animal Health, Compton, Newbury, Berkshire, RG20 7NN, UK; e-mail: fred.davison@dunelm.ac.uk*

R.A.H.M. ten Doeschate, *ABNA Ltd, ABNA House, Oundle Road, Peterborough, PE2 9PW, UK; e-mail: rtendoeschute@abn.co.uk*

A.M. Donoghue, *Poultry Production and Product Safety Research Unit, ARS, USDA, Fayetteville, Arkansas, USA*

D.J. Donoghue, *Department of Poultry Science, University of Arkansas, Fayetteville, Arkansas 72701, USA*

R. Ducatelle, *Department of Pathology, Bacteriology and Avian Disease, Research Centre on Veterinary Food Safety and Zoonoses, Faculty of Veterinary Medicine, Ghent University, Salisburylaan, 133, B-9820, Merelbeke, Belgium*

S.H. Duncan, *Microbial Ecology Group, Gut Health Division, Rowett Research Institute, Greenburn Road, Bucksburn, Aberdeen, AB21 9SB, UK*

V.S. Elbrønd, *Department of Animal and Veterinary Basic Science, The Royal Veterinary and Agricultural University, DK-1870, Frederiksberg C, Denmark*

M.B. Farnell, *Poultry Production and Product Safety Research Unit, ARS, USDA, Fayetteville, Arkansas, USA*

H.J. Flint, *Microbial Ecology Group, Gut Health Division, Rowett Research Institute, Greenburn Road, Bucksburn, Aberdeen, AB21 9SB, UK; email: h.flint@rowett.ac.uk*

I. Gantois, *Department of Pathology, Bacteriology and Avian Disease, Research Centre on Veterinary Food Safety and Zoonoses, Faculty of Veterinary Medicine, Ghent University, Salisburylaan, 133, B-9820, Merelbeke, Belgium*

J.S. Guy, *College of Veterinary Medicine, North Carolina State University, 4700 Hillsborough Street, Raleigh, North Carolina 27606, USA; fax: 919-513-6464; e-mail: jim_guy@ncsu.edu*

F. Haesebrouck, *Department of Pathology, Bacteriology and Avian Disease, Research Centre on Veterinary Food Safety and Zoonoses,*

Faculty of Veterinary Medicine, Ghent University, Salisburylaan, 133, B-9820, Merelbeke, Belgium

I. Hautefort, *Molecular Microbiology Group, Institute of Food Research, Norwich Research Park, Norwich, NR4 7UA, UK*

J. Hinton, *Molecular Microbiology Group, Institute of Food Research, Norwich Research Park, Norwich, NR4 7UA, UK*

S. Idei, *European Commission, Health and Consumer Protection Directorate-General, Directorate D – Food Safety: Production and Distribution Chain; D2 – Biological Risks,Commission Européenne ; B-1049 Bruxelles/ Europese Commissie, B-1049 Brussels,Belgium ; tel: (32–2) 299 11 11; e-mail: Sarolta.Idei@cec.eu.int*

A. Kettunen, *Alimetrics Ltd, Höyläämötie 14, FIN-00380, Helsinki, Finland*

M.T. Kidd, *Department of Poultry Science, Mississippi State University, Box 9665, Mississippi 39762, USA; fax: 662 325 8292; e-mail: mkidd@poultry.msstate.edu*

K.C. Klasing, *Department of Animal Science, University of California Davis, Davis, California 95616, USA; tel: 530-752-1901; fax: 530-752-0175; e-mail: kcklasing@ucdavis.edu*

G. Laverty, *Department of Biological Sciences, University of Delaware, Newark, Delaware 19716, USA; e-mail: laverty@udel.edu*

E.C.M. Leitch, *Microbial Ecology Group, Gut Health Division, Rowett Research Institute, Greenburn Road, Bucksburn, Aberdeen, AB21 9SB, UK*

S.A. Lister, *Crowshall Veterinary Services, Attleborough, Norfolk, UK; (address for communication: SAC, Ayr Campus, Ayr, KA6 5HW); tel: +44 (0)1953 455454; fax: +44 (0)1953 455661; e-mail: salister@crowshall.co.uk*

P. Louis, *Microbial Ecology Group, Gut Health Division, Rowett Research Institute, Greenburn Road, Bucksburn, Aberdeen, AB21 9SB, UK*

M.A. Mitchell, *Scottish Agricultural College, Bush Estate, Midlothian, EH16 AA, UK; e-mail: malcolm.mitchell5@btinternet.com*

M. Moretó, *Departament de Fisiologia, Facultat de Farmàcia, Av. de Joan XXIII s/n, 08028 Barcelona, Spain*

J.E. Olsen, *Department of Veterinary Pathobiology, The Royal Veterinary and Agricultural University, Stigbojlen 4, DK-1870, Frederiksberg C, Denmark*

F. Pasmans, *Department of Pathology, Bacteriology and Avian Disease, Research Centre on Veterinary Food Safety and Zoonoses, Faculty of Veterinary Medicine, Ghent University, Salisburylaan, 133, B-9820, Merelbeke, Belgium*

A.J. Patterson, *Microbial Ecology Group, Gut Health Division, Rowett Research Institute, Greenburn Road, Bucksburn, Aberdeen, AB21 9SB, UK*

C. Powers, *Division of Immunology and Immunopathology, Institute for Animal Health, Compton, Newbury, Berkshire, RG20 7NN, UK*

H. Raine, *ABNA Ltd, ABN House, Oundle Road, Peterborough, PE2 9PW, UK*

M.T. Rincon, *Microbial Ecology Group, Gut Health Division, Rowett Research Institute, Greenburn Road, Bucksburn, Aberdeen, AB21 9SB, UK*

G.D. Rosen, *Holo-Analysis Services Ltd., 66 Bathgate Road, Wimbledon, London, SW19 5PH, UK; tel./fax 020 8946 9575; e-mail: g.d.rosen@btinternet.com*

C. Schneitz, *Orion Corporation Orion Pharma, Animal Health, PO Box 425, FI-20101 Turku, Finland; fax: 2358-9-4521764; e-mail: carita.schneitz@orionpharma.com*

K.P. Scott, *Microbial Ecology Group, Gut Health Division, Rowell Research Institute, Greenburn Road, Bucksburn, Aberdeen, AB21 9SB, UK*

M.W. Shirley, *Eimerian Genomics Group, Institute for Animal Health, Compton, Newbury, Berkshire, RG20 7NN, UK; tel: 01635 577293; fax: 01635 577263*

E. Skadhauge, *Department of Animal and Veterinary Basic Science, The Royal Veterinary and Agricultural University, DK-1870, Frederiksberg C, Denmark*

A.L. Smith, *Enteric Immunology Group, Institute for Animal Health, Compton, Newbury, Berkshire, RG20 7NN, UK; tel: 44 (0)1635 577293; fax: 44 (0)1635 577263; e-mail: adrian.smith@bbsrc.ac.uk*

B. Svihus, *Department of Animal and Aquacultural Sciences, Norwegian University of Life Sciences, PO Box 5003, N-1432 Ås, Norway; e-mail: birger.svihus@umb.no*

L. Timbermont, *Department of Pathology, Bacteriology and Avian Disease, Research Centre on Veterinary Food Safety and Zoonoses, Faculty of Veterinary Medicine, Ghent University, Salisburylaan, 133, B-9820, Merelbeke, Belgium*

Z. Uni, *Department of Animal Sciences, Faculty of Agricultural, Food and Environmental Quality Sciences, The Hebrew University of Jerusalem, PO Box 12, Rehovot 76-100, Israel; e-mail: uni@agri.huji.ac.il*

F. Van Immerseel, *Department of Pathology, Bacteriology and Avian Diseases, Research Centre on Veterinary Food Safety and Zoonoses, Faculty of Veterinary Medicine, Ghent University, Salisburylaan, 133, B-9820, Merelbeke, Belgium; e-mail: filip.vanimmerseel@UGent.be*

A.W. Walker, *Microbial Ecology Group, Gut Health Division, Rowett Research Institute, Greenburn Road, Bucksburn, Aberdeen, AB21 9SB, UK*

PREFACE

The proceedings which comprise this volume are taken from the 28th Symposium in a series organized by the UK branch of the World's Poultry Science Association. The chosen topic attracted a large audience, members of which came from all continents except South America. This probably demonstrates the importance that poultry scientists and industries worldwide attach to avian gut function.

Since the late 1960s, when the Swann Committee in the UK investigated the use of antibiotics in animal husbandry and veterinary medicine, pressure on their use as growth promoters has grown. Swann recommended that only antibiotics which 'have little or no application as therapeutic agents in man or animals and will not impair the efficacy of a prescribed therapeutic drug or drugs through the development of resistant strains of organisms' should be used for growth promotion. The UK government accepted these recommendations and, subsequently, the European Union adopted this principle, although it had to resist pressure from Sweden for a complete ban on antibiotic growth promoters.

Political pressure continued and, since the mid-1990s, several antibiotics have been banned and withdrawn from the market. Concerns have also been expressed over the use of certain coccidiostats, with the result that some of the earlier products have now also been withdrawn.

These restrictions have taken place because of concerns over drug resistance and possible carry-over to consumers of poultry products. However, little consideration has been given to the consequences of their withdrawal to poultry health and performance. The worldwide interest in this symposium, therefore, demonstrates the concern of poultry scientists about these challenges, but also highlights the dearth of sound scientific knowledge about avian gut function in healthy birds.

Scientists of international standing were invited to present their views at the symposium. The starting point was a description of the history of feed additives. This was followed by a paper describing the likely consequences of the withdrawal of prescription-free use of antimicrobials, which concluded with a brief consideration of likely alternatives.

Four papers dealt with the development of the gastrointestinal tract, its function related to growth, uptake of nutrients and immune development.

This was followed by three papers concerned with gut microbiology, including the analysis of the microbial ecosystems, a description of the more commonly encountered microbes and a discussion of mechanisms of pathogen control.

Six papers covered nutritional effects. These included the effect of nutrients, anti-nutrients, non-nutrients or toxins and microflora on gut health and bird growth. Unfortunately, due to health reasons, the main author was unable to complete his review paper so it has not been included in this volume. The effect of non-starch polysaccharides on gut function were covered, as well as amino acid and protein supply on nutrition and health. The last three papers in this section dealt with the role of feed processing on gut function, micronutrient supply and its effect on immunity and, lastly, the association between nutrition and wet litter.

Three papers dealt with pathology. A description of virally induced gastrointestinal diseases of chickens and turkeys was followed by a paper highlighting the intestinal tract as a portal of entry of bacteria. The final paper in this section described a new approach to the investigation of the host–parasite relationship in which parasite genetics, DNA fingerprinting and selection by immunity can be used.

The immunological and pathogen control session contained papers dealing with feed acidification, competitive exclusion, the influence of vaccines and campylobacters and their bacteriophage. The author of the paper describing the influence of some specific vaccines on the intestinal health of broilers was unable to provide a review chapter for these proceedings. This is an unfortunate omission, since the use of vaccines is clearly an important practical approach to the control of gut pathogens. The concluding paper in this session discussed breeding for disease resistance and the role of gene mapping and microarray studies.

The concluding session included a description of the legislative role of the EU in monitoring zoonoses and zoonotic infections, and the final paper brought the discussions back to the farm level by asking: what does it mean to the farmer?

In addition to the formal presentations, 47 poster abstracts were offered by scientists working in the field of avian gut function.

I am indebted to the Organizing Committee for their contributions and support during the preparation and conduct of the symposium. I hope that readers of these proceedings will feel that the subject was adequately covered and brought up to date by the distinguished panel of speakers.

The efficiency with which the Symposium was run was due entirely to the efforts of Rita and Christine, who worked tirelessly to ensure that the interests of the sponsors and the needs of speakers and delegates were well catered for. I owe them my gratitude for making my responsibilities so much easier to handle.

Organizing Committee

G.C. Perry (Chairman), J.A. Parsons (Secretary), K.J. McCracken (ex-officio, Treasurer), T. Acamovic, A. Ball, P. Barrow, F. Davison, P. Hocking (ex-officio, President, WPSA UK Branch), T. Humphrey, G. Mead, M. Mitchell and G. Rosen.

Administrative secretaries

Rita Hinton and Christine Rowlings.

PART I
Introduction

CHAPTER 1

History and current use of feed additives in the European Union: legislative and practical aspects

R.A.H.M. ten Doeschate* and H. Raine

*ABNA Ltd., Peterborough, UK; *e-mail: rtendoeschate@abagri.com*

ABSTRACT

Whilst the term feed additives encompasses a variety of products, this chapter will concentrate on product groups such as antibiotic growth promoters, coccidiostats and enzymes. These products have been, and are, subject to scrutiny and licensing at European Union (EU) and national level.

The history of antibiotic growth promoters serves as a case study for the development of a product group, ensuing legislation, pressure group activity and the subsequent demise of the product group.

EU legislation has evolved over the years from a situation of limited scope and relatively relaxed rules to the current system where the scope is being extended to more products and with stricter rules. The current EU regulation (1831/2003) is based on the precautionary principle related to human health, animal health and the environment. At present there are signs that legislation has tightened to such a level that certain products are not available on the EU market, maybe as a result of high costs or delays, due to the registration process. A balance needs to be found between effective legislation regulating the safe use of additives and allowing new product and concept development.

From a practical nutritionist's point of view the whole area is very complicated, with several products on the market, and it is unclear whether they are additives in the legal sense or not, but there is a responsibility on the nutritionist to ensure compliance with the law. With antibiotic growth promoters disappearing, several 'replacement' products need to be evaluated, which is another big challenge for nutritionists.

INTRODUCTION

Feed additives encompass a variety of products. According to the currently applicable legislation (EC 1831/2003, Art. 2 (2a)), 'feed additives' means substances, microorganisms or preparations – other than feed materials and premixtures – which are intentionally added to feed or water to perform, in particular, one or more of the functions mentioned in Article 5 (3). Article 5 (3) can be summarized as follows: a feed additive should favourably affect one of the following:

- characteristics of feed or animal products;
- colour of ornamental fish or birds;
- the environment;
- animal production, performance or welfare through positive effects at gut level,

or satisfy the nutritional needs of animals or have a coccidiostatic or histo-monostatic effect.

This, potentially, gives a huge range of possible additives but, for the scope of this chapter only product groups will be covered such as antibiotic growth promoters, coccidiostats and enzymes. The reason for this is that these products have been, and are, subject to a fair degree of scrutiny and approval/authorization at EU and national level.

This chapter is intended to give an overview of the history of both feed additives and applicable EU legislation, which will inevitably be incomplete but will hopefully serve to give an understanding of the subject. Further, it will show how current use of additives is impacted by legislation, and how this and commercial practice interact.

HISTORY

For antibiotic growth promoters the story began in the 1940s, with a 1946 (Moore *et al.*) and/or a 1949 (Stokestad *et al.*) paper being widely cited as the first scientific paper(s) to note a growth-promoting effect of feeding antibiotics. The interest in antibiotics actually stemmed from a quest to find substitutes for the animal protein growth factor. Vitamin B_{12} was partly identified as a factor promoting growth in the absence of animal protein, but it was found that either antibiotics or antibiotic growth medium improved growth over and above the animal protein effect.

Subsequently, antibiotics quickly found a place in poultry nutrition, which can be illustrated by the following quote from Heuser (1955):

> One of the most spectacular recent developments in nutrition science is the discovery that antibiotics have growth-stimulating properties when fed to chickens, turkeys and swine. Few nutritional discoveries have been so quickly and universally applied to feeding practice.

The nutritional community at that time was easily convinced of the benefits, and subsequently the principle has been used for as long as it was allowed.

Initially, several antibiotics were used for growth-promoting purposes but, by the late 1960s, the Swann Committee investigated their use in the light of the potential for the development of antibiotic resistance in bacteria. They concluded that: 'The administration of antibiotics to farm livestock, particularly at sub-therapeutic levels, poses certain hazards to human and animal health.' In particular, it had led to resistance in enteric bacteria of animal origin. This resistance was transmissible to other bacteria (it had been the discovery that this might be so, following an epidemic of resistant *Salmonella typhimurium* in 1963–1965, which prompted the UK government minister to appoint the Swann Committee). It had also been shown that enteric bacteria were transferable from animals to man (House of Lords, 1998). The Swann Committee recommended that only antibiotics which 'have little or no application as therapeutic agents in man or animals and will not impair the efficacy of a prescribed therapeutic drug or drugs through the development of resistant strains of organisms' (Swann, 1969 as quoted by the House of Lords, 1998) should be usable for growth promotion. Their report named the following antibiotics, which were then in use for growth promotion, as unsuitable for such use: chlortetracycline, oxytetracycline, penicillin, tylosin (a macrolide related to erythromycin) and the sulphonamides. The UK Government largely accepted these recommendations (House of Lords, 1998).

Over the following years pressure on the use of antibiotics as growth promoters has increased steadily. Sweden was the first country in Europe to completely ban the use of antibiotic growth promoters in 1985. This ban was actually requested by the organization of Swedish farmers, and was subsequently enforced by the Swedish Government. When Sweden joined the EU it was granted a derogation allowing the ban to continue in Sweden whilst other EU countries continued the use of antibiotic growth promoters (SOU, 1997). Sweden lobbied for an EU-wide ban, and during the late 1990s in Europe the political pressure was at such a level that since 1996 several antibiotics have disappeared completely.

Table 1.1 shows the full list of active ingredients banned or taken off the market, with their year of disappearance. Avoparcin is the only one of these that has disappeared globally. Whilst all other products have been banned in the EU, they can still be used in other parts of the world.

Before the loss of four products in 1999, it was common practice in some companies to switch between different ingredients on a regular basis, say every 6–9 months. It was commonly thought that this helped maintain the efficacy of the various antibiotic growth promoters. However, after 1999, when only two active ingredients remained, there was no real opportunity to do this and as a result most companies tended to use the same product all the time.

As a result of the public pressure on the use of antibiotic growth promoters in 1999, one UK integrator decided to satisfy the perceived demand for poultry produced without an antibiotic growth promoter, and stopped their use. Other producers quickly followed and at present chickens destined for the retail channel are fed on Non-growth Promoter (NGP) feed. The only chickens still fed antibiotic growth promoters in the UK are those destined for the not insignificant remaining part of the broiler industry that supplies birds to non-retail markets.

Table 1.1. Antibiotic growth promoters for use in poultry feeds withdrawn from the EU market since 1996.

Active ingredient	Year of withdrawal
Avoparcin	1996
Spiramycin	1999
Tylosin	1999
Virginiamycin	1999
Zinc bacitracin	1999
Avilamycin	2006
Flavomycin	2006

For coccidiostats, the story started later than that for antibiotic growth promoters but on the other hand the story has not finished yet. In the 1950s the sulpha-drugs started to be used to control coccidiosis as a disease, but development really started with the chemical coccidiostats in the 1960s. The availability of coccidiostats is generally considered to be one of the main factors that allowed the development of the intensive poultry industry. Ionophore coccidiostats became available in the 1970s, after which a typical programme in the UK was to use a chemical anticoccidiostat (ACS) in the starter feeds, followed by an ionophore later on.

Pure chemical products had a habit of breaking down, i.e. resistance to them developed and caused significant problems. Ionophores and chemical/ionophore mixtures seemed to enable more stable production levels, yet coccidiosis remains ever present and any lapse in hygiene or management may result in mini-outbreaks. As a consequence of the development of resistance to the pure chemical products, some of these are no longer on the market whereas others, whilst available, are used very sporadically and then only in a tactical manner. Some products have been banned from use in Europe on other than strictly technical grounds. Nicarbazin as a pure product, for example, had its licence withdrawn as a result of an incomplete renewal dossier, whereas some of the generic versions of other active ingredients have been caught in the brand-specific approvals process and are subsequently no longer available to the industry.

Table 1.2 shows some of the products that have disappeared recently (since 2002).

The reduction of available products and the high costs involved in maintaining registration no doubt mean that the cost to the industry of coccidiosis control is higher than it could have been. In particular, the loss of generic versions of active ingredients means that the price paid may increase due to lack of competition.

Currently the use of coccidiostats is controlled under the EU additive legislation, which allows feed suppliers to decide which products to use and when. There is a proposal to change this situation in that the control of product choice would move from nutritionists to veterinarians in 2012. Given that this transfer of control means that individual farms may have individual coccidiostat programmes, feed manufacturers are worried about this proposal as it could lead to huge inefficiencies and additional costs.

Table 1.2. Coccidiostats that have disappeared from the market since 2002.

Active ingredient	Brand name
Amprolium	Amprol
Amprolium/ethopabate	Amprol Plus
Clopidol	Coyden
Clopidol/methylbenzoquate	Lerbek
Nicarbazin	Various
Salinomycin	Generic versions
Monensin	Generic versions

Product cycles

The history of products described earlier illustrates a common cycle of development. The cycle starts with the scientific discovery, for example, that a low level of antibiotic improves growth. To enable the marketing of this idea, one would expect to pass legislative hurdles, but if the idea is novel the legislator may not have thought of it and, as such, there would be no hurdles to clear initially. The product would then be marketed and, if successful, would be accepted on the market. As the market is using a new product or concept, the regulatory framework catches up and controls the products via an authorization system, resulting in an increase in legislative hurdles.

During use of a product either real problems turn up or pressure groups campaign against products based on perceived problems, resulting in ever-increasing legislative hurdles. Eventually these may/will become so difficult to overcome that the product is either banned outright or the company decides not to defend the dossier, resulting in the withdrawal of that product from the market. One by one, products thus disappear until finally the whole product group will have gone. A new group of products then appears on the market to fill the now vacant niche or solve the problem, which starts the whole cycle again.

Currently, everything evolves faster and there is a belief that these product cycles are also speeding up. As an example, one could compare the cycles of antibiotic growth promoters (with bacitracin as an example) with those of non-starch polysaccharide (NSP) enzymes, as shown in Table 1.3.

Based on this, it could be predicted that some of the products we now consider to be normal may, in the not too distant future, be no longer available, unless action is taken to defend them. Of course this is only a hypothesis, but nevertheless it is an interesting one to consider.

History of legislation

Not only is the history of products or product groups of interest but the history of legislation is also of interest, especially if one looks at the shift in its emphasis. At the EU level, Directive 70/524 was really the first one that tried to regulate the use of feed additives across the EU Member States. The aim was expressed thus:

Table 1.3. Product cycles of Bacitracin and NSP enzymes.

	Bacitracin	NSP enzymes
Scientific development	1950s	1980s
Initial legislative hurdles	None	None
Market product	1958	1980s
Acceptance in market	1960s	1990s
Increase in legislative hurdles	1990s	Late 1990s
EU ban	1998	1831/2003: some products for some species

Whereas the provisions laid down by law, regulation or administrative action concerning additives in feedingstuffs, insofar as they exist, differ as regards their basic principles; whereas it follows that they directly affect the establishment and functioning of the common market and should therefore be harmonized.

However:

Whereas because of special situations of certain Member States, and in particular because of their different systems of animal feeding, it is necessary in certain cases to allow derogations and Member States should also retain the power to suspend the use of certain additives or to lower the maximum levels if animal or human health is endangered; whereas Member States should not, however, be able to have recourse to that power in order to hinder the free movement of the various products.

This in effect meant that even though the directive was intended to lead to consistent legislation across the EU, it did not achieve this. As seen earlier, the fact that the Scandinavian countries banned anti-growth promoters ahead of the rest of Europe resulted in a campaign to get these products banned in the whole of the EU.

According to Directive 70/524, authorization for an additive shall only be given if it is efficacious and does not adversely affect human or animal health or the environment, nor harm the consumer by impairing the characteristics of animal products (CONSLEG, 2003). So, both efficacy and protection are mentioned, but protection was based more on known negatives rather than on the precautionary principle seen now in Regulation 1831/2003.

Directive 70/524 appears complicated, with many annexes and different authorization periods for different products.

The industry worked with the Directive and the country-specific implementation of the Directive into law as well as possible. Regulation 1831/2003 states, however:

Experience with the application of Council Directive 70/524/EES of 23 November 1970 concerning additives in feedingstuffs has shown that it is necessary to review all the rules on additives in order to take into account the need to ensure a greater degree of protection of animal and human health and of the environment. It is also necessary to take into account the fact that technological progress and scientific developments have made available new types of additives, such as those used on silage or in water.

(OJ L 268, 2003)

Regulation 1831/2003 thus sets out to do this, with a change of emphasis towards the protection of human health, animal health and the environment, based on the precautionary principle.

The replacement of Directive 70/524 with regulation 1831/2003 has had a number of consequences, one of which is that as it is a Regulation it applies directly in each Member State, and there should thus be less opportunity for country-specific rules and derogations. This should make life easier for an additive supplier once the product is approved but may create difficulties for a supplier with a product that may be considered not to be an additive in one country but could be considered an illegal additive in another.

Regulation 1831/2003 was published on 18 October 2003 and applied from 18 October 2004. There was a transitional period during which any additive previously authorized under Directive 70/524 and already on the market needed to be notified for evaluation to prevent its removal from the market. This period was 1 year and started from the date the Regulation came into force, which was 20 days after the publication in the *Official Journal of the European Union*, i.e. 7 November.

Any additive that has missed this deadline has to be considered as a new application and cannot thus be marketed until the application process has been completed successfully. Any additives that did meet the deadline and were authorized under 70/524 now require a full application to be submitted as well, either 1 year before expiration of the time-limited authorization based on 70/524 or, in the case of old, permanently approved additives, within 7 years of the entry into force of Regulation 1831/2003. All additives authorized under regulation 1831/2003 will be given time-limited authorizations to allow technological progress and scientific developments to be taken into account in the review of the product authorizations.

CURRENT LEGISLATIVE PRINCIPLES

The basic principle of Regulation 1831/2003 is that only those additives approved may be placed on the market. Approval is based on the presence of positive effects within one of the categories of additives and the absence of negative effects on human health, animal health and welfare, environment and users' and consumers' interests.

Positive effects need to be proved in appropriate trial work showing statistically significant benefits. Absence of negative effects is, of course, almost impossible to prove, but toxicity studies and residue studies need to be part of the application. For the evaluation of the data one needs to remember that it is clearly stated that action by the Community relating to human health, animal health and the environment should be based on the precautionary principle and, as such, an additive will not receive the benefit of the doubt.

The categories of additives identified in 1831/2003 are:

1. Technological additives: any substance added to feed for a technological purpose.

2. Sensory additives: any substance, the addition of which improves or changes the organoleptic properties of the feed, or the visual characteristics of the food derived from animals.

3. Nutritional additives (such as amino acids).

4. Zootechnical additives: any additive used to affect favourably the performance of animals in good health or used to affect favourably the environment.

5. Coccidiostats and histomonostats.

Within these categories there is then a subdivision into functional groups, and the applicant has to suggest into which category and functional group an additive should be classified. This classification is important, as some of the other rules depend on which category the additive is in. For example, post-marketing monitoring is required for additives in categories 3, 4 and 5, and for those falling under the Genetically Modified Organisms (GMO) regulations, but not for additives in categories 1 and 2 (OJ L268, 2003).

Given all this, and the presence of all sorts of products on the market which look like additives but are not (yet) authorized under regulation 1831/2003, one could ask the question: when is an additive not an additive?

Some 'additives' are classified as feed materials, and as long as no additive claims are made they can be used as feed materials. Merely making a performance claim does not turn a feed material into an additive but it does increase the risk that an ingredient may be considered an additive by an authority. However, some of the products being marketed as replacement products for antibiotic growth promoters (identified as 'magic potions' by some people in the industry) appear more like additives than feed materials, yet are not authorized as such.

Also, some additives might be promoted for a use other than the one for which they are authorized, e.g. when zootechnical claims are made for a technological additive. It is not necessarily so that such a claim should mean it should be authorized as a zootechnical additive, but some authorities may say so. If one were to make any medicinal claims however, then a new set of rules would apply, and one would probably prefer to avoid the medicinal additives.

Currently this is a 'grey' area, and it will be necessary to find out how the legislators and enforcement agencies will deal with the issue in the future.

PUTTING LEGISLATION INTO PRACTICE

Understanding the principles of the current regulations is, of course, necessary and useful, but it is also important to apply the legislation correctly. A nutritionist in the compound feed business has many things to consider in evaluating any of the available additives. The first question to ask is whether the product is legal for the purpose under consideration. This is not as easy as it should be, as suppliers are not always totally sure themselves. Secondly, one needs to consider whether the product would be allowed to be used under Health and Safety rules, the interpretation of which is company specific. Some of the additives which would be legal to use from the point of view of Regulation 1831/2003 have given some concern with respect to Health and Safety and thus could either not be used or would need feedmill modifications to allow use.

After answering these questions, the nutritionist then needs to make the commercial decisions about whether to use a particular product. The first questions in this respect would be: does the product work under the particular conditions in which it will be used, and are any interactions with other diet components likely? If a product works it is necessary to consider whether it is the best possible choice and whether its use would be cost-effective.

Most nutritionists faced with these questions would prefer to conduct their own comparative studies. To evaluate products to replace antibiotic growth promoters, some companies tend to run pen trials comparing a range of products with both a positive and a negative control. As the removal of antibiotic growth promoters has taken place, our company has conducted several trials of this type and it is interesting to note that in nearly all of them the positive control (with antibiotic growth promoter) outperformed the negative control. These studies were run in a pen trial facility, but to replicate a commercial environment the birds were started on a mixture of new shavings and some used litter from a previous crop. The trials are statistically evaluated and if products show a positive response and all the other questions can be answered positively, then consideration is given to the use of those products in the field.

Complications of legislation in practice

With the demise of antibiotic growth promoters there is, naturally, huge interest in products that could potentially fill the now vacant niche in the market. However, some of these 'magic potion' products look like they may be in the 'grey' area mentioned earlier, where it is not clear whether their use is legal. One could consider that some of the feed materials, technological products or sensory additives actually would require authorization as zootechnical additives. If not, then maybe some of the claims made are difficult to justify and it might leave a nutritionist exposed if he or she decided to use the products with the aim of improving zootechnical performance.

Another issue to consider is that the EU regulations are quite demanding and involved, both in terms of the cost of getting a product authorized and in the duration of the authorization process. As a result, international additive suppliers may either ignore the EU altogether or, alternatively, a novel product may not become available in the EU for quite some time after it has been launched in other markets. This would result in a lower level of competitiveness of animal production in the EU, which is a bad outcome if one is trying to do business within the EU, but a good outcome if one is trying to compete with the EU.

Based on the (perceived) lack of a level playing field for animal production between the EU and elsewhere, producers often request that if they cannot use antibiotic growth promoters then imported poultry should also not have been fed antibiotic growth promoters. Whereas that makes sense in terms of trade, it does not make sense if one considers the original reasoning behind the ban of

AGPs. The hypothesis on which the ban is based is that the use of AGPs could induce the development of antibiotic-resistant pathogenic microbes. If this hypothesis is correct, then AGPs should be globally banned. It does not matter where in the world AGPs are used if, as a consequence, resistant pathogens occur. Obviously, contaminated poultry meat could be a major vector in the transfer of resistant pathogens, but it is not the only possible route.

If the EU thesis is that there is a real danger from the use of AGPs, then the EU should have no physical contact whatsoever with countries that are still using them, as there is a risk that resistant pathogens could enter the EU. As this is clearly impossible, there are only two logical conclusions: (i) if the use of AGPs poses a threat to human and animal health and the environment, their use should stop across the globe; or (ii) if there is no threat from their use producers in the EU are simply being forced down a route that leads to their becoming less competitive.

CONCLUSIONS

The use of additives and the legislation associated with them has evolved over the years and, no doubt, will continue to evolve in the future. Regulation 1831/2003 is still relatively new and, as a result, it is not always clear how to interpret the rules, but the future will show how the interaction of authority and industry will work out. The legislation needs to regulate the use of additives in such a way that safety is ensured, whilst still allowing development of new products and concepts. Practical nutritionists have to make decisions on the use of additives and need to understand and work with legislation.

REFERENCES

CONSLEG (2003) Consolidated text of: *Council Directive of 23 November 1970 concerning additives in feeding-stuffs (70/524/EEC)*. Office for Official Publications of the European Communities (accessed via Internet).

Heuser, G.F. (1955) *Feeding Poultry.* J. Wiley and Sons, Inc., New York; Chapman and Hall Ltd., London.

House of Lords (1998) *Select Committee on Science and Technology – Seventh Report.* HMSO, London.

Moore, P.R., Evenson, A., Luckey, T.D., McCoy, C., Elvehjem, C.A. and Hart, E.B. (1946) Use of sulfasuxidine, streptothricin, and streptomycin in nutritional studies with the chick. *Journal of Biological Chemistry* 165, 437–441.

OJ L 268 (2003) Regulation (EC) No 1831/2003 of the European Parliament and of the Council of 22 September 2003 on additives for use in animal production. OJ L 268, 18 October 2003, p. 29.

SOU (Swedish Ministry of Agriculture) (1997) Antimicrobial feed additives. *Report from the Commission on Antimicrobial Feed Additives.* SOU 1997:132, Stockholm.

Stokstad, E.L.R., Jukes, T.H., Pierce, J., Page, A.C. Jun. and Franklin, A.L. (1949) The multiple nature of the animal protein factor. *Journal of Biological Chemistry* 180, 647–654.

Swann, M.M. (1969) Report: *Joint Committee on the Use of Antibiotics in Animal Husbandry and Veterinary Medicine.* HMSO, London.

CHAPTER *2*
Poultry nutrition without pronutrient antibiotics

G.D. Rosen

Holo-Analysis Services Ltd., London, UK; e-mail: g.d.rosen@btinternet.com

ABSTRACT

Ongoing bans around the world on the veterinary prescription-free use of antimicrobials in poultry production evoke several questions. What are the results of a ban? Are there effective replacements? Which are the candidates? How is efficacy best assessed? What are the specific effects of chronological, diet ingredient content, environmental, genetic, geographical, managemental and nutrient content independent variables? Are admixtures useful?

Current nomenclature in feed additives can be off-putting and non-transparent for consumers. The scientific use of terms such as additives, non-nutrients, probiotics and prebiotics as descriptors is inapt. Better terminology includes pronutrients, microbials and saccharides. The routine use of duplex generic descriptors is recommended, jointly specifying function and nature of a product with or without any relevant brand name.

The net result of a ban for a producer is reduced productivity and profit. As yet there are no proven, fully effective replacement products amongst the plethora of diverse candidates on offer, as well as nutrients per se. Large variations in nutritional responses pose major problems, evidenced by coefficients of variation of feed intake, liveweight gain, egg production, feed conversion and mortality effects, respectively, of 203–1604, 101–240, 141–301, 109–261 and 304–1783%.

Cogent efficacy assessments relate to the setting and the meeting of standards for efficient replacement. Comprehensive holo-analytical mathematical models can be elaborated to provide predictive algebraic equations and software – based on all available negatively controlled published test data – for quantifying feed intake, liveweight, egg production, feed conversion, mortality and economically dependent variable responses with confidence limits in terms of all available independent variables. Basic terms studied are: (i) level of control performance; (ii) duration; (iii) dosage; (iv) test year; and (v) country. Some of the other independent variables required to relate research conditions to praxis include: (i) sex; (ii) housing; (iii) feed processing; (iv) disease; (v) part-purified diet; (vi) mode of action/metabolic test; (vii) individual product brands; and (viii) several dietary ingredient and nutrient contents.

Currently, exogenous enzymes appear to be leading contenders, with saccharides and microbials in pursuit. The potentials of nutrients, acids and botanicals in poultry nutrition are topics for future holo-analyses.

INTRODUCTION

Advancing worldwide progressive withdrawals of the use of pronutrient antibiotics in poultry feeds pose a challenge to food animal producers and their associated supply industries of how best to compensate for consequent loss of productivity (Rosen, 2006). Several questions arise, with most immediate urgency in the European Union, which will have finalized its progressive bans on the use of veterinary prescription-free feed antibiotics with effect from 1 January 2006. This poses the following questions:

- What are the results of a ban?
- Are there effective replacements?
- Which are the candidates?
- How is efficacy assessed?
- What are the specific effects of genetic, managemental, chronological, geographical and dietary variables?
- Are admixtures useful?

There is already widespread concern that the net result of a ban for producers will be reduced productivity and/or profitability, comporting a strong likelihood of consumer price inflation. As yet there is little or no agreement on whether or not any fully-effective replacement products are available amongst the plethora of diverse candidates on offer. This stems in part from inadequate research as yet in this field and also from problems posed by the manifest substantial variations in nutritional responses to feed antibiotics. The latter have coefficients of variation in their respective effects on feed intake, liveweight gain, egg production, feed conversion and mortality of 203–1604, 101–240, 140–301, 109–261 and 304–1783% (Rosen, 2005, unpublished results).

In essence, we are faced by the need firstly to set standards whereby the efficiency of replacements can be assessed and, thereafter, with the task of how best to meet such standards, in part at least at the outset and wholly as soon as possible thereafter.[1]

NOMENCLATURE

Substandard, ill-chosen and inaccurate nomenclature has long been an undesirable feature in this field. The very term additive has little or no

---

[1] Symbols used in the text and tables are defined in the Appendix.

consumer appeal in the sense that it may be vaguely accurate but it carries auras of minority, subsidiarity, afterthought and optional extra (Rosen, 1996). Garland (1995) considered that the term 'additive' had unhealthy connotations for current products from the poultry industry. The well-known use of additives in locomotive fuels also precipitates an unhealthy consumer reaction about their use in both foods and feeds. Such a descriptor of invaluable feed components is ill-conceived, non-transparent and relatively non-descriptive.

As yet there is no legal definition of a feed additive in the USA. However, its definition in the European Union is somewhat abstruse and over-complicated:

- 'feed additives' refers to substances, microorganisms or preparations – other than feed material and premixtures – which are intentionally added to feed or water in order to perform, in particular, one or more of the functions mentioned in Article 5 (3);
- 'feed materials' refers to various products of vegetable or animal origin – in their natural state, fresh or preserved, and to products derived from the industrial processing thereof and to organic or inorganic substances, whether or not containing additives, which are intended for use in oral animal feeding either directly as such or, after processing, in the preparation of compound feedingstuffs or as carriers of premixtures;
- 'premixtures' refers to mixtures of feed additives or mixtures of one or more feed additives with feed materials or water used as carriers, not intended for direct feeding to animals.

Article 5(3) states that:

The feed additive shall:

- favourably affect the characteristics of feed;
- favourably affect the characteristics of animal products;
- favourably affect the characteristics of the colour of ornamental fish and birds;
- satisfy the nutritional needs of animals;
- favourably affect the environmental consequences of animal production;
- favourably affect animal production, performance or welfare, particularly by affecting the gastrointestinal flora or digestibility of feedingstuffs; or
- have a coccidiostatic or histomonostatic effect.

In stark contrast, the European Commission in 2002 rejected, without explanation, adoption of the following proposed simplified definition that: 'A feed additive is a product to be placed on the market for use in animal nutrition, excluding feed materials, feedingstuffs, premixtures, processing aids and veterinary medicinal products.'

Whilst we are stuck legislatively in Europe with 'feed additive', in nutrition science it is more meaningful and precise to term these valuable dietary components 'pronutrients'. A pronutrient is simply defined as a substance which improves the value of nutrients (Rosen, 1997). In a random test with 100 consumers worldwide, none of whom had ever heard the term pronutrient, 86% responded more or less with 'something good for nutrition' when asked what they thought a pronutrient might be. Pronutrient is a logical semantic counterpart to the traditional, widely accepted and equally meaningful term 'antinutrient', as follows:

$+$

$\uparrow$

Better performance

|

PRONUTRIENT

|

NUTRIENT MIX

|

ANTINUTRIENT

|

Worse performance

$\downarrow$

$-$

Three other malapropisms in this field are non-nutrient feed additives, probiotics and prebiotics. The descriptor 'non-nutrient' used by the US National Research Council may be apposite as to nature, but it is incorrect and misleading as to nutritive value.

The term 'probiotic' was originally and correctly introduced by Winter (1955) in his studies on the significance for therapy and diet of antibiotic substances from flowering plants (with special reference to nasturtiums, cress and horseradish). He correctly averred that: 'We can possibly therefore call these substances, due to their local function, probiotics; they influence life without having the character of vitamins: they are antibiotic against pathogenic microbes and they are therefore probiotic for the infected organism.' In this way Winter also corrected the unfortunate anomaly introduced in 1941 by Waksman (1954), when he referred to actinomycin and other life-saving drugs as antibiotics, true though this appellation may be in microbiology. In fact, it was a marine geographer (Maury, 1869) who first correctly used the term antibiotic with reference to his inclination to an 'antibiotic hypothesis', concerning his disbelief in the presence or possibility of life existing in the earth's as-yet-unexplored deep ocean beds.

Ten years later the word probiotic was used by Lilley and Stillwell (1965) for a stimulatory protozoal growth factor produced by another protozoan. Parker (1974) used the term probiotic to describe 'organisms which contribute to the intestinal balance', thereby covering up the identity of a microbial agent. The use of the term probiotic was continued when Fuller (1989) defined it as 'a live microbial feed supplement which beneficially affects the host animal by improving its intestinal balance'. Neither the FDA nor the European Commission have concurred in the use of this flexible and loose terminology. The FDA (together with AAFCO) specifies (direct-fed) microbials. The EU Commission specifies microorganisms. One may conclude therefore that any pronutrient which improves animal health and/or performance could be dubbed as probiotic.

A 'prebiotic' is correctly defined, with a historical and traditional usage in chemistry, as a substance believed to have been involved in the origin of life, i.e. before life. Gibson and Roberfroid (1995) used this descriptor, defining it as a 'nondigestible food ingredient that beneficially affects the host by selectively stimulating the growth and/or activity of one or a limited number of bacteria in the colon and thus improves host health'. Such substances are, however, synthesized by living organisms. They act on live microorganisms that inhabit live host animals. In this context they might more appropriately and correctly have been termed as postbiotic. In nutrition such substances should be described as nutrient sources for gut microbiota, intestinal immuno-modulators, pathogen binders, etc., according to their modes of action.

The routine use of duplex descriptors in nutrition would be more transparent, jointly defining both nature and function as, for example, in pronutrient xylanase, diformate, oregano, *B. cereus*, etc.; prophylactic narasin, *L. salivarius*, *B. acidilactici*, etc.; therapeutic penicillin, tylosin phosphate, lincomycin hydrochloride, etc.; and pro-environmental phytase, protease, etc. Whenever relevant in this field, individual brand names should be used with their duplex descriptor.

HOLO-ANALYSIS IN ANIMAL NUTRITION

The setting of standards for the efficient replacement of antibiotics in poultry nutrition can best be achieved by the use in praxis of holo-analytical empirical antibiotic models, which quantify their nutritional responses for individual circumstances of use (Rosen, 2004a). The meeting of such standards is then effected by the deployment of analogous models for proposed replacements, such as acids, botanicals (including essential oils), enzymes, microbials, saccharides, etc. These pronutrient types with their various modes of action have a common function sparing the limiting nutrient in a diet. There is a need to also include nutrients per se as potential antibiotic replacements in our research and development programmes.

In praxis, holo-analytical models are utilized in software designed to determine the specific nutritive and economic values in each and every situation. For some pronutrients, it may be necessary to provide software targeted on statistically significant subgroups. For example, regarding copper in pigs – models derived from 628 publications (1928–1995) using 789 start-to-finish tests on 32,000 pigs (Rosen and Roberts, 1996) – separate models were used for six subclasses within and outside North America, for weaners, slaughter pigs to target weight and slaughter pigs to target date, as exemplified in Table 2.1.

The nutritional responses with confidence limits are used to assess the equivalence or significance of any difference between a pair of candidates. Different dosages can also be assessed in order to optimize net profit for any user's situation.

A holo-analysis integrates *all available* test results in a multiple regression empirical model, quantifying a dependent variable nutritional response in terms of *all available* genetic, chronological, environmental, geographical,

Table 2.1. Input–output software for practical application of holo-analytical models for slaughter pigs to target liveweight outside North America.

Required input values		Output effects, standard errors, confidence limits (95%) and economics		
Copper (mg/kg feed)	200		LWGeff (g)	FCReff
Feeding system (enter 1 if *ad lib* or 0 other)	0			
Control daily liveweight gain (kg)	0.75	*Estimates*	0.0147	0.0010
Control feed conversion ratio	2.8	Percentage of control	1.96	0.04
Initial liveweight (kg)	25	Standard error	0.0052	0.0166
Final liveweight (kg)	100	Confidence limits	0.0104	0.0328
Terminal rate of gain (kg/day)	1.0			
Control feed cost (MU[a]/kg)	0.16	*Economics*		
Liveweight value (MU/kg)	0.96	Net profit due to copper (MU/pig), 0.28		
Cost of copper (MU/g)	0.0028	Return on investment ratio, 2.4		
Other costs (MU/pig/day)	0.2			

[a] Monetary unit (£ in this example).

managemental, dietary ingredient and nutrient content independent variables. The basic structure and scope of holo-analysis is summarized in Table 2.2.

Key points are: (i) elimination of repeats due to multiple publication of a test; (ii) use of start-to-finish effects; (iii) exclusion of outlier (3 × SD) test results; (iv) elaboration of separate models for statistically significantly different subclasses; and (v) economic integration of partial nutritional effects of all available feed intake, liveweight gain, feed conversion, mortality, carcass and depollution responses.

Table 2.2. Structure and scope of holo-analysis.

1. Collection of all available, published, negatively controlled, feeding test reports
2. Abstraction and computer filing of numeric and descriptive variables
3. Elimination of repeats and errors; collection from authors and/or calculation of missing values
4. Multiple regression analysis, relating start-to-finish dependent variable nutritional effects to all available significant independent variables and important interactions
5. Determination of best-fit model for the effect, having maximum correlation coefficient square, minimum standard deviation and statistically significant partial regression coefficients[a]
6. Exclusion of outliers and computation of confidence limits
7. Elaboration of models for significantly different subclasses
8. Integration of partial nutritional responses (feed, gain, conversion, mortality, carcass, depollution) as an economic effect

[a] Draper and Smith (1981).

Table 2.3 contains a guideline to the number of tests used in the progression of holo-analysis (Rosen, 2004b).

Exploratory and preliminary models based on 20–100 tests could be invaluable in guiding the course of future research and development. Working models normally require 100–300 tests. Successive increments of 30–60 tests thereafter are used to update models. Expansion beyond 300 tests normally has only marginal effects on partial regression coefficients.

Model structures

The basic nature and structure of holo-analytical models is illustrated for present purposes with a large test resource of the five most utilized antibiotics in broiler feeds (Table 2.4). The four models for 'Bromycin' shown in Table 2.5 incorporate key variables, level of control performance, duration, year of test, logarithmic dosage, anticoccidial feed and disease. This set of models accounts only for 5–69% of the variations in the four nutritional responses.

Partial regression coefficients for valid models must have logical algebraic signs. In Table 2.5 the control performance terms are all negative showing, as would be expected, that the magnitude of response at a given dosage is reduced with superior negative control performances. Within the dosage range 0–100 ppm in this data set, the control performance and dosage partial regression coefficients are critical for optimal posology. As expected, gain, conversion and mortality effects are greater in the presence of diagnosed or endemic disease.

Table 2.3. Numbers of negatively controlled tests used in holo-analysis.

Number of tests	Applications
3	EU efficacy minimum for first registration
10–20	Guide to the use of an average dosage for an average response
20–50	Exploratory models
50–100	Preliminary models
100–300	Working models
>300	Updated models

Table 2.4. 'Bromycin' resource for broiler feed pronutrient antibiotics.

	Number of tests	
Antibiotic	FDIeff/LGWeff/FCReff	MORTeff
All	1709	708
Bacitracin MD	194	82
Bacitracin Zn	640	346
Chlortetracycline	299	88
Oxytetracycline	247	76
Virginiamycin	329	116

Table 2.5. 'Bromycin' holo-analytical models.

FDIeff   = − 79.5 − 0.0380 FDIC   + 2.98 DUR   + 1.02 EXD
R^2 0.052   − 24.3 COC
SD 126

LWGeff   = − 72.4 − 0.0114 LWGC   + 1.02 DUR   + 0.871 EXD + 19.5 log(ABP+1)
R^2 0.171   − 16.8 COC + 76.1 VET
SD 47.4

FCReff   = 0.301 − 0.161 FCRC   + 0.00306 DUR − 0.00159 EXD − 0.0286 log(ABP+1)
R^2 0.287   − 0.116 VET
SD 0.100

MORTeff   = − 1.13 − 0.647 MORTC + 0.0363 DUR   + 1.081 log(ABP+1)
R^2 0.685   − 2.12 VET
SD 3.12

A large variety of other independent variables must, however, be investigated for the elaboration of comprehensive working models for use in praxis. Many of these are listed in Table 2.6.

Inclusion of all available significant variables raises R^2 and SD reduces in working models. Normally, chronological and geographical variables disappear with the inclusion of additional significant variables such as those in Table 2.6 or, alternatively, they regress to minor contributors to response variations. Examples of two working models are detailed in Table 2.7 for phytase in broilers.

These larger models account for 64 and 72%, respectively, of the variation in feed intake and liveweight gain responses. Further model development, however, beyond 75% is largely restricted by the unavailability in published reports of key variables such as temperature, altitude and/or the absence of key dietary factors. Nevertheless, models such as those in Table 2.7 are already useful in the appraisal of effective antibiotic replacements.

Table 2.6. Independent variables used in holo-analytical nutritional models.

control performance	feed process	maize[b]	gross energy[c]
duration	antibiotic	sorghum	net energy[c]
year of test	anticoccidial	wheat	crude protein
dose	antihistomonial	barley	crude fat
initial age	metabolic test	oats	crude fibre
not day-old	diet marker	rye	calcium
sex	part-purified diet	animal fat	phosphorus
phased dose	disease challenge	vegetable oil	lysine
factor 2 dose[a]	supplier test	animal protein	methionine
selected weight birds	institute test	vegetable protein	methionine + cystine
housing	country	wheat offal	threonine
stocking density	brand	rice bran	tryptophan

[a] Second antibiotic/enzyme/acid/microbial/other pronutrient/nutrient.
[b] As dietary concentrations (columns 3 and 4).
[c] Digestible or metabolizable energy are alternatives.

Table 2.7. Phytase broiler holo-analytical feed intake and liveweight gain models.

FDIeff = 232 − 0.136FDIC + 20.0 DUR + 226 log(PHY+1) − 514 log(P+1) − 78.9 CAG + 93.0 NDO

R^2 0.641	SE	56.0	0.014	1.83	46.3	54.2	13.2	18.2
SD 62.1	p	0.000	0.000	0.000	0.000	0.000	0.000	0.000

+ 65.2 COC + 7.02 Ca − 150 NAT − 222 NOV − 207 FIN − 0.573 MZP + 71.0 AOF

	65.2 COC	7.02 Ca	150 NAT	222 NOV	207 FIN	0.573 MZP	71.0 AOF
SE	12.0	2.38	14.5	23.7	21.3	0.018	19.9
p	0.000	0.004	0.000	0.000	0.000	0.002	0.000

+ 12.1 ROP − 14.7 PFP − 9.13 AFP − 1.66 VOP

	12.1 ROP	14.7 PFP	9.13 AFP	1.66 VOP
SE	0.264	0.311	0.269	0.352
p	0.000	0.000	0.000	0.001

LWGeff = 118 − 0.231 LWGC + 16.4 DUR + 168 log(PHY+1) − 339 log(P+1) − 49.6 CAG + 54.2 NDO

R^2 0.717	SE	33.3	0.017	1.08	26.0	29.0	7.67	8.85
SD 35.4	p	0.000	0.000	0.000	0.002	0.000	0.000	0.000

+ 48.1 COC − 86.3 NAT − 142 NOV − 122 FIN − 0.716 MZP − 0.662 SOP − 1.97 BAP

	48.1 COC	86.3 NAT	142 NOV	122 FIN	0.716 MZP	0.662 SOP	1.97 BAP
SE	6.72	7.71	13.1	12.5	0.012	0.021	0.055
p	0.000	0.000	0.006	0.000	0.000	0.000	0.000

+ 105 AOF + 5.58 ROP − 7.41 PFP

	105 AOF	5.58 ROP	7.41 PFP
SE	13.0	0.156	0.174
p	0.000	0.000	0.000

Model applications

Holo-analytical models can be used to compare classes of pronutrients (e.g. antibiotics versus enzymes) and also specific generic or branded individual products (e.g. zinc bacitracin versus phytase as generics or Albac zinc bacitracin versus Natuphos phytase as brands).

The mean characteristics of the class resources in Table 2.8 manifest striking differences between the conditions used in their test programmes during the course of research and development, e.g. mean enzyme test duration is 17 days less than antibiotic using 59 versus 39% caged birds, with only 15 versus 40% anticoccidial in 31 versus 17% mash feeds. Models for these two classes have been used to compare gain and conversion responses in praxis conditions at a current broiler performance level (Rosen, 2001). At respective mean dosages of 18 ppm and 7.5 units/g feed there is no significant difference in antibiotic and enzyme responses. The enzyme model also predicts that the mean one-point difference in feed conversion effect could be eliminated by the use of a higher enzyme dosage of 9.2 u/g feed, i.e. +1.7 u/g.

In Table 2.9, zinc bacitracin and phytase are compared for equal investment and for equal feed conversion improvement. At equal investment zinc bacitracin (80 ppm) is five conversion points better than phytase, which is ineffective at 500 units/kg feed, a dosage commonly deployed for depollution and phosphorus economy. These models also show that a 3.2-point conversion improvement by 42 ppm zinc bacitracin would require a 20-fold higher phytase dosage (10,000 u/kg feed).

Table 2.8. Pronutrient class comparison of antibiotics and exogenous enzymes in broilers.

	n	LWGC (g)	LWGeff (g)	FCRC	FCReff	MORTC	MORTeff	DUR (days)	CAG	MAL	COC	MAS
Antibiotics	1709/708	1235	38.0	2.274	−0.0650	4.32	−0.41	47.7	0.39	0.30	0.40	0.17
Enzymes	1869/365	1133	57.0	1.946	−0.100	6.83	−1.80	30.5	0.59	0.51	0.15	0.31

For 42-day-old, as-hatched on the floor, pellet feeds, with anticoccidial LWGC 2326 g; FCRC 1.777

	LWGeff (g)	FCReff
Antibiotics (17.9 ppm)	33.7	−0.041
Enzymes (7.5 u/g)	33.8	−0.030

Table 2.9. Iso-input cost and iso-feed conversion effect broiler comparisons of zinc bacitracin and first-generation phytases.

Pronutrient	Dosage	Basis	LWGeff (g)[a]	FCReff[a]
Zinc bacitracin (1165 tests)	80 ppm	iso-cost	59	−0.054
	42 ppm	iso-FCReff	47	−0.032
Phytase (296 tests)	500 u/kg	iso-cost	−1	+0.001
	10,000 u/kg	iso-FCReff	159	−0.032

[a] For as-hatched 56-day-old broilers of LWGC 3266 g and FCRC 2.079.

PRONUTRIENT ADMIXTURES

As yet this is virtually uncharted territory, though there appears to be a keen and growing interest in the potential for admixtures. This is an area in which the terminology is also confused and often incorrect, particularly for 'additive' and 'synergistic'. Admixture terminology is precisely defined in Table 2.10.

Comprehensive tests of admixtures are keenly awaited and much needed. As background, a design of pronutrient admixture tests to guide admixture strategies is summarized in Table 2.11.

Table 2.10. Admixture terminology.

Product A = +2
Product B = +3

	A + B
Sub-additive	= +4
Additive	= +5
Synergistic	= +6
Valueless	= +2 or +3
Antagonistic	= +1

Table 2.11. Design of pronutrient admixture tests.

Treatment	Feed concentration (ppm)	Investment cost (MU[a])
(i) Basal feed	0	0
Pronutrient A	a	4
Pronutrient B	b	8
A + B	a + b	12
(ii) Pronutrient A	3a	12
Pronutrient B	1.5b	12
(iii) A + B	0.33a + 0.33b	4
A + B	0.67a + 0.67b	8

[a] Money units.

Table 2.11 encompasses the following: (i) basal 2×2 factorial tests; (ii) iso-mixture cost tests for higher single dosages; and (iii) iso-mixture costs for each factorial component.

Concerning pronutrient admixtures the literature contains several examples of diverse interactions in broiler nutrition. In the total number of 2×2 factorial tests (17) on phytases and xylanases reported up to 2004, mainly (15) for wheat-based diets, interactions ranged from antagonistic to synergistic, related, at least in part, to the presence or absence of substantial declared or undeclared enzyme side activities in phytase and xylanase products (Rosen, 2004c).

DISCUSSION AND CONCLUSION

The task of setting standards for the efficient replacement of pronutrient antimicrobials in poultry nutrition is greatly facilitated by the availability in the literature of thousands of negatively-controlled test results, which provide large data resources for the elaboration of holo-analytical models (Rosen, 1995). The setting of specific standards using such models for the assessment of efficacy in each and every individual circumstance of use is well advanced in the case of enzymes.

Preliminary models are also available for a pronutrient saccharide product (Rosen, 2005). The studies to date have indicated the likelihood that enzymes represent the current best antibiotic replacement prospect. In the use of enzymes, however, care is needed in distinguishing between nutritional and depollution responses and their economic implications.

A simple Seven Question Test has been devised as a starting point in the appraisal of the efficacy and modelling status of the many hundreds of candidates on offer as antibiotic replacements. Illustrative answers are given in Table 2.12.

The answers shown to Questions 1–3 counsel a cautious approach to Product X with its need to expand its data base as soon as possible. In such cases where own tests are preferred the best policy is to investigate Product X at

Table 2.12. Seven Question Test for pronutrient antimicrobial replacement candidates (illustrative answer).

Question	Reply
1. How many properly controlled feeding tests do you have on the efficacy of Product X?	20
2. How many of these have no negative controls?	2
3. Can you supply a bibliography for questions 1 and 2?	yes
4. How many times out of ten does Product X improve liveweight gain and feed conversion?	7/10
5. What are the coefficients of variation in the gain and conversion responses?	100–200%
6. What dosage of Product X will maximize return on my investment?	x ppm because…
7. Can you supply me with a model to predict responses to Product X under my specific conditions?	yes

the average dosage used in tests to date with half and double doses, at least in a dose-response trial. The answers to Questions 4–6 are very satisfactory, particularly if supporting data justify the dosage advocated in reply to Question 6. A positive answer to Question 7 with provision of software to utilize the model for Product X is optimal.

The next priorities in modelling programmes are for the modelling of organic acids used individually or in admixtures, followed by microbials and botanicals. The role of nutrients as antibiotic replacements also needs to be defined.

REFERENCES

Draper, N.R. and Smith, H. (1981) *Applied Regression Analysis,* 2nd edn. John Wiley and Sons Inc., New York, 709 pp.

Fuller, R. (1989) Probiotics in man and animals. *Journal of Applied Bacteriology* 66, 365–378.

Garland, P.W. (1995) Range of substances currently available and problems to be addressed for the future. In: Senkoylu, N. (ed.) *Proceedings of the 10th European Symposium on Poultry Nutrition,* 15–19 October, Antalya, Turkey, pp. 203–207.

Gibson, G.R. and Roberfroid, M.B. (1995) Dietary modulation of the human colonic microbiota: introducing the concept of prebiotics. *Journal of Nutrition* 125, 1401–1412.

Lilley, D.M. and Stillwell, R.H. (1965) Probiotics: growth-promoting factors produced by micro-organisms. *Journal of Bacteriology* 89, 747–48.

Maury, N.F. (1869) *The Physical Geography of the Sea and its Meteorology,* 14th edn. Sampson Low, Son and Marston, London, 324 pp.

Parker, R.B. (1974) Probiotics, the other half of the antibiotic story. *Animal Nutrition and Health* 29, 4–8.

Rosen, G.D. (1995) Antibacterials in poultry and pig nutrition. In: Wallace, R.J. and Chesson, A. (eds) *Biotechnology in Animal Feeds and Animal Feeding,* VCH Verlagsgesellschaft mbH, D69451 Weinheim, Germany, pp. 143–172.

Rosen, G.D. (1996) Feed additive nomenclature. *World's Poultry Science Journal* 54, 53–57.

Rosen, G.D. (1997) Future prospects for pronutrients in poultry production. In: *Proceedings of the 16th Scientific Day,* 16 October, University of Pretoria, Pretoria, Republic of South Africa, pp. 60–79.

Rosen, G.D. (2001) Multi-factorial efficacy evaluation of alternatives to antimicrobials in pronutrition. *British Poultry Science* 42S, S104–S105.

Rosen, G.D. (2004a) Holo-analysis in animal nutrition. *Poultry International* 25 (12), 17–18, 21.

Rosen, G.D. (2004b) How many broiler feeding tests are required to elaborate working models for the prediction of nutritional effects of pronutrients in praxis? *British Poultry Science* 45S1, S8–S9.

Rosen, G.D. (2004c) Admixture of exogenous phytases and xylanases in broiler nutrition. In: *Abstracts and Proceedings, XXII World's Poultry Congress,* 13–17 June, Istanbul (CD-ROM).

Rosen, G.D. (2005) Holo-analysis of the effects of genetic, managemental, chronological and dietary variables on the efficacy of a pronutrient mannanoligosaccharide in broilers. *British Poultry Abstracts* 1 (1), 27–29.

Rosen, G.D. (2006) Setting and meeting standards for the efficient replacement of pronutrient antibiotics in poultry and pig nutrition. In: Barung, D., de Jong, J., Kies, A.K. and Verstegan, M.W.A. (eds) *Antimicrobial Growth Promoters: Where do We Go From Here?* Wageningen Academic Publishers, Wageningen, Netherlands, pp. 381–397.

Rosen, G.D. and Roberts, P. (1996) *A Comprehensive Survey of the Response of Growing Pigs to Supplementary Copper in Feed.* Field Investigations and Nutrition Services Ltd., London, p. 94.

Waksman, S.A. (1954) In: *My life with the Microbes*. Robert Hale Ltd., London, p. 210.
Winter, A.G. (1955) The significance for therapy and diet of antibiotic substances from flowering plants (with special reference to nasturtiums, cress and horseradish). *Die Medizinische 2*, 73–80.

APPENDIX. SYMBOLS, VARIABLES AND UNITS

Symbol	Variable (unit)	Symbol	Variable (unit)
ABP	antibiotic (ppm)	MORTC	control mortality (%)
AFP	animal fat (%)	MORTeff	effect on MORTC (%)
AOF	added oil/fat (0 or 1)	MU	money unit
BAP	barley (%)	MZP	maize (%)
Ca	calcium (g/kg)	n	number of tests
CAG	cage housing (0 or 1)	NAT	Natuphos (0 or 1)
COC	anticoccidial feed (0 or 1)	NDO	not day-old (0 or 1)
CV	coefficient of variation (%)	NOV	Novo phytase (0 or 1)
DUR	test duration (days)	p	probability
EXD	test year – 1900	P	phosphorus (g/kg)
FCRC	control feed conversion ratio	PFP	poultry fat (%)
FCReff	effect on FCRC	PHY	phytase (u/g feed)
FDIC	control feed intake[a]	R^2	multiple correlation coefficient square
FDIeff	effect on FDIC[a]	ROP	rape oil (%)
FIN	Finase (0 or 1)	SD	standard deviation
log	$\log_{10}$	SE	standard error
LWGC	control liveweight gain[a]	SOP	sorghum (%)
LWGeff	effect on LWGC[a]	u	units
MAL	male (0 or 1)	VET	presence of disease (0 or 1)
MAS	mash feed (0 or 1)	VOP	vegetable oil (%)
MD	methylene disalicylate	Zn	zinc

[a] broiler and turkey (g), pig (kg/day).

PART II
Gastrointestinal structure and functional development

CHAPTER 3
Early development of small intestinal function

Z. Uni

Department of Animal Sciences, Faculty of Agricultural, Food and Environmental Quality Sciences, The Hebrew University of Jerusalem, Israel; e-mail: uni@agri.huji.ac.il

Dedicated to the memory of David Sklan. A unique individual and scientist – my partner in exploring the chicken intestine.

ABSTRACT

To accommodate the rapid transition to external nutrient sources, the chicken small intestine goes through morphological, cellular and molecular changes towards the end of incubation. The weight of the intestine, as a proportion of embryonic weight, increases from approximately 1% at 17 days of incubation (17E) to 3.5% at hatch. At this time, the embryonic small intestinal villi are divided into two main developmental stages, which differ in both length and shape. Mucin-producing cells can be observed from 17E and at that time contain only acidic mucin. The activity and RNA expression of brush-border enzymes, which digest disaccharides and small peptides, and of major transporters (sodium–glucose transporter and ATPase) which are found at 15E, begin to increase at 19E (2 days prior to hatch), and increase further on the day of hatch. Although the digestive capacity begins to develop a few days before hatch, most of the development occurs post-hatch when the neonatal chick begins consuming feed. During the post-hatch period, the weight of the small intestine increases at a faster rate than body mass. Rapid enterocyte proliferation and differentiation occur. In addition the intestinal crypts, which begin to form at hatch, are clearly defined several days post-hatch, increasing in both cell number and size. Goblet cells produce acidic and neutral mucins in similar proportions. Studies have shown that feeding immediately post-hatch or even pre-hatch accelerates the functional development of the small intestine, while delayed access to external feed arrests the development of the small intestine's mucosal layer and changes mucin dynamics.

INTRODUCTION

The investigation of intestinal cell dynamics is crucial to understanding both digestive physiology and the efficiency of animal production. The epithelial layer of the intestine, the site of nutrient digestion and absorption, is composed of a continuously renewable population of cells in which stem cells, located in the crypt region, give rise to enterocyte and goblet cells. These cells migrate up the villi where they exhibit changes in functional characteristics.

The developing intestine undergoes morphogenesis and cyto-differentiation. Enzymes, transporters and glycosidation patterns change dramatically during the transition from embryo to independent chick, accompanying the crucial transition that occurs in the nutrient source of the hatchling. The yolk sac consists primarily of lipids and proteins, while the external feed consists of carbohydrates, lipids and proteins. This major change demands suitable redeployment of the cellular mechanisms that catalyse the digestion or hydrolysis and absorption of food components.

As the chicks make the metabolic and physiological transition from egg nutrition (i.e. the yolk) to exogenous feed, the yolk lipids – such as oleic acid – are quickly absorbed (this occurs close to hatch), whereas dietary carbohydrates and amino acids are not well utilized because of limitations in digestive and absorptive capacity. Therefore, the digestive system plays a major role in determining the development potential of the hatchling. Nutrient digestion, absorption, assimilation and incorporation into developing and growing tissues, including skeletal muscle, depend directly on the functional capabilities of the intestinal epithelial layer.

To achieve optimal muscle growth rates – and ultimately muscle and meat yields in poultry – it is essential to maximize enterocyte development, function and performance. Early functions of the gastrointestinal tract (GIT) are vital for the chicken's growth performance. Hence, it is crucial to achieve optimal intestinal development and functional capacity. This chapter examines the morphological and functional development of the small intestine in chickens.

GROSS AND STRUCTURAL ASPECTS OF THE DEVELOPING SMALL INTESTINE

As incubation progresses, the body weight of the embryo increases, as does the weight of the small intestine. However, small intestinal weight increases at a much greater rate than does body weight, the latter showing only a small increase close to hatch. The enhanced growth rate of the small intestine can clearly be seen by calculating the weight of the intestine as a percentage of embryo weight. During the last 3 days of incubation this increases from approximately 1 on day 17 of embryonic age (17E) to 3.5 at hatch.

The morphology of the small intestine also changes rapidly. Histology indicates that the intestine – including the external muscular layers and the villi – are growing rapidly. Villi at 15E are rudimentary; however, at 17E, villi at different stages of development are observed.

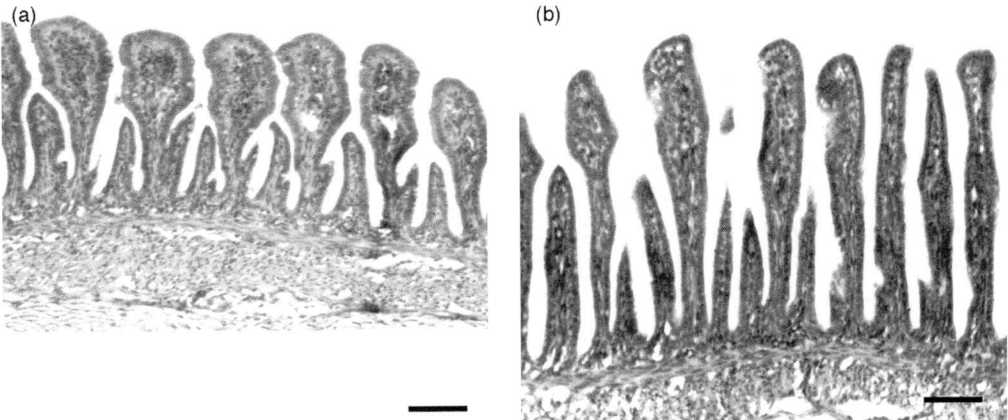

Fig. 3.1. Representative light microscopy (× 200) of intestinal villi from the jejunum of the broiler embryo at 17 days (17E) of incubation (a) and at 20 days (20E) of incubation (b). Sections show the pear-shaped and rocket-shaped villi. Villi in different stages of development can be observed. Staining was with haematoxylin-eosin. Bar = 10 μm.

Two main developmental stages can be seen, differing in length and shape: the larger villi are often pear-shaped and the smaller ones (approximately 65% of the size of the large ones) are narrower, with a rocket-like shape. Adjacent to each larger villus there appears to be one smaller one. A similar pattern was observed at 18E and 19E, with villi at both developmental stages growing, mainly in length, while maintaining similar shapes and distribution ratios. At 19E there was some 'budding' at the base of existing villi. These buds developed considerably and, by 20E, an additional 'wave' of small villi was found, comprising some 30% of the total.

At this embryonic stage, one day before hatch, the earlier waves of villi continue to grow and the third series of developing villi are approximately 65% of the size of the next largest ones. Villus growth rate between 19E and 20E was 30–40%; between 20E and day of hatch, however, growth was twofold for villi from all three developmental stages (Uni *et al.*, 2003b).

Further rapid changes are observed post-hatch, with significant morphological development of the small intestine occurring immediately after hatch. The intestines of hatchlings increase in weight more rapidly than does body mass. Growth rate of the intestines relative to body weight was greatest at 5–7 days of age in chicks (Uni, 1999). A two- to fourfold increase in intestinal length was observed up to 12 days of age, while the weight of the three intestinal segments (duodenum, jejunum and ileum) increased seven- to tenfold.

Villi increase in size and number, giving them a greater absorptive surface per unit of intestine. This rapid morphological development immediately after hatch involves differing rates of increase in villus volume in the duodenum, jejunum and ileum. Although duodenal villus growth is almost complete by day 7, jejunum and ileum development continues beyond 14 days of age (Uni *et al.*, 1995, 1999).

Enterocyte density ranges from 200,000–280,000 cells/cm^2 in all three segments of the small intestine and changes little with age. Nevertheless, an increase in the total number of enterocytes per villus is observed with age, resulting from the dramatic increase in villus length (Uni *et al.*, 1998a).

Crypt formation and enterocyte proliferation and migration in the developing small intestine

A continuous renewal process occurs in the intestinal mucosa as proliferating cells in the mucosal crypts differentiate, predominantly to enterocytes, which migrate up the villus and are sloughed into the lumen from the villus tip. During this migration process, the enterocytes acquire differentiated functions in terms of digestion, absorption and mucin secretion (Traber *et al.*, 1991, 1992; Ferraris *et al.*, 1992).

The site of enterocyte proliferation and the rate of their migration vary among vertebrates. In most mammals, proliferation is restricted to the crypts at the base of the villi, and their progeny migrate up the villus where they specialize and lose their capacity to divide (Cheng and Leblond, 1974a,b). However, in lower vertebrates such as amphibians (McAvoy *et al.*, 1975; McAvoy and Dixon, 1978a,b) and turtles (Wurth and Musacchia, 1964), the proliferating zones lie in troughs between mucosal folds and not in clearly defined zones of proliferation as in the higher orders of vertebrates.

In most mammals, the small intestinal crypts develop in the prenatal period from the flat intervillus epithelium (Marshall and Dixon, 1978; Quaroni, 1985a). DNA synthesis and cell proliferation are common over the entire length of the villus until birth, after which proliferation becomes restricted to the crypt area (Quaroni, 1985a, b).

In poultry, at hatch, all small intestinal enterocytes are proliferating; with time, the proportion of proliferating cells decreases rapidly, reaching approximately 50% in the crypt 2–3 days post-hatch. Interestingly, by day 10 post-hatch, proliferating enterocytes are not restricted to the crypt region but also exist along the villi. Their proportion along the villi decreases quickly in the first days post-hatch and then more slowly, reaching approximately 10% (Uni *et al.*, 1998b). Lack of access to feed increases the proportion of proliferating enterocytes, both in the crypt and along the villus (Geyra *et al.*, 2001a, b).

In hatching birds, a single crypt is observed per villus with a relatively small number of cells per crypt. During the initial days post-hatch, crypt numbers increase by branching and, at the same time, a process of hyperplasia occurs in the crypt region. After 4–5 days, the rate of growth declines and by the end of the second week post-hatch, three to four crypts are observed per villus.

Enterocyte migration from crypt to villus tip takes approximately 72 h in 4-day-old chicks and 96 h in older birds (Uni *et al.*, 1998b; Geyra *et al.*, 2001b). At hatch, small intestinal enterocytes are round and non-polar; from 24–48 h post-hatch, these cells increase rapidly in length and develop pronounced polarity and a defined brush border.

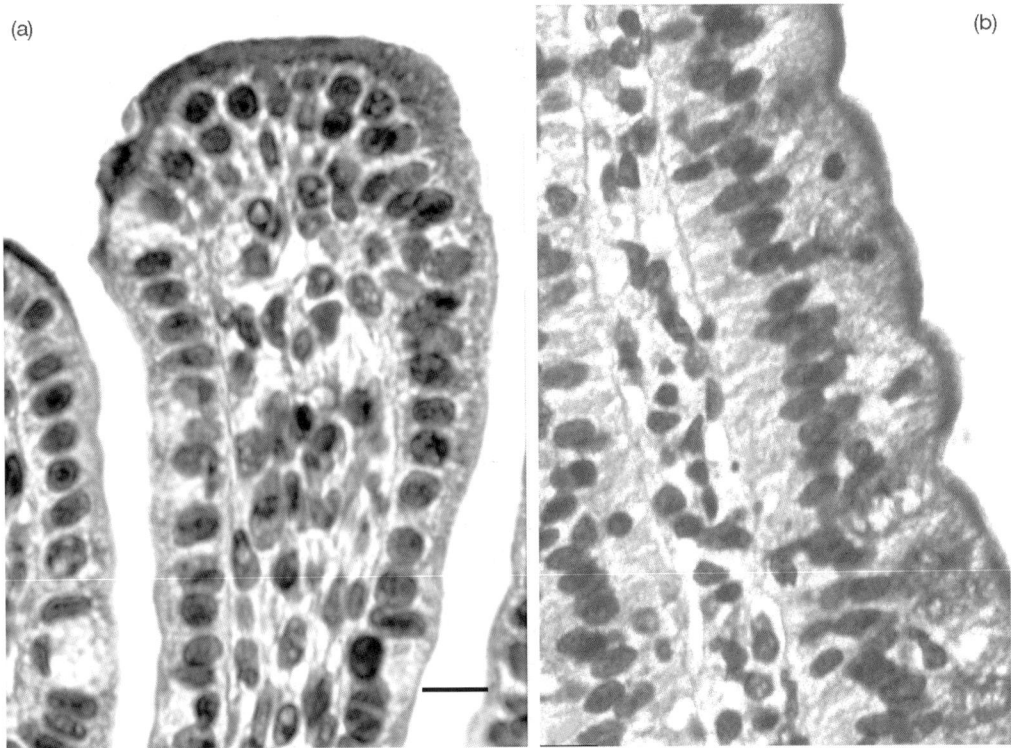

Fig. 3.2. Light microscope section of jejunal sections at hatching (a); round and non-polar enterocytes can be observed. At day 7 post-hatch (b) enterocytes are polar. Bar = 10 μm.

Thus, the small intestine of a newly hatched chick is immature and undergoes dramatic changes during the first few days post-hatch. This is demonstrated by the immaturity of the crypts and the absence of a defined zone of enterocyte proliferation. During the initial 4–5 days post-hatch, the crypt–villus axis is established with a distinct area of proliferation and constant enterocyte migration. The extensive changes in the morphological development of the small intestine close to and immediately post-hatch include the differentiation of enterocytes and the definition of crypts, as well as a many-fold enlargement of the intestine's absorptive surface.

DEVELOPMENT OF DIGESTIVE AND ABSORPTIVE FUNCTIONS OF THE INTESTINAL MUCOSA

In addition to luminal digestion of feed-derived macromolecules, the final stages of digestion are carried out by membrane-anchored enzymes at the brush-border region of the enterocyte cells. These include sucrase-isomaltase (SI), peptidases and phosphatases (Semenza, 1986).

Chickens have a considerable capacity to hydrolyse disaccharides via the brush-border enzyme SI immediately post-hatch and even pre-hatch (Uni, 1999; Uni *et al.*, 2003b). SI expression increases rapidly just before hatch and continues to increase post-hatch (Sklan *et al.*, 2003). The pre-hatch expression is intriguing because intestinal enterocytes have not yet been exposed to carbohydrates. This was explained by the homeobox gene *CDX* which is involved in inducing SI expression in mammals (Ferraris, 2001) and in chickens (Sklan *et al.*, 2003).

The major transporter for glucose is the sodium–glucose transporter (SGLT-1), which is expressed during the prenatal period. Expression of SGLT-1 increases 2 days pre-hatch and continues to increase post-hatch (Sklan *et al.*, 2003; Uni *et al.*, 2003b).

Another enzyme which plays a major role in protein digestion, via the hydrolysis of peptides, is the brush-border enzyme aminopeptidase. This enzyme is expressed at a relatively high constant level (compared to the SGLT and SI genes) from 15E until hatch (Uni *et al.*, 2003b).

For absorption to occur, sodium is required in the lumen and, once absorbed, must be removed from enterocytes to maintain ionic equilibrium. Sodium transport occurs via the enterocyte's basolateral Na^+/K^+-ATPase. The expression of this enzyme is detected 2 days prior to hatching (Uni *et al.*, 2003b).

An examination of the activities of the jejunal brush-border membrane enzymes maltase, aminopeptidase, SGLT-1 and ATPase during the last period of embryonic development revealed low activity of both maltase and aminopeptidase at 15E and 17E. Activities of all the enzymes and transporters examined began to increase at 19E and further increased on the day of hatch.

The relative mRNA expressions of the different enzymes and transporters were correlated, as were their activities ($r = 0.75$–0.96); however, expression was not correlated with the enzymatic activities.

In general, at 19E, the mRNA levels of the apical and basolateral digestive enzymes and nutrient transporters have significantly increased relative to their levels at 15E and 17E. This elevation in gene expression is followed by an increase in the biochemical activities of the enterocyte enzymes and transporters on the day of hatch.

The intestinal mucosal layer appears to change rapidly during the late pre-hatch period, and the late-term embryonic intestine seems to have the capacity to digest and absorb disaccharides and short peptides 2 days prior to hatch. These major changes in the expression and localization of the functional brush-border proteins prepare the framework for ingestion of carbohydrate- and protein-rich exogenous feed post-hatch.

Calculation of the total mucosal enzyme activity on a regional basis yields activity curves for disaccharides, gamma glutamyl transferase and alkaline phosphatase that increase curvilinearly with age. Regional activity of mucosal enzymes is related to the digestive capacity in the three different regions of the intestine: the jejunum has the highest regional capacity to digest disaccharides, the duodenum has the lowest capacity and the ileum is intermediate (Uni *et al.*, 1998a). From the end of the first week post-hatch, mucosal enzyme activity per mass of intestine is closely correlated to the number of enterocytes per villus, suggesting that the amount of enzyme activity expressed per enterocyte does not change greatly with age.

Development and function of mucous-secreting goblet cells

The epithelium of the GIT is covered by a mucous layer that acts as a medium for protection, lubrication and transport between the luminal contents and the epithelial cells. The mucous layer is composed predominantly of mucin glycoproteins produced by mucous-secreting goblet cells. Other components of the mucous layer are water, various serum and cellular macromolecules, electrolytes, microorganisms and sloughed cells.

Mucin-type glycoprotein is believed to be capable of aggregating several bacterial species and preventing the attachment of pathogenic bacteria by modulating their adherence to the intestinal epithelium. The mucous layer is a component of the innate host response which is regulated in response to inflammation and infection. A number of protective mucous layer-associated proteins are either co-secreted with mucins or interact with mucous environments to perform their protective action (Allen and Kent, 1968; Allen, 1981, 1989; Allen *et al.*, 1993).

Mucin production for maintenance of the mucous layer is the responsibility of the goblet cells. The goblet cells arise by mitosis from pluripotential stem cells at the base of the crypt or from poorly differentiated cells in the lower crypt region – referred to as oligomucous cells.

Mucins are categorized into three distinct families: gel-forming, soluble and membrane-bound. Secreted gel-forming mucins are characterized by their large size, viscoelastic behaviour and high content of carbohydrates. Secretory gel-forming mucin genes have a large central region made up of multiple tandem repeats which vary in length and sequence, but all have been found to contain threonine and/or serine residues. The amino- and carboxy-terminal regions (D-domains) are very lightly glycosylated, have a high cysteine content and are important to the mucin multimerization process. Mucin glycoproteins exhibit a high level of heterogeneity resulting from the diversity in length, composition, branching and degree of sulphation and acetylation of the oligosaccharide chains attached to the peptide backbone.

Mucins are classified into neutral and acidic subtypes, and are also distinguished by sulphated or non-sulphated groups. Expression of different mucins, defined by differences in either their protein backbones or their glycosylation patterns, has been shown to vary both between and within tissues, and during intestinal ontogenesis. Mucin molecules are assembled and stored in large, membrane-bound granules before secretion from the goblet cells. Secretory mucins are secreted from the apical surface of the goblet cells by either baseline secretion or compound exocytosis. Degradation of the mucous layer is part of the balance between synthesis secretion and breakdown of the mucous layer (Forstner, 1978; Forstner and Forstner, 1985, 1994; Forstner *et al.*, 1995).

An investigation of the ontogenesis and development of mucin-producing cells in the chicken small intestine (Uni *et al.*, 2003a) showed that mucin-producing cells can be detected from 17E, containing only acidic mucin. After hatch and until day 7 post-hatch, the small intestine contains similar proportions of goblet cells producing acidic and neutral mucins. An increasing gradient of goblet cell density was observed along the duodenal–ileal axis.

Thus, the development of small intestinal mucous-secreting cells in chicks occurs in the late embryonic and immediate post-hatch period.

An analysis of the expression of the chicken intestinal mucin gene (Smirnov *et al.*, 2004) revealed a sustained increase in mucin mRNA expression from 17E to day 3 post-hatch, at which point expression stabilized.

This increase in mucin mRNA levels parallels the increase in the number of goblet cells occurring from the late incubation stage through the first week post-hatch. Western-blot analysis showed the lowest amount of mucin glycoprotein to be in the duodenum, and an increase in content distally along the small intestine.

Once mucin is synthesized in the goblet cells and secreted to the intestinal surface, it forms a layer which undergoes continuous degradation and renewal. In 3-week-old chickens, the adhering mucous is of similar thickness throughout all segments of the mature small intestine.

Studies in chickens and turkeys have confirmed that feeding conditions, dietary composition and microbial flora influence the different parameters of mucin dynamics in the avian small intestine (Smirnov *et al.*, 2004, 2005). However, the role of mucin in absorption and protection against pathogens, and the age at which the mucin layer acts as a functional barrier, are not yet fully understood.

EARLY FEEDING AND GUT DEVELOPMENT

At hatch, the immature small intestine is physiologically active and primed for the activities of food assimilation (Uni *et al.*, 2003a). Delayed access to feed after hatch creates an abnormal physiological situation and leads to the negative consequences of a lack of enteral nutrition. Nutrient supply immediately after hatch is a critical factor for small intestinal development in chicks. The common practice in which feed is first made available to chicks more than 48 h post-hatch may depress subsequent development (Nunez *et al.*, 1996; Noy *et al.*, 2001).

Feeding immediately post-hatch led to an acceleration in intestinal morphological development, whereas late access to external feed resulted in delayed development of the small intestine's mucosal layer (Uni *et al.*, 1998a; Geyra *et al.*, 2001a; Uni and Ferket, 2003). Furthermore, birds denied access to first feed for 24 to 48 h exhibited decreased villus length (Yamauchi *et al.*, 1996), decreased crypt size and fewer crypts per villi, as well as decreased enterocytic migration rate (Geyra *et al.*, 2001b). This finding was accompanied by a decrease in the number of enterocytes and an increase in the density of goblet cells producing acid and neutral mucins in the jejunal and ileal villi at the end of the fasting period (Uni *et al.*, 2003a).

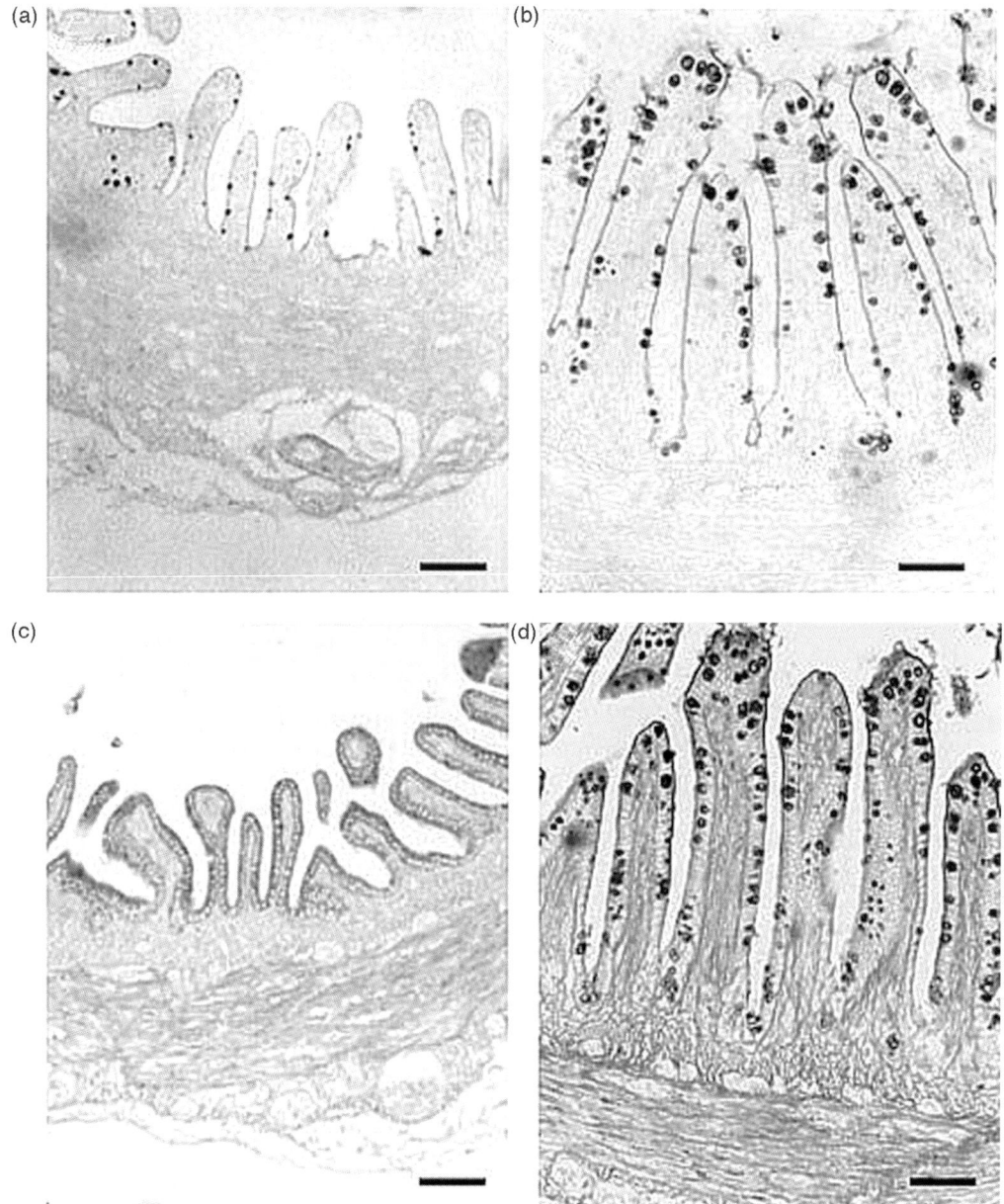

Fig. 3.3. Representative light micrographs of jejunum stained with Alcian blue (a, b) or periodic acid-Schiff (c, d). Goblet cells producing only acid mucin can be observed at 18 days (18E) of incubation (a, c). On the day of hatch (b, d) goblet cells producing both acidic and neutral mucin are observed. Crypts are not yet developed and goblet cells are distributed all along the villi. (Magnification × 200; bar = 50 μm at both stages).

Furthermore, the size of the cross-sectional area of an individual goblet cell in the small intestinal segments was significantly greater in fasting chicks than in controls (Uni *et al.*, 2003a). This accumulation of mucin in the goblet cells

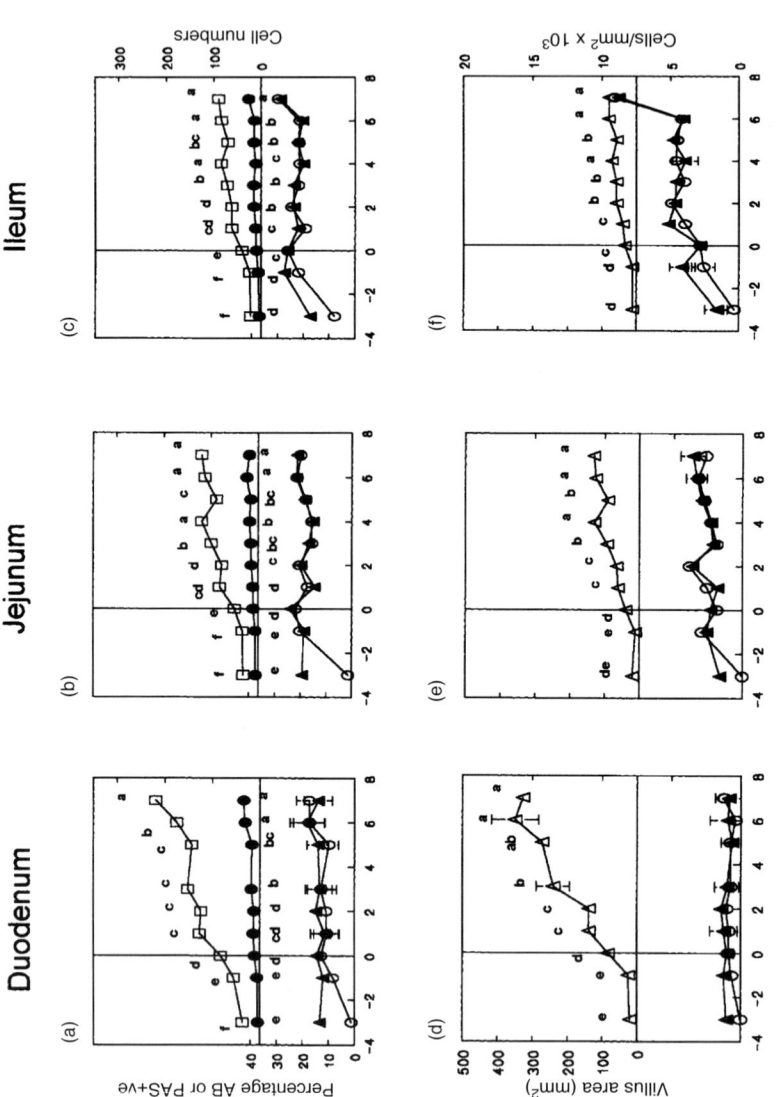

Fig. 3.4. a–c: number of enterocytes (□) and goblet cells (●) (right axis) and the percentage of cells staining positive for Alcian blue (AB+ve, ▲) or periodic acid-Schiff (PAS+ve, o) (left axis) per villus column in various regions of the small intestine of broiler embryos and chicks; d–f: villus surface area (△) (right axis) and density of goblet cell populations (AB+ve, ▲; PAS+ve, o) (right axis) in various regions of small intestine of broilers embryos and chicks. Values are means with standard errors represented by vertical bars (when they do not fall within the symbols). Differences between the percentage of AB+ve and PAS+ve were significant ($p < 0.05$) at 3 days before hatch in the duodenum, jejunum and ileum.

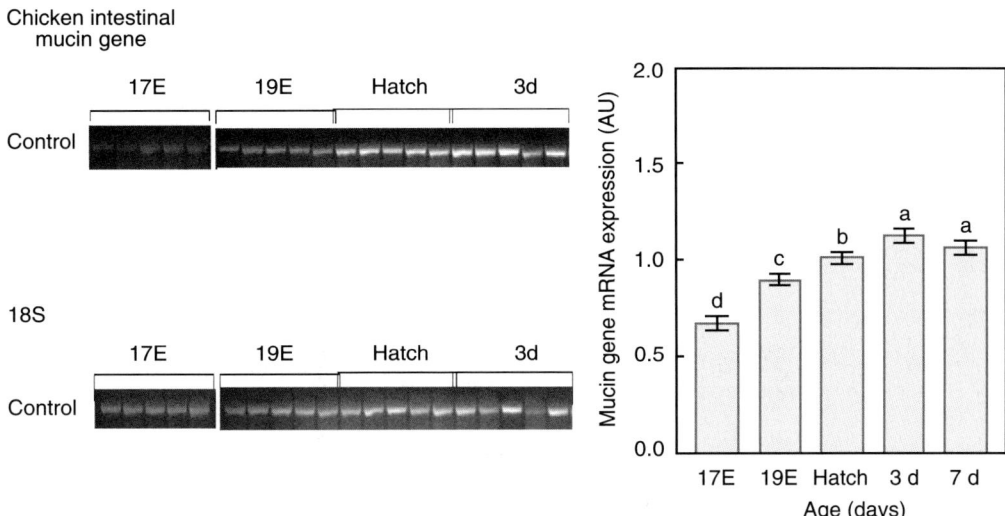

Fig. 3.5. Mucin gene expression in the chicken small intestine. Amplification of chicken intestinal mucin mRNA and 18 S RNA. Changes with age in mucin mRNA expression were measured by semi-quantitative RT-PCR and expressed relative to the expression of 18 S rRNA. Values are means ± SEM, $n = 5$. Means without a common letter are significantly different ($p < 0.05$).

might be due to its increased synthesis and/or decreased secretion. Mucin biosynthesis changes with fasting-associated alterations in the migration rate of epithelial cells from the proliferating crypt zones (Shea-Donohue *et al.*, 1985) and with perturbations in the differentiation rates of precursor cells into mature goblet cells (De Ritis *et al.*, 1975; Wattel *et al.*, 1979; Shub *et al.*, 1983).

Since the mucous layer has both protective and transport functions, time of access to feed in the first hours post-hatch is important to the normal development of the mucous cover in the chicken small intestine. Thus, since early access to feed is critical to intestinal development, nutrient supply to the intestine by early feeding (at the hatchery, in the delivery boxes or to the embryo before hatch by *in ovo* feeding methods) (Uni and Ferket, 2003) is expected to enhance development of the small intestine.

Tako *et al.* (2004, 2005) demonstrated that administration of exogenous nutrients and minerals into the amnion enhances intestinal development by increasing villus size and by generating early activation of the enterocyte brush-border membrane.

The intra-amniotic administration of carbohydrates, β-hydroxy-β-methylbutyrate (HMB), Na^+Cl^- ions and zinc-methionine led to elevated mRNA levels of the genes encoding nutrient enzymes and transporters, followed by their increased biochemical activities in the embryo. Research performed with early-feeding methods indicates that the small intestines of *in ovo*-fed hatchlings are at a functional stage of development equivalent to that of conventionally fed 2-day-old chicks.

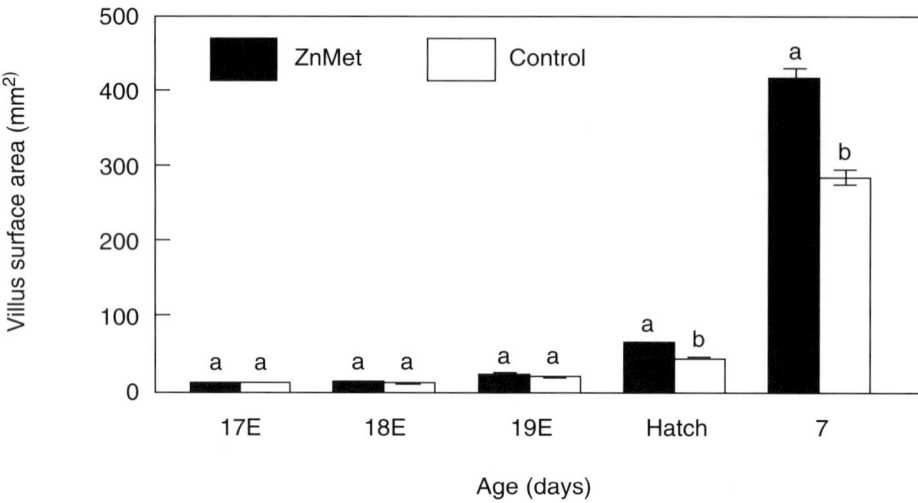

Fig. 3.6. Effect of intra-amniotic administration of Zinc-Methionine (ZnMet) on days 17E, 18E and 19E, day of hatch and 7 days post-hatch on the jejunal small intestinal villus surface area. Values are means ± SEM, $n = 4$. Means with different letters differ ($p < 0.05$).

SUMMARY

All aspects of the early development of the small intestine have important influences on later growth and development of the whole bird. The effects of early feeding and the immediate environment of the chick upon the initiation and development of GIT morphology, digestive and absorptive function and immunological and bioprotective barrier activities must all be considered in relation to husbandry and nutritional regimes and in the context of optimization of commercial production strategies and methods.

REFERENCES

Allen, A. and Kent, P.W. (1968) Biosynthesis of intestinal mucins. *Biochemical Journal* 106, 301–309.

Allen, A. (1981) Structure and function of gastrointestinal mucus. In: Johnson, L.R. (ed.) *Physiology of the Gastrointestinal Tract*. Raven Press, New York, pp. 617–639.

Allen, A. (1989) Gastrointestinal mucus. In: *Handbook of Physiology. The Gastrointestinal System. Salivary, Gastric, Pancreatic, and Hepatobiliary Secretion*. American Physiology Society, Bethesda, Maryland, pp. 359–382.

Allen, A., Flemstrom, G., Garner, A. and Kivilaakso, E. (1993) Gastroduodenal mucosal protection. *Physiological Reviews* 73, 823–857.

Cheng, H. and Leblond, C.P. (1974a) Origin, differentiation and renewal of the four main epithelial cell types in the mouse small intestine. I: columnar cell. *American Journal of Anatomy* 141, 461–479.

Cheng, H. and Leblond, C.P. (1974b) Origin, differentiation and renewal of the four main epithe-lial cells in the mouse small intestine. IV: unitarian theory of the origin of the four epithelial cell types. *American Journal of Anatomy* 141, 537–561.

De Ritis, G., Falchuk, Z.M. and Trier, J.S. (1975) Differentiation and maturation of cultured fetal rat jejunum. *Developmental Biology* 45, 304–317.

Ferraris, R.P. (2001) Dietary and developmental regulation of intestinal glucose transport. *Physiology Review* 77, 257–302.

Ferraris, R.P., Villenas, S.A., Hirayama, B.A. and Diamond, J. (1992) Effect of diet on glucose transporter site density along the intestinal crypt–villus axis. *American Journal of Physiology* 262, G1060–G1068.

Forstner, G.G. and Forstner, J.F. (1985) Structure and function of gastrointestinal mucus. In: Desnuelle, P., Sjostrom, H. and Noren, O. (eds) *Molecular and Cellular Biology of Digestion.* Vol. II, Elsevier Science Publishers B.V. (Biomedical Division), Amsterdam.

Forstner, J.F. (1978) Intestinal mucins in health and disease. *Digestion* 17, 234–263.

Forstner, J.F. and Forstner, G.G. (1994) Gastrointestinal mucus. In: Johnson, L.R. (ed.) *Physiology of the Gastrointestinal Tract*, 3rd edn. Raven Press, New York, pp. 1255–1284.

Forstner, J.F., Oliver, M.G. and Sylvester, F.A. (1995) Production, structure and biologic relevance of gastrointestinal mucins. In: Blaser, M.J., Smith, P.D., Ravdin, J.I., Greenberg, H.B. and Guerrant, R.L. (eds) *Infections of the Gastrointestinal Tract*, Raven Press, New York, pp. 7– 88.

Geyra, A., Uni, Z. and Sklan, D. (2001a) The effect of fasting at different ages on growth and tissue dynamics in the small intestine of the young chick. *British Journal of Nutrition* 86, 53–61.

Geyra, A., Uni, Z. and Sklan, D. (2001b) Enterocyte dynamics and mucosal development in the posthatch chick. *Poultry Science* 80, 776–782.

Marshall, J.A. and Dixon, K.E. (1978) Cell specialization in the epithelium of the small intestine of feeding *Xenopus laevis* tadpoles. *Journal of Anatomy* 126, 133–144.

McAvoy, J.W. and Dixon, K.E. (1978a) Cell specialization in the small intestinal epithelium of adult *Xenopus laevis*: functional aspects. *Journal of Anatomy*. 125, 237–245.

McAvoy, J.W. and Dixon, K.E. (1978b) Cell specialization in the small intestinal epithelium of adult *Xenopus laevis*: structural aspects. *Journal of Anatomy* 125, 155–169.

McAvoy, J.W., Dixon, K.E. and Marshall, J.A. (1975) Effects of differences in mitotic activity, stage of cell cycle, and degree of specialization of donor cells on nuclear transplantation in *Xenopus laevis*. *Developmental Biology* 45, 330–339.

Noy, Y., Geyra, A. and Sklan, D. (2001) The effect of early feeding on growth and small intestinal development in the posthatch poult. *Poultry Science* 80, 912–919.

Nunez, M.C., Bueno, J.D., Ayudarte, M.V., Almendros, A., Suarez, M.D. and Gol, A. (1996) Dietary restriction-induced biochemical and morphometric changes in the small intestine of nursing piglets. *Journal of Nutrition* 126, 933–944.

Quaroni, A. (1985a) Crypt cell development in newborn rat small intestine. *Journal of Cellular Biology* 100, 1601–1610.

Quaroni, A. (1985b) Pre- and postnatal development of differentiated functions in rat intestinal epithelial cells. *Developmental Biology* 111, 280–292.

Semenza, G. (1986) Anchoring and biosynthesis of stalked brush border membrane proteins: gly-cosidases and peptidases of enterocytes and renal tubuli. *Annual Review of Cell Biology* 2, 2255–2313.

Shea-Donohue, T., Dorval, T.E.D., Montcalm, E., El-Bayer, H., Durakovich, A., Conklin, J.J. and Dubois, A. (1985) Alterations in gastric mucus secretion in rhesus monkeys after exposure to ionizing radiation. *Gastroenterology* 88, 685–690.

Shub, M.D., Pang, K.Y., Swann, D. and Walker, W.A. (1983) Age-related changes in chemical composition and physical properties of mucus glycoprotein from rat small intestine. *Biochemical Journal* 215, 405–411.

Sklan, D., Geyra, A., Tako, E., Gal-Gerber, O. and Uni, Z. (2003) Ontogeny of brush border carbohydrate digestion and uptake in the chick. *British Journal of Nutrition* 89, 747–753.

Smirnov, A., Sklan, D. and Uni, Z. (2004) Mucin dynamics in the chick small intestine are altered by starvation. *Journal of Nutrition* 134, 736–742.

Smirnov, A., Perez, R., Amit-Romach, E., Sklan, D. and Uni, Z. (2005) Mucin dynamics and microbial populations in chicken small intestine are changed by dietary probiotic and antibiotic growth promoter supplementation. *Journal of Nutrition* 135, 187–192.

Tako, E., Ferket, P.R. and Uni, Z. (2004) The effects of *in ovo* feeding of carbohydrates and beta-hydroxy-beta-methylbutyrate on the development of chicken intestine. *Poultry Science* 83, 2023–2028.

Tako, E., Ferket, P.R. and Uni, Z. (2005) Changes in chicken intestinal Zinc exporter (ZnT1) mRNA expression and small intestine functionality following an intra-amniotic Zinc-Methionine (ZnMet) administration. *Journal of Nutritional Biochemistry* 16, 339–346.

Traber, P.G., Gumucio, D.L. and Wang, W. (1991) Isolation of intestinal epithelial cells for the study of differential gene expression along the crypt–villus axis. *American Journal of Physiology* 260, 895–903.

Traber, P.G., Yu, L., We, G. and Judge, J. (1992) Sucrase-Isomaltase gene expression along the crypt-villus axis of human small intestine is regulated at the level of mRNA abundance. *American Journal of Physiology* 262, G123–G130.

Uni, Z. (1999) Functional development of the small intestine in domestic birds: cellular and molecular aspects. *Poultry and Avian Biology Reviews* 10, 167–179.

Uni, Z. and Ferket, P.R. (2003) Enhancement of development of oviparous species by *in ovo* feeding. *Patent number US 6,592,878.*

Uni, Z., Noy, Y. and Sklan, D. (1995) Posthatch changes in morphology and function of the small intestines in heavy- and light-strain chicks. *Poultry Science* 74, 1622–1629.

Uni, Z., Ganot, S. and Sklan, D. (1998a) Posthatch development of mucosal function in the broiler small intestine. *Poultry Science* 77, 75–82.

Uni, Z., Platin, R. and Sklan, D. (1998b) Cell proliferation in chicken intestinal epithelium occurs both in the crypt and along the villus. *Journal of Comparative Physiology* (B) 168, 24– 247.

Uni, Z., Noy, Y. and Sklan, D. (1999) Posthatch development of small intestinal function in the poult. *Poultry Science* 78, 215–222.

Uni, Z., Smirnov, A. and Sklan, D. (2003a) Pre- and posthatch development of goblet cells in the broiler small intestine: effect of delayed access to feed. *Poultry Science* 82, 320–327.

Uni, Z., Tako, E., Gal-Garber, O. and Sklan, D. (2003b) Morphological, molecular, and functional changes in the chicken small intestine of the late-term embryo. *Poultry Science* 82, 1747–1754.

Wattel, W., Van Huis, G.A., Kramer, M.F. and Geuze, J.J. (1979) Glycoprotein synthesis in the mucous cell of the vascularly perfused rat stomach. *American Journal of Anatomy* 156, 313–320.

Wurth, M.A. and Musacchia, X.J. (1964) Renewal of intestinal epithelium in the freshwater turtle, *Chrysemys picta. Anatomical Record* 148, 427–439.

Yamauchi, K., Kamisoyama, H. and Isshiki, Y. (1996) Effects of fasting and re-feeding on structures of the intestinal villi and epithelial cells in White Leghorn hens. *British Poultry Science* 37, 909–921.

CHAPTER *4*
Absorptive function of the small intestine: adaptations meeting demand

M.A. Mitchell[1]* and M. Moretó[2]

[1]*Scottish Agricultural College, Bush Estate, Midlothian, UK;*
[2]*Departament de Fisiologia, Facultat de Farmàcia, Barcelona, Spain;*
**e-mail: malcolm.mitchell5@btinternet.com*

ABSTRACT

The mucosal epithelium of the small intestine constitutes a highly dynamic interface with the external environment through the delivery, digestive processing and absorption of nutrients. The intestinal mucosa is capable of rapid and extensive morphological and functional adaptation in response to evolutionary changes, genetic selection, altered dietary composition or food availability and pathological processes. The function of the intestine and adaptations therein may be studied at the molecular, cellular, organic or whole animal level, employing a variety of *in vitro* and *in vivo* techniques. This review addresses the mechanisms mediating adaptations in mucosal structure and function in chickens in response to selection for production traits and in the face of thermal stress and of changes in dietary sodium intake. The data presented draw upon the outputs of studies using membrane vesicles, isolated enterocytes, *in vitro* sheets and *in vivo* perfusion preparations. The findings of the different methodologies and approaches are compared and contrasted and their contribution to characterization of digestive and absorptive function in the whole animal is considered. These studies help elucidate the basic mechanisms involved in nutrient absorption from the small and large intestine in the fowl and an understanding of the adaptive potential of the mucosa and the underlying mechanisms may provide the sound scientific basis for nutritional strategies that exploit the physiological adaptations induced by the requirement for improved growth rates or efficiency of production under a wide range of climatic conditions.

INTRODUCTION

The mucosa of the gastrointestinal tract is a functional interface between the environment and the internal physiological compartments of the organism. As such, the mucosal and associated cells constitute a dynamic and metabolically active barrier possessing selective permeability (Baumgart and Dignass, 2002). This barrier has multiple functions that involve the digestion, transport and uptake of specific substances and nutrients and exclusion of microorganisms and toxins. The processes of translocation (motility or peristalsis), digestion and absorption occur in a micro-environment modified by the intestinal mucosa (e.g. secretion) and the ancillary organs (pancreas, liver). The importance of 'the intestinal barrier' in understanding gut function and gut health in poultry has been recently reviewed (Hughes, 2005).

Optimal digestive and absorptive functions are essential for growth, development and health of the animal. In addition, the intestine must act as a physical barrier to pathogenic organisms and toxins and play a role in both innate and acquired immunity. The integration of the digestive, absorptive and immune function of the gastrointestinal tract and the genetic regulation of these processes are central to animal production and health.

This chapter focuses upon absorptive function of the small intestine and on its capacity for adaptive responses in the face of varying demands and challenges. Areas of interest that have not been addressed herein but which have been discussed recently elsewhere include the effects of specific pathologies upon absorptive function, e.g. 'malabsorption syndrome' (Rebel *et al.*, 2006) and the effects of antibiotics and growth promoters on intestinal structure and function (Miles *et al.*, 2006).

Whilst information relating to fundamental mechanisms of absorption and adaptation may be drawn from research upon many species, particular attention is paid to examples in avian species and, where possible, specific findings in poultry. This is pertinent because of the central role of nutrient assimilation and processing in the full expression of the genetic potential in modern, highly selected lines of poultry. Thus optimum production in both meat birds and egg layers is dependent upon the efficient digestion and absorption of nutrients and their provision in the required amounts and forms to the various biochemical pathways underpinning the desirable production traits. Whilst laying hens have a high requirement for absorption of nutrients including carbohydrate, protein, lipid and calcium in sustaining high levels of egg production, the demands of very rapid growth rate in meat-type birds present unique challenges for the development and function of the intestinal absorptive epithelium.

Artificial genetic selection has proved extremely successful in the improvement of growth rate and food conversion efficiency in the commercial broiler chicken. Many of the physiological mechanisms mediating these responses and their consequences have yet to be fully characterized (Scanes, 1987; Siegel and Dunnington, 1987). It is clear, however, that in order to support the increased growth rate of the 'demand organs' such as muscle, bone, fat, skin and feathers, appropriate adaptations must occur in the 'supply organs' including the intestine, liver, cardiovascular and respiratory systems (Lilja, 1983; Lilja *et al.*, 1985; McNabb *et al.*, 1989).

It has been postulated that failure of the supply systems to meet the demands of a disproportionately large growth rate may underlie many of the current pathologies and welfare problems encountered in commercial broiler production (Savory, 1995). It has also been suggested that the apparent increased susceptibility to 'stress' of the modern broiler fowl may also be attributable to the exploitation of the genetic potential for growth in these birds in the absence of compensatory development of the corresponding homeostatic and regulatory mechanisms (Teeter, 1994).

The mucosal epithelium of the small intestine constitutes a highly dynamic interface with the external environment through the delivery, processing and absorption of nutrients. The intestinal mucosa is capable of rapid and extensive morphological and functional adaptation in response to evolutionary, genetic and ontogenetic development demands (Ferraris, 2001), as well as to environmental and nutritional challenges (Julian, 2005).

In the domestic fowl, intensive artificial genetic selection has resulted in commercial meat breeds (broiler chickens) whose growth rates and food conversion efficiencies greatly exceed those of their genetic predecessors (Griffin and Goddard, 1994). Central to the balance of physiological supply and demand in the rapidly growing broiler chicken is the availability and uptake of essential nutrients. There is evidence that genetic selection for growth rate has resulted in both anatomical and physiological adaptations in the small intestine of the broiler which are consistent with improved efficiency of digestion and absorption of dietary substrates under normal production conditions (Mitchell and Smith, 1990, 1991; Smith *et al.*, 1990).

As it has been proposed, however, that the capacity for the intestine to absorb and assimilate nutrients may pose a proximal constraint upon the rate of growth in precocial species – including the domestic fowl (Obst and Diamond, 1992) – it is essential that feeding regimes and dietary composition are optimized to take full advantage of the available absorptive mechanisms. Changes in dietary formulation may be necessary if nutrient requirements, for optimum performance, are modified in the face of environmental stressors but, in turn, may be further influenced by any concomitant adaptive response in intestinal function.

It may be proposed that the increased growth of 'demand' tissues such as skeletal muscle should be accompanied by appropriate adaptations in structure and function of 'support tissues' such as the gastrointestinal tract (Rance *et al.*, 2002). Indeed, previous studies have demonstrated such adaptations in the small intestine of the highly selected broiler chicken in terms of crypt cell dynamics and enterocyte migration rates (Smith *et al.*, 1990). Such birds, however, appear to have a small intestinal mucosal compartment (relatively smaller, compared to body size, than that in genetically unselected lines (Mitchell and Smith, 1991)), and it is proposed that functional adaptations in terms of nutrient absorption at the enterocyte level have supported the increased demands of elevated growth rate (Mitchell and Smith, 1990). It is now considered that nutrient absorption and adaptations therein may represent the rate-limiting step in further genetic improvements in broiler chicken growth rates (Croom *et al.*, 1998, 1999).

ADAPTATIONS IN INTESTINAL ABSORPTIVE FUNCTION

Adaptive regulation of nutrient (glucose and amino acid) transportation in the vertebrate intestine has been reviewed in the past (Karasov and Diamond, 1983, 1987; Jenkins and Thomson, 1994; Ferraris and Diamond, 1997). It is proposed that many mechanisms have been involved in adaptation, including mucosal mass, specific transport systems and the sodium gradient, and the signals for adaptive responses included the effects of dietary solutes, starvation, hyperphagia, dietary bulk, diabetes, intestinal position, intestinal resection, hibernation, lactation and ageing, and also differences between species.

Major conclusions (e.g. Karasov and Diamond, 1983) were that increased metabolic requirements should be met by increased absorption involving increased mucosal mass, whereas nutritionally essential and non-essential solutes used as caloric sources should, respectively, repress and induce their own transport. A later review addressed the nature of the mechanisms of adaptation in intestinal sugar transport (Ferraris and Diamond, 1997).

Non-specific adaptations are described that lead to parallel changes in transport of different nutrients involving alterations in mucosal surface area and the ratio of absorbing to non-absorbing cells. Specific adaptations, for single substrates, considered include changes in transporter site densities and affinity constants. The effects upon sugar transport of changes in diet, energy budgets and environmental salinity, as well as intestinal resection, starvation, age and stress, are discussed. This work also emphasizes the importance of different routes of substrate uptake, including the transporter-mediated pathway and uptake by the non-specific paracellular route. Clearly many, but not all, of the physiological conditions and challenges described above are pertinent to avians and specifically to poultry, and where information is available these have been addressed.

Evolutionary adaptation

The function of the intestinal epithelial cells or enterocytes may change or adapt in response to a wide number of demands or challenges. These include the following: (i) evolutionary forces; (ii) genetic selection (natural or artificial); (iii) responses to altered physiological status; (iv) environmental challenges or constraints upon nutrient availability; (v) dietary composition; (vi) exposure to anti-nutrients or toxins; (vii) disease states; and (viii) external human factors such as surgical interventions.

The most essential and basic nutrients whose absorptive processes are most likely to exhibit adaptive regulation are sugars and amino acids (Karasov and Hume, 1997) and, in particular, the transport of hexose sugars has received a great deal of attention as glucose absorption is a fundamental major pathway in the intestines of all species. Glucose absorption is the most studied topic in this field and many major insights into the understanding of transport mechanisms have evolved from these studies (Reuss, 2000). Glucose transport mechanisms represent the key model for understanding nutrient absorption systems. Adaptations in this process have been studied widely (Karasov, 1988).

The specific nutrient requirements of the different classes and species of vertebrates have resulted in the development of matched digestive and absorptive functions. Higher vertebrates may be divided into two groups: those which eat diets with a high proportion of carbohydrates (herbivores and omnivores) and those that eat relatively carbohydrate-free diets (carnivores). Vertebrates differ dramatically in the amounts of food they eat daily, and daily metabolic rates and food requirements of endotherms (e.g. mammals and birds) are generally of an order of magnitude higher than those of ectothermic fish, amphibians and reptiles (Karasov, 1988).

Thus the uptake of glucose and amino acids in the small intestine of mammals and birds is around 13-fold greater than in reptiles and fish. This is achieved by changes in intestinal morphology and functional capacity of the absorptive mechanisms (Karasov, 1987, 1988). Differences in morphological features such as villus size and microvillus length have been demonstrated between mammals and reptiles and correlated with turnover numbers for glucose transporter sites and transporter density, representing a major contribution to regional and species differences in glucose uptake (Ferraris *et al.*, 1989). The intestinal absorptive epithelium exhibits an unusual feature related to this evolutionary development of physiological function (Diamond and Hammond, 1992).

Generally, natural selection eliminates unutilized capacities because of their costs. In the case of intestinal absorptive capacity this often exceeds natural nutrient loads by an order of two- to threefold. This apparent uneconomic imbalance may well underpin the adaptive capacity of the small intestine and thus acts as a buffer against changing demand, availability and requirement. Such spare capacities may play a very important role in the responses of animals to their changing environments and physiological status (Karasov and McWilliams, 2005). It may be argued that in the development of flight, birds may have required a highly efficient gut to maximize absorption for the minimum organ weight (Karasov, 1987, 1988). In the selection of modern poultry for high production efficiency these evolutionary processes may have been exploited to optimize digestive absorptive efficiency (Mitchell and Smith, 1990, 1991).

Whilst the mechanisms of carbohydrate and amino acid absorption are fundamentally similar in mammals and avians and have been well characterized (Miller *et al.*, 1973; Boorman, 1976; Levin, 1976, 1984; Lerner, 1984; Austic, 1985; Vinardell, 1992), adaptive responses in avian absorption have received relatively little attention (Levin and Mitchell, 1982). It may be safely assumed, however, that many adaptations exhibited by mammals will also be observed in birds. In the domestic fowl the capacity to absorb nutrients depends mainly on the mucosal surface area of the small and large intestine, as well as on the functional properties of the specific nutrient transporters present in the brush border and the basolateral membranes.

There are specific transport systems for the major dietary hexoses. Glucose is absorbed across the apical sodium-dependent SGLT-1 system, expressed along the small and large intestine (Ferrer *et al.*, 1994; Garriga *et al.*, 1999b; Barfull *et al.*, 2002); fructose is taken up by the apical-facilitated GLUT5-type system (Garriga *et al.*, 2004) and both sugars are transported to the interstitial compartment through the basolateral GLUT2 transporter (Garriga *et al.*, 1997, 2000). Detailed studies, however, of the responses by these systems during adaptations in poultry intestinal function are in their infancy.

Ontogenetic adaptation

The efficient development of the small intestine, its structure and digestive and absorptive functions immediately prior to and around birth or hatch and during the early stages of growth is vital to the further development of the organism. In the case of commercial poultry it is recognized that early intestinal function is related to later full expression of growth potential and other production traits. This topic has been reviewed by Sklan (2001) and is described elsewhere in this volume (Uni, Chapter 3, this volume).

Ontogenetic development of absorptive function has been examined in turkeys (Sklan and Noy, 2003; Applegate *et al.*, 2005) and in ducks (King *et al.*, 2000; Watkins *et al.*, 2004). It is concluded that absorptive capacity increases in proportion to growth rate and body weight in order to sustain these functions, but that absorption rate (depending upon the parameter upon which this is calculated) may decrease after the first day or few days following hatch. In the chicken, monosaccharide absorption measured *in vivo* and expressed per unit body weight decreases with age (Gonzalez *et al.*, 1996). Sodium-dependent glucose uptake of brush border membrane vesicles prepared from jejunal tissue of the chicken is greatest during the first week post-hatch and then declines, as does the density of the SGLT-1 brush border membrane glucose transporter (Vasquez *et al.*, 1997; Barfull *et al.*, 2002).

Interestingly, similar developmental changes have been observed in the red jungle fowl, a wild genetic precursor of the domesticated chicken, which exhibited a decreased specific absorption of glucose with age post-hatch but an increased absorptive capacity primarily related to body weight, which was mainly attributed to a corresponding increase in gut mass and the size of the absorptive compartment (Jackson and Diamond, 1995).

Of particular interest to commercial broiler production is the fact that the digestive/absorptive mechanisms of the chick may be stimulated towards more rapid maturation by early feeding (see Uni, Chapter 3, this volume). Delayed access to feed and water can impede development of the absorptive functions of turkey poults (Corless and Sell, 1999), whereas early provision of substrates promotes development of absorptive function (Noy and Pinchasov, 1993; Sklan, 2001). Early food deprivation, which is common in commercial practice, reduces crypt depth, crypt cell proliferation, crypts per villus, villus height and crypt cell migration rates in comparison to fed chicks, and clearly would therefore depress subsequent development and absorptive function (Geyra *et al.*, 2001).

The important transporter systems are present in the pre-hatch embryo. Expression of the mRNA for SGLT-1 can be demonstrated from days 15 or 17 of incubation and the activity of the transporter and other brush border membrane enzymes has been shown to increase at day 19 (Sklan *et al.*, 2003; Uni *et al.*, 2003).

It is possible that manipulation of the transporter expression and function *in ovo* may facilitate increased absorptive activity from hatch in rapidly growing birds, thus improving nutrient assimilation and early growth and efficiency. In general it may be concluded that the elevated requirements for nutrient absorption required in the rapidly growing broiler chicken are met by an elevated specific transporter activity (which declines marginally with age), coupled with a major increase in the size of the transport compartment in terms of active mature enterocytes and total absorptive surface area.

Fasting and food restriction

Fasting, restricted feeding and re-alimentation possibly represent the most studied stimuli for intestinal adaptation. These treatments have been employed to elucidate the mechanisms involved in whole gut and mucosal adaptations in many species (Ferraris and Carey, 2000). Early studies demonstrated that fasting may increase absorption per unit of active mucosa or gut weight, although total absorption by the animal might be reduced (Levin, 1974). In mammals it is postulated that apparently increased glucose absorption during fasting might represent an adaptation supporting cerebral function in the face of inanition, but the associated more generalized reductions in glucose utilization (Das *et al.*, 2001). Caloric restriction also stimulates sugar absorption in a manner similar to that of short-term fasting (Casirola *et al.*, 1996).

Most recent studies have developed an integrated model of the mechanisms mediating the cellular adaptations occurring during fasting and re-feeding (Habold *et al.*, 2005). These studies have examined the effects of these treatments upon gene and protein abundance and cellular localization of glucose transporters (SGLT-1, GLUT-2, GLUT-5) as well as those of key gluconeogenic enzymes in enterocytes from rats. It was concluded that increased expression of the SGLT-1 during fasting may allow efficient absorption of glucose from low concentrations as soon as food became available. Reductions in basolateral GLUT-2 abundance may be compensated for by transfer of glucose to the blood through membrane trafficking, involving enhanced glucose-6-phosphatase activity. Fasting-induced increased passive permeability of the mucosa also increases the potential for glucose uptake via the paracellular route. Finally, re-feeding triggered synthesis of GLUT-2 and GLUT-5 and apical (brush border) recruitment of GLUT-2 in order to facilitate greater glucose absorption as the substrate became available.

Responses in hexose absorption in the domestic fowl during fasting have also been studied in detail (Levin and Mitchell, 1976, 1979a, b, 1980, 1982). These studies used transport kinetic analysis of *in vivo* absorption data corrected for the presence of an unstirred layer of water and reported increased transport function at the cellular level, as indicated by increased maximum transport capacities (V_{max} or J_{max}) and/or increased transporter affinity or reduced K_ms. These studies did not quantify the transporter numbers or measure the expression of the associated genes but it may be proposed that similar effects to those observed in mammals may responsible for the adaptations in transport.

More recent studies indicate that the morphological changes in the intestine and mucosa in response to fasting for 1, 2 or 3 days are similar to those observed in mammals in terms of villus height, surface cell area and mitotic numbers (Yamamauchi and Tarachai, 2000). *In vivo* studies have confirmed the increase in duodenal absorption of glucose in acutely or intermittently starved chicks (Rayo *et al.*, 1992). Indeed, these procedures were able to demonstrate the effect of age in young birds, which is generally the reduction in uptake rates (*vide infra*). Gal-Garber *et al.* (2000) described a lower transport affinity but higher transporter activity (SGLT-1) in brush border membrane vesicles from food-deprived chicks compared to that in fed controls. Increased expression of transporter mRNA was thought to reflect up-regulation by low substrate concentrations. These findings are consistent with the corresponding knowledge in mammalian systems.

Most recently, Maneewan and Yamauchi (2005) have described the restoration of histological appearance of villi and apparent absorptive function by the re-feeding of single nutrients (protein, fat or carbohydrate) in previously fasted chickens compared to birds receiving 'formula diets'. It was concluded that a balanced or complete formulation was required to restore all aspects of structure and function to the absorptive compartment.

Dietary composition

The concept of dietary modulation of absorptive function has been explored extensively in mammalian systems and a number of hypotheses relating to the underlying mechanisms have been proposed (Karasov, 1992; Ferraris, 2001). In wild birds seasonal changes in diet result in adaptive responses in the absorption rates of individual substrates (Afik *et al.*, 1997). The effects of different protein/carbohydrate dietary compositions upon intestinal function have been described in wild and domestic turkeys and the findings suggest that the commercially produced strains are able to regulate the digestive and absorptive functions of protein in response to altered composition (McCann *et al.*, 2000; Foye and Black, 2003).

The effects of dietary fibre have also been demonstrated in domestic fowl. The absorption of both hexose and pentose sugars has been shown to respond to the inclusion of grass to increase the fibre content of standard diets, and differential responses were observed in different regions of the intestine (Savory and Mitchell, 1991; Savory, 1992). In turkey poults early feeding of high dietary carbohydrate up-regulates the glucose transport system in the small intestine and stimulates mucosal carbohydrate digestion enzyme activity, e.g. maltase (Suvarna *et al.*, 2005). This, of course, is an important finding in relation to the design of dietary formulations for early feeding and enhanced maturation of the digestive/absorptive systems. Glucose absorption in the fowl is also known to respond to altered dietary sodium (Barfull *et al.*, 2002;

Gal-Garber *et al.*, 2003) and the mechanisms mediating these adaptive responses have been studied in some detail, as described below.

Adaptations to dietary sodium intake

In chickens, glucose is absorbed in the small intestine as well as in some regions of the large intestine. Indeed, the proximal caecum (Ferrer *et al.*, 1994) and the rectum or colon (Amat *et al.*, 1996) can also take up sugars against a concentration gradient. Although the total sugar uptake capacity of the distal regions is low, due to its relatively small surface compared to that of the small intestine, it role is not negligible as it may account for up to 5% of total intestinal absorption capacity.

The uptake of sugars by distal regions may serve two functions: (i) to absorb hexose not absorbed in the small intestine or present in the contents emptied by the caecum; and (ii) to reabsorb glucose excreted in the urine and appearing the rectum and caecum by retro-peristalsis from the coprodeum. In the chicken intestinal distal regions, glucose is transported by an SGLT-1 isoform that is very similar (if not identical) to the protein present in the small intestine (Garriga *et al.*, 1999b). SGLT-1 is located in the apical or brush border membrane, and transports glucose and galactose along with Na^+ from the lumen to the cytosol.

Sugar absorption is a very well-regulated process in most vertebrates. In mammals, SGLT-1 responds to changes in luminal glucose concentration, it follows a circadian rhythm and can adapt to several physiological and pathological conditions (Ferraris, 2001). However, in some wild birds, dietary carbohydrate does not affect its uptake and this has been attributed to the high predominance of the passive paracellular absorption compared to mediated uptake in these species (Chediak *et al.*, 2001).

Ferraris (2001) suggested that SGLT-1 can be regulated by luminal Na^+, as dietary Na^+ can affect the expression of SGLT-1. However, the current results strongly indicate that these effects are mediated by the renin–angiotensin–aldosterone (RAA) system, rather than by direct effects of Na^+ at the sugar absorption site. It is interesting to point out that a low-Na^+ diet reduces the amount of SGLT-1 in the apical membrane of the proximal caecum and rectum, as well as of the the distal ileum. This is therefore an effect that is restricted to distal regions of the chicken intestine, while the absorptive function in the duodenum and jejunum is unaltered. This means that the basic needs of glucose are still fulfilled, even in strictly long-term low-Na^+ conditions.

The role of the RAA system in the regulation of glucose transport capacity of the chicken intestinal distal regions is well established (see Fig. 4.1).

First, there is a close correlation between Na^+ intake, plasma aldosterone concentration and SGLT-1 expression (Garriga *et al.*, 2000); and, secondly, the low-Na^+ effects can be mimicked by the administration of subcutaneous aldosterone to chickens fed a high-Na^+ diet (Garriga *et al.*, 2001). Although an aldosterone-mediated mechanism can apparently explain all or most of the effects of low-Na^+ dietary intake on glucose uptake, other hormones may be

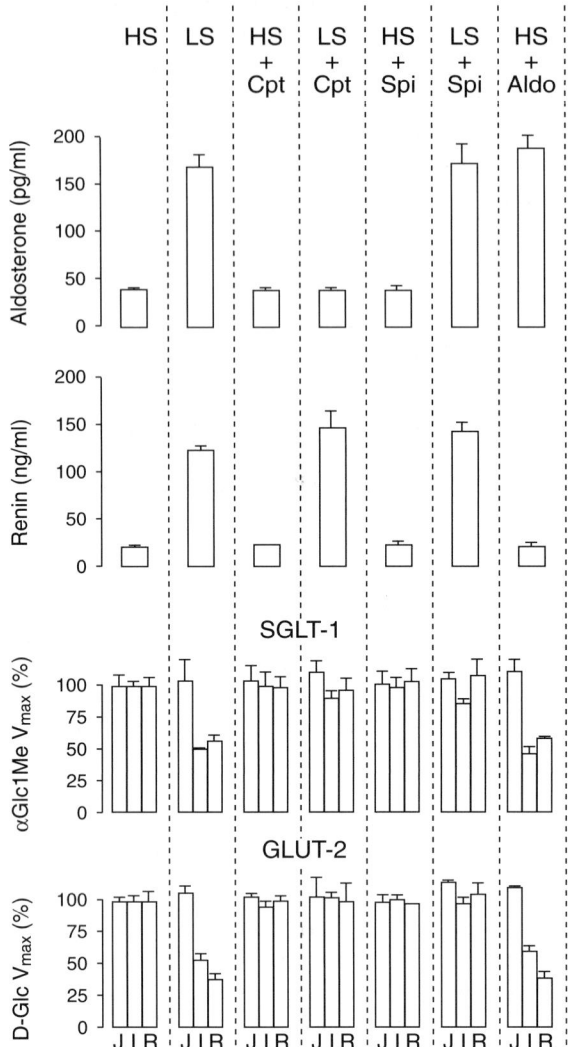

Fig. 4.1. Effects of low- and high-sodium diets on plasma renin and angiotensin concentrations and relative uptakes of hexoses by the SGLT-1 brush border and GLUT-2 basolateral membrane system. HS, high sodium intake; LS, low sodium intake; Cpt, Captopril, angiotensin-converting enzyme inhibitor; Spi, Spironolactone, mineralcorticoid receptor inhibitor; Aldo, aldosterone administered subcutaneously by osmotic mini-pump; SGLT-1, apical glucose transporter; GLUT-2, basolateral glucose transporter; D-Glc, D-glucose; αGlc1Me, glucose analogue, specific for SGLT-1; V_{max}, maximal transport capacity; J, jejunum; I, ileum; R, rectum (Garriga *et al.*, 2001).

involved (De la Horra *et al.*, 2001). The response affects both apical and basolateral glucose transporters of distal intestine, the effect being more dramatic in the rectum that in the distal ileum (Garriga *et al.*, 1999a) and, once established, it is reversed within 4 h after resalination (Garriga *et al.*, 2000), suggesting that regulation involves the control of vesicle trafficking to the plasma membrane of mature enterocytes.

Further studies demonstrated that the reduction in the number of Na^+-dependent D-glucose co-transporters by aldosterone does not affect the amount of SGLT-1 mRNA, supporting the view that the effects of dietary Na^+ on intestinal hexose transport involve post-transcriptional regulation of SGLT-1 (Barfull *et al.*, 2002).

The reduction of SGLT-1 expression is paralleled by up-regulation of the Na^+, H^+ exchanger and the amiloride-sensitive Na^+ channel in the distal intestine (Donowitz *et al.*, 1998; Laverty *et al.*, 2006). This indicates that the adaptive mechanisms in response to low-Na^+ intakes give priority to the homeostasis of Na^+, probably because the variable plasma Na concentration occupies a relevant position in homeostatic hierarchy ranking. Suppression of Na^+–glucose uptake at the distal intestine suggests that this mechanism is not efficient enough to fulfil the supply of Na^+ during low-Na^+ intakes and that other systems with higher capacity to scavenge luminal Na^+ are required.

Heat stress

It is well established that chronic heat stress reduces growth rate and the feed conversion efficiency in broiler chickens (Geraert *et al.*, 1996a; Temim *et al.*, 2000). Whilst these effects are in part attributable to the hyperthermia-induced decrease in food intake, growth depression may also be mediated directly by the associated metabolic and endocrine responses (Geraert *et al.*, 1996b; Lin *et al.*, 2004), as indicated by paired feeding studies. Broiler chickens are more susceptible to heat stress than are slower growing domestic fowl, and both their adaptive and pathological responses to extended thermal challenge involve multiple organs and systems (Altan *et al.*, 2003). It is therefore possible that both inanition and changes in metabolic and endocrine status will induce adaptive responses in intestinal absorptive function.

The ability of the vertebrate intestine to adapt during altered nutrient availability and thermal challenges is well recognized (Karasov and Diamond, 1983; Diamond and Karasov, 1987; Mitchell and Carlisle, 1992; Levin, 1994), and characterization of the mechanisms mediating the responses in the absorption of individual nutrients during exposure to specific stressors will underpin the design of appropriate nutritional strategies.

Adaptations in absorption of substrates from the small intestine in response to chronic heat stress have received detailed examination. One series of studies has characterized the effects of elevated environmental temperatures upon *in vivo* absorption of a hexose sugar and an amino acid *in vivo*, which was followed by elucidation of the cellular mechanisms using isolated enterocyte preparations and, finally, the full characterization of the alterations in transporter functions employing preparations of membrane vesicles. These studies are described in detail below.

In chickens, elevated environmental temperature reduces food intake but apparently increases the capacity of the intestinal epithelium to take up sugars.

Thus moderate chronic heat stress may induce an increased potential for hexose and amino acid absorption per unit of tissue mass (e.g. dry weight) in the small intestine of the broiler chicken, and this response may be mediated by changes in both active uptake of substrate by the enterocytes and an elevated passive transfer through the absorptive epithelium (Mitchell and Carlisle, 1992).

This apparently enhanced absorption capacity was confirmed in *in vitro* studies where isolated enterocytes from chronically heat-adapted birds showed a 50% increase in galactose accumulation ratio when compared to cells from control chickens (Mitchell *et al.*, 1995). However, the precise mechanisms of this adaptation and the contributions of reduced food intake associated with heat stress and hyperthermia *per se* were not clear. It was suggested that the enhanced uptake of galactose and methionine observed during heat stress reflected an adaptation to optimize nutrient absorption in reponse to reduced luminal nutrition and decreases in the size of the absorptive compartment.

Further mechanistic studies elucidated the systems mediating the altered hexose absorption and established whether the effects were attributable to lower food intake or to heat stress per se (Garriga *et al.*, 2006). Glucose transport kinetics were examined in preparations of apical and basolateral membranes of the jejunum of broiler chickens maintained in climatic chambers for 2 weeks. Morphological analysis of the intestinal tissue was also undertaken. Experimental groups were: (i) control *ad libitum* group (CAL), fed *ad libitum* and in thermoneutral conditions (20°C, 50% RH); (ii) heat stress *ad libitum* group (HSAL), kept in a heated environment (30°C, 70% RH); and (iii) control pair-fed group (CPF), maintained in thermoneutral conditions and fed the same amount of food as consumed by the HSAL group.

This approach facilitated discrimination between adaptive responses induced by decreased food intakes and those caused by heat stress or hyperthermia. CPF and HSAL groups showed reduced body weight gain; only HSAL birds exhibited marked hyperthermia. Morphometric studies showed that the fresh weight and length of the jejunum were reduced in the HSAL group and that this effect is independent of food restriction. Distal jejunum showed shorter villi in CPF and HSAL groups while microvilli in both groups were significantly longer than in CAL birds. The kinetics of hexose absorption measured in membrane vesicles are presented in Fig. 4.2.

It is clear that whilst no major differences between the CAL and CPF groups exist, there are substantial changes in the absorptive process in the HSAL group. The kinetic constants for these transport processes and the abundance of the SGLT-1 (measured as phloridzin binding) and GLUT-2 (measured as cytochlasin-B binding) in the brush border and basolateral membranes, respectively, are shown in Table 4.1.

The activity and expression of apical SGLT-1 were increased by about 50% in HSAL chickens, without any effects in the CPF group. No changes in basolateral GLUT-2 were induced by heat stress or food restriction. During prolonged exposure to elevated thermal loads the chicken jejunum increased its apical microvillous surface to compensate for the reduction in villous size due to food restriction. The increased apical hexose transport capacity is entirely dependent on adaptations of SGLT-1 expression to heat stress and not on reduced food intake.

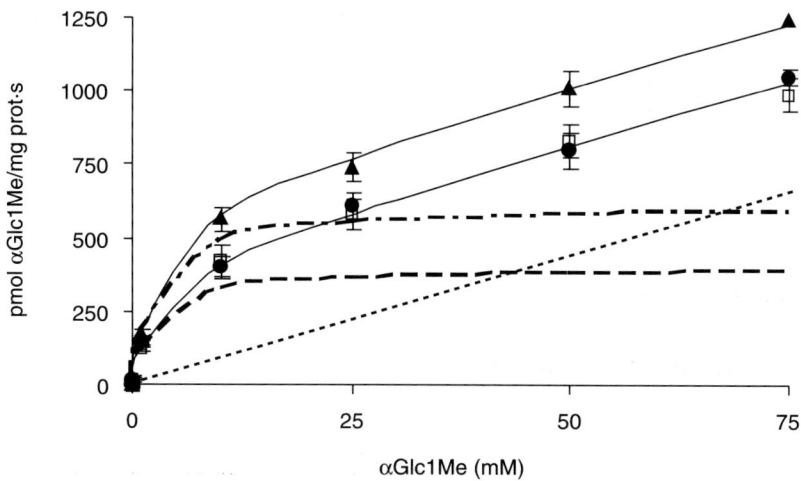

Fig. 4.2. Kinetics of αGlc1Me uptake by BBMV of jejunum. Total αGlc1Me influx (pmol/mg prot·s) was determined in vesicles from CAL (●), CPF (□) and HSAL (△) chickens. Values are the means ± SEM of five separate experiments. Total influx can be broken down into a non-saturable linear component and a saturable component. The saturable components of control groups CAL and CPF are grouped thus (– – – – –), which is different to that of the HSAL group (·– ·– ·– ·–). The non-saturable components of all groups are grouped thus (--------). The calculated kinetic constants are shown in Table 4.1.

This series of studies are the first to fully characterize the mechanisms of hexose absorption in birds and adaptations in the process at the *in vivo* level, in cells using *in vitro* preparations and then finally at the membrane level using vesicles. The work has demonstrated that the mechanisms underlying the apparent increase in substrate uptake in perfused intestine from heat-stressed chickens may be explained, at least in part, by an increase in transport at the brush border membrane through the SGLT-1 system caused by heat stress and not by the associated reduced food intake. The gene regulation and endocrine processes mediating the adaptations are now under investigation.

Luminal nutrient concentrations

Whilst the major focus of studies upon adaptations in nutrient transport has involved characterization of uptake of individual substrates using experimental preparations, few of these have considered the actual concentrations of these substrates in the intestinal lumen *in vivo*. The intestinal epithelium acts as the interface between the ingesta, the intercellular fluids and the circulation.

Absorption of individual nutrients is mediated by a number of systems (Alpers, 1987; Hopfer, 1987). These include both active transporter-mediated and passive mediated and non-mediated components (Pappenheimer, 1990; Buddington, 1992; Stevens, 1992; Sadowski and Medding, 1993; Karasov and Cork, 1994). Experimental determinations of nutrient transport or transfer rates and capacities are based upon exposure of intestinal tissue, cells or membrane

Table 4.1. Kinetic parameters of: (i) αGlc1Me uptake in BBMV (brush border membrane vesicles) and specific phlorizin binding; and (ii) D-Glc uptake in BLMV (basolateral membrane vesicles) and specific cytochalasin B binding. Values are the means ± SEM of five separate experiments. No differences were found between experimental groups (Garriga *et al.*, 2006).

	CAL	CPF	HSAL
BBMV			
V_{max} (pmol αGlc1Me/mg prot·s)	392 ± 12	402 ± 67	604 ± 57*
K_m (mM)	2.4 ± 0.1	2.2 ± 0.3	2.4 ± 0.2
K_d (nL/mg prot·s)	8.8 ± 0.8	8.8 ± 1.1	8.2 ± 1.7
$B_{50\ Phz}$ (pmol/mg prot)	153 ± 5.0	143 ± 3.0	238 ± 7.0*
BLMV			
V_{max} (pmol D-Glc/mg prot·s)	1210 ± 76	1125 ± 101	1177 ± 84
K_m (mM)	11.6 ± 1.2	11.7 ± 0.5	12.0 ± 2.0
K_d (nL/mg protcs)	20.0 ± 1.8	21.3 ± 1.0	18.9 ± 1.3
$B_{1\ cyto\ B}$ (pmol/mg prot)	40.4 ± 0.8	42.1 ± 1.7	39.7 ± 1.3

CAL, control *ad libitum* group; CPF, control pair-fed group; HSAL, heat stress *ad libitum* group.

preparations either *in vivo* or *in vitro* to a given range of substrate concentrations in the perfusates or incubation media.

The basis for the selection of these concentrations is often unspecified and they are frequently matched to the characteristics of the transport component to be studied. Whilst this approach provides a sound mechanistic understanding of the individual transport systems and adaptations therein to physiological state, environmental challenge and altered dietary composition, it is difficult to relate the findings to absorption in the whole animal during normal feeding. Indeed, it has been proposed that 'uncertainty about luminal concentration makes it difficult to calculate the relative contributions of mediated versus non-mediated transport from measures made *in vivo* or *in vitro*' (Karasov and Cork, 1994).

A first step towards achieving a more integrated model of whole animal absorption is to measure the composition of the luminal fluid in the small intestine. This fluid not only contains the bulk phase concentrations of the nutrients to which the digestive/absorptive surface is exposed, but represents a major determinant of the micro-environment in which digestion and absorption take place. More specifically, the luminal fluid determines the ionic, pH and osmolarity micro-environment of the unstirred layer at the brush border membrane of active enterocytes.

Surprisingly few attempts have been made to measure comprehensively the luminal concentrations of substrates and electrolytes in the luminal bulk phase, and those studies which do exist report either limited data or conflicting results, the latter being partly attributable to the different species studied and to different locations of measurement within the small intestine (Cole, 1961; Nasset, 1964; Steiner and Gray, 1969; Nixon and Mawer, 1970; Mongin, 1976; Murakami *et al.*, 1977; Ferraris *et al.*, 1990; Savory and Mitchell, 1991; Karasov and Cork, 1994; Barfull *et al.*, 2002). Specific questions may therefore be posed with regard to luminal substrates and intestinal function in a range of species – including the domestic fowl, which has been the subject of numerous mechanistic studies relating to intestinal substrate absorption and adaptation (e.g. Mitchell and Levin, 1981; Levin *et al.*, 1983; Mitchell and Smith, 1990; Mitchell and Carlisle, 1992; Mitchell and Hunter, 1996; Hunter *et al.*, 2000).

Thus, it may be asked: 'what substrate and electrolyte concentrations are available to the absorptive compartment throughout the small intestine and do these conditions change with dietary composition, food intake, environment and physiological status?' Intestinal luminal fluid contains the bulk phase concentrations of the nutrients to which the digestive/absorptive surface is exposed and represents a major determinant of the micro-environment in which digestion and absorption take place.

A recent study (Mitchell *et al.*, 2006) has determined the concentrations of glucose, methionine, sodium and potassium concentrations and the pH and osmolality of luminal fluid from different regions of the small intestine of broiler chickens. Comparisons of these parameters were made in birds maintained in thermoneutral conditions and groups exposed to chronic heat stress; this latter condition is reported to alter absorption of nutrients as well as food intake and growth (Mitchell and Carlisle, 1992; Mitchell *et al.*, 1999). It was demonstrated that luminal glucose and methionine concentrations exhibited a proximal–distal gradient (Table 4.1) and that for most of the small intestine these concentrations were considerably higher than the concentrations associated with the kinetic characteristics of the known transport systems.

This suggests that passive paracellular absorption must play a significant part in nutrient uptake under normal conditions of food intake and even during reduced intake such as in periods of heat stress. It also supports the hypothesis that adaptations in absorptive mechanisms may be mediated by changes in both the active transporter-dependent component and the passive paracellular route. Luminal fluid also contains less sodium and more potassium than plasma and interstitial fluid and has a higher osmolarity (Table 4.2).

Luminal fluid also exhibits a proximal–distal pH gradient, increasing from around 6.4 in the upper small intestine to almost 8.0 in the distal ileum. Heat stress reduced the proximal–distal gradient for luminal glucose and reduced absolute concentrations, possibly as a result of reduced food intake. Methionine concentrations were also decreased, whilst luminal sodium increased. Luminal osmolarity also increased during heat stress.

It is clear that the composition of the luminal fluid presented to the absorptive compartment or mucosa in control and heat-stressed birds differs substantially and this must influence interpretation of the measurements of substrate absorption in the context of whole-animal nutrient uptake under different environmental conditions. All these features described above will affect the absorptive process for specific nutrients and must be taken into account in the design of experimental techniques for the characterization of the absorptive process and adaptations therein. Also, these considerations must be included in the interpretation of current and previously published findings.

Table 4.2. Luminal concentrations of glucose, methionine, sodium, potassium and super-natant pH and osmolarity in the duodenum, jejunum and ileum of control (C, thermoneutral, 20°C) and heat-stressed (HS, 30°C) broiler chickens (values are presented as the mean ± one SEM, $n = 12$) (Mitchell *et al.*, 2006).

	Temperature regime	Duodenum	Jejunum	Ileum
Glucose	C	68.1 ± 6.4	58.9 ± 4.0	7.6 ± 1.6
(mmol/l)	HS	56.1 ± 6.8	44.4 ± 3.3	12.4 ± 2.4
Methionine	C	11.0 ± 3.5	6.8 ± 1.9	3.0 ± 1.6
(mmol/l)	HS	1.9 ± 1.2	1.3 ± 0.5	1.3 ± 0.3
Sodium	C	67.3 ± 5.0	68.5 ± 4.0	83.5 ± 6.5
(mmol/l)	HS	118.4 ± 2.8	83.2 ± 7.6	71.3 ± 4.0
Potassium	C	19.8 ± 1.3	13.3 ± 1.0	26.8 ± 2.4
(mmol/l)	HS	12.2 ± 0.5	14.5 ± 1.4	37.1 ± 3.0
pH	C	6.4 ± 0.06	6.6 ± 0.16	7.7 ± 0.09
	HS	6.6 ± 0.06	6.8 ± 0.16	8.2 ± 0.26
Osmolarity	C	389 ± 16.7	430 ± 7.8	342 ± 4.9
(mOsm/l)	HS	463 ± 6.9	452 ± 22.2	413 ± 13.6

It was concluded that the luminal fluid of the small intestine of the domestic fowl may be described as a hyperosmolar, low-sodium, high-potassium solution of non-physiological pH. The associated concentrations of major nutrients vary significantly with anatomical location and may differ substantially from those assumed from known characteristics of transport systems or used in many experimental protocols. It is apparent that the composition of luminal fluid in the small intestine has important implications for the design of physiologically meaningful nutrient absorption studies and their interpretation in relation to absorptive mechanisms, capacities and adaptations.

SUMMARY

The small intestine – and in particular the digestive/absorptive compartment or epithelial mucosa – is capable of extensive adaptation in the face of a number of nutritional and environmental challenges. The responses in digestive and absorptive function are often accompanied by alterations in mucosal morphology, villus and microvillus surface area and structure and changes in crypt cell dynamics, migration and maturation. The adaptations usually favour more efficient absorption of nutrients in the face of the imposed stress or challenge and the intestine appears to regulate a reserve capacity in absorptive function to optimize nutrient uptake and assimilation both during and following any dietary, nutritional or environmental stress.

The control or regulation of these processes is important in the context of poultry production, as optimization of nutrient digestion and absorption underpins efficient growth and development and the more complete expression of genetic potential. Understanding of the mechanisms mediating adaptive responses in absorptive function and how they relate to bird growth and development may facilitate the design of more effective nutritional strategies and formulations for use in commercial chicken production, particularly in conditions where environmental stressors may otherwise compromise efficient production.

REFERENCES

Afik, D., Darken, B.W. and Karasov, W.H. (1997) Is diet shifting facilitated by modulation of intestinal nutrient uptake? Test of an adaptational hypothesis in yellow-rumped warblers. *Physiological Zoology* 70, 213–221.

Alpers, D.H. (1987) Digestion and absorption of carbohydrates and proteins. In: Johnson, L.R. (ed.) *Physiology of the Gastrointestinal Tract* 2nd edn. Raven Press, New York, pp. 1469–1488.

Altan, O., Pabuccuoglu, A., Altan, A., Konyalioglu, S. and Bayraktar, H. (2003) Effect of heat stress on oxidative stress, lipid peroxidation and some stress parameters in broilers. *British Poultry Science* 44, 545–550.

Amat, C., Planas, J.M. and Moretó, M. (1996) Kinetics of hexose uptake by the small and large intestine of the chicken. *American Journal of Physiology* 271, R1085–R1089.

Applegate, T.J., Karcher, D.M. and Lilburn, M.S. (2005) Comparative development of the small intestine in the turkey poult and Pekin duckling. *Poultry Science* 84, 426–431.

Austic, R.E. (1985) Development and adaptation of protein digestion. *Journal of Nutrition* 115, 686–697.

Barfull, A., Garriga, C., Tauler, A. and Planas, J.M. (2002) Regulation of SGLT-1 expression in response to Na(+) intake. *American Journal of Physiology* 282, R738–R743.

Baumgart, D.C. and Dignass, A.U. (2002) Intestinal barrier function. *Current Opinions in Clinical Nutrition and Metabolic Care* 5, 685–694.

Boorman, K.N. (1976) Digestion and absorption of protein. In: Boorman, K.N. and Freeman, B.M. (eds) *Digestion in the Fowl.* British Poultry Science Ltd., Edinburgh, UK, pp. 27–61.

Buddington, R.K. (1992) Intestinal nutrient transport during ontogeny of vertebrates. *American Journal of Physiology* 263, R503–R509.

Casirola, D.M., Rifkin, B., Tsai, W. and Ferraris, R.P. (1996) Adaptations of intestinal nutrient transport to chronic caloric restriction in mice. *American Journal of Phsyiology* 271, G192–G200.

Chediak, J.G., Caviedes-Vidal, E. and Karasov, W.H. (2001) Passive absorption of hydrophilic carbohydrate probes by the house sparrow *Passer domesticus*. *Journal of Experimental Biology* 204, 723–731.

Cole, A.S. (1961) Soluble material in the gastrointestinal tract of rats under normal feeding conditions. *Nature* 191, 502–503.

Corless, A.B. and Sell, J.L. (1999) The effects of delayed access to feed and water on the physical and functional development of the digestive system of young turkeys. *Poultry Science* 78, 1158–1169.

Croom, W.J., McBride, B., Bird, A.R., Fan, K.-Y., Odle, J., Froetschel, M. and Taylor, I.L. (1998) Regulation of intestinal glucose absorption: a new issue in animal science. *Canadian Journal of Animal Science* 78, 1–13.

Croom, W.J., Brake, J., Havenstein, G.B., Christensen, V.L., McBride, B., Peebles, E.D. and Taylor, I.L. (1999) Is intestinal absorption capacity rate-limiting for performance in poultry? *Journal of Applied Poultry Research* 8, 242–252.

Das, S., Yadav, R.K. and Nagchoudhuri, J. (2001) Effect of fasting on the intestinal absorption of D-glucose and D-xylose in rats *in vivo*. *Indian Journal of Physiology and Pharmacology* 45, 451–456.

De la Horra, M.C., Cano, M., Peral, M.J., Calonge, M.L. and Ilundáin, A.A. (2001) Hormonal regulation of chicken intestinal NHE and SGLT-1 activities. *American Journal of Physiology* 280, R655–R660.

Diamond, J. and Hammond, K. (1992) The matches achieved by natural selection between biological capacities and their natural loads. *Experientia* 15, 551–557.

Diamond, J.M. and Karasov, W.H. (1987) Adaptive regulation of intestinal nutrient transporters. *Proceedings of the National Academy of Sciences* 84, 2242–2245.

Donowitz, M., De la Horra, C., Calonge, M.L., Wood, I.S., Dyer, J., Grible, S.M., Sanchedz de Medina, F., Tse, C.M., Shirazi-Beechey, S.P. and Ilundáin, A.A. (1998) In birds, NHE2 is major brush border Na/H exchanger in colon and is increased by a low-NaCl diet. *American Journal of Physiology* 274, R1659–R1669.

Ferraris, R.P. (2001) Dietary and developmental regulation of intestinal sugar transport. *Biochemical Journal* 360, 265–276.

Ferraris, R.P. and Carey, H.V. (2000) Intestinal transport during fasting and malnutrition. *Annual Review of Nutrition* 20, 195–219.

Ferraris, R.P. and Diamond, J. (1997) Regulation of intestinal sugar transport. *Physiological Reviews* 77, 257–302.

Ferraris, R.P., Lee, P.P. and Diamond, J.M. (1989) Origin of regional and species differences in intestinal glucose uptake. *American Journal of Physiology* 257, G689–G697.

Ferraris, R.P., Sasan Yasharpour, K.C., Lloyd, K., Mirzayan, R. and Diamond, J.M. (1990) Luminal glucose concentrations in the gut under normal conditions. *American Journal of Physiology* 259, G822–G837.

Ferrer, R., Gil, M., Moretó, M., Oliveras, M. and Planas, J.M. (1994) Hexose transport across the apical and basolateral membrane of enterocytes from different regions of the chicken intestine. *Pflügers Archive* 426, 83–88.

Foye, O.T. and Black, B.L. (2006) Intestinal adaptation to diet in the young domestic and wild turkey (*Melleagris gallopavo*). *Comparative Biochemistry and Physiology* A143, 184–192.

Gal-Garber, O., Mabjeesh, S.J., Sklan, D. and Uni, Z. (2000) Partial sequence and expression of the gene for and activity of the sodium-glucose transporter in the small intestine of fed, starved and refed chickens. *Journal of Nutrition* 130, 2174–2179.

Gal-Garber, O., Mabjeesh, S.J., Sklan, D. and Uni, Z. (2003) Nutrient transport in the small intestine: Na+,K+-ATPase expression and activity in the small intestine of the chicken as influenced by dietary sodium. *Poultry Science* 82, 1127–1133.

Garriga, C., Moretó, M. and Planas, J.M. (1997) Hexose transport across the basolateral membrane of the chicken jejunum. *American Journal of Physiology* 272, R1330–R1335.

Garriga, C., Moretó, M. and Planas, J.M. (1999a) Hexose transport in the apical and basolateral membranes of enterocytes in chickens adapted to high and low NaCl intakes. *Journal of Physiology* 514 (1), 189–199.

Garriga, C., Rovira, N., Moretó, M. and Planas, J.M. (1999b) Expression of Na$^+$-glucose cotransporter in brush border membrane of the chicken intestine. *American Journal of Physiology* 276, R627–R631.

Garriga, C., Moretó, M. and Planas, J.M. (2000) Effect of resalination on intestinal glucose transport in chickens adapted to low Na$^+$ intakes. *Experimental Physiology* 85, 371–378.

Garriga, C., Planas, J.M. and Moretó, M. (2001) Aldosterone mediates the changes in hexose transport induced by low sodium intake in chicken distal intestine. *Journal of Physiology* 535 (1), 197–205.

Garriga, C., Barfull, A. and Planas, J.M. (2004) Kinetic characterization of apical D-fructose transport in chicken jejunum. *Journal of Membrane Biology* 197, 71–76.

Garriga, C., Hunter, R.R., Amat, C., Planas, J.M., Mitchell, M.A. and Moretó, M. (2006) Heat stress increases apical glucose transport in the chicken jejunum. *American Journal of Physiology* 290, R195–R201.

Geraert, P.A., Padilha, J.C.F. and Guillaumin, S. (1996a) Metabolic and endocrine changes induced by chronic heat exposure in broiler chickens: growth performance, body composition and energy retention. *British Journal of Nutrition* 75, 195–204.

Geraert, P.A., Padilha, J.C.F. and Guillaumin, S. (1996b) Metabolic and endocrine changes induced by chronic heat exposure in broiler chickens: biological and endocrine variables. *British Journal of Nutrition* 75, 205–216.

Geyra, A., Uni, Z. and Sklan, D. (2001)The effect of fasting at different ages on growth and tissue dynamics in the small intestine of the young chick. *British Journal of Nutrition* 86, 53–61.

Gonzalez, E., Marti, T. and Vinardell, M.P. (1996) Comparative study of the small intestinal monosaccharide and amino acid transport as a function of age. *Nutrition Research* 16, 865–868.

Griffin, H.D. and Goddard, C. (1994) Rapidly growing broiler (meat type) chickens: their origin and use for comparative studies of the regulation of growth. *International Journal of Biochemistry* 26, 19–28.

Habold, C., Foltzer-Jourdainne, C., Le Maho, Y., Ligno, J.H. and Oudart, H. (2005) Intestinal gluconeogenesis and glucose transport according to body fuel availability in rats. *Journal of Phsyiology* 566, 575–586.

Hopfer, U. (1987). Membrane transport mechanisms for hexoses and amino acids in the small intestine. In: Johnson, L.R. (ed.) *Physiology of the Gastrointestinal Tract*, 2nd edn. Raven Press, New York, pp. 1499–1527.

Hughes, R.J. (2005) An integrated approach to understanding gut function and gut health of chickens. *Asia Pacific Journal of Clinical Nutrition* 14, S27.

Hunter, R.R., Mitchell, M.A., Garriga, C., Mitjans, M., Amat, C., Planas, J.M. and Moreto, M. (2000) Chronic heat stress induces an adaptive increase of SGLT-1 expression in broiler chickens. *Poultry Science* 79 (1), 100.

Jackson, S. and Diamond, J. (1995) Ontogenic development of gut function, growth and metabolism in a wild bird, the red jungle fowl. *American Journal of Physiology* 269, R1163–R1173.

Jenkins, A.P. and Thompson, R.P.H. (1994) Mechanisms of small intestinal adaptation. *Digestive Disease* 12, 15–27.

Julian, R.J. (2005) Production and growth related disorders and other metabolic diseases of poultry: a review. *The Veterinary Journal* 3, 350–369.

Karasov, W.H. (1987) Nutrient requirements and design and function of guts in fish, reptiles and mammals. In: Dejours, P., Bolis, L., Taylor, C.R. and Weible, E.R. (eds) *Comparative Physiology: Life in Water and on Land*. Fidia Research Series IX, Liviana Press, Padua, Italy, pp. 181–199.

Karasov, W.H. (1988) Nutrient transport across vertebrate intestine. In: *Advances in Comparative and Environmental Physiology*, Vol. 2, Chapter 4. Springer Verlag, Berlin and Heidelberg, pp. 131–171.

Karasov, W.H. (1992) Tests of the adaptive modulation hypothesis for dietary control of intestinal nutrient transport. *American Journal of Phsyiology* 263, R496–R502.

Karasov, W.H. and Cork, S.J. (1994) Glucose absorption by a nectarivorous bird: the passive pathway is paramount. *American Journal of Physiology* 267, G18–G26.

Karasov, W.H. and Diamond, J.M. (1983) Adaptive regulation of sugar and amino acid transport by vertebrate intestine. *American Journal of Physiology* 245, G443–G462.

Karasov, W.H. and Diamond, J. (1987) Adaptation of intestinal nutrient transport. In: Johnson, L.R. (ed.) *Physiology of the Gastrointestinal Tract*, 2nd edn. Raven Press, New York, pp. 1489–1497.

Karasov, W.H. and Hune, I.D. (1997) Vertebrate gastrointestinal system. In: Dantzler, W. (ed.) *Handbook of Physiology*, Vol. 1, Section 13: *Comparative Physiology*, pp. 409–480.

Karasov, W.H. and McWilliam, S.R. (2005) Digestive constraints in mammalian and avian ecology. In: Starck, J.M. and Wang, T. (eds) *Physiological and Ecological Adaptations to Feeding in Vertebrates*. Science Publishers, Enfield, New Hampshire, pp. 246–267.

King, D.E., Asem, E.K. and Adeola, O. (2000) Ontogenetic development of intestinal digestive functions in White Pekin ducks. *Journal of Nutrition* 130, 57–62.

Laverty, G., Elbrond, V.S., Árnason, S.S. and Skadhauge, E. (2006) Endocrine regulation of ion transport in the avian lower intestine. *General and Comparative Endocrinology*, http://www.sciencedirect.com

Lerner, J. (1984) Cell membrane amino acid transport processes in the domestic fowl (*Gallus domesticus*). *Comparative Biochemistry and Physiology* 78A, 205–215.

Levin, R.J. (1974) A unified theory for the action of partial or complete reduction of food intake on jejunal hexose absorption *in vivo*. In: Dowling, R.H. and Riecken, E.O. (eds) *Intestinal Adaptation*. FK-Schattauer Verlag, Stuttgart, Germany, pp. 125–133.

Levin, R.J. (1976) Digestion and absorption of carbohydrate: from embryo to adult. In: Boorman, K.N. and Freeman, B.M. (eds) *Digestion in the Fowl*. British Poultry Science Ltd., Edinburgh, UK, pp. 63–116.

Levin, R.J. (1984) Absorption from the alimentary tract. In: Freeman, B.M. (ed.) *Physiology and Biochemistry of the Domestic Fowl, Vol. 5*. Academic Press, London, pp. 1–37.

Levin, R.J. (1994) Digestion and absorption of carbohydrates – from molecules and membranes to humans. *American Journal of Clinical Nutrition* 59, 690S–698S.

Levin, R.J. and Mitchell, M.A. (1976) The differential effects of starvation on the kinetics of electrogenic glucose absorption in the jejunum and ileum of the domestic fowl. *Journal of Physiology* 257, 31P–32P.

Levin, R.J. and Mitchell, M.A. (1979) The effects of fasting on real kinetic parameters for L-valine absorption measured in jejunum and ileum *in vivo* in fowls. *Journal of Physiology* 295, 20P–21P.

Levin, R.J. and Mitchell, M.A. (1979) Adaptation of neutral amino acid transfer systems in the avian small intestine during starvation. *Journal of Physiology* 298, 47P.

Levin, R.J. and Mitchell, M.A. (1980) Three types of kinetic response in amino acid intestinal absorption *in vivo* to fasting. *Journal of Physiology* 306, 42P.

Levin, R.J. and Mitchell, M.A. (1982) Intestinal adaptations to fasting – use of corrected parameters to assess responses of jejunal and ileal absorption *in vivo*. In: Robinson, J.W.L., Dowling, R.H. and Riecken, E.O. (eds) *Mechanisms of Intestinal Adaptation*. MTP Press Ltd., Lancaster, UK.

Levin, R.J., Mitchell, M.A. and Barber, D.C. (1983) Comparison of jejunal and ileal absorptive functions for glucose and valine *in vivo* – a technique for estimating real K_m and J_{max} in the domestic fowl. *Comparative Biochemistry and Physiology* A, 74 (4), 961–966.

Lilja, C. (1983) A comparative study of postnatal growth and organ development in some species of bird. *Growth* 47, 317–339.

Lilja, C., Sperber, I. and Marks, H.L. (1985) Postnatal growth and organ development in Japanese Quail selected for high growth rate. *Growth* 49, 51–62.

Lin, H., Malheiros, R.D., Moraes, V.M.B., Careghi, C., Decuypere, E. and Buyse, J. (2004) Acclimation of broiler chickens to chronic high environmental temperatures. *Archives fur Geflugelkunde* 68, 39–46.

Maneewan, B. and Yamauchi, K. (2005) Recovery of duodenal villi and cells in chickens refed protein, carbohydrate and fat. *British Poultry Science* 46, 415–423.

McCann, R., Foye, O. and Black, B. (2000) Effect of diet on intestinal glucose and alanine transport rates in wild *vs* domestic turkey. *FASEB Journal* 14, A364.

McNabb, F.M.A., Dunnington, E.A., Freeman, T.B. and Siegel, P.B. (1989) Thyroid hormones and growth patterns of embryonic and post-hatch chickens from lines selected for high and low juvenile body weight. *Growth, Development and Ageing* 53, 87–92.

Miles, R.D., Butcher, G.D., Henry, P.R. and Littell, R.C. (2006) Effect of antibiotic growth promoters on broiler performance, intestinal growth parameters, and quantitative morphology. *Poultry Science*, 85, 476–485.

Miller, D.S., Houghten, D., Burrill, P., Herzberg, G.R. and Lerner, J. (1973) Specificity characteristics of the intestinal absorption of model amino acids in domestic fowl. *Comparative Biochemistry and Physiology* 44A, 17–34.

Mitchell, M.A. and Carlisle, A.J. (1992) The effects of chronic exposure to elevated environmental temperature on intestinal morphology and nutrient absorption in the domestic fowl (*Gallus domesticus*). *Comparative Biochemistry and Physiology* A 101A, 137–142.

Mitchell, M.A. and Hunter, R.R. (1996) Effects of chronic heat stress upon intestinal absorption of DL-methonine and methionine hydroxy analogue *in vivo* in the broiler chicken. *British Poultry Science* 37 (supplement), S88–S89.

Mitchell, M.A. and Levin, R.J. (1981) Amino acid-absorption in jejunum and ileum *in vivo* – a kinetic comparison of function on surface area and regional bases. *Experientia* 37 (3), 265–266.

Mitchell, M.A. and Smith, M.W. (1990) Jejunal alanine uptake and structural adaptation in response to genetic selection for growth rate in the domestic fowl (*Gallus domesticus*) *in vitro*. *Journal of Physiology* 424, 7P.

Mitchell, M.A. and Smith, M.W. (1991) The effects of genetic selection for increased growth rate on mucosal and muscle weights in the different regions of the small intestine of the domestic fowl (*Gallus domesticus*). *Comparative Biochemistry and Physiology* A 99A, 251–258.

Mitchell, M.A., Carlisle, A.J. and Hunter, R. (1995) Increased enterocyte hexose accumulation in response to chronic heat stress. *Italian Journal of Gastroenterology* 27, 162.

Mitchell, M.A., Carlisle, A.J. and Hunter, R.R. (1999) Cellular mechanisms of adaptation in intestinal hexose absorption in the broiler chicken during chronic heat stress. *Poultry Science* 78 (supplement 1), 56 (248).

Mitchell, M.A., Hunter, R.R., Sandercock, D.A. and Lemme, A. (2006) Composition of the luminal fluid in the small intestine of the domestic fowl. *British Poultry Science*

Mongin, P. (1976) Ionic constituents and osmolality of the small intestinal fluids of the laying hen. *British Poultry Science* 17, 383–392.

Murakami, E., Saito, M. and Suda, M. (1977) Contribution of diffusive pathway in intestinal absorption of glucose in rat under normal feeding condition. *Experientia* 33, 1469.

Nasset, E.S. (1964) The role of the digestive tract in protein metabolism. *American Journal of Digestive Diseases (new series)* 9, 175–190.

Nixon, S.E. and Mawer, G.E. (1970) The digestion and absorption of protein in man. *British Journal of Nutrition* 24, 241–258.

Noy, Y. and Pinchasov, Y. (1993) Effect of a single post-hatch intubation of nutrients on subsequent early performance of broiler chickens and turkey poults. *Poultry Science* 72, 1861–1866.

Obst, B.S. and Diamond, J. (1992) Ontogenesis of intestinal nutrient transport in domestic chickens (*Gallus gallus*) and its relation to growth. *The Auk* 109, 451–464.

Pappenheimer, J.R. (1990) Paracellular intestinal absorption of glucose, creatinine and mannitol in normal animals: relation to body size. *American Journal of Physiology* 259, G290–G299.

Rance, K.A., McEntee, G.M. and McDevitt, R.M. (2002) Genetic and phenotypic relationships between and within support and demand tissues in a single line of broiler chickens. *British Poultry Science* 43, 518–527.

Rayo, J.M., Esteban, S. and Tur, J.A. (1992) Effect of starvation on the *in vivo* intestinal absorption of sugars and amino acids in young chickens (*Gallus domesticus*). *Archive Internationales de Physiologie de Biochemie et de Biophysique* 100, 155–158.

Rebel, J.M.J., Balk, F.R.M., Post, J., Van Hemert, S., Zekarias, B. and Stockhofe, N. (2006) Malabsorption syndrome in broilers. *World's Poultry Science Journal* 62, 17–29.

Reuss, L. (2000) One hundred years of inquiry: the mechanism of glucose absorption int the intestine. *Annual Review of Physiology* 62, 939–946.

Sadowski, D.C. and Medding, J.B. (1993) Luminal nutrients alter tight-junction permeability in the rat jejunum: an in vivo perfusion model. *Canadian Journal of Physiology and Pharmacology* 71, 835–839.

Savory, C.J. (1992) Gastrointestinal morphology and absorption of monosaccharides in fowls conditioned to different types and levels of dietary fibre. *British Journal of Nutrition* 67, 77–89.

Savory, C.J. (1995) Broiler welfare: problems and prospects. *Archives fur Geflugelkunde*, Sonerheft 1, 48–52.

Savory, C.J. and Mitchell, M.A. (1991) Absorption of hexose and pentose sugars in vivo in perfused intestinal segments in the fowl. *Comparative Biochemistry and Physiology* 101A (4), 969–974.

Scanes, C.G. (1987) The physiology of growth, growth hormone and other growth factors in poultry. In: Dietert, R.R. (ed.) *Critical Reviews in Poultry Biology 1*. CRC Press, Florida, pp. 51–105.

Siegel, P.B. and Dunnington, E.A. (1987) Selection for growth in chickens. In: Dietert, R.R. (ed.) *Critical Reviews in Poultry Biology 1*. CRC Press, Florida, pp. 1–24.

Sklan, D. (2001) Development of the digestive tract of poultry. *World's Poultry Science Journal* 57, 415–428.

Sklan, D. and Noy, Y. (2003) Functional development and intestinal absorption in the young poult. *British Poultry Science* 44, 651–658.

Sklan, D., Geyra, A., Tako, E., Gal-Gerber, O. and Uni, Z. (2003) Ontogeny of brush border carbohydrate digestion and uptake in the chick. *British Journal of Nutrition* 89, 747–753.

Smith, M.W., Mitchell, M.A. and Peacock, M.A. (1990) Effects of genetic selection on growth rate and intestinal structure in the domestic fowl (*Gallus domesticus*). *Comparative Biochemistry and Physiology* A 97, 57–63.

Steiner, M. and Gray, S.J. (1969) Effect of starvation on intestinal amino acid absorption. *American Journal of Physiology* 217, 747–752.

Stevens, B.R. (1992) Vertebrate intestine apical membrane mechanisms of organic nutrient transport. *American Journal of Physiology* 263, R458–R463.

Suvarna, S., Christensen, V.L., Ort, D.T. and Croom, W.J. (2005) High levels of dietary carbohydrate increase glucose transport in poult intestine. *Comparative Biochemistry and Physiology A – Molecular and Integrative Physiology* 141, 257–263.

Teeter, R.G. (1994) Optimising production of heat-stressed broilers. *Poultry Digest*, May, 10–16.

Temim, S., Chagneau, A.M., Peresson, R. and Tesseraud, S. (2000) Chronic heat exposure alters protein turnover of three different skeletal muscles in finishing broiler chickens fed 20 or 25% protein diets. *Journal of Nutrition* 130, 813–819.

Uni, Z., Tako, E., Gal-Garber, O. and Sklan, D. (2003) Morphological, molecular and functional changes in the chicken small intestine of the late-term embryo. *Poultry Science* 82, 1747–1754.

Vasquez, C., Rovira, N., Ruiz-Gutierrez, V. and Planas, J.M. (1997) Developmental changes in glucose transport, lipid composition and fluidity of jejunal BBM. *American Journal of Physiology* 273, R1086–R1093.

Vinardell, M.P. (1992) Age influences on amino acid intestinal transport. *Comparative Biochemistry and Physiology* 103A, 169–171.

Watkins, E.J., Butler, P.J. and Kenyon, B.P. (2004) Post-hatch growth of the digestive system in wild and domesticated ducks. *British Poultry Science* 45, 331–341.

Yamamauchi, K. and Tarachai, P. (2000) Changes in intestinal villi, cell area and intracellular auotphagic vacuoles realted to intestinal function in chickens. *British Poultry Science* 41, 416–423.

Chapter 5
Epithelial structure and function in the hen lower intestine

G. Laverty,[1]* V.S. Elbrønd,[2] S.S. Árnason[3] and E. Skadhauge[2]

[1]Department of Biological Sciences, University of Delaware, Newark, Delaware, USA; [2]Department of Basic Animal and Veterinary Sciences, The Royal Veterinary and Agricultural University, Frederiksberg, Denmark; [3]Department of Physiology, University of Iceland, Reykjavik, Iceland; *e-mail: laverty@udel.edu

ABSTRACT

In birds, transport processes in the lower intestine mediate absorption of ions, water and a variety of organic substrates, including significant amounts of glucose, amino acids derived from protein associated with urate spheres, and short-chain fatty acids derived from fermentation processes. These transport pathways contribute to both osmoregulation and energy homeostasis. Although birds lack a urinary bladder, evidence has shown that ureteral urine, entering the distal lower intestine, is forced into the colon, caecae and even distal portions of the small intestine. Further, substrates also enter the lower intestine from the small intestine. Thus, the lower intestine serves as an 'integrating segment' for mixing and modification of both urinary and intestinal inputs. Of particular interest is that much of the transport activity of the lower intestine is regulated by the hormone aldosterone, which in turn varies inversely with dietary salt content. Numerous studies have shown that following acclimation to low-salt diets, sodium-linked organic substrate co-transporters are largely suppressed, while electrogenic sodium channels (ENaCs) are induced. These changes can be manipulated, at least in part, by resalination of low-salt hens, or by aldosterone administration to high-salt-acclimated birds. In the coprodeum, the changes in transport are paralleled by extensive remodelling of the mucosal surface, with low-salt acclimation increasing cell numbers, microvillus density and length and the proportion of 'mitochondrial-rich' (MR) cells. The latter may be the sites of proton secretion in the lower intestine. A cAMP-activated chloride secretion pathway is also present in both colon and coprodeum, and may be mediated by a CFTR-like Cl⁻ channel. There are still a number of unresolved issues, including whether other hormones contribute to the regulation of transport activity.

INTRODUCTION

The lower intestine of the domestic fowl includes the cloaca (proctodeum, urodeum and coprodeum), the colon and the paired caeca. The colon, coprodeum and caeca are known to be highly active, transporting epithelia that contribute to both energy homeostasis and osmoregulation. Birds lack a urinary bladder and ureteral urine enters the cloaca, where it is moved by retroperistaltic muscle contractions into the more proximal regions of the colon and caeca where it mixes with chyme entering from the small intestine.

It is well established that the lower intestine controls mixing, retrograde peristalsis and transport activities in order to regulate the conservation of nutrient substrates, ions and water. The retrograde movement of urine anteriorly (Skadhauge, 1968; Goldstein and Braun, 1986), for example, has been shown to be influenced by dietary nitrogen levels (Waldenstedt and Björnjag, 1995), and urine osmolarity and hydration state modulate colonic retroperistaltic muscle contractions (Brummermann and Braun, 1995). Thus, colon and caeca are the final regulators of nutrient uptake and post-renal adjustments to water and electrolyte excretion.

Microbial fermentation, particularly in the caeca, produces large amounts of volatile fatty acids (Annison *et al.*, 1968; Titus and Ahearn, 1992), and degradation of urinary uric acid spheres results in the production of peptides and amino acids, derived from protein associated with these spheres (Boykin and Braun, 1996; Casotti and Braun, 2004). Transport of hexoses, peptides, amino acids and volatile fatty acids – as well as of ions and water by these tissues – has been established and demonstrates the marked transport capacity of the lower intestine (see Thomas, 1982; Skadhauge, 1993; Laverty and Skadhauge, 1999). Furthermore, the production of ammonia from urate degradation may provide a mechanism for nitrogen recycling via microbial incorporation into amino acids and subsequent epithelial absorption of these amino acids (Karasawa, 1999).

This chapter will focus on the morphology and transport processes found in the lower intestine, with an emphasis on dietary and hormonal regulation of these systems.

MORPHOLOGY AND TRANSPORT PROCESSES IN THE LOWER INTESTINE

The basic morphology of the transporting segments of the lower intestine consists of a mucosal surface amplified by either villi, as in the colon and *basis caecum* (closest to the ileocaecal junction) (Dantzer, 1989; Clauss *et al.*, 1991; Laverty *et al.*, 2001), or by flatter mucosal folds or rugae, as in the coprodeum and *corpus caecum* (Clauss *et al.*, 1988; Dantzer, 1989; Elbrønd *et al.*, 1991, 1998; Mayhew *et al.*, 1992) (see Fig. 5.1). Crypts are found in all three structures. The innermost layer is a simple columnar epithelium, which in both colon and caeca has an extensively developed microvillus brush border. The coprodeum has relatively sparse and short microvilli on normal or high-salt diets, but undergoes a marked development during acclimation to low-salt diets, including increases in microvillus density and height (*see* below). The *lamina epithelialis* is composed of three major cell types that have been well described in the colon and coprodeum, but less so in the caecum (*see* Figs 5.2a, b).

The major transporting cell has been designated the 'absorptive epithelial cell' (AEC). In addition there are goblet cells and mitochondria-rich (MR) cells, which have been particularly well described in the coprodeum (Elbrønd *et al.*, 1998, 1999, 2004). The MR cells – sometimes in the past referred to as 'brush cells' and perhaps identical to 'chloride cells' (in reference to their reaction with $AgNO_3$, Clauss *et al.*, 1988) – have been the subject of recent investigation (Elbrønd *et al.*, 2004). In the coprodeum these cells are localized to the upper

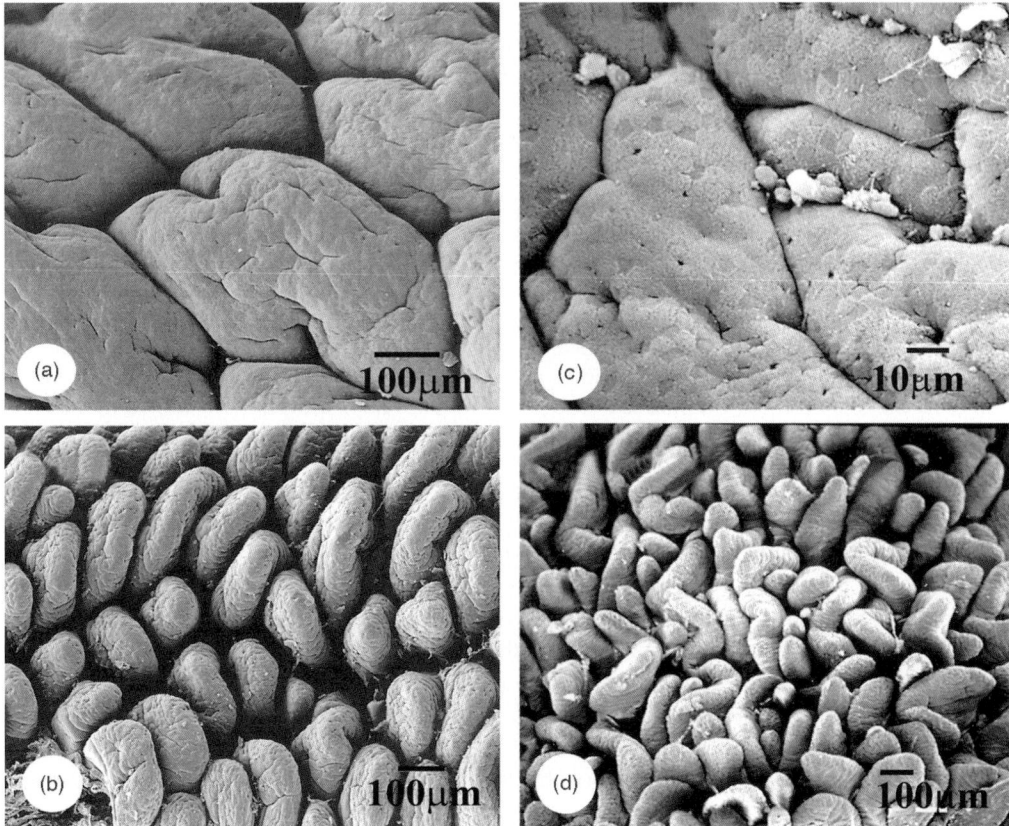

Fig. 5.1. Scanning electron micrographs (SEMs) of hen lower intestine showing how different regions amplify the absorptive mucosal surface area. The upper left panel (a) presents the coprodeum with foldings and openings to crypts in between. The lower left panel (b) is from the colon and illustrates a mucosa with densely packed and leaf-shaped villi with crypt openings in between. The right-hand panels (c and d) present, respectively, high and low magnification views of the *basis caeci* in which densely packed villi and crypts are seen (similar to those of the colon). Figures 5.1c and 5.1d were kindly provided by Dr V. Dantzer, The Royal Veterinary and Agricultural University, Frederiksberg, Denmark.

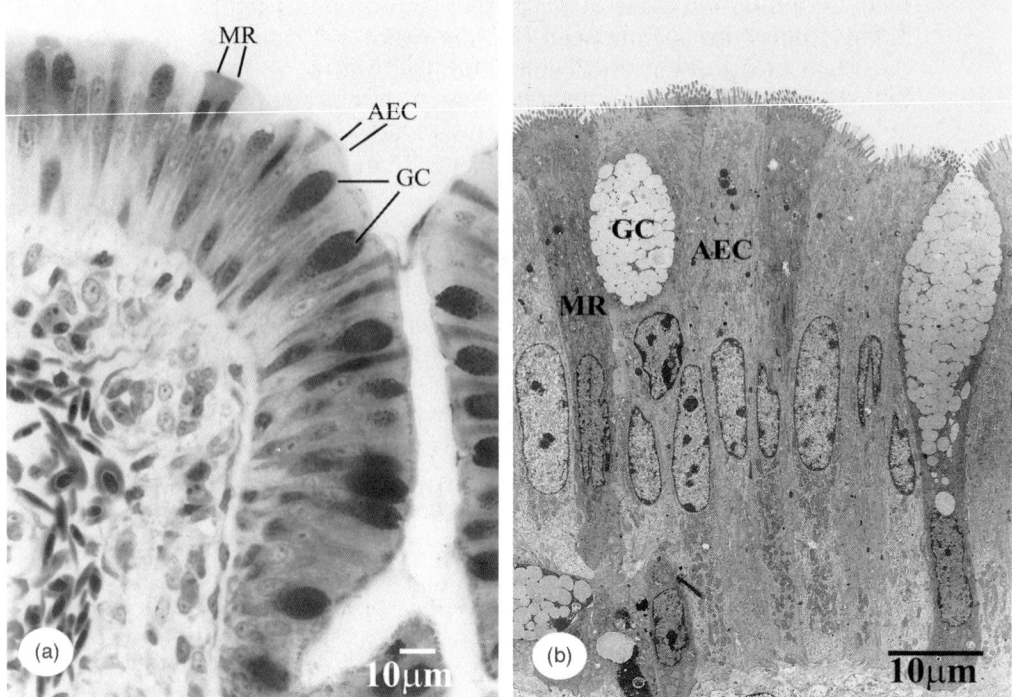

Fig. 5.2. Light micrograph (LM) of epon-embedded, toluidine blue-stained tissue from the hen coprodeum. The three major epithelial cell types: absorptive intestinal epithelial cells (AEC), goblet cells (GC) and mitochondria-rich cells (MR) are present in the epithelium on the mucosal folds (a). Figure 5.2b is a transmission electron micrograph of colonic epithelium which, with respect to the cell types, is similar to that of the coprodeum. To increase the cellular surface area the cells are amplified apically with microvilli. The figure also illustrates more clearly the ultrastructural differences between the three cell types, such as: (i) electron density of the cytoplasm and nucleus; (ii) number of mitochondria; (iii) length and packing density of microvilli; and (iv) cytoplasmic granules.

lateral margins of mucosal folds, and express both carbonic anhydrase (CA) and the epithelial isoform of the sodium hydrogen exchanger, NHE-2, suggesting a possible role in proton secretion. Both colon and coprodeum have an acidic microclimate on the mucosal surface, presumably arising from proton secretion and unstirred layer effects (Holtug *et al.*, 1992; Laverty *et al.*, 1994). Thus, MR cells may be important in proton secretion by these epithelia, which in turn may provide a mechanism for absorption of volatile fatty acids (Holtug, 1989).

There have been numerous studies on the transport properties of the hen lower intestine, both *in vivo* and *in vitro*. The epithelia of these segments display a high capacity for NaCl absorption and solute-linked water absorption against osmotic gradients (Bindslev and Skadhauge, 1971a, b; Choshniak *et al.*, 1977; Lind *et al.*, 1980 a, b; Árnason and Skadhauge, 1991). In addition, there is secretion of both H^+ and K^+ ions (Skadhauge and Thomas, 1979; Lind *et al.*, 1980a; Rice and Skadhauge, 1982a; Grubb and Bentley, 1990; Laverty *et al.*, 1994) and a cAMP-stimulated chloride secretion pathway (Clauss *et al.*, 1988; Árnason and Skadhauge, 1991; Árnason, 1997).

In general, these properties are not unlike those of the mammalian colon. However, unlike the mammalian colon, the hen colon also expresses Na^+-linked co-transporters for hexoses and amino acids, transport characteristics more akin to those of the small intestine (Lind *et al.*, 1980a, b; Clauss *et al.*, 1984; Árnason and Skadhauge, 1991). Phloridzin-sensitive hexose transport has also been demonstrated in isolated cells from the proximal caecum (Ferrer *et al.*, 1986). However, another study found that, while both glucose and acetate stimulated transport metabolically, there did not appear to be a hexose-stimulated co-transporter in the caecal bulb (Grubb, 1991).

Most of the studies of hen lower intestinal transport have focused on the regulation of transport activity by dietary salt and aldosterone. Numerous studies have documented the need to optimize dietary Na^+ and Cl^- intake levels in fowl, as these affect growth rates, bone development and excreta moisture content, among other variables (Smith *et al.*, 2000; Murakami *et al.*, 2001). Additionally, there has been some work related to the effects of water deprivation and osmoregulatory hormones, such as arginine vasotocin (AVT) and prolactin. These areas will be the subject of the following sections. What has become very clear from both morphological and physiological studies is that the transporting segments of the lower intestine represent a high-capacity system for both nutrient and electrolyte absorption as well as water conservation.

Effects of dietary salt and aldosterone on the structure and function of the lower intestine

The most widely studied aspect of lower intestinal function is the striking change in transport activity that accompanies acclimation of hens to varying levels of dietary salt. Much of this work has been done *in vitro*, using the Ussing chamber and voltage clamping to measure currents associated with net ion flux. Under voltage clamp conditions, these short circuit currents (I_{SC}) are nearly equal to active net Na^+ absorption, as shown by isotopic flux measurements (Choshniak *et al.*, 1977; Lind *et al.*, 1980a, b).

When hens are maintained on high-salt (HS) or medium-salt diets (including commercial feeds), Na^+ transport and I_{SC} in the coprodeum are close to zero, with a high epithelial resistance. However, when acclimated to low-salt (LS) diets (< 1 mmol Na^+/kg BW/d intake), this tissue markedly increases Na^+ transport, with net fluxes and I_{SC} values as high as 14 μEq/cm^2/h (380 μA/cm^2) (Choshniak *et al.*, 1977; Clauss *et al.*, 1988; Árnason and Skadhauge, 1991). Nearly all of this current can be inhibited by the diuretic amiloride, indicating that the net Na^+ transport conforms to a model of electrogenic entry of Na^+ at the apical membrane via amiloride-sensitive sodium channels (ENaCs) and active basolateral extrusion via Na^+/K^+ ATPase pumps. This change in transport alone represents one of the largest ranges in transport capacity for any tissue, and is clearly an adaptation for maximal salt conservation.

In the colon, the changes that accompany dietary salt acclimation are more complex. Again, with LS acclimation, transport and I_{SC} are characterized by electrogenic, ENaC-mediated sodium absorption. However, on HS, ENaC channel activity, as measured by amiloride-sensitive I_{SC}, almost completely disappears. Sodium transport continues in this tissue, but apical Na^+ entry now occurs through a variety of Na^+-linked substrate co-transporters. These include Na^+-hexose co-transport (Lind *et al.*, 1980 a, b; Árnason and Skadhauge, 1991; Clauss *et al.*, 1991), attributed to: (i) sodium glucose luminal transporter (SGLT) family symporter (Bindslev *et al.*, 1997; Laverty *et al.*, 2001); (ii) Na^+ amino acid co-transport (Lind *et al.*, 1980 a, b; Árnason and Skadhauge, 1991); and (iii) Na^+/H^+ exchange, mediated primarily by a sodium hydrogen exchanger-2 (NHE-2) isoform (Donowitz *et al.*, 1998; De la Horra *et al.*, 2001). Thus, under HS conditions, maximal Na^+ transport requires the presence of hexoses and amino acids to stimulate these transporters.

This transition in colonic transport pattern has been confirmed by uptake studies with isolated brush border membrane vesicles (BBMV). Looking at rates of Na^+ gradient-dependent glucose uptake, Garriga and co-workers (1999a) have shown markedly reduced V_{max} values for glucose uptake in BBMV prepared from LS-acclimated hens (see Mitchell and Moretó, Chapter 4, this volume). It is considered that this co-transporter pattern seen with HS acclimation not only helps to conserve nutrient substrates derived from both urinary and intestinal inputs, but also functions in water conservation.

Indeed, there is recent evidence that Na^+-linked symporters, such as SGLT, can drive large fluxes of water (Loo *et al.*, 2002), and a number of *in vivo* perfusion studies have shown that the hen lower intestine and that of other species can generate solute-linked water absorption sufficient to offset imposed osmotic pressure gradients (lumen > blood) of almost 200 mOsm/kg (Bindslev and Skadhauge, 1971b; Skadhauge and Thomas, 1979; Rice and Skadhauge, 1982a; Goldstein and Braun, 1986; Goldstein *et al.*, 1986). Thus, one function of HS colonic co-transporter activity may be to compensate for osmotic water fluxes into the hyperosmotic lumen. We therefore propose that both colon and coprodeum of the hen can regulate transport activities to either optimize salt conservation or to minimize osmotic water loss.

There have been some *in vivo* and *in vitro* studies of caecal transport. The main segment (caecal bulb, or *corpus caeci*) also displays high rates of mainly amiloride-sensitive sodium absorption (Rice and Skadhauge, 1982b; Thomas and Skadhauge, 1989a, b; Grubb, 1991; Grubb and Bentley, 1992). *In vivo* perfusion studies show that this transport is accompanied by high rates of solute-linked water transport which, taken together with the retrograde flow of urine into the caecal lumen, indicates that the caeca may be major contributors to overall water balance (Rice and Skadhauge, 1982b; Thomas and Skadhauge, 1989a, b). Ion and water fluxes were shown to be enhanced by inclusion of glucose or acetate in the perfusate (Rice and Skadhauge, 1982b), but this effect is probably metabolic rather than co-transporter mediated (Grubb, 1991).

Both LS acclimation and aldosterone (ALDO) treatment enhanced net sodium transport (Thomas and Skadhauge, 1989a, b; Grubb and Bentley,

1992), as did dehydration (Rice and Skadhauge, 1982b; Thomas and Skadhauge, 1989b). As mentioned above, phloridzin-sensitive glucose uptake, presumably reflecting co-transporter activity, has been demonstrated in isolated cells from the proximal caecum (*basis caeci*), the region closest to the ileocaecal junction, while other regions apparently lose this transport capacity during the first 2 weeks of age (Ferrer *et al.*, 1986).

It is interesting to look in more detail at the dietary salt levels at which the LS to HS transitions in the colon and coprodeum take place. Figure 5.3 summarizes data, derived from logistical regression analysis, of hexose and amino acid-stimulated I_{SC}, amiloride-sensitive I_{SC}, and circulating ALDO levels, for colons of hens chronically acclimated to one of six different salt intake levels (Árnason and Skadhauge, 1991). As stated earlier, with increasing dietary salt hexose and amino acid-stimulated I_{SC} increase whereas amiloride-sensitive I_{SC} and ALDO levels decline but, interestingly, all the major changes take place over a relatively small range of Na^+ intake, between 1 and 10 mmol Na^+/kg BW/d. Plasma ALDO levels show a particularly steep decline, with a half maximal intake level of 0.9 mmol Na^+/kg BW/d and a detection threshold at about 2 mmol Na^+/kg BW/day.

In contrast, the half maximal dietary levels for the transport activities of ENaCs and co-transporters lie between 3.8 and 5.9 mmol Na^+/kg BW/day. Thus, while there is a strong correlation between plasma ALDO levels and transport pattern at the extremes of dietary intake (very low and very high), this relation breaks down somewhat at intermediate levels. While it has generally been assumed that ALDO accounts for most or all of the changes induced by dietary salt manipulation, these studies raise the possibility that other hormones at factors may also be involved (see below).

When hens are initially placed on LS diets, it takes 5–10 days for full expression of the chronic, steady-state LS transport pattern (complete amiloride sensitivity) (Thomas and Skadhauge, 1982; Skadhauge, 1983). The response is faster in the colon than in the coprodeum (Fig. 5.4). There seem to be two or more distinct phases to the long-term acclimation. In the first few days there is presumably a recruitment or activation of ENaC channels, as well as suppression of co-transporter activity in the colon, as plasma ALDO levels rise.

As noted earlier (Fig. 5.3), the highest levels of plasma ALDO require very low levels of salt intake, but this probably takes a period of days to develop, as whole animal electrolyte balances shift. This early phase may involve a mechanism for relatively rapid genomic responses, as recently described for other aldosterone-sensitive tissues such as the intercalated cells of the renal distal tubule. Briefly, elevated ALDO has been shown to transcriptionally activate serum and glucocorticoid-activated kinase (SGK), which then phosphorylates and inactivates the ubiquitin ligase Nedd4–2 (Bhargava *et al.*, 2004; McCormick *et al.*, 2005). Nedd4–2 normally targets membrane ENaC units for proteolytic destruction. Thus, by inhibiting this ubiquitination system, ALDO indirectly increases surface expression of ENaCs. No experiments have yet been performed to investigate this signalling pathway in hen lower intestine.

Over a longer time course in the coprodeum, there is additionally a major remodelling of the mucosal surface. This includes a near doubling of total cell

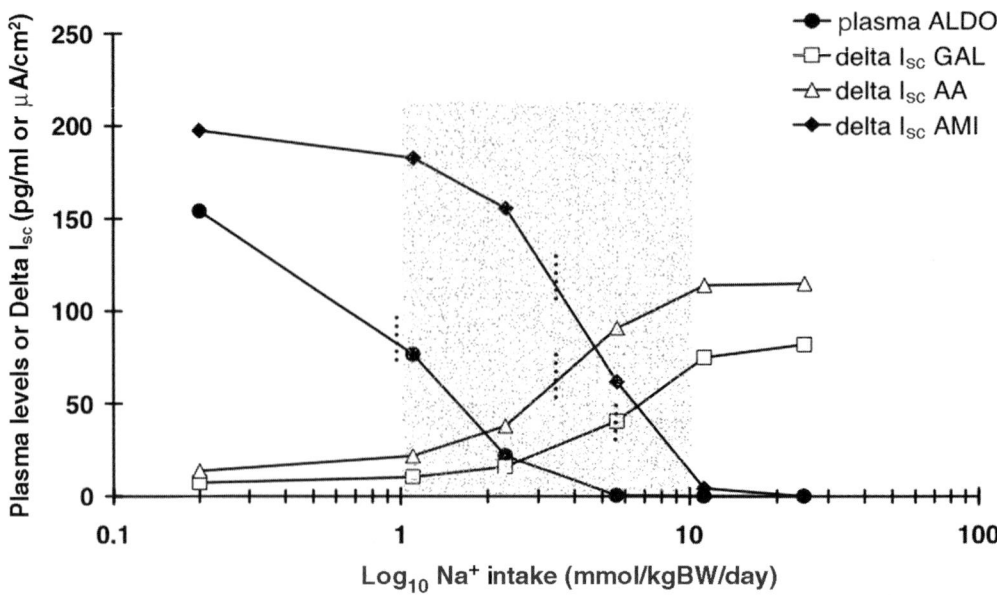

Fig. 5.3. Logistic regression curves showing changes in plasma aldosterone and in short-circuit current (I_{SC}) responses to administration of amiloride, hexose (galactose) or amino acids (leucine + lysine), as a function of the log of steady-state dietary salt (Na^+) intake (data from Árnason and Skadhauge, 1991). Hens were acclimated to one of six different diets resulting in Na^+ intake that ranged from 0.2 to 24.7 mmol/kg BW/day. Data shown are for colonic tissues; there is no hexose- or amino acid-stimulated current in the coprodeum. Note that most of the changes take place within the short range of 1–10 mmol/kg BW/day of dietary Na^+ intake, emphasized by the shaded area. Cross-hatches indicate the dietary Na^+ intake levels at which each parameter is half maximal. Delta I_{SC} amiloride (AMI) and delta I_{SC} amino acids (AA) had identical half-maximal values of 3.8 mmol/kg BW/day. For delta I_{SC} galactose (GAL) the half-maximal value was at 5.9 and for plasma aldosterone (ALDO) it was at 0.9 mmol/kg BW/day.

number and increases in microvillus length and density (Fig. 5–5a) (Elbrønd *et al.*, 1991, 1998; Mayhew *et al.*, 1992). Together, these changes more than double the total apical surface area in the fully LS-acclimated coprodeum, thus providing for increased transport capacity. There does not appear to be any such tissue remodelling in the colon (unpublished observations). Measurements of cell turnover rate in the coprodeum, using bromodeoxyuridine labelling, showed that 14–16 d were required for newly labelled cells to move from the crypts to the tops of the mucosal folds (Elbrønd *et al.*, 1998). This low turnover rate supports the idea of a slower component to the LS acclimation, as compared to the colon.

LS acclimation also caused an increase in the relative proportion of MR cells in the coprodeum, from about 2–5% of total cell number in HS hens to as high as 26% in LS hens (Elbrønd *et al.*, 1999). This may correlate with an increased H^+ ion secretion rate in LS versus HS coprodeum (Laverty *et al.*, 1994).

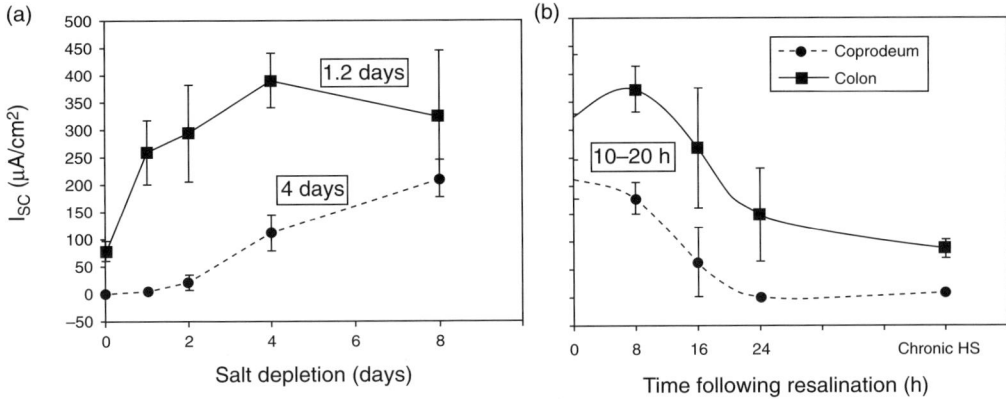

Fig. 5.4. Time courses of changes in initial short-circuit current (I_{SC}) during NaCl depletion (a) and acute resalination (b) in the hen colon and coprodeum. Note different abscissa scales (days versus hours) for the two panels. HS, high-salt diet. Ordinate axis scales are the same for both panels. Numbers in boxes refer to the estimated half-times for depletion and resalination for the two tissues. Resalination half-times are the same for both colon and coprodeum. The zero value (a) and chronic HS value (b) are derived from fully acclimated HS hens. Means ± SEM. Data are derived and replotted from Thomas and Skadhauge, 1982 and from Skadhauge, 1983.

When LS-acclimated hens are acutely resalinated, the transport patterns in both colon and coprodeum revert back to those seen with HS over a much shorter period of 24–72 h (Fig. 5.4), corresponding to a rapid reduction in plasma ALDO (Thomas and Skadhauge, 1982; Clauss *et al.*, 1988, 1991; Elbrønd *et al.*, 1998). There is a slower decrease in both microvillus length and density in the coprodeum, as seen in Fig. 5–5b (Elbrønd *et al.*, 1998), but no apparent changes in the colon (Laverty *et al.*, 2001).

If aldosterone is administered to HS hens by injection (three doses/day), ENaC activity in the colon (amiloride-sensitive I_{SC}) rises to high levels, similar to those of LS, after only 24 h, although amino acid-stimulated I_{SC} is not suppressed at this time (Thomas *et al.*, 1980; Clauss *et al.*, 1991). However, the 24-h response in the coprodeum represents only a small fraction of the LS activity and even 6 d of ALDO treatment induced only 40% of the chronic LS amiloride-sensitive I_{SC} (Thomas *et al.*, 1980; Clauss *et al.*, 1988; Elbrønd *et al.*, 1998). During this period of stimulation, new microvilli appear and increases in both mean microvillus height and MR cell proportion are observed (Elbrønd *et al.*, 1998). These results again point to tissue remodelling in the coprodeum as a major part of the acclimation to low dietary salt.

On the other hand, if ALDO is administered to LS hens in combination with acute resalination, the hormone is able to maintain LS transport patterns in both colon and coprodeum during a period when resalination would otherwise cause transport to revert to the HS patterns. In one study a 24-h ALDO regimen – given at different times after the beginning of resalination – showed a gradual reduction in the ability of the coprodeum to sustain the high rates of amiloride-sensitive I_{SC}, with almost complete insensitivity to ALDO by 5 days post-resalination (Clauss *et al.*, 1984). The same protocol, however,

Fig. 5.5. Scanning electron-micrographs (SEMs) of the epithelial surface of hen coprodeum with insets of transmission electron-micrographs (TEMs) from birds on, respectively, a low-salt (LS, a) and a high-salt (HS, b) diet. Figure 5.5a presents the three major cell types and their localization. The cell surfaces are amplified with long and well-developed microvilli (inset), and the mitochondria-rich cells (MR) are the longest and most densely packed when compared to the absorptive intestinal epithelial cells (AEC). In Fig. 5.5b the epithelial cells have only a few and very short microvilli. The number of MR cells is very limited and only AEC and goblet cells (GC) are presented in the figures.

resulted in full maintenance of colonic amiloride sensitivity for 5 days (Clauss *et al.*, 1984). If ALDO treatment is started at the same time as resalination and continued over a period of days, the high amiloride-sensitive I_{SC} and low co-transporter activity levels are maintained for up to 7 days (Árnason, 1997; Laverty *et al.*, 2001).

The data from these various studies indicate that high levels of ALDO alone can both induce ENaC channel activity and suppress co-transporter activity. On the other hand, it is not known whether other hormones contribute to the induction of co-transporters when ALDO levels drop. Thus, the regression data in Fig. 5.3 show that co-transporter activity is only half maximal at steady-state dietary salt levels that result in undetectable ALDO concentrations (Árnason and Skadhauge, 1991). Similarly, Fig. 5.3 also shows that the half maximal dietary level for amiloride-sensitive I_{SC} is close to the detection threshold for ALDO and considerably higher than its half maximal level. Thus, there may be either changes in the ALDO sensitivity of tissues or other factors that interact with ALDO at medium salt intakes.

The BBMV methods mentioned earlier have confirmed and supported the major role of ALDO in suppressing co-transporter activity. Thus, LS diets and ALDO administration result in low V_{max} values for sodium gradient-driven glucose uptake, whereas resalination of LS hens increases V_{max} (Garriga *et al.*, 1999a, 2000, 2001; Barfull *et al.*, 2002). Even more convincing was the finding that both captopril – a blocker of angiotensin converting enzyme – and spironolactone – a blocker of mineralocorticoid receptors – prevented the reduction in V_{max} seen with LS acclimation (Garriga *et al.*, 2001). These data support a role for the renin–angiotensin–aldosterone axis in control of colonic co-transporter activity.

Several studies have also looked at co-transporter suppression at the protein expression level. These studies have focused on a colonic isoform of SGLT, which appears to be different from the hen small intestinal transporter, based on antibody specificity (Donowitz *et al.*, 1998; Garriga *et al.*, 1999b). In both whole cell extracts (Bindslev *et al.*, 1997) and BBMV preparations (Barfull *et al.*, 2002) Western blotting results show a decrease in colonic SGLT protein expression with LS acclimation. At the same time, Northern blotting failed to detect any changes in SGLT mRNA between LS and HS, suggesting post-transcriptional regulation (Barfull *et al.*, 2002). We have further shown that SGLT protein expression can be maintained at the suppressed LS level with ALDO administered along with resalination, while resalination with vehicle induces the activity and expression of the co-transporter (Laverty *et al.*, 2001).

Interestingly, while glucose-stimulated I_{SC} was significantly elevated after only 1 day of resalination (above both LS controls and 1 day ALDO plus resalination), the SGLT protein expression did not increase until after 7 days of HS. This suggests either the presence of a pool of available SGLT that can be shuttled to the plasma membrane, or control of the transport activity of membrane-expressed proteins during acute changes in salt intake.

Recent studies have begun to look at a possible role for the actin cytoskeleton in controlling ENaC activity in the hen lower intestine. In other systems, studies have provided evidence for multiple functions of actin in ENaC regulation. In renal epithelial cells, for example, submembranous actin-binding proteins such as fodrin and ankyrin bind to the alpha subunit of ENaC channels and stabilize them within the apical membrane domain, thus maintaining cell polarity (Smith *et al.*, 1991; Rotin *et al.*, 1994).

Aside from this structural role, studies in the A6 kidney cell line (Cantiello *et al.*, 1991) and in artificial membranes (Berdiev *et al.*, 1996) have shown that short actin filaments increase ENaC activity, possibly mediated by actin-binding proteins. Furthermore, protein kinase A (PKA) was shown to stimulate ENaC activity by phosphorylating actin and thereby inhibiting elongation of actin filaments. The stimulatory effects of arginine vasopressin may also be mediated by PKA effects on actin (Prat *et al.*, 1993a, b).

Finally, actin also plays a role in the aldosterone-dependent trafficking of ENaC-containing membrane vesicles to and from the apical membrane (Verrey *et al.*, 1995; Rehn *et al.*, 1998). Preliminary studies in the hen lower intestine have compared Na^+ transport and plasma ALDO with the amount and distribution of globular (G) actin monomers and filamentous actin, under different dietary conditions. These findings provide support for both structural and regulatory roles of actin in ENaC-mediated Na^+ transport in these tissues (V.S. Elbrønd, personal observations).

Transport of other electrolytes

In vivo perfusion studies of the hen lower intestine revealed a net secretory flux for K^+ ions and absorptive fluxes of NH_4^+ and PO_4^{2-} (Skadhauge and Thomas, 1979; Rice and Skadhauge, 1982a). Of these, phosphate flux was considered to be passive, while those for ammonium and potassium ions appeared to be coupled to Na^+ ion fluxes (Skadhauge and Thomas, 1979). There have been additional *in vitro* studies of K^+ secretion; under voltage clamp conditions, net secretion is consistently demonstrated, indicating an active transport pathway for this ion, although accounting for only 5–10% of total ion current (Lind *et al.*, 1980a; Grubb and Bentley, 1990; Munck and Munck, 1990). Net secretion was higher in colons from HS- than LS-acclimated hens (Lind *et al.*, 1980a; Munck and Munck, 1990) and could be enhanced by manipulations that increased Na^+ absorptive flux, probably due to increased activity of the Na^+/K^+ ATPase (Munck and Munck, 1990).

Under open-circuit conditions and in the presence of physiologically high mucosal concentrations of K^+ (50 mM), net secretion was abolished, although both unidirectional fluxes remained high (Grubb and Bentley, 1990). This study also demonstrated a significant enhancement of K^+ unidirectional absorptive flux in colons from hens acclimated to a low K^+ dietary intake, which effectively abolished net secretion (voltage clamp conditions). Thus, the colon appears to have a highly regulated and adaptive transport pathway for K^+. The components of K^+ transport in the hen colon have not been well studied, although one report demonstrated the presence of an ouabain-sensitive K^+-activated ATPase on apical membranes of both colon and caecum (Calonge *et al.*, 1997). This activity was sodium independent. In the mammalian colon a H^+/K^+-ATPase activity is thought to drive proton secretion, but inhibitors of this enzyme failed to block activity of the hen colonic K^+-ATPase activity.

As noted earlier, proton secretion is also seen in the hen lower intestine (Lind *et al.*, 1980a; Laverty *et al.*, 1994). This activity may be a function, at least in part, of the carbonic anhydrase-positive MR cells (Elbrønd *et al.*, 2004) and may play a role in the absorption of volatile fatty acids (Holtug *et al.*, 1992). HS acclimation completely eliminated H^+ secretion in the coprodeum, but increased secretion rates in colonic tissues (Laverty *et al.*, 1994). The colonic H^+ secretion was insensitive to both Na^+ ion replacement and ouabain. This study and others (Holtug *et al.*, 1992) also demonstrated an acidic microclimate, a near-surface extracellular pH lower than that of the bulk phase, in both colon and coprodeum. This microclimate probably develops in association with an unstirred mucous layer on the mucosal surface, and may help to neutralize the charge on volatile fatty acids, thus making them membrane permeant.

A well-documented transport activity of the hen lower intestine is that of chloride secretion. In some cases a small flux into the lumen is seen under basal conditions (Lind *et al.*, 1980a), but this secretory transport can be markedly stimulated by agents that raise intracellular cAMP. Thus theophylline, a phosphodiesterase inhibitor, has been extensively used to activate this transport pathway (Clauss *et al*., 1988, 1991; Árnason and Skadhauge, 1991; Árnason, 1997). Chloride secretion in the coprodeum is consistently increased with LS acclimation, whereas responses to LS acclimation in the colon have been mixed. The electrogenic Cl⁻ secretion pathway appears to be similar to that of the mammalian intestine, and may drive net water flow into the lumen under stimulated conditions. In this regard, the Cl⁻ secretion pathway is probably relevant in the pathogenesis of secretory diarrhoea in the fowl.

Besides theophylline, Cl⁻ secretion can be activated by the adenylate cyclase agonist forskolin (unpublished observations) and by vasoactive intestinal peptide (VIP, Munck *et al*., 1984) and serotonin (Hansen and Bindslev, 1989). These latter responses imply physiological regulation. It has also been demonstrated that the avian intestine, including the colon, expresses receptors for guanylin-like endogenous compounds (Krause *et al*., 1995). These compounds are known to activate Cl⁻ secretion via cyclic GMP activation (Forte, 2004). To our knowledge, however, there have not been any functional studies of these compounds in the hen lower intestine.

Secretion of chloride in the hen lower intestine is sensitive to basolateral bumetanide, a diuretic that inhibits Na^+-$2Cl^-$-K^+ symport (Clauss *et al*., 1988, 1991). Thus, as in the mammalian intestine, this probably represents the major uptake step for net Cl⁻ secretion. Apical exit in the mammalian colon is thought to be mediated by cAMP/PKA-activated cystic fibrosis transmembrane regulator (CFTR) chloride channels (Greger, 2000), although other chloride conductances may also play a role. In isolated cells from the hen colon one study (Fischer *et al*., 1991) identified, by patch clamp methods, two chloride channel types, both of which were blocked by NPPB, a commonly used but non-specific blocker of CFTR channels (Schultz *et al*., 1999). It would be of interest to determine whether CFTR-like activity accounts for the cAMP-activated Cl⁻ secretion in the hen lower intestine.

Effects of other hormones

The logistical regression analysis of Fig. 5.3, and other findings, indirectly implicate other hormones in the regulation of hen lower intestinal transport. The most likely candidates are corticosterone, prolactin and arginine vasotocin (AVT). Non-stressed plasma corticosterone, like aldosterone, decreases with increasing dietary salt, although only at the highest levels of intake (Árnason and Skadhauge, 1991). It is also elevated with water deprivation (Árnason *et al*., 1986). Prolactin and AVT are also elevated by water deprivation (Árnason *et al*., 1986), and are significantly increased at the highest levels of dietary salt intake (Árnason and Skadhauge, 1991). The changes in plasma hormone levels with dietary salt intake take place at levels well above those at which the major changes in transport take place, thus calling into question their role in the control of these transport systems. Nevertheless, several studies provide some evidence for effects on lower intestinal transport under certain conditions.

Both corticosterone and the synthetic analogue dexamethasone enhanced amiloride-sensitive I_{SC} in the colon when administered by injection as little as

4–5 h prior to tissue removal (Grubb and Bentley, 1989). With regard to prolactin, although this hormone is generally regarded as a vertebrate osmoregulatory hormone (Loretz and Bern, 1982), there have been very few studies of its effects in birds. The increases in plasma levels with dehydration and HS acclimation (Árnason *et al.*, 1986; Árnason and Skadhauge, 1991) suggest that this hormone might function both in the minimization of water excretion and in increased Na^+ excretion.

On the other hand, prolactin was shown to inhibit water absorption in the rat colon (Krishnamra *et al.*, 2001), thus making any interpretation more complicated. In hens, one study found decreased Na^+ and water absorption by everted jejunal and colonic sacs after 7 days of prolactin injections (Morley *et al.*, 1981). Prolactin infusions were also shown to increase renal excretion of Na^+ and Cl^- in feral chickens (Roberts, 1998). Thus, these studies support the hypothesis of prolactin decreasing Na^+ transport under HS conditions, perhaps by decreasing ENaC activity. However, the inhibitory effect on water absorption does not fit well with elevated prolactin levels associated with water deprivation.

There is more information regarding the possible role of AVT in lower intestinal transport. Most of this comes from a series of studies on colonic BBMV preparations from hens acclimated to different conditions. Looking at two transport systems, the SGLT glucose transporter and the NHE-2 sodium hydrogen exchanger, these studies confirmed the reduction in Na^+ gradient-driven glucose transport activity (decreased V_{max} for glucose uptake; see also Garriga *et al.*, 1999a) and also found increased NHE-2 activity with LS acclimation (Donowitz *et al.*, 1998; De la Horra *et al.*, 2001). When hens were water deprived for 4 days, BBMV glucose transport and NHE-2 activity were both significantly increased in the colon, as well as in both ileum and jejunum (De la Horra *et al.*, 1998). Under these conditions both AVT and ALDO are elevated above baseline (Árnason *et al.*, 1986).

When tissues were removed and pre-incubated directly with hormones for 3 h prior to BBMV preparation, AVT exposure resulted in increased BBMV glucose uptake in the colon, ileum and jejunum (De la Horra *et al.*, 2001). Pre-incubation with ALDO, on the other hand, had no effect on glucose uptake, although it did increase colonic NHE-2 activity. The authors suggest, based on these findings, that colonic SGLT transport activity is primarily regulated by AVT rather than by aldosterone.

However, there are several factors that should also be considered in this interpretation. First, direct exposure of lower intestinal tissues to aldosterone, *in vitro*, has previously been shown to be without effect – even for up to 8 h – on amiloride-sensitive I_{SC} (Elbrønd and Skadhauge, 1992). Secondly, while elevated aldosterone in water-deprived hens was associated with increased – rather than decreased – SGLT activity (De la Horra *et al.*, 1998) , the actual level of plasma ALDO under these conditions was not nearly as high as that seen with LS acclimation (Árnason *et al.*, 1986; Árnason and Skadhauge, 1991).

Moreover, this interpretation does not explain the suppression of SGLT activity under LS acclimation. On the other hand, these are interesting studies, because they might suggest that water deprivation – with increases in both AVT and ALDO – ultimately activates SGLT co-transporter activity and, as mentioned earlier, associated solute-linked water absorption. In this case, elevated AVT would override the suppressive effects of modest increases in ALDO and provide a physiologically appropriate response to aid in water conservation. The possible opposing effects of ALDO and AVT on glucose (and possibly other) co-transporter activity needs to be confirmed.

SUMMARY

The lower intestine of the domestic fowl has been the subject of extensive investigation for several decades. These studies have provided a wealth of information about the transport properties and morphology of these tissues (coprodeum, colon and caeca). It is now clear that the epithelia of these tissues maintain a diverse array of substrate transporters that contribute to both osmotic and nutritional homeostasis. The lower intestine is positioned in such a way as to allow for final recovery and adjustment to both urine and chyme. Moreover, retroperistaltic contractions force these materials into the highly developed caeca, where both fermentative breakdown and further absorption of nutrients, ions and water can take place.

Particularly intriguing are the changes in transport capacity and patterns that are associated with variations in the dietary Na^+ intake. Thus, on low-salt diets, the lower intestine provides a high-capacity transport system for salt conservation. On the other hand, with high salt intake levels, and possibly in response to water deprivation, the colon and coprodeum transport systems switch to a pattern that seems adapted to minimizing osmotic water loss into the intestinal lumen. In this regard, the hen lower intestine has also become a valuable model for the study of actions of aldosterone, particularly the suppression of co-transporter (e.g. SGLT) activity and the extensive tissue remodelling seen in the coprodeum with LS diets.

Ongoing studies are continuing to look at these aspects, as well as at the contributions of other hormones in transport regulation, the characteristics of chloride secretion by these tissues and the role of the actin cytoskeleton in regulating sodium channel activity. These tissues may also provide novel systems to study the signal transduction mechanisms and cellular responses of aldosterone and other hormones.

ACKNOWLEDGEMENTS

Some of the work reported here was supported by the National Science Foundation, NSF IBN0343478 (GL), by the Icelandic Research Council, by The Research Fund and Sattmalasjodur of The University of Iceland (SSA) and by SJVF (VSE). Technical assistance was also provided by Inge Bjerring-Nielsen, Hanne Holm and Gunnel Holden.

REFERENCES

Annison, E.F., Hill, K.J. and Kenworthy, R. (1968) Volatile fatty acids in the digestive tract of the fowl. *British Journal of Nutrition* 22, 207–216.

Árnason, S.S. (1997) Aldosterone and the control of lower intestinal Na^+ absorption and Cl^- secretion in chickens. *Comparative Biochemistry and Physiology* 118A, 257–259.

Árnason, S.S. and Skadhauge, E. (1991) Steady-state sodium absorption and chloride secretion of colon and coprodeum, and plasma levels of osmoregulatory hormones in hens in relation to sodium intake. *Journal of Comparative Physiology* B 161, 1–14.

Árnason, S.S., Rice, G.E., Chadwick, A. and Skadhauge, E. (1986) Plasma levels of arginine, vasotocin, prolactin, aldosterone and corticosterone during prolonged dehydration in the domestic fowl: effects of dietary NaCl. *Journal of Comparative Physiology* B 156, 383–397.

Barfull, A., Garriga, C., Tauler, A. and Planas, J.M. (2002) Regulation of SGLT1 expression in response to Na^+ intake. *American Journal of Physiology. Regulatory, Integrative Comparative Physiology* 282, R738–R743.

Berdiev, B.K., Prat, A.G., Cantiello, H.F., Ausiello, D.A., Fuller, C.M., Jovov, B., Benos, D.J. and Ismailov, I.I. (1996) Regulation of epithelial sodium channels by short actin filaments. *Journal of Biological Chemistry* 271, 17704–17710.

Bhargava, A., Wang, J. and Pearce, D. (2004) Regulation of epithelial ion transport by aldosterone through changes in gene expression. *Molecular Cellular Endocrinology* 217, 189–196.

Bindslev, N. and Skadhauge, E. (1971a) Salt and water permeability of the epithelium of the coprodeum and large intestine in the normal and dehydrated fowl (*Gallus domesticus*). *In vivo* perfusion studies. *Journal of Physiology* 216, 735–751.

Bindslev, N. and Skadhauge, E. (1971b) Sodium chloride absorption and solute-linked water flow across the epithelium of the coprodeum and large intestine in the normal and dehydrated fowl (*Gallus domesticus*). In vivo perfusion studies. *Journal of Physiology* 216, 753–768.

Bindslev, N., Hirayama, B.A. and Wright, E.M. (1997) Na/D-glucose co-transport and SGLT1 expression in hen colon correlates with dietary Na^+. *Comparative Biochemistry and Physiology* 118A, 219–227.

Boykin, S.L.B. and Braun, E.J. (1996) A role for serum albumin in the excretion of uric acid. *Federation of American Socities for Experimental Biology Journal* 10, A389.

Brummermann, M. and Braun, E.J. (1995) Effect of salt and water balance on colonic motility of white leghorn roosters. *American Journal of Physiology* 268, R690–R698.

Calonge, M.L., De la Horra, C., Cano, M., Sanchez-Aguayo, I. and Ilundain, A.A. (1997) Apical ouabain-sensitive K^+-activated ATPase activity in colon and caecum of the chick. *Pflügers Archiv European Journal of Physiology* 433, 330–335.

Cantiello, H.F., Stow, J.L., Prat, A.G. and Ausiello, D.A. (1991) Actin filaments regulate epithelial Na channel activity. *American Journal of Physiology: Cell Physiology* 261, C882–C888.

Casotti, G. and Braun, E.J. (2004) Protein location and elemental composition of urine spheres in different avian species. *Journal of Experimental Zoology* 301A, 579–587.

Choshniak, I., Munck, B.G. and Skadhauge, E. (1977) Sodium chloride transport across the chicken coprodeum. Basic characteristics and dependence on sodium chloride intake. *Journal of Physiology* 271, 489–503.

Clauss, W., Árnason, S.S., Munck, G.B. and Skadhauge, E. (1984) Aldosterone-induced sodium transport in lower intestine: effects of varying NaCl intake. *Pflügers Archiv European Journal of Physiology* 401, 354–360.

Clauss, W., Dantzer, V. and Skadhauge, E. (1988) A low-salt diet facilitates Cl secretion in hen lower intestine. *Journal of Membrane Biology* 102, 83–96.

Clauss, W., Dantzer, V. and Skadhauge, E. (1991) Aldosterone modulates electrogenic Cl secretion in the colon of the hen (*Gallus domesticus*). *American Journal of Physiology Regulatory and Integrative Comparative Physiology* 261, R1533–R1541.

Dantzer, V. (1989) Ultrastructural differences between the two major components of chicken caeca. *Journal of Experimental Zoology* Supplement 3, 21–31.

De la Horra, M.C., Calonge, M.L. and Ilundáin, A.A. (1998) Effect of dehydration on apical Na⁺–H⁺ exchange activity and Na⁺-dependent sugar transport in brush-border membrane vesicles isolated from chick intestine. *Pflügers Archiv European Journal of Physiology* 436, 112–116.

De la Horra, M.C., Cano, M., Peral, M.J., Calonge, M.L. and Ilundain, A.A. (2001) Hormonal regulation of chicken intestinal NHE and SGLT-1 activities. *American Journal of Physiology. Regulatory, Integrative Comparative Physiology* 280, R655–R660.

Donowitz, M., De la Horra, C., Calonge, M.L., Wood, I.S., Dyer, J., Gribble, S.M., Sanchez de Medina, F., Tse, C.M., Shirazi-Beechey, S.P. and Ilundáin, A.A. (1998) In birds, NHE2 is major brush-border Na⁺/H⁺ exchanger in colon and is increased by a low-NaCl diet. *American Journal of Physiology. Regulatory, Integrative Comparative Physiology* 274, R1659–R1669.

Elbrønd, V.S. and Skadhauge, E. (1992) Na transport during long-term incubation of the hen lower intestine: no aldosterone effect. *Comparative Biochemistry Physiology* 101A, 203–208.

Elbrønd, V.S., Dantzer, V., Mayhew, T.M. and Skadhauge, E. (1991) Avian lower intestine adapts to dietary salt (NaCl) depletion by increasing transepithelial sodium transport and micro-villous membrane surface area. *Experimental Physiology* 76, 733–744.

Elbrønd, V.S., Skadhauge, E., Thomsen, L. and Dantzer, V. (1998) Morphological adaptations to induced changes in transepithelial sodium transport in chicken lower intestine (coprodeum): a study of resalination, aldosterone stimulation, and epithelial turnover. *Cell and Tissue Ressearch* 292, 543–552.

Elbrønd, V.S., Dantzer, V. and Skadhauge, E. (1999) Differences in epithelial morphology corre-late to Na⁺ transport: a study of the proximal, mid-, and distal regions of the coprodeum from hens on high and low NaCl diet. *Journal of Morphology* 239, 75–86.

Elbrønd, V.S., Jones, C.J.P. and Skadhauge, E. (2004) Localization, morphology and function of the mitochondria-rich cells in relation to transepithelial Na⁺ transport in chicken lower intestine (coprodeum). *Comparative Biochemistry and Physiology* 137A, 683–696.

Ferrer, R., Planas, J.M. and Moretó, M. (1986) Characteristics of the chicken proximal caecum hexose transport system. *Pflügers Archiv European Journal of Physiology* 407, 100–104.

Fischer, H., Kromer, W. and Clauss, W. (1991) Two types of chloride channels in hen colon epithelial cells identified by patch-clamp experiments. *Journal of Comparative Physiology* B 161, 333–338.

Forte, L.R. Jr. (2004) Uroguanylin and guanylin peptides: pharmacology and experimental therapeutics. *Pharmacology and Therapeutics* 104, 137–162.

Garriga, C., Moretó, M. and Planas, J.M. (1999a) Hexose transport in the apical and basolateral membranes of enterocytes in chickens adapted to high and low NaCl intakes. *Journal of Physiology (Lond.)* 514 (1), 189–199.

Garriga, C., Rovira, N., Moretó, M. and Planas, J.M. (1999b) Expression of Na⁺-D-glucose co-transporter in brush-border membrane of the chicken intestine. *American Journal of Physiology. Regulatory, Integrative Comparative Physiology* 276, R627–R631.

Garriga, C., Moretó, M. and Planas, J.M. (2000) Effects of resalination on intestinal glucose transport in chickens adapted to low Na⁺ intakes. *Experimental Physiology* 85, 371–378.

Garriga, C., Planas, J.M. and Moretó, M. (2001) Aldosterone mediates the changes in hexose transport induced by low sodium intake in chicken distal intestine. *Journal of Physiology* 535 (1), 197–205.

Goldstein, D.L. and Braun, E.J. (1986) Lower intestinal modification of ureteral urine in hydrated house sparrows. *American Journal of Physiology* 250, R89–R95.

Goldstein, D.L. Hughes, M.R. and Braun, E.J. (1986) Role of the lower intestine in the adapta-tion of gulls (*Larus glaucescens*) to seawater. *Journal of Experimental Biology* 123, 345–357.

Greger, R. (2000) Role of CFTR in the colon. *Annual Review of Physiology* 62, 467–491.

Grubb, B.R. (1991) Avian caecum: role of glucose and volatile fatty acids in transepithelial ion transport. *American Journal of Physiology* 260, Gastrointestinal and Liver Physiology G703–G710.

Grubb, B.R. and Bentley, P.J. (1989) Avian colonic ion transport: effects of corticosterone and dexamethasone. *Journal of Comparative Physiology* B 159, 131–138.

Grubb, B.R. and Bentley, P.J. (1990) Potassium transport across the intestines of the fowl, *Gallus domesticus*. *Journal of Comparative Physiology* B 160, 17–22.

Grubb, B.R. and Bentley, P.J. (1992) Effects of corticosteroids on short-circuit current across the caecum of the domestic fowl, *Gallus domesticus*. *Journal of Comparative Physiology* B 162, 690–695.

Hansen, M.B. and Bindslev, N. (1989) Serotonin-induced chloride secretion in hen colon. Possible second messengers. *Comparative Biochemistry and Physiology* 94A, 315–321.

Holtug, K. (1989) Mechanisms of absorption of short-chain fatty acids coupling to intracellular pH regulation. *Acta Veterinaria Scandinavica* (Supplement) 86, 126–133.

Holtug, K., McEwan, G.T.A. and Skadhauge, E. (1992) Effects of propionate on the acid microclimate of hen (*Gallus domesticus*) colonic mucosa. *Comparative Biochemistry and Physiology* 103A, 649–652.

Karasawa, Y. (1999) Significant role of the nitrogen recycling system through the caeca occurs in protein-depleted chickens. *Journal of Experimental Zoology* 283, 418–425.

Krause, W.J., Freeman, R.H., Eber, S.L., Hamra, F.K., Fok, K.F., Currie, M.G. and Forte, L.R. (1995) Distribution of *Escherichia coli* heat-stable enterotoxin/guanylin/uroguanylin receptors in the avian intestinal tract. *Acta Anatomica* 153, 210–219.

Krishnamra, N., Ousingsawat, J. and Limiomwongse, L. (2001) Study of acute pharmacological effects of prolactin on calcium and water transport in the rat colon by an *in vivo* perfusion technique. *Canadian Journal of Physiology and Pharmacology* 79, 415–421.

Laverty, G. and Skadhauge, E. (1999) Physiological roles and regulation of transport activities in the avian lower intestine. *Journal of Experimental Zoology* 283, 480–494.

Laverty, G., Holtug, K., Elbrønd, V.S., Ridderstråle, Y. and Skadhauge, E. (1994) Mucosal acidification and an acid microclimate in the hen colon *in vitro*. *Journal of Comparative Physiology* B 163, 633–641.

Laverty, G., Bjarnadóttir, S., Elbrønd, V.S. and Árnason, S.S. (2001) Aldosterone suppresses expression of an avian colonic sodium–glucose co-transporter. *American Journal of Physiology. Regulatory, Integrative and Comparative Physiology* 281, R1041–R1050.

Lind, J., Munck, B.G., Olsen, O. and Skadhauge, E. (1980a) Effects of sugars, amino acids and inhibitors on electrolyte transport across hen colon at different sodium chloride intakes. *Journal of Physiology* 305, 315–325.

Lind, J., Munck, B.G. and Olsen, O. (1980b) Effects of dietary intake of sodium chloride on sugar and amino acid transport across isolated hen colon. *Journal of Physiology* 305, 327–336.

Loo, D.D.F., Wright, E.M. and Zeuthen, T. (2002) Water pumps. *Journal of Physiology* 542 (1), 53–60.

Loretz, C.A. and Bern, H.A. (1982) Prolactin and osmoregulation in vertebrates. *Neuroendocrinology* 35, 292–304.

Mayhew, T.M., Elbrønd, V.S., Dantzer, V. and Skadhauge, E. (1992) Quantitative analysis of factors contributing to expansion of microvillous surface area in the coprodeum of hens transferred to a low-NaCl diet. *Journal of Anatomy* 181, 73–77.

McCormick, J.A., Bhalla, V., Pao, A.C. and Pearce, D. (2005) SGK1: a rapid aldosterone-induced regulator of renal sodium reabsorption. *Physiology* 20, 134–139.

Morley, M., Scanes, C.G. and Chadwick, A. (1981) The effect of ovine prolactin on sodium and water transport across the intestine of the fowl (*Gallus domesticus*). *Comparative Biochemistry and Physiology* 68A, 61–66.

Munck, B.G., Andersen, V. and Voldsgård, P. (1984) Chloride transport across the isolated hen colon. In: Skadhauge, E. and Heintze, K. (eds) *Intestinal Absorption and Secretion*. MTP Press, Lancaster, UK, pp. 373–386.

Munck, L.K. and Munck, B.G. (1990) Intestinal transport of potassium. Effects of changing apical and basolateral influx of sodium in the isolated mucosa of the hen (*Gallus domesticus*) colon. *Comparative Biochemistry and Physiology* 96A, 181–186.

Murakami, A.E., Oviedo-Rondón, E.O., Martins, E.N., Pereira, M.S. and Scapinello, C. (2001) Sodium and chloride requirements of growing broiler chickens (twenty-one to forty-two days of age) fed corn–soybean diets. *Poultry Science* 80, 289–294.

Prat, A.G., Ausielleo, D.A. and Cantiello, H.F. (1993a) Vassopressin and proteinkinase A activate G Protein-sensitive epithelial Na channels. *American Journal of Physiology: Cell Physiology* 265, C218–C223.

Prat, A.G., Bertorello, A.M., Ausielleo, D.A. and Cantiello, H.F. (1993b) Activation of the epithelial Na channels by protein kinase A requires actin filaments. *American Journal of Physiology: Cell Physiology* 265, C224–C233.

Rehn, M., Weber, W.M. and Clauss, W. (1998) Role of the cytoskeleton in stimulation of Na channels in A6 cells by changes in osmolality. *European Journal of Physiology* 146, 270–279.

Rice, G.E. and Skadhauge, E. (1982a) The *in vivo* dissociation of colonic and coprodeal transepithelial transport in NaCl-depleted domestic fowl. *Journal of Comparative Physiology* B 146, 51–56.

Rice, G.E. and Skadhauge, E. (1982b) Caecal water and electrolyte absorption and the effects of acetate and glucose in dehydrated, low-NaCl diet hens. *Journal of Comparative Physiology* B 147, 61–64.

Roberts, J.R. (1998) The effect of acute or chronic administration of prolactin on renal function in feral chickens. *Journal of Comparative Physiology* B 168, 25–31.

Rotin, D., Bar-Sagi, D., O'Brodovich, H., Merilainen, J., Lehto, V.P., Canessa, C.M., Rossier, B.C. and Downey, G.P. (1994) An SH3 binding region in the epithelial Na channel (αrENaC) mediates its localization at the apical membrane. *The European Molecular Biology Organization Journal* 13, 4440–4450.

Schultz, B.D., Singh, A.K., Devor, D.C. and Bridges, R.J. (1999) Pharmacology of CFTR chloride channel activity. *Physiological Reviews* 79 (1), S109–S144.

Skadhauge, E. (1968) The cloacal storage of urine in the rooster. *Comparative Biochemistry and Physiology* 24, 7–18.

Skadhauge, E. (1983) Temporal adaptation and hormonal regulation of sodium transport in the avian intestine. In: Gilles-Baillien, M. and Gilles, R. (eds) *Intestinal Transport.* Springer-Verlag, Berlin, pp. 284–294.

Skadhauge, E. (1993) Basic characteristics and hormonal regulation of ion transport in avian hindguts. In: Clauss, W. (ed.) *Advances in Comparative and Environmental Physiology*, Vol. 16. Springer-Verlag, Berlin, pp. 67–93.

Skadhauge, E. and Thomas, D.H. (1979) Transepithelial transport of K^+, NH_4^+, inorganic phosphate and water by hen (*Gallus domesticus*) lower intestine (colon and coprodeum) perfused luminally *in vivo*. *Pflügers Archiv European Journal of Physiology* 379, 237–243.

Smith, A., Rose, S.P., Wells, R.G. and Pirgozliev, V. (2000) Effect of excess dietary sodium, potassium, calcium and phosphorus on excreta moisture of laying hens. *British Poultry Science* 41, 598–607.

Smith, P.R., Saccomani, E.H., Joe, E.H., Angelides, K.J. and Benos, D.J. (1991) Amiloride-sensitive sodium channel is linked to the cytoskeleton in the renal epithelial cells. *Proceedings of National Academy of Natural Sciences of the USA* 88, 6971–6975.

Thomas, D.H. (1982) Salt and water excretion in birds: the lower intestine as an integrator of renal and intestinal excretion. *Comparative Biochemistry and Physiology* 71A, 527–535.

Thomas, D.H. and Skadhauge, E. (1982) Time course of adaptation to low- and high-NaCl diets in the domestic fowl. Effects on electrical behaviour of isolated epithelia from the lower intestine. *Pflügers Archiv European Journal of Physiology* 395, 165–170.

Thomas, D.H. and Skadhauge, E. (1989a) Water and electrolyte transport by the avian caeca. *Journal of Experimental Zoology* Supplement 3, 95–102.

Thomas, D.H. and Skadhauge, E. (1989b) Function and regulation of the avian caecal bulb: influence of dietary NaCl and aldosterone on water and electrolyte fluxes in the hen (*Gallus domesticus*) perfused *in vivo*. *Journal of Comparative Physiology* B 159, 51–60.

Thomas, D.H., Jallageas, M., Munck, B.G. and Skadhauge, E. (1980) Aldosterone effects on electrolyte transport of the lower intestine (coprodeum and colon) of the fowl (*Gallus domesticus*) *in vitro*. *General and Comparative Endocrinology* 40, 44–51.

Titus, E. and Ahearn, G.A. (1992) Vertebrate gastrointestinal fermentation: transport mechanisms for volatile fatty acids. *American Journal of Physiology. Regulatory, Integrative and Comparative Physiology* 262, R547–R553.

Verrey, F., Groscurth, P. and Bolliger, U. (1995) Cytoskeletal disruption in A6 kidney cells: impact on endo/exocytosis and NaCl transporter regulation by antidiuretic hormone. *Journal of Membrane Biology* 145, 193–204.

Waldenstedt, L. and Björnhag, G. (1995) Retrograde flow of urine from cloaca to caeca in laying hens in relation to different levels of nitrogen intake. *Deutsche Tierärztliche Wochenschrift* 102, 168–169.

CHAPTER 6
Immunological development of the avian gut

R.K. Beal,[1] C. Powers,[1] T.F. Davison[2]* and A.L. Smith[1]

[1]*Division of Immunology, Institute for Animal Health, Newbury, UK;*
[2]*Formerly of the above;* *e-mail: fred.davison@dunelm.org.uk*

ABSTRACT

The gut is a highly specialized site for absorption of nutrients but also represents an important barrier between the external and internal environments. It is the site of replication and/or portal of entry for several important pathogens and requires an array of different immune effector cells that are organized in specialized tissues such as the bursa of Fabricius, caecal tonsils, Meckel's diverticulum and Peyers' patches, or more broadly distributed in the gut epithelium or the sub-epithelial *lamina propria*.

Both the innate and adaptive immune effector mechanisms play important roles in protection from gut pathogens.

The period after hatching is especially crucial for the development of the gut immune system. With the onset of food intake and microbial colonization the immune system must differentiate between innocuous food components and potentially harmful pathogens. At this stage the immune system is immature and the chick must rely on innate effector mechanisms and maternal antibodies, mainly IgY transmitted from the hen via the yolk. This makes the chick especially vulnerable to a number of pathogens, until the adaptive immune system has developed sufficiently to produce effective immune responses. Development of the adaptive immune system depends on the arrival and replication of specialized leukocytes at key locations in the gut, especially the B and T lymphocytes.

Investigations on the influence of age at primary infection with *Salmonella enterica* have shown that clearance from the gut is associated with the ability to produce strong adaptive immune responses. *Salmonella* persists in the gut beyond 8–9 weeks of age, irrespective of the age at first exposure, indicating the importance of a fully mature gut immune system to control infection.

INTRODUCTION

The gut, being a specialized organ concerned with digestion, requires a large mucosal surface that allows intimate contact for the absorption of nutrient molecules. This is achieved through millions of villi lined with columnar epithelial cells which undergo continual renewal and are produced from stem cells lining the crypts. The gut is a crucial barrier between the internal milieu and the external environment. In addition, it provides a residence and site for the replication of many commensal organisms, some of which break down cellulose and uric acid in the caecum and make important contributions to the digestive process. The gut is also an established site of infection for viruses, bacteria and parasites. Although some pathogens remain within the gut lumen others, such as *Salmonella* and *Eimeria*, invade the lining to gain access to internal tissues where their replication can cause disease. To protect the host against disease it is essential that defence mechanisms be employed along the mucosal surfaces. These defences include physical, chemical and cellular barriers to pathogen entry, as well as a highly specialized gut immune system. The gut-associated lymphoid tissue (GALT) comprises lymphoid cells residing in the epithelial lining and distributed in the underlying *lamina propria*, as well as specialized lymphoid structures located at strategic sites along the gut.

Gut immune responses include both chemical and cellular responses that are either non-specific or broadly specific and, lacking memory, are referred to as innate responses. In addition the adaptive immune system which is mediated by both B and T lymphocytes has both specificity and memory and makes important contributions to pathogen control through its populations of specialized inducer, effector and immuno-regulatory cells. The gut B and T cells share much in common with counterparts in the systemic immune system, but also have some very distinct differences.

An important component of the immune system is a capacity for different cell populations to interact, which is achieved by co-stimulatory molecules (direct cell to cell contact) and by the release of soluble mediators such as cytokines. A broad range of cell types can produce different cytokines and these are involved in recruiting cells to infected sites and in inducing cell proliferation, activation and suppression (for a review of avian cytokines see Kaiser *et al.*, 2005).

The GALT is a key immunological system estimated to have more lymphocytes than any of the other lymphoid tissues (Kasahara *et al.*, 1993b). Like the spleen, the GALT provides a major system involved in the development of immune responses (Bar-Shira and Friedman, 2005). The gut immune system, in contrast to the other luminal immune systems such as those protecting the respiratory and reproductive tracts, must be capable of differentiating between two types of foreign materials. Harmless nutrient molecules in the gut need to be tolerated, whereas antigens derived from pathogens must evoke appropriate protective immune responses.

A fine balance between tolerance and response is essential for gut health and to provide optimal conditions for the absorption of nutrients, and also to act as an efficient barrier to the entry of pathogens. Moreover, normal gut structure is very sensitive to perturbation and the level of a local immune response can often result in gross pathology. Classical signs of an ongoing immune response in the gut include: villus atrophy and hyperplasia of enterocytes in the crypts.

ORGANIZATION OF THE AVIAN GUT IMMUNE SYSTEM

Lymphocyte populations in the epithelial lining, and also in the *lamina propria*, are markedly affected by age, genotype, pathogen status and diet. However, in the mature healthy chicken, the epithelial cells lining the apical regions of the villi are heavily interspersed with lymphocytes (intra-epithelial lymphocytes, IEL; see Fig. 6.1).

This IEL population principally comprises the following: (i) natural killer (NK) cells (Gobel *et al.*, 2001); and (ii) γδ-T cells and αβ-T cells, the latter principally CD8$^+$ with a minor population of CD4$^+$ cells (Lillehoj *et al.*, 2004). More than 50% of the TCR-positive IEL population have been found to express CD8αα homodimers, rather than the αβ heterodimer observed elsewhere (Imhof *et al.*, 2000) and, in addition, small numbers of B cells (Vervelde and Jeurissen, 1993) and heterophils (Bar-Shira and Friedman, 2005) have been identified. The cell populations distributed in the gut will be discussed in greater detail later in this review.

Lymphocytes are also distributed throughout the *lamina propria* underlying the epithelial lining. Although NK, B and T lymphocytes are present, the composition of this population is considerably different from the IEL population. In terms of lymphocyte numbers, B and T cells predominate (~90%), with the remainder comprising the NK cell phenotype. In contrast to the IEL population, the T cell population of the *lamina propria* contains only a small proportion of γδ-T cells (~ 10%), with the much larger αβ-T cell population dominated by CD4$^+$ T cells and only a minor population of CD8$^+$ cells.

Chickens lack highly structured lymph nodes such as those found in mammals, but have a number of distinct lymphoid aggregates that line the length of the gut. In their simplest form these consist of: (i) a specialized lymphoepithelium containing M cells which are phagocytic; and (ii) sample antigens from the gut lumen which transport these cells to underlying macrophages and dendritic cells (Burns and Maxwell, 1986; Gallego *et al.*, 1995; Jeurissen *et al.*, 1999). In close proximity to these cells, and with varying degrees of organization, are follicles rich in B and T lymphocytes, usually undergoing cell division.

The chicken foregut is considered to have relatively little organized lymphoid tissue compared with the hindgut (Befus *et al.*, 1980), although oesophageal tonsils have been described, located at the junction of the oesophagus and the proventriculus (Olah *et al.*, 2003). Up to 8 tonsillar units have been described; each consisting of a crypt lined with lymphoepithelium and surrounded by dense lymphoid tissue organized into B- and T-cell

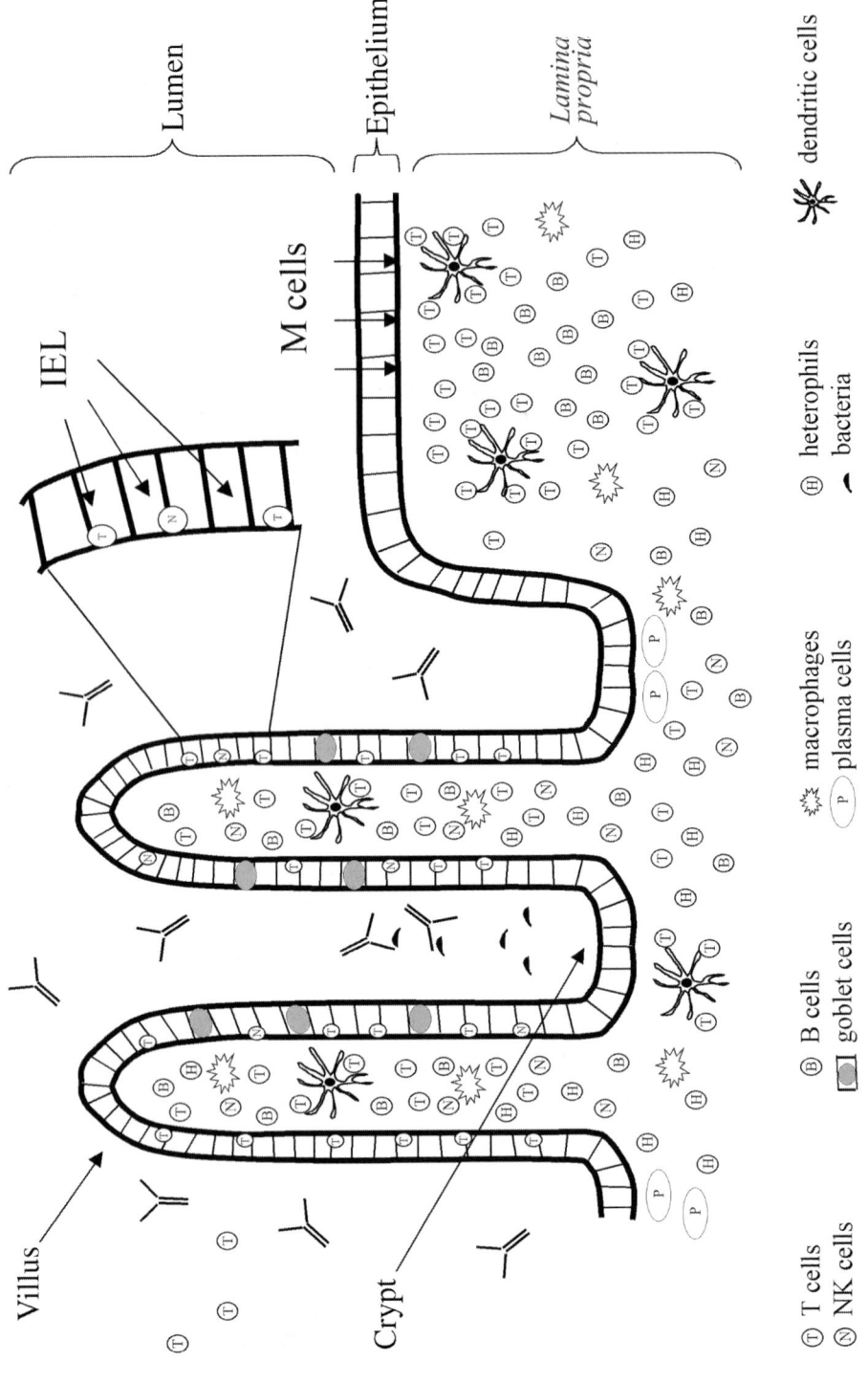

Fig. 6.1. Schematic drawing of a section through the wall of the avian intestine showing the villus structure, epithelial layer and underlying *lamina propria*, with the different types of cells found in these regions.

T cells   NK cells   B cells   goblet cells   macrophages   plasma cells   dendritic cells   heterophils   bacteria regions. Organized lymphoid tissue is also located throughout the entire *lamina propria* of the proventriculus above the glandular units (Jeurissen *et al.*, 1994).

A specialized lymphoid structure, Meckel's diverticulum, develops at the junction of the duodenum and jejunum in the young chick (Olah and Glick, 1984). The importance of this lymphoid aggregate is not fully understood but, since it is formed from the lining of the vitelline duct, in the newly hatched chick it may help prevent pathogens gaining entry to or from the yolk sac whilst it is being re-adsorbed in the abdomen. Several lymphoid structures with the characteristics of mammalian Peyer's patches (PP) can be found scattered throughout the intestine (Befus *et al.*, 1980).

Although not evident at hatching, chicken PP become visible to the naked eye by the second week (Befus *et al.*, 1980) and can increase in number to about five by 16 weeks of age. The PP anterior to the ileocaecal junction is the most obvious and this becomes especially prominent after infection. Caecal tonsils (CT) at the mouth of each of the caeca are prominent, and the most well-studied structures of the GALT; however, numerous lymphoid follicles are present throughout the caecal wall, especially towards the apical region of the caecum (del Cacho, *et al.* 1993; Kitagawa *et al.*, 1996, 1998).

The colon is generally lacking in lymphoid follicles but a few solitary ones have been observed in the proctodeal and urodeal regions of the cloaca (Befus *et al.*, 1980), and they are particularly abundant around the bursal duct (Friedman *et al.*, 2003).

The bursa of Fabricius, a lymphoepithelial outgrowth from the proctodeum, is one of the primary lymphoid organs of birds where B lymphocyte development occurs during the late embryonic stage and in the young chick. The role of the bursa as a primary lymphoid organ essential for the development of B cells is discussed in a later section. However, it has been suggested that the bursa could have a role as a secondary lymphoid organ (Friedman *et al.*, 2003).

There is good evidence that antigens from the gut are transported from the cloacal region up the bursal duct and into the bursal lumen (Schaffner *et al.*, 1974; Sorvari *et al.*, 1975). These antigens cross the lining of the bursal follicles through specialized phagocytic epithelial cells (similar to M cells) and then enter the bursal medulla, where they are taken up by resident macrophages – and probably by dendritic cells.

The process of retrograde peristalsis, usually considered as a means of increasing water conservation from kidney secretions entering the cloaca, could also be the means by which bursal follicles sample hindgut bacteria (Friedman *et al.*, 2003). It has been reported that the introduction of antigens into the bursa results in increased B-cell responses in the peripheral tissues after subsequent systemic challenge (Ekino *et al.*, 1979), but these introduced antigens can only induce B-cell priming and do not appear to induce antibody responses directly (Ratcliffe *et al.*, 1996).

Bramburger and Friedman (see Abstracts, this volume) have suggested that the bursa functions as a secondary lymphoid tissue as it has the capacity to express immunoglobulin (Ig) transporter genes and actively to secrete antibodies. Whilst antibody production by follicles within, or in the vicinity of, the bursa could be valuable in the young chick it is not clear that this function is maintained in older birds after the bursa has regressed.

IMMUNE EFFECTOR CELLS

Innate immune cell populations

Macrophages

In avian species, as in mammals, the principal roles of macrophages are in phagocytosis and in antigen presentation. These cells express a number of pattern recognition molecules on their surface in order to detect invading pathogens, including mannose receptors and scavenger receptors.

They also express a range of Toll-like receptors (TLR) that recognize a variety of compounds derived from bacteria, fungi, viruses and parasites. Expression in the chicken of TLR1/6/10, TLR 2 (type 2), TLR4, TLR5 and TLR7 have been demonstrated (Dil and Qureshi, 2002; Iqbal *et al.*, 2005a, b; Philbin *et al.*, 2005). Macrophages also express Fc receptors that bind antibody and complement receptors that bind opsonized pathogens (reviewed in Qureshi *et al.*, 2000). Activated macrophages increase in size, produce a variety of cytokines and up-regulate phagocytic and cytotoxic activity (Qureshi, 2003).

NK cells

The function and characterization of mammalian NK cells is relatively well defined by comparison with avian equivalents. These cells are large and lymphoid, with prominent intracellular granules containing perforin and granzymes, but do not express the T cell receptor (TCR) or B cell receptor (BCR). In mammals NK cells can recognize infecting pathogens coated with antibody via Fc receptors that bind IgG_1 and IgG_3, thus allowing antibody-dependent cell cytotoxicity to take place.

Recognition of carbohydrate molecules on the surface of infected cells by C-type lectins also enables NK cells to recognize and kill intracellular pathogens. Additionally, these cells recognize the levels of MHC Class I molecules on the surface of cells and prevent the killing of healthy self cells expressing normal levels of MHC Class I molecules (Janeway *et al.*, 1999). Although NK cell biology is still not fully understood, the properties of NK cells have led some to suggest that they provide a first line of defence and are particularly important in responses to viruses and tumours.

In the chicken early studies were limited to assaying NK cell-like activity (Sharma and Okazaki, 1981; Chai and Lillehoj, 1988; Lillehoj and Chai, 1988). More recently, NK cell populations have been characterized in the chicken on the basis of the classic NK cell phenotype; these cells were CD8[+] but lacked specific markers for the T- or B-cell lineage (Gobel *et al.*, 2001). Within the spleen and peripheral blood the proportions of these cells are very low (< 1%), but the proportion present in the IEL poplulation is much greater (~30%).

Heterophils

These are the most abundant polymorphonuclear leukocytes in the chicken and play an important role in the phagocytosis and destruction of invading pathogens. Chicken heterophils lack myeloperoxidase and do not produce oxygen radicals; their killing mechanisms mainly rely on the production of defensins (Brune *et al.*, 1972; Penniall and Spitznagel, 1975; Evans *et al.*, 1994). Heterophils are associated with inflammation and are recruited early after infection; for example, large numbers of heterophils infiltrate the *lamina propria* following *Salmonella* infections (Henderson *et al.*, 1999; Van Immerseel *et al.*, 2002).

Chicken heterophils express a range of pattern recognition molecules similar to those expressed by macrophages, including Fc and complement receptors. Heterophils express a range of TLR molecules including: TLR1/6/10, both forms of TLR2, TLR3, TLR4, TLR5 and TLR7 (Iqbal *et al.*, 2005a; Kogut *et al.*, 2005). Moreover, heterophils have the capacity to degranulate and to produce a range of pro-inflammatory cytokines in response to specific TLR agonists.

Paneth cells

These are specialized epithelial cells located in the crypts of the small intestine and have been described in detail in mammals (reviewed by Ganz, 2000). Paneth cells are rich in secretory granules that contain lysozyme, secretory phospholipase A_2 and α- and β-defensins, which all have anti-microbial activity *in vitro* against bacteria (Selsted *et al.*, 1984), fungi (Selsted *et al.*, 1985; Lehrer *et al.*, 1988) and some enveloped viruses (Lehrer *et al.*, 1985; Daher *et al.*, 1986). It is considered that these cells help maintain the gastrointestinal barrier in the crypt lumen (Ayabe *et al.*, 2000) and contribute to the host's defence against pathogenic microorganisms.

Paneth cells have not yet been described and may be scarce in birds (Bezuidenhout and Van Aswegen, 1990). β-defensins have been identified in the chicken (Zhao *et al.*, 2001; Higgs *et al.*, 2005) but have been associated with heterophils and macrophages (Evans *et al.*, 1994; Zhao *et al.*, 2001; Sugiarto and Yu, 2004), leading some to speculate that these cells could have an important antimicrobial role in the avian gut.

Adaptive immunity

T cells

As in mammals, chicken T cells are classified on the basis of their T-cell receptors either as TCR$\alpha\beta^+$ ($\alpha\beta$-T cells) or TCR$\gamma\delta^+$ ($\gamma\delta$-T cells). Chickens have only two families of TCR$\alpha\beta^+$ cells expressing either Vβ1 (recognized by the monoclonal antibody TCR2) – referred to as the TCR$\alpha\beta$1 subfamily – and those expressing Vβ2 (recognized by the monoclonal antibody TCR3) –

referred to as the TCRαβ2 subfamily. The tissue distributions of TCR2[+] and TCR3[+] cells differ. Within the IEL population very few TCR3[+] cells are present in the intestine and TCR2[+] cells are relatively abundant in the intestinal *lamina propria* (in contrast, both populations are abundant in the spleen).

In a series of elegant experiments Chen *et al.* (1989) and Cihak *et al.* (1991) investigated functional differences between TCR2[+] and TCR3[+] cells. Injection of anti-TCR2 monoclonal antibody into the embryo followed by post-hatch thymectomy resulted in depletion of TCR2[+] cells and elevated proportions of TCR3[+] T cells. The resulting chickens responded normally to a range of different antigens (both T cell-dependent and -independent) and had normal levels of circulating IgG and IgM, but were severely compromised in their capacity to produce IgA.

In the chicken, γδ-T cells are present in higher proportions in the peripheral blood and spleen than in mice and humans (Cooper *et al.*, 1991). In the thymus and peripheral blood the majority of γδ-T cells express neither the CD8 nor the CD4 marker on their cell surface, whereas those in the gut are predominantly CD8[+] (Sanchez-Garcia and McCormack, 1996). Whilst the biological functions of γδ-T cells are still unclear, *in vitro* studies have demonstrated that they are capable of cytotoxicity (Chen *et al.*, 1994) and may also have immunoregulatory functions (Quere *et al.*, 1990). Chicken γδ-T cells respond poorly to stimulation by either mitogens or to TCR crosslinking (Sowder *et al.*, 1988; Cooper *et al.*, 1991); however, the addition of αβ-T cells to cultures restored the capacity of γδ-T cells to proliferate *in vitro* (Arstila *et al.*, 1993).

Kasahara *et al.* (1993a) demonstrated that these γδ-T cells require exogenous sources of the cytokine interleukin (IL)-2, to potentiate stimulation by mitogen or TCR cross-linking. Studies using mice have also demonstrated that γδ-T cells are involved in the regulation of αβ-T cells. When antibodies were used to modulate γδ-T cells *in vivo*, both an increase in IL-2 production by CD4[+] T cells and increased cytotoxicity by CD8[+] T cells were observed (Kaufmann *et al.*, 1993), although this has not yet been tested in the chicken.

B cells

The chicken provided the first evidence for the differentiation of lymphocyte populations into those derived from the bursa of Fabricius (B cells) and those derived from the thymus (T cells) (Glick *et al.*, 1956; Cooper *et al.*, 1965). The major function of B cells is the production of antibodies. Classically, activation of B cells is dependent on the specific antigen and to assistance from CD4[+] T cells (T cell help).

T cell-dependent antigens bind to surface antibody on the B cell, are internalized, processed and then re-expressed as peptides in association with MHC Class II molecules before recognition by helper T cells; these helper cells then deliver activating signals to the B cell via cytokines and co-stimulatory molecules. Activation of B cells can occur independently of T cell help, usually by highly repetitive antigenic structures that cross-link many BCRs (Toivanen *et al.*, 1972; Toivanen and Toivanen, 1973; Dil and Qureshi, 2002).

Once activated, B cells differentiate to form plasma cells that produce and secrete antibodies into the local environment. Repeated exposure to an antigen results in class switching from IgM to IgG and IgA (see below). It has been suggested that peripheral B cells are pre-programmed for switching to a particular class before they leave the bursa (Kincade and Cooper, 1973; Ivanyi, 1975).

Antibodies

Only three classes of chicken antibodies have been described and characterized: (i) IgM; (ii) IgY (the avian homologue of mammalian IgG); and (iii) IgA. IgM is the first antibody produced in response to primary challenge and is expressed on the surface of all B cells prior to class switching. The structure and properties of chicken IgM are very similar to those of mammalian IgM (reviewed in Higgins, 1975). IgM is present in the serum predominantly in either pentameric (Leslie and Clem, 1969) or tetrameric configurations; however, a small amount can also be present as a monomer (Lebacq-Verheyden et al., 1974; Higgins, 1976). The role of IgM in immune responses in the gut is limited; it is present in bile and intestinal washes only in very small quantities (Leslie et al., 1971; Bienenstock et al., 1973b).

IgY is the predominant antibody active against systemic infection, and it also plays a major role as the vehicle of maternal immunity, passing to the chick via the yolk sac (Van Meter et al., 1969).

The predominant secretory antibody in the chicken, as in mammals, is IgA (Leslie et al., 1971). It is present at high concentrations in intestinal fluids and bile at 3.5–12 mg/ml, compared with IgM at $< 100\mu g/ml$ (Lebacq-Verheyden et al., 1972; Bienenstock et al., 1973b; Mockett, 1986). In avian serum IgA is predominantly present as a monomer, whereas in secretions it is polymeric (Bienenstock et al., 1973b). The majority of antibody-forming plasma cells in the intestine produce IgA, which is transported across the gut epithelium and into the gut lumen by a process similar to that described in mammals (Peppard et al., 1986). Polymeric IgA binds to epithelial cells or hepatocytes using the secretory component (SC); it is then internalized and transported across the cells and into the lumen (Bienenstock et al., 1973a; Parry and Porter, 1978; Rose et al., 1981; Peppard et al., 1986).

DEVELOPMENT OF THE GUT IMMUNE SYSTEM

After hatching, considerable changes take place in the structure and organization of the GALT. During the first 5 days epithelial cells in the crypts and villi are rapidly proliferating; however, after this stage, proliferation is limited to defined proliferative zones in the crypts (Uni et al., 2000). Epithelial cells themselves also undergo rapid change: from round, non-polar enterocytes to elongated cells with typical enterocyte morphology by 24 h post-hatch (Geyra et al., 2001).

Staining for proliferating cell nuclear antigen (PCNA) has revealed that most enterocytes in the crypts and villi are proliferating at hatch, but that only those enterocytes in the crypts are still proliferating by 3 days post-hatch (Geyra *et al.*, 2001).

Development of goblet cells in the chicken in the immediate post-hatch period has been studied by Uni *et al.* (2003); these are apparent in the intestinal epithelium as early as 3 days before hatching. Goblet cells present before hatching stained mainly for acidic mucins, but after hatching there was an increase in neutral mucins. Chicks whose feed was withheld for 48 h had higher levels of intracellular mucin, which could be due to either impaired mucin secretion or to enhanced mucin production (Uni *et al.*, 2003).

Change in the structure of the intestinal epithelium could have profound implications for the ability of the newly hatched chick to resist enteric infection: low mucous production, enterocyte immaturity and a general lack of villus and crypt development reduces the physical barriers within the intestine for preventing invasion of enteric pathogens.

Pre-/post-hatch population of the gut with immune cells

Increases in the numbers and diversity of different cell types in the intestinal epithelium and *lamina propria* and development of the specialized lymphoid tissues such as the PP and CT (Befus *et al.*, 1980; Vervelde and Jeurissen, 1993; Gomez Del Moral *et al.*, 1998) are a consequence of the export of thymic T cells and bursal B cells.

Dunon *et al.* (1997) monitored T cell export from the thymus and proposed that the process occurred in a series of three waves, each wave beginning with the export of TCR1[+] cells, followed 2–3 days later by the export of TCR2[+] and TCR3[+] cells. Thus TCR1[+] cells are exported to the periphery at 15–17 E, around the time of hatching and at 6–8 days post-hatch. TCR2[+] and TCR3[+] cells migrate to the periphery at 18–20 E, at 2–4 days post-hatch and at 9–11 days. Studies tracking bursal cells using ^{3}H-thymidine demonstrated that B cells begin to be exported from the bursa around day 18 of embryonic development (Linna *et al.*, 1969; Hemmingsson and Linna, 1972).

At hatching, histological studies have indicated that the intestinal tract is very low in lymphoid cell content (Bar-Shira and Friedman, 2005). In the epithelium the number of lymphocytes increased with age as cells were exported from the thymus, with peak numbers present by approximately 8 weeks post-hatch (Vervelde and Jeurissen, 1993).

In recent studies by Bar-Shira *et al.* (2003), measurements of the expression of specific mRNA produced by different cell subsets was used to monitor age-related changes in lymphocyte populations in different parts of the gut. These workers observed a large increase in CD3[+] cells at 4 days post-hatch in all parts of the gut. The number of B cells also increased significantly, initially in the small intestine (4 days post-hatch) and then in the large intestine and caecal tonsils (6 days post-hatch). However, at hatching, the level of B cell-specific mRNA expression was greater in the CT than any of the other parts of the gut. These authors suggested that GALT lymphocytes are functionally immature at hatching and measurements of mRNA for cytokines, such as IL-2 and interferon-γ (IFN- γ), indicated that functional maturation of both B and T cells was biphasic, with the first stage occurring during the first week after hatching and the second stage during the second week (Bar-Shira *et al.*, 2003).

During the first 4 weeks post-hatch there are also important changes in the proportions of different lymphocyte subsets within the different gut compartments. For example, there is a shift from TCR2$^+$ to TCR1$^+$ cells in the intestinal epithelium, and also an increase in the percentage of IEL expressing CD8. In the *lamina propria* there are increases in the percentage of lymphocytes expressing TCR2 (Lillehoj and Chung, 1992). The *lamina propria* of 1-day-old chicks is poorly developed, containing little stroma (capillaries, lacteals, reticular and muscle fibres) and few lymphocytes (Yason *et al.*, 1987). By 17 days post-hatch the *lamina propria* has changed significantly, with increases in stromal tissue, lymphocytes, mononuclear cells and eosinophils.

Cells of the innate immune system develop earlier than T and B cells, with large numbers of heterophils and monocytes being present in the blood at hatching (Burton and Harrison, 1969; Wells *et al.*, 1998). During the first 2 weeks post-hatch there are further increases in polymorphonuclear cells in all parts of the intestine (Bar-Shira and Friedman, 2005).

DEVELOPMENT OF IMMUNE RESPONSES TO MODEL ANTIGENS

Systemic responses

There have been few reports in the literature of changes in immune responses to model antigens. Mast and Goddeeris (1999) reported changes in antibody responses following *in ovo* (E16 or E18) or subcutaneous immunizations (1-, 7- or 12-day-old chicks) of bovine serum albumin (BSA) administered with an adjuvant. Antigen-specific IgM and IgY were produced only in those chicks immunized at 12 days, peaking at 7 and 10 days post-immunization for IgM and IgY, respectively.

The chickens immunized at 7 and 10 days were capable of an anamnestic response after a second immunization; however, those immunized at 1 day old were not. These data suggest that young chicks (< 1 week old) are not capable of mounting an effective antibody response to antigen.

Responses to oral antigens

In a similar series of experiments, Bar-Shira *et al.* (2003) investigated the capacity of gut antigens to evoke antibody responses in differently aged birds. When chickens were immunized orally with BSA, or cloacally with haemocyanin, only those aged 8 days or older at the time of immunization were capable of mounting specific antibody responses. Other studies by Klipper *et al.* (2001, 2004) demonstrated that oral antigens induced tolerance in chicks less than 3 days of age.

Whilst this appears to be a good mechanism for preventing inappropriate immune responses to food antigens later in life, it could also risk the development of tolerance to antigens derived from pathogens. However, such chickens become capable of developing immune responses to these enteric pathogens as their GALT develops.

Klipper *et al.* (2004) hypothesized that the presence of maternal antibodies at hatching may help prevent tolerance against antigens derived from pathogens. They suggested that maternal antibodies block interactions between the immune cells involved in generating tolerance and pathogen-derived antigens. The progeny of hens immunized with BSA were compared with those from non-immunized hens and their capacity to produce BSA-specific antibodies measured after oral immunisation at days 1–6 post-hatch and an oral booster at 2 weeks of age.

Chicks exhibiting high levels of maternal anti-BSA antibodies produced high levels of their own anti-BSA antibodies after this booster immunization. In comparison, oral immunization at 1–6 days induced tolerance in the progeny of non-immunized hens. Due to the paucity of information on oral tolerance in the chicken, this phenomenon is still poorly understood and would benefit from further investigation.

DEVELOPMENT OF RESPONSES TO PATHOGENS

Importance of maternal antibodies for preventing infection

The young chick is usually endowed with high levels of maternal antibodies, mainly IgY derived from the yolk (see earlier). These antibodies have a transient, but essential, role in protecting chicks from pathogens such as: *Eimeria*, *Cryptosporidium*, *Salmonella enterica* and *Campylobacter* (Rose, 1972; Smith *et al.*, 1994; Hassan and Curtiss, 1996; Hornok *et al.*, 1998; Sahin *et al.*, 2003).

In challenge experiments, the progeny of immunized – or infected – hens were compared with the progeny of control hens. This represents a somewhat contrived system, for levels of specific maternal antibodies against the infecting pathogen were elevated due to the recently acquired infection/immunization (Wallach *et al.*, 1995), but it demonstrates the capacity of maternal antibodies to confer protection to the progeny. The transitory nature of maternal protection has been demonstrated with *E. maxima*, where significant protection occurred only in chicks from the eggs laid between 17 and 30 days post-infection (Rose, 1972).

Development of protection against infection with age

Studies into the development of the immune response – and subsequent protection against enteric pathogens – with age have not been undertaken in many experimental systems.

Chickens infected with the broad host strains of *Salmonella enterica* (such as Typhimurium and Enteritidis) show marked age-related differences in their disease biology. Those infected soon after hatch (< 3 days) develop a severe systemic disease that results in high mortality, whilst those infected when older develop very few clinical symptoms, with infection predominantly limited to the intestine (Barrow *et al.*, 1988; Gast and Beard, 1989). The differences in infection biology at these two stages may simply be due to more *Salmonella* crossing the gut epithelium and gaining access to systemic sites, as a consequence of the immaturity of the gut structure (Uni *et al.*, 2000; Geyra *et al.*, 2001). Studies in our laboratory have examined more subtle differences in the capacity of chicks infected at 1, 3 and 6 weeks to clear an oral *S.* Typhimurium infection (Beal *et al.*, 2004).

Regardless of age at infection, none of the chicks exhibited clinical signs of disease and all cleared the infection at a similar age (approximately 9–11 weeks). The chicks infected at 1 week old also had reduced antigen specific antibody and T cell proliferation compared with those infected at 3 and 6 weeks. Furthermore, chickens infected at 1 week of age were less protected against re-challenge than those initially infected at 3 and 6 weeks.

These results have important implications to the producers of broiler chickens. First, at whatever age chicks become infected they are still likely to remain infected at the age of slaughter (6–8 weeks). Secondly, vaccination is unlikely to be successful in protecting these chicks from infection, as protective immune responses to enteric *Salmonella* infection do not appear to develop until chickens reach 9–11 weeks of age.

Interestingly, the infection biology of the intestinal parasite *E. tenella* does not appear to alter with age (Lillehoj, 1988). Regardless of the age at infection (1, 7, 14 and 21 days of age), all chickens produced similar numbers of oocysts; unfortunately, immune responses were not measured in this study. Clearly, age-related effects on immunity are dependent on both the nature of the invading pathogen and the mechanisms employed by the host.

The age of infection has a major impact on the capacity of chickens to respond to enteric viruses, such as reoviruses which infect predominantly via the faecal/oral route. Studies carried out recently by Mukiibi-Muka and Jones (1999) demonstrated an increased resistance of chickens to reovirus infection with increasing age, and an associated increase in their capacity to produce virus specific antibodies. Chicks infected on the day of hatch were incapable of producing significant virus specific IgA, whilst those infected at 3 weeks exhibited a significant increase in intestinal IgA.

Implications for vaccination

In recent years there has been pressure from many quarters to reduce, or even to eliminate, the use of antibiotics and other drug-based feed additives in poultry diets. As a consequence the poultry industry has placed increasing reliance on vaccination as protection against pathogens. The gut immune system is poorly developed in the neonate; this has important implications for developing vaccine strategies against enteric pathogens. A better understanding of the development of gut immune responses in the neonate is a prerequisite for the development of new vaccination strategies against enteric pathogens.

REFERENCES

Arstila, T.P., Toivanen, P. and Lassila, O. (1993) Helper activity of CD4+ alpha beta T cells is required for the avian gamma delta T cell response. *European Journal of Immunology* 23, 2034–2037.

Ayabe, T., Satchell, D.P., Wilson, C.L., Parks, W.C., Selsted, M.E. and Ouellette, A.J. (2000) Secretion of microbicidal alpha-defensins by intestinal Paneth cells in response to bacteria. *Nature Immunology* 1, 113–118.

Bar-Shira, E. and Friedman, A. (2005) Ontogeny of gut associated immune competence in the chick. *Israel Journal of Veterinary Medicine* 60, 42–50.

Bar-Shira, E., Sklan, D. and Friedman, A. (2003) Establishment of immune competence in the avian GALT during the immediate post-hatch period. *Developmental and Comparative Immunology* 27, 147–157.

Barrow, P.A., Simpson, J.M. and Lovell, M.A. (1988) Intestinal colonization in the chicken of food-poisoning *Salmonella* serotypes; microbial charactreistics associated with faecal excretion *Avian Pathology* 17, 571–588.

Beal, R.K., Wigley, P., Powers, C., Hulme, S.D., Barrow, P.A. and Smith, A.L. (2004) Age at primary infection with *Salmonella enterica* serovar Typhimurium in the chicken influences persistence of infection and subsequent immunity to re-challenge. *Veterinary Immunology and Immunopathology* 100, 151–164.

Befus, A.D., Johnston, N., Leslie, G.A. and Bienenstock, J. (1980) Gut-associated lymphoid tissue in the chicken. I. Morphology, ontogeny, and some functional characteristics of Peyer's patches. *Journal of Immunology* 125, 2626–2632.

Bezuidenhout, A.J. and Van Aswegen, G. (1990) A light microscopic and immunocytochemical study of the gastrointestinal tract of the ostrich (*Struthio camelus* L.). *Onderstepoort Journal of Veterinary Research* 57, 37–48.

Bienenstock, J., Gauldie, J. and Perey, D.Y. (1973a) Synthesis of IgG, IgA, IgM by chicken tissues: immunofluorescent and 14C amino acid incorporation studies. *Journal of Immunology* 111, 1112–1118.

Bienenstock, J., Perey, D.Y., Gauldie, J. and Underdown, B.J. (1973b) Chicken A: physico-chemical and immunochemical characteristics. *Journal of Immunology* 110, 524–533.

Brune, K., Leffell, M.S. and Spitznagel, J.K. (1972) Microbicidal activity of peroxidaseless chicken heterophile leukocytes. *Infection and Immunity* 5, 283–287.

Burns, R.B. and Maxwell, M.H. (1986) Ultrastructure of Peyer's patches in the domestic fowl and turkey. *Journal of Anatomy* 147, 235–243.

Burton, R.R. and Harrison, J.S. (1969) The relative differential leucocyte count of the newly hatched chick. *Poultry Science* 48, 451–453.

Chai, J.Y. and Lillehoj, H.S. (1988) Isolation and functional characterization of chicken intestinal intra-epithelial lymphocytes showing natural killer cell activity against tumour target cells. *Immunology* 63, 111–117.

Chen, C.H., Sowder, J.T., Lahti, J.M., Cihak, J., Losch, U. and Cooper, M.D. (1989) TCR3: a third T-cell receptor in the chicken. *Proceedings of the National Academy of Sciences of the USA of America* 86, 2351–2355.

Chen, C.H., Gobel, T.W., Kubota, T. and Cooper, M.D. (1994) T cell development in the chicken. *Poultry Science* 73, 1012–1018.

Cihak, J., Ziegler-Heitbrock, H.W., Stein, H. and Losch, U. (1991) Effect of perinatal anti-TCR2 treatment and thymectomy on serum immunoglobulin levels in the chicken. *Journal of Veterinary Medicine*, Series B, 38, 28–34.

Cooper, M.D., Peterson, R.D. and Good, R.A. (1965) Delineation of the thymic and bursal lymphoid systems in the chicken. *Nature* 205, 143–146.

Cooper, M.D., Chen, C.L., Bucy, R.P. and Thompson, C.B. (1991) Avian T cell ontogeny. *Advances in Immunology* 50, 87–117.

Daher, K.A., Selsted, M.E. and Lehrer, R.I. (1986) Direct inactivation of viruses by human granulocyte defensins. *Journal of Virology* 60, 1068–1074.

del Cacho, E., Gallego, M., Sanz, A. and Zapata, A. (1993) Characterization of distal lymphoid nodules in the chicken caecum. *The Anatomical Record* 237, 512–517.

Dil, N. and Qureshi, M.A. (2002) Differential expression of inducible nitric oxide synthase is associated with differential Toll-like receptor-4 expression in chicken macrophages from different genetic backgrounds. *Veterinary Immunology and Immunopathology* 84, 191–207.

Dunon, D., Courtois, D., Vainio, O., Six, A., Chen, C.H., Cooper, M.D., Dangy, J.P. and Imhof, B.A. (1997) Ontogeny of the immune system: gamma/delta and alpha/beta T cells migrate from thymus to the periphery in alternating waves. *Journal of Experimental Medicine* 186, 977–88.

Ekino, S., Matsuno, K., Harada, S., Fujii, H., Nawa, Y. and Kotani, M. (1979) Amplification of plaque-forming cells in the spleen after intracloacal antigen stimulation in neonatal chicken. *Immunology* 37, 811–815.

Evans, E.W., Beach, G.G., Wunderlich, J. and Harmon, B.G. (1994) Isolation of antimicrobial peptides from avian heterophils. *Journal of Leukocyte Biology* 56, 661–665.

Friedman, A., Bar-Shira, E. and Sklan, D. (2003) Ontogeny of gut-associated immune competence in the chick. *World's Poultry Science Journal* 59, 209–219.

Gallego, M., del Cacho, E. and Bascuas, J.A. (1995) Antigen-binding cells in the cecal tonsil and Peyer's patches of the chicken after bovine serum albumin administration. *Poultry Science* 74, 472–479.

Ganz, T. (2000) Paneth cells – guardians of the gut cell hatchery. *Nature Immunology* 1, 99–100.

Gast, R.K. and Beard, C.W. (1989) Age-related changes in the persistence and pathogenicity of *Salmonella* typhimurium in chicks. *Poultry Science* 68, 1454–1460.

Geyra, A., Uni, Z. and Sklan, D. (2001) Enterocyte dynamics and mucosal development in the posthatch chick. *Poultry Science* 80, 776–782.

Glick, B., Chang, T.S. and Jaap, R.G. (1956) The bursa of Fabricius and antibody production. *Poultry Science* 35, 224–225.

Gobel, T.W., Kaspers, B. and Stangassinger, M. (2001) NK and T cells constitute two major, functionally distinct intestinal epithelial lymphocyte subsets in the chicken. *International Immunology* 13, 757–762.

Gomez Del Moral, M., Fonfria, J., Varas, A., Jimenez, E., Moreno, J. and Zapata, A.G. (1998) Appearance and development of lymphoid cells in the chicken (*Gallus gallus*) caecal tonsil. *The Anatomical Record* 250, 182–189.

Hassan, J.O. and Curtiss, R., 3rd (1996) Effect of vaccination of hens with an avirulent strain of *Salmonella* typhimurium on immunity of progeny challenged with wild-Type *Salmonella* strains. *Infection and Immunity* 64, 938–944.

Hemmingsson, E.J. and Linna, T.J. (1972) Ontogenetic studies on lymphoid cell traffic in the chicken. I. Cell migration from the bursa of Fabricius. *International Archives of Allergy and Applied Immunology* 42, 693–710.

Henderson, S.C., Bounous, D.I. and Lee, M.D. (1999) Early events in the pathogenesis of avian salmonellosis. *Infection and Immunity* 67, 3580–3586.

Higgins, D.A. (1976) Fractionation of fowl immunoglobulins. *Research in Veterinary Science* 21, 94–99.

Higgs, R., Lynn, D.J., Gaines, S., McMahon, J., Tierney, J., James, T., Lloyd, A.T., Mulcahy, G. and O'Farrelly, C. (2005) The synthetic form of a novel chicken beta-defensin identified in silico is predominantly active against intestinal pathogens. *Immunogenetics* 57, 90–98.

Hornok, S., Bitay, Z., Szell, Z. and Varga, I. (1998) Assessment of maternal immunity to *Cryptosporidium baileyi* in chickens. *Veterinary Parasitology* 79, 203–212.

Imhof, B.A., Dunon, D., Courtois, D., Luhtala, M. and Vainio, O. (2000) Intestinal CD8 alpha alpha and CD8 alpha beta intraepithelial lymphocytes are thymus derived and exhibit subtle differences in TCR beta repertoires. *Journal of Immunology* 165, 6716–6722.

Iqbal, M., Philbin, V.J. and Smith, A.L. (2005a) Expression patterns of chicken Toll-like receptor mRNA in tissues, immune cell subsets and cell lines. *Veterinary Immunology and Immunopathology* 104, 117–127.

Iqbal, M., Philbin, V.J., Withanage, G.S., Wigley, P., Beal, R.K., Goodchild, M.J., Barrow, P., McConnell, I., Maskell, D.J., Young, J., Bumstead, N., Boyd, Y. and Smith, A.L. (2005b) Identification and functional characterization of chicken toll-like receptor 5 reveals a fundamental role in the biology of infection with *Salmonella enterica* serovar Typhimurium. *Infection and Immunity* 73, 2344–2350.

Ivanyi, J. (1975) Immunodeficiency in the chicken. I. Disparity in suppression of antibody responses to various antigens following surgical bursectomy. *Immunology* 28, 1007–1013.

Janeway, C.A., Travers, P., Walport, M. and Capra, J.D. (1999) *Immunobiology: the Immune System in Health and Disease*, 4. Elsevier, London.

Jeurissen, S.H.M., Vervelde, L. and Janse, M. (1994) Structure and function of lymphoid tissues in the chicken. *Poultry Science Reviews* 5, 183–207.

Jeurissen, S.H., Wagenaar, F. and Janse, E.M. (1999) Further characterization of M cells in gut-associated lymphoid tissues of the chicken. *Poultry Science* 78, 965–972.

Kaiser, P., Poh, T.Y., Rothwell, L., Avery, S., Balu, S., Pathania, U.S., Hughes, S., Goodchild, M., Morrell, S., Watson, M., Bumstead, N., Kaufman, J. and Young, J.R. (2005) A genomic analysis of chicken cytokines and chemokines. *Journal of Interferon and Cytokine Research* 25, 467–484.

Kasahara, Y., Chen, C.H. and Cooper, M.D. (1993a) Growth requirements for avian gamma delta T cells include exogenous cytokines, receptor ligation and *in vivo* priming. *European Journal of Immunology* 23, 2230–2236.

Kasahara, Y., Chen, C.L., Gobel, T.W.F., Bucy, R.P. and Cooper, M.D. (1993b) Intraepithelial lymphocytes in birds. In: Kiyono, H. and McGhee, J.R. (eds) *Mucosal Immunology: Intraepithelial Lymphocytes*. Raven Press, New York.

Kaufmann, S.H., Blum, C. and Yamamoto, S. (1993) Crosstalk between alpha/beta T cells and gamma/delta T cells *in vivo*: activation of alpha/beta T-cell responses after gamma/delta T-cell modulation with the monoclonal antibody GL3. *Proceedings of the National Academy of Sciences of the USA of America* 90, 9620–9624.

Kincade, P.W. and Cooper, M.D. (1973) Immunoglobulin A: site and sequence of expression in developing chicks. *Science* 179, 398–400.

Kitagawa, H., Imagawa, T. and Uehara, M. (1996) The apical caecal diverticulum of the chicken identified as a lymphoid organ. *Journal of Anatomy* 189 (3), 667–672.

Kitagawa, H., Hiratsuka, Y., Imagawa, T. and Uehara, M. (1998) Distribution of lymphoid tissue in the caecal mucosa of chickens. *Journal of Anatomy* 192 (2), 293–298.

Klipper, E., Sklan, D. and Friedman, A. (2001) Response, tolerance and ignorance following oral exposure to a single dietary protein antigen in *Gallus domesticus*. *Vaccine* 19, 2890–2897.

Klipper, E., Sklan, D. and Friedman, A. (2004) Maternal antibodies block induction of oral tolerance in newly hatched chicks. *Vaccine* 22, 493–502.

Kogut, M.H., Iqbal, M., He, H., Philbin, V., Kaiser, P. and Smith, A. (2005) Expression and function of Toll-like receptors in chicken heterophils. *Developmental and Comparative Immunology* 29, 791–807.

Lebacq-Verheyden, A.M., Vaerman, J.P. and Heremans, J.F. (1972) A possible homologue of mammalian IgA in chicken serum and secretions. *Immunology* 22, 165–175.

Lebacq-Verheyden, A.M., Vaerman, J.P. and Heremans, J.F. (1974) Quantification and distribution of chicken immunoglobulins IgA, IgM and IgG in serum and secretions. *Immunology* 27, 683–692.

Lehrer, R.I., Daher, K., Ganz, T. and Selsted, M.E. (1985) Direct inactivation of viruses by MCP-1 and MCP-2, natural peptide antibiotics from rabbit leukocytes. *Journal of Virology* 54, 467–472.

Lehrer, R.I., Ganz, T., Szklarek, D. and Selsted, M.E. (1988) Modulation of the *in vitro* candidacidal activity of human neutrophil defensins by target cell metabolism and divalent cations. *The Journal of Clinical Investigation* 81, 1829–1835.

Leslie, G.A. and Clem, L.W. (1969) Phylogen of immunoglobulin structure and function. 3. Immunoglobulins of the chicken. *Journal of Experimental Medicine* 130, 1337–1352.

Leslie, G.A., Wilson, H.R. and Clem, L.W. (1971) Studies on the secretory immunologic system of fowl. I. Presence of immunoglobulins in chicken secretions. *Journal of Immunology* 106, 1441–1446.

Lillehoj, H.S. (1988) Influence of inoculation dose, inoculation schedule, chicken age, and host genetics on disease susceptibility and development of resistance to *Eimeria tenella* infection. *Avian Diseases* 32, 437–444.

Lillehoj, H.S. and Chai, J.Y. (1988) Comparative natural killer cell activities of thymic, bursal, splenic and intestinal intraepithelial lymphocytes of chickens. *Developmental and Comparative Immunology* 12, 629–643.

Lillehoj, H.S. and Chung, K.S. (1992) Postnatal development of T-lymphocyte subpopulations in the intestinal intraepithelium and *lamina propria* in chickens. *Veterinary Immunology and Immunopathology* 31, 347–360.

Lillehoj, H.S., Min, W. and Dalloul, R.A. (2004) Recent progress on the cytokine regulation of intestinal immune responses to *Eimeria*. *Poultry Science* 83, 611–623.

Linna, T.J., Hemmingsson, E. and Back, R. (1969) A method for local labelling of the thymus and of the bursa of Fabricius of the chicken with tritiated thymidine. *Acta Pathologica et Microbiologica Scandinavica* 77, 557–558.

Mast, J. and Goddeeris, B.M. (1999) Development of immunocompetence of broiler chickens. *Veterinary Immunology and Immunopathology* 70, 245–256.

Mockett, A.P.A. (1986) Monoclonal antibodies used to isolate IgM from chicken bile and avian sera and to detect specific IgM in chicken sera. *Avian Pathology* 15, 337.

Mukiibi-Muka, G. and Jones, R.C. (1999) Local and systemic IgA and IgG responses of chicks to avian reoviruses: effects of age of chick, route of infection and virus strain. *Avian Pathology* 28, 54–60.

Olah, I. and Glick, B. (1984) Meckel's diverticulum. I. Extramedullary myelopoiesis in the yolk sac of hatched chickens (*Gallus domesticus*). *The Anatomical Record* 208, 243–252.

Olah, I., Nagy, N., Magyar, A. and Palya, V. (2003) Esophageal tonsil: a novel gut-associated lymphoid organ. *Poultry Science* 82, 767–770.

Parry, S.H. and Porter, P. (1978) Characterization and localization of secretory component in the chicken. *Immunology* 34, 471–478.

Penniall, R. and Spitznagel, J.K. (1975) Chicken neutrophils: oxidative metabolism in phagocytic cells devoid of myeloperoxidase. *Proceedings of the National Academy of Sciences of the USA of America* 72, 5012–5015.

Peppard, J.V., Hobbs, S.M., Jackson, L.E., Rose, M.E. and Mockett, A.P. (1986) Biochemical characterization of chicken secretory component. *European Journal of Immunology* 16, 225–229.

Philbin, V.J., Iqbal, M., Boyd, Y., Goodchild, M.J., Beal, R.K., Bumstead, N., Young, J. and Smith, A.L. (2005) Identification and characterization of a functional, alternatively spliced

Toll-like receptor 7 (TLR7) and genomic disruption of TLR8 in chickens. *Immunology* 114, 507–521.

Quere, P., Bhogal, B.S. and Thorbecke, G.J. (1990) Characterization of suppressor T cells for antibody production by chicken spleen cells. II. Comparison of CT8+ cells from con-canavalin A-injected normal and bursa cell-injected agammaglobulinaemic chickens. *Immunology* 71, 523–529.

Qureshi, M.A. (2003) Avian macrophage and immune response: an overview. *Poultry Science* 82, 691–698.

Qureshi, M.A., Heggen, C.L. and Hussain, I. (2000) Avian macrophage: effector functions in health and disease. *Developmental and Comparative Immunology* 24, 103–119.

Rose, M.E. (1972) Immunity to coccidiosis: maternal transfer in *Eimeria maxima* infections. *Parasitology* 65, 273–282.

Rose, M.E., Orlans, E., Payne, A.W. and Hesketh, P. (1981) The origin of IgA in chicken bile: its rapid active transport from blood. *European Journal of Immunology* 11, 561–564.

Sahin, O., Luo, N., Huang, S. and Zhang, Q. (2003) Effect of *Campylobacter*-specific maternal antibodies on *Campylobacter jejuni* colonization in young chickens. *Applied and Environmental Microbiology* 69, 5372–5379.

Sanchez-Garcia, F.J. and McCormack, W.T. (1996) Chicken gamma delta T cells. *Current Topics in Microbiology and Immunology* 212, 55–69.

Schaffner, T., Mueller, J., Hess, M.W., Cottier, H., Sordat, B. and Ropke, C. (1974) The bursa of Fabricius: a central organ providing for contact between the lymphoid system and intestinal content. *Cellular Immunology* 13, 304–312.

Selsted, M.E., Szklarek, D. and Lehrer, R.I. (1984) Purification and antibacterial activity of anti-microbial peptides of rabbit granulocytes. *Infection and Immunity* 45, 150–154.

Selsted, M.E., Szklarek, D., Ganz, T. and Lehrer, R.I. (1985) Activity of rabbit leukocyte peptides against *Candida albicans*. *Infection and Immunity* 49, 202–206.

Sharma, J.M. and Okazaki, W. (1981) Natural killer cell activity in chickens: target cell analysis and effect of antithymocyte serum on effector cells. *Infection and Immunity* 31, 1078–1085.

Smith, N.C., Wallach, M., Miller, C.M., Braun, R. and Eckert, J. (1994) Maternal transmission of immunity to *Eimeria maxima*: Western blot analysis of protective antibodies induced by infection. *Infection and Immunity* 62, 4811–4817.

Sorvari, T., Sorvari, R., Ruotsalainen, P., Toivanen, A. and Toivanen, P. (1975) Uptake of environ-mental antigens by the bursa of Fabricius. *Nature* 253, 217–219.

Sowder, J.T., Chen, C.H., Ager, L.L., Chan, M.M. and Cooper, M.D. (1988) A large sub-population of avian T cells express a homologue of the mammalian T gamma/delta receptor. *Journal of Experimental Medicine* 167, 315–322.

Sugiarto, H. and Yu, P.L. (2004) Avian antimicrobial peptides: the defence role of beta-defensins. *Biochemical and Biophysical Research Communications* 323, 721–727.

Toivanen, P. and Toivanen, A. (1973) Bursal and postbursal stem cells in chicken. Functional characteristics. *European Journal of Immunology* 3, 585–595.

Toivanen, P., Toivanen, A. and Good, R.A. (1972) Ontogeny of bursal function in chicken. 3. Immunocompetent cell for humoral immunity. *Journal of Experimental Medicine* 136, 816–831.

Uni, Z., Geyra, A., Ben-Hur, H. and Sklan, D. (2000) Small intestinal development in the young chick: crypt formation and enterocyte proliferation and migration. *British Poultry Science* 41, 544–551.

Uni, Z., Smirnov, A. and Sklan, D. (2003) Pre- and posthatch development of goblet cells in the broiler small intestine: effect of delayed access to feed. *Poultry Science* 82, 320–327.

Van Immerseel, F., De Buck, J., De Smet, I., Mast, J., Haesebrouck, F. and Ducatelle, R. (2002) Dynamics of immune cell infiltration in the caecal *lamina propria* of chickens after neonatal infection with a *Salmonella* Enteritidis strain. *Developmental and Comparative Immunology* 26, 355–364.

Van Meter, R., Good, R.A. and Cooper, M.D. (1969) Ontogeny of circulating immunoglobulin in normal, bursectomized and irradiated chickens. *Journal of Immunology* 102, 370–374.

Vervelde, L. and Jeurissen, S.H. (1993) Postnatal development of intra-epithelial leukocytes in the chicken digestive tract: phenotypical characterization *in situ*. *Cell and Tissue Research* 274, 295–301.

Wallach, M., Smith, N.C., Petracca, M., Miller, C.M., Eckert, J. and Braun, R. (1955) *Elmeria maxima* gametocyte antigens: potential use in a subunit maternal vaccine against coccidiosis in chickens. *Vaccine* 13, 347–354.

Wells, L.L., Lowry, V.K., DeLoach, J.R. and Kogut, M.H. (1998) Age-dependent phagocytosis and bactericidal activities of the chicken heterophil. *Developmental and Comparative Immunology* 22, 103–109.

Yason, C.V., Summers, B.A. and Schat, K.A. (1987) Pathogenesis of rotavirus infection in various age groups of chickens and turkeys: pathology. *American Journal of Veterinary Research* 48, 927–938.

Zhao, C., Nguyen, T., Liu, L., Sacco, R.E., Brogden, K.A. and Lehrer, R.I. (2001) Gallinacin-3, an inducible epithelial beta-defensin in the chicken. *Infection and Immunity* 69, 2684–2691.

PART III
Gastrointestinal flora

CHAPTER 7
Molecular approaches to the analysis of gastrointestinal microbial ecosystems

H.J. Flint,* E.C.M. Leitch, S.H. Duncan, A.W. Walker, A.J. Patterson, M.T. Rincon, K.P. Scott and P. Louis

*Microbial Ecology Group, Gut Health Division, Rowett Research Institute, Aberdeen, UK; * e-mail: h.flint@rowett.ac.uk*

ABSTRACT

Molecular analyses of ribosomal sequence diversity in mammalian and avian gut ecosystems reveal that a high proportion of gut bacteria do not correspond to known, cultured species. Many of the dominant microbial colonizers of the large intestine and rumen, of which the majority are related to the *Bacteroides* or *Firmicutes* groups, are highly sensitive to oxygen or are fastidious in their growth requirements. A wide range of 'culture-independent' approaches is now available for tracking the microbial composition of gut samples, ranging from specific probes to profiling techniques that exploit ribosomal sequence information. It remains crucially important, however, to be able to relate particular phylogenetic groups to their roles in the gut, and to their impact upon the host. This can be achieved through the isolation of abundant, but previously little-studied, groups of gut microbes, through molecular analysis of shifts in microbial communities *in vivo* and *in vitro* and, in some cases, through the targeting of functionally significant genes. Examples of these approaches, as applied to the breakdown of dietary carbohydrates, the metabolism of short-chain fatty acids (SCFAs) and the spread of antibiotic resistance genes, will be used to illustrate their potential for advancing understanding of complex gut ecosystems.

INTRODUCTION

The microorganisms that colonize the gut of mammals and birds have a major influence on the nutrition and health status of their hosts. Some species, notably many pathogens, have the ability to survive in the external environment and in alternative host species. For this reason, and because the microbial ecosystems represented by different host species share many common features, it is important to consider gut microbial ecology as a discipline that crosses host–species boundaries, including comparison of the mammalian and avian gut.

In both types of host, the large intestine and caecum develop a dense, predominantly anaerobic, microbial community, while the small intestine supports a less dense community that is dominated by lactic acid bacteria. The roles of these communities in supplying energy and nutrients to the host, in harbouring or resisting colonization by pathogens and in the development of the host's immune system continue to be major topics for research.

Molecular approaches offer new prospects for answering some of the most basic questions in gut microbial ecology. These cover the following: (i) the true extent of microbial diversity, including detection of uncultivated organisms; (ii) the influence of diet and inter-host variation upon species composition of the microbiota; (iii) the sequence of early colonization; and (iv) the existence of specialized communities within particular micro-niches and in different regions of the gut.

INFORMATION FROM 16S RRNA SEQUENCES

Ribosomes are universally present in living cellular organisms, and the genes that encode ribosomal rRNA molecules have provided a vast amount of valuable information on microbial phylogeny, diversity and evolution (Woese, 1987). The small subunit rRNA molecule (16S in bacteria), in particular, has proved valuable, as it has been possible to identify regions that are highly conserved in sequence, e.g. identical in all bacteria, as well as regions that vary in sequence between different phylogenetic groups (Weisburg *et al.*, 1991).

PCR amplification, using primers that target conserved regions from DNA that is extracted from a gut or faecal sample, yields products that should comprise all the 16S rRNA coding sequences present in the extracted DNA. Simple library construction, followed by sequencing of the PCR products, has been widely used to analyse the complexity of the bacterial and, in some cases, also archaean and eukaryotic, microbial colonizers of the gut in mammals, birds and insects.

Diversity in gut microbial communities

A number of clone library analyses of 16S TRNA gene sequence diversity are available for human faeces and large intestine (Wilson and Blitchington, 1996; Suau *et al.*, 1999; Hayashi *et al.*, 2002; Hold *et al.*, 2002; Eckburg *et al.*, 2005). There is general agreement that around 75% of the eubacterial sequences detected do not correspond to known cultured organisms and represent species that are yet to be described (cut-offs used vary from 97–99% for the degree of sequence identity corresponding to a known species).

In all cases, the dominant groups found are: (i) the CFB (*Cytophaga, Flavobacterium, Bacteroides*) group of Gram-negative bacteria; (ii) the *Clostridium* cluster XIVa group of Gram-positive bacteria; and (iii) the *Clostridium* cluster IV group of Gram-positive bacteria. Interestingly, the same three groups are also found to be predominant in the rumen of cattle (Whitford *et al.*, 1998; Tajima *et al.*, 1999), in the large intestines of pigs (Pryde *et al.*, 1999; Leser *et al.*, 2002) and in horses (Daly *et al.*, 2001).

In the case of pigs, *Streptococcus/Lactobacillus* are also numerically dominant. Several studies have used 16S clone libraries to investigate the distribution of microorganisms in different regions of the gastrointestiinal tract (GIT) (e.g. Wang *et al.*, 2004). These have confirmed earlier information from cultural studies, indicating a major shift between the stomach, small intestine and large intestine in monogastric mammals and man, with the more dense and complex anaerobically dominated communities occurring in the large intestine.

Studies in poultry (Apajalahti *et al.*, 2001; Gong *et al.*, 2002; Zhu *et al.*, 2002; Lu *et al.*, 2003) have detected a high proportion of sequences corresponding to *Clostridium*-related anaerobes (*Firmicutes*) in the caecum. In particular, relatives of the *Cl. leptum* (cluster IV) and *Cl. coccoides* (cluster XIVa) groups were found, as in the mammalian large intestine (Table 7.1) and rumen. Zhu *et al.* (2002) also reported a high frequency of the *Clostridium* cluster IX (*Sporomusa*) group and of Gram-negative, enteric, coliform bacteria. The *Sporomusa* group, which includes major propionate producers, has also been detected in the pig and human intestine (Pryde *et al.*, 1999; Walker *et al.*, 2005). None of these studies reported high proportions of CFB (*Bacteroides*) or *Bifidobacterium* in the chicken gut, although these groups were detected in previous cultural studies. A clear difference was reported between the small intestine – which is dominated by relatives of *Lactobacillus*, *Streptococcus* and *Enterococcus* (up to 80%) – and

Table 7.1. Estimates of the eubacterial composition of gut microbial communities based on sequencing of 16S rRNA clone libraries.

	Chicken	Human		Pig
	Caecum	Colon	Stool	Intestine
	(Percentage sequences or OTUs recovered)			
Clostridium cluster XIVa (*Cl. coccoides* group)	27[a]	46[b]	44[c]	25[d]
Clostridium cluster IV (*Cl. leptum* group)	20	15	20	19
Other Gram-positives	25	5	2	33
CFB (*Bacteroides*)	2	26	31	11
Other (inc. proteobacteria)	21	7	3	8

[a] Zhu *et al.* (2002).
[b] Hold *et al.* (2002).
[c] Suau *et al.* (1999).
[d] Leser *et al.* (2002).

the caecum, in agreement with earlier cultural studies (Barnes 1972; Salanitro *et al.*, 1974).

It should be stressed that 16S clone library analyses are by no means free from potential biases and pitfalls. Large numbers of PCR cycles can lead to over-representation of certain groups, making it desirable to keep the cycle number low (Wilson and Blitchington, 1996). The method used for nucleic acid extraction is also important in equalizing release of DNA from different groups of microbes, and in minimizing degradation. Certain widely used 'universal' primer combinations may fail to amplify some eubacterial groups. In particular, bifidobacteria (which are high G+C content, Gram-positive bacteria) amplify poorly with certain 'universal' primer pairs, accounting for their apparent absence from many clone libraries. Extensive discussions of 16S methodologies are available elsewhere (von Wintzingerode *et al.*, 1997; Zoetendal *et al.*, 2004) and their relative merits are summarized in Table 7.2.

Exploiting ribosomal gene sequences: specific probes and primers

The ribosomal database now includes more than 100,000 16S sequences (Zoetendal *et al.*, 2004; Ribosomal Database Project II at http://rdp.cme.msu.

Table 7.2. 16S rRNA-based approaches for studying diversity in microbial communities.

	Identification	Sample capacity	Quantification	Sensitivity	Comment
Sequence analysis of PCR amplicons	species/ strain	limited	good[a]	low[b]	costly, laborious
Oligonucleotide probing					
Quantitative hybridization	targeted groups	high	good	low	laborious
FISH (quantitative)		limited	good	low[c]	in
Arrays		high	semi-quant.	low	development
Specific PCR amplification					
Competitive PCR	targeted groups	limited	good	high	little used
Real-time PCR		limited	good	high	costly
Profiling approaches					
DGGE, TGGE	limited[d]	good	poor	low/high[e]	rapid,
T-RFLP	limited	good	poor	low/high	comparative
SSCP	limited	good	poor	low/high	
Localization studies					
FISH	targeted groups	limited	qualitative	moderate[c]	direct visualization

[a] Subject to possible biases in DNA extraction, PCR amplification.
[b] Limited by numbers of clones that can be sequenced.
[c] FISH methods detect single cells, but microscopic methods limited by number of fields that can be searched; sensitivity can be improved, e.g. by FACS sorting.
[d] Amplification of targeted group; definitive identification of bands within that group requires cloning and sequencing.
[e] Depending on specificity of the primer combinations used.

edu/index.jsp), and variable regions are widely used to design oligonucleotide probes that specifically recognize particular phylogenetic groups. Probes can be labelled radioactively or with fluorescent tags. Binding of radio-labelled probes to extracted ribosomal RNA has been used extensively to quantify the abundance of particular groups in gut samples (Stahl *et al.*, 1988).

An increasingly attractive option, however, is hybridization of fluorescent probes to whole cells (FISH) (Langendijk *et al.*, 1995; Franks *et al.*, 1998), accompanied by automated microscopic counting or recovery and enumeration by fluorescence-activated cell sorting (FACS) (Amann *et al.*, 1990; Zwirglmaier 2005).

FISH-based approaches have the advantage of providing information on morphology and localization, but enumeration is essentially limited to cells in the liquid phase and to relatively abundant groups (> 0.1% of total cells). Nevertheless, panels of FISH probes that cover the major microbial groups can provide valuable information on individual variation and on shifts occurring within the gut microbial community (e.g. Franks *et al.*, 1998; Duncan *et al.*, 2003; Walker *et al.*, 2005).

A logical extension of the use of probes is to construct array-based detection using a wide range of specific oligonucleotides. Although technically challenging, this may provide at least semi-quantitative descriptions of microbial communities (Small *et al.*, 2001; Wilson *et al.*, 2002). A high-throughput checkerboard DNA–DNA hybridization approach based on total genomic DNA has also been reported for microbial community analysis (Socransky *et al.*, 2004).

As an alternative approach, specific primer pairs can be used in conjunction with real-time PCR to quantify specific microbial groupings (Tajima *et al.*, 2001). This provides greater sensitivity of detection. As with probing approaches, great care must be taken to validate specificity of detection under the conditions used. The DNA extraction method used for real-time-based approaches is also a crucial factor in obtaining reliable data.

Rapid profiling approaches

Rapid profiling techniques in microbial ecology have been developed to study species diversity and bacterial community dynamics where other methods, such as cloning, are unsuitable due to their time-consuming and costly nature. A wide range of methods has been explored for 'fingerprinting' microbial communities based on chemical, e.g. lipid, analysis, but nucleic acid-based methods that combine PCR amplification of ribosomal sequences with a sequence-sensitive separation technique are now dominant.

In denaturing gradient gel electrophoresis (DGGE) (Fig. 7.1) or variants that employ temperature gradients (TGGE), one of the primers used for amplification carries a GC-rich 'clamp' to prevent complete denaturation of the amplified, double-stranded DNA (Muyzer and Smalla, 1998; Zoeteandal *et al.*, 1998). Amplicons disassociate in accordance with their sequence-dependent melting properties, decreasing their electrophoretic mobility through an

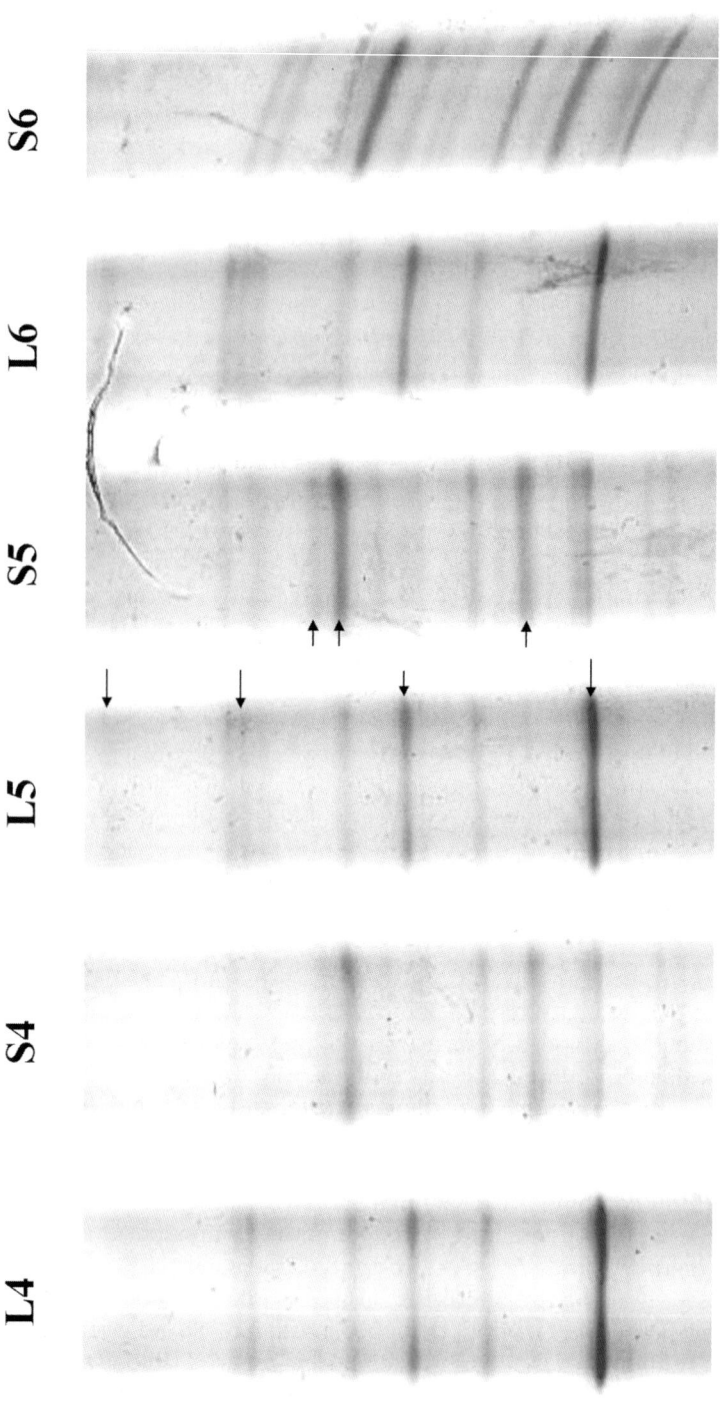

Fig. 7.1. DGGE analysis of bacteria attached to starch particles (S) or present in the liquid medium (L) of an anaerobic fermentor inoculated with a human faecal sample. Samples were analysed at different times after inoculation (4–6 h), as indicated. Consistent differences in banding pattern are evident between the attached and non-attached communities (E.C. Leitch, unpublished results).

acrylamide gel containing a linearly increasing chemical (usually urea or formamide) or thermal denaturant.

Although, in theory, a single base change can be detected, in practice, amplicons produced from gut communities using universal eubacterial primers yield highly complex patterns, which make it difficult to identify all bands present, even when they are recovered for sequencing (Kocherginskaya *et al.*, 2001). Primers targeted at narrower groups provide patterns that are more easily interpreted (e.g. Knarreborg *et al.*, 2002).

DGGE-based amplification from DNA broadly reflects genomic diversity (although the rRNA gene copy number can vary widely between bacteria: Coenye and Vandamme, 2003). It is also possible to amplify extracted ribosomal RNA, however, and comparison with the DNA-based DGGE profile gives some indication of the physiological state of bacteria *in situ*, since bacterial rRNA/DNA content is known to be related to growth rate (Zoetendal *et al.*, 1998).

An alternative profiling approach is to cleave PCR-amplified 16S sequences with particular restriction enzymes, and perform size separation of cleavage products by capillary electrophoresis. In the T-RFLP method, one primer is end-labelled either radioactively or, more usually, fluorescently (Wood *et al.*, 1998; Nagashima *et al.*, 2003; Wang *et al.*, 2004) and only the terminal fragment is detected. In principle, the sizes of cleavage products for any given bacterium are predictable from the sequence database, and the abundance of different fragments can be quantified using an automated DNA sequencer. A further approach is provided by single-strand, conformational polymorphism (SSCP), which is based on variations in the migration of single-stranded DNA due to secondary structure.

Variability in the intergenic regions within ribosomal operons has also been used for microbial community analysis (e.g. Larue *et al.*, 2005).

RELATING STRUCTURE TO FUNCTION

The rationale behind many fingerprinting studies is that the observed responses of individual phylotypes, e.g. as defined by DGGE bands, to diet or growth stage will help to define their function, and this can be followed by sequence-based identification. Likewise, FISH probes can provide information on localization and morphology of organisms ahead of their identification and isolation. Other approaches are available, however, that address the issue of functionality more directly.

Stable isotope probing

Certain substrates are utilized only by specialized microorganisms within a complex community. If these substrates can be labelled isotopically in the intact system, and the label becomes incorporated into microbial DNA, it is possible to recover this DNA by density gradient ultra-centrifigation. Subsequent amplification and sequencing of 16S rRNA sequences can then reveal those species that had incorporated the labelled substrate (Radajewski *et al.*, 2000).

This technique, and variations of it – e.g. using RNA – is referred to collectively as stable isotope probing (SIP). While, in principle, providing a powerful means of relating sequence to function, the rapid dissemination of label from many substrates, e.g. due to metabolic cross-feeding, may complicate analysis in situations as complex as those found in the gut.

Targeting of functionally relevant genes

Molecular studies have also targeted microbial genes that are of key interest in relation to disease and host nutrition, e.g. those encoding toxins, antibiotic resistance and adhesion factors. Generic PCR primers and probes can be designed in some cases where sequence conservation is high. Examples include antibiotic resistance genes (Aminov *et al.*, 2001; Lu *et al.*, 2002) and amplification of the APS reductase gene to track sulphate-reducing bacteria (Deplancke *et al.*, 2000).

In general, however, the degree of sequence divergence between functionally similar genes from different bacteria precludes the design of truly generic, gene-specific oligonucleotides.

Metagenomics

The ability to prepare gene libraries from mixed microbial communities, rather than from cultured organisms, means that genes can be studied from hitherto uncultured microorganisms, either by selecting inserts that encode functions of interest or by random sequencing. Thus, the approach can identify novel genes and provide information on members of the gut community that may be under-represented or absent in culture collections (Steele and Streit, 2005).

If metagenome libraries are constructed so as to maximize insert sizes, using BAC vectors, this increases the chance of recovering whole operons and of identifying the origin of the insert from the presence of a 16S rRNA gene. Smaller insert libraries are, however, technically easier to construct. The possibility that vast sequencing projects can lead to resolution of whole chromosomes is suggested by recent work (Tyson *et al.*, 2004) but, for gut communities, the extent of the diversity would make this an enormous task.

Cultivation

In many environments, e.g. soils, oceans, a high proportion (> 99%) of microbes are thought to be very slow growing and these are seldom, if ever, recovered in culture. This is not the case in the gut, where a certain minimum growth rate is a likely prerequisite for survival. In practice, careful anaerobic procedures appear capable of revealing most of the gut bacterial diversity, but work of this type usually pre-dated the introduction of 16S rRNA methodologies.

Thus, it seems likely that the majority of gut bacteria in humans and farm animals have been cultured at some time or other, e.g. Barnes (1972) and Finegold *et al.* (1983), but many cultures have been lost and, unfortunately, cannot be related to the wealth of ribosomal sequence data that is now emerging.

The value of cultured isolates in providing functional information is obvious, and these still offer the best opportunities for exploiting the progress in molecular methodologies. It is possible for isolated organisms to be subjected to genetic analysis, e.g. through gene knockouts or transposon mutagenesis, thus allowing genes to be related to their functions (Handfield and Levesque, 1999). Gene transfer methodologies still require development for most gut anaerobes, but some progress is being made (Accetto *et al.*, 2005).

SOME APPLICATIONS OF MOLECULAR MICROBIAL ECOLOGY

Regulation of SCFA metabolism in the human colon

Anaerobic fermentation of diet- and host-derived substrates provides most of the energy for microbial growth and metabolism in the large intestine of mammals and birds. The products of microbial fermentation influence the host in many different ways, acting as energy sources, regulators of gene expression and cell differentiation, inhibitors of pathogens or, in some cases, as toxins and pro-carcinogens.

In the human large intestine, there has been much interest in the proposed role of butyrate, one of the three main SCFA products of microbial fermentation, in maintaining gut health and in preventing colitis and colorectal cancer (McIntyre *et al.*, 1993; Mortensen and Clausen, 1996). The microorganisms responsible for many of the key metabolic transformations, including butyrate formation (Pryde *et al.*, 2002), have, however, remained little understood until recently. In fact, butyrate-producing bacteria were found to be readily isolated, although most are strictly anaerobic and their 16S rRNA sequences reveal that the majority are *Firmicutes*, related to *Eubacterium rectale/Roseburia* spp. or to *Faecalibacterium prausnitzii* (Barcenilla *et al.*, 2000; Duncan *et al.*, 2002a). Related butyrate-producing bacteria are reported to form a significant fraction of the microbiota of the chicken caecum (Gong *et al.*, 2002).

In view of their phylogenetic diversity, the possibility was considered of basing the enumeration of butyrate-producing bacteria on genes involved in butyrate synthesis. Degenerate primers were designed that recognize butyrate kinase/phosphotransbutyrylase genes from a wide range of *Clostridium*-related anaerobes. Butyrate kinase sequences, however, proved to be undetectable in 34 out of 38 human colonic butyrate producers, correlating with a lack of butyrate kinase activity (Louis *et al.*, 2004).

It appears that the alternative CoA transferase enzyme (Diez-Gonzalez *et al.*, 1999; Duncan *et al.*, 2002b) provides the major route for butyrate synthesis in the human colon (Louis *et al.*, 2004). The physiological significance of this finding is not yet established, but may be connected to the high acetate concentrations found in the colonic environment. The genes for the central pathway of butyrate synthesis, although present in butyrate-producing strains, have proved less amenable as targets for the design of consensus primers.

Specific 16S rRNA-based FISH probes have now been designed for the two most dominant groups of butyrate producers, enabling their enumeration in faecal or colonic samples (Suau *et al.*, 2001; Hold *et al.*, 2003; Walker *et al.*, 2005). Meanwhile, additional probes are available for other phylogenetic groups that correspond to particular metabolic outputs (Harmsen *et al.*, 2002; Loy *et al.*, 2003). Thus the CFB- (*Bacteroides*) and *Selenomonas*-related groups should account for most of the major producers of propionate, while *Bifidobacterium* and *Lactobacillus* probes account for many lactic acid-producing bacteria.

A panel of FISH probes was used to investigate the influence of pH upon *in vitro* fermentor communities established from human faecal inocula and supplied with a mixture of polysaccharide substrates (Walker *et al.*, 2005) Fig. 7.2. A pH value of 6.5 was shown to select strongly for the CFB group (Bac 303), while pH 5.5 favoured butyrate-producing bacteria (RTec, FpTan). As a result, butyrate concentrations were some fourfold higher at pH 5.5 compared with those at 6.5, while propionate and acetate concentrations rose at pH 6.5. Since fermentable dietary carbohydrates that escape digestion in the small intestine tend to reduce the pH of the proximal colon, they may be expected to boost butyrate formation.

Such a butyrogenic effect has been widely reported, for example, for resistant starches (Topping and Clifton, 2001; Pryde *et al.*, 2002). The correlation that is evident from Fig. 7.2 between system SCFA outputs and key functional groups of bacteria targeted by FISH probes goes some way towards establishing ecological roles for several of the major phylogenetic groups of human gut bacteria.

The ability to target and accurately enumerate significant, functionally defined groups with probes or PCR primers provides an important approach that has also been applied, for example, to quercitin-degrading bacteria from the human colon (Simmering *et al.*, 2002) and to lactic acid bacteria.

Defining the bacteria that colonize insoluble substrates in the gut

Much of the dietary carbohydrate that survives digestion in the small intestine is insoluble, in the form of plant cell wall fragments. Such substrate particles, along with host-produced mucous and mucosal surfaces, provide potential sites for the development of specialized microbial biofilms, although these will necessarily be limited by rapid gut transit and cell turnover. It seems likely that the microbiota within these biofilms include the key primary metabolizers of the underlying substrate.

Molecular methodologies allow the identification of such biofilm-associated populations. Using a combination of PCR (cloning and DGGE) and non-PCR

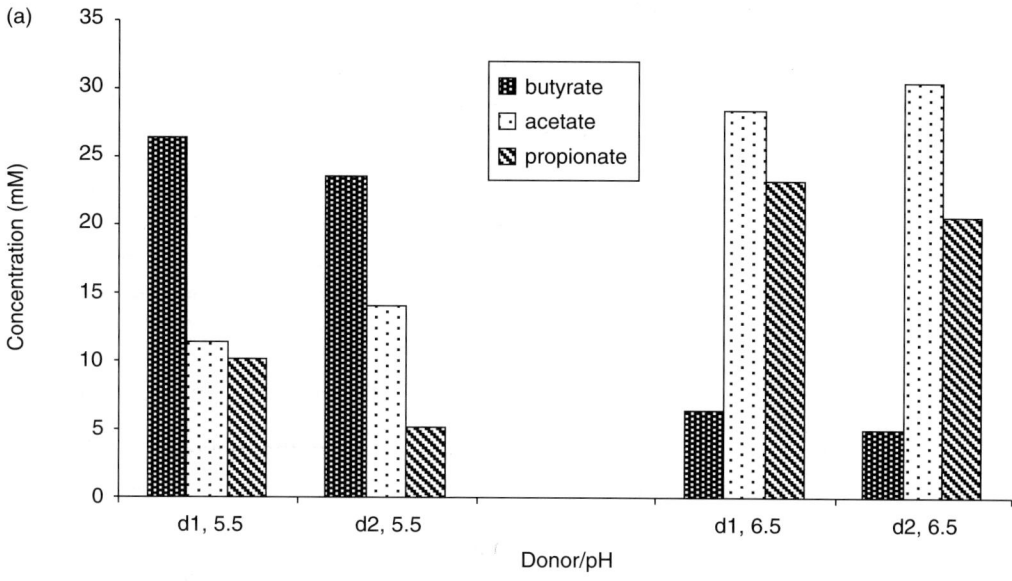

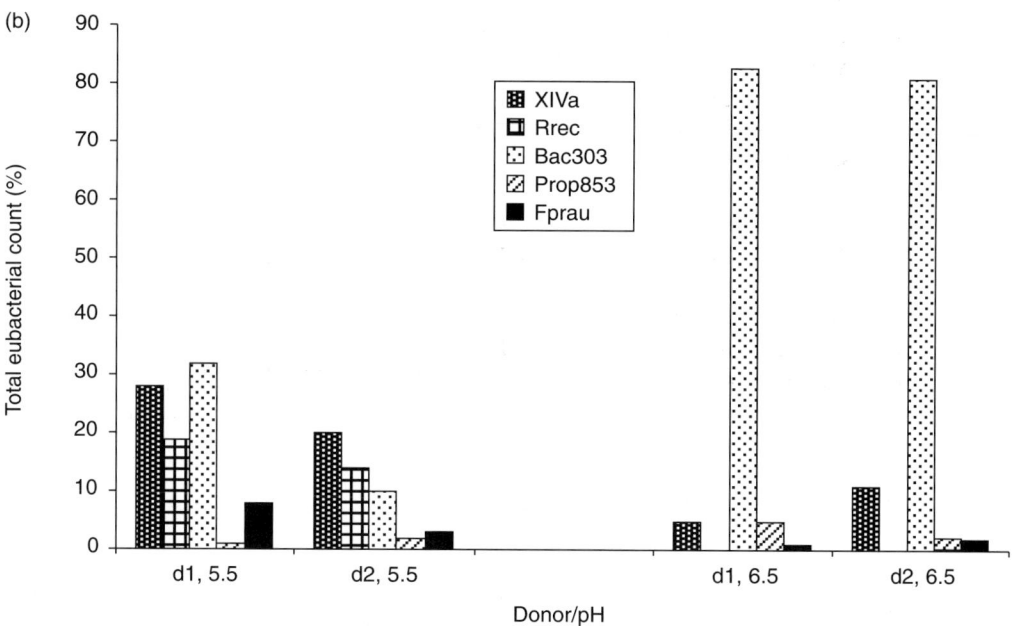

Fig. 7.2. pH-dependent changes in short-chain fatty acid concentration (a) and bacterial numbers detected by FISH probes (see text) (b) in an anaerobic fermentor system inoculated with human faecal bacteria. Data are from Walker *et al.* (2005). Separate experiments are shown for two faecal communities (d1, d2) in which pH was shifted from 5.5 to 6.5. Samples are for time points taken after extensive periods (around 150 h) of equilibration of the community at each pH value.

(FISH)-based technologies, Leitch *et al.* (2004) found preliminary evidence to suggest that, under conditions found in the human gut, a specialized biofilm community can form on insoluble substrates (Fig. 7.1). Furthermore, biofilm composition is greatly influenced by the nature of the substrate. Similarly, the substrate-associated bacterial fractions in the ovine rumen differ between sheep fed a diet of grass or starch, as determined by cloning and DGGE (Larue *et al.*, 2005).

Tracking and detecting new classes of antibiotic resistance genes in gut bacteria

The increase in microbial resistance to clinically important antibiotics and the presumed impact that veterinary or agricultural use of antibiotics has on the spread of resistance genes is of widespread concern. Past research has focused largely on pathogenic microorganisms, but it is clear that non-pathogenic gut microbes are equally exposed to antibiotic selection and represent potential reservoirs of antibiotic resistance genes.

Indeed, new resistance genes have been identified in commensal gut microbes, including *tet(W)* (Barbosa *et al.*, 1999), which has proved subsequently to be among the most abundant of all resistance genes, and is also present in pathogens such as *Arcanobacterium pyogenes* (Billington *et al.*, 2002).

Taking advantage of the generally high degree of sequence conservation found within classes of resistance genes, the distribution of known resistance gene sequences in environmental samples can be tracked by PCR approaches (Aminov *et al.*, 2001; Lu *et al.*, 2002; Villedieu *et al.*, 2003) or by macro-array-based methods that use a range of gene-targeted probes (Patterson *et al.*, 2004) (Fig. 7.3). These studies have shown that the distribution of bacterial resistance genes between gut samples from different hosts is more similar than between oral and gut samples from the same host species.

It remains possible that many types of resistance gene are still unknown, and it has become apparent recently that selective pressure from antibiotics drives the formation of mosaic resistance genes that arise by homologous recombination between existing types of resistance genes (Stanton *et al.*, 2005). This has to be taken into account when interpreting data obtained from targeted probes and primers. An important approach to searching for these and other novel genes is to screen metagenome libraries, selecting for inserts that confer resistance to a given antibiotic.

CONCLUSIONS

Molecular techniques have revolutionized the study of microbial ecology in recent years by revealing previously suspected, but often undescribed, microbial diversity, while also providing powerful culture-independent methods for studying the composition of microbial communities. These approaches are proving informative when applied to avian gut communities. To be of practical

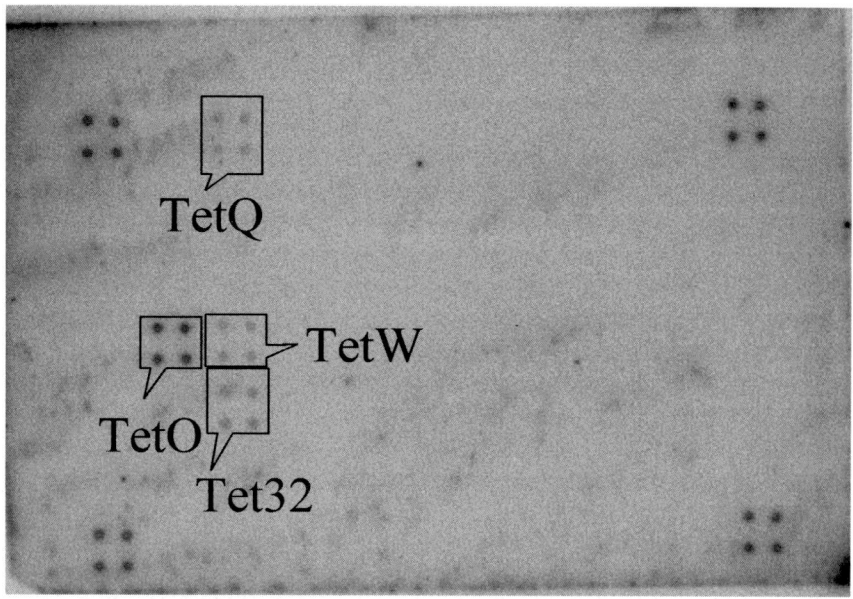

Fig. 7.3. Macro-array detection of antibiotic resistance genes in a human faecal sample. The macro-array was spotted (in quadruplicate) with amplified DNA sequences that recognized 23 different classes of tetracycline resistance genes and ten different classes of erythromycin resistance genes, and hybridized against DNA extracted from a faecal sample, labelled with [32]P. 16S rRNA sequences were spotted on to the four corners as controls (Patterson *et al.*, 2004).

value, however, molecular descriptions of the microbial community must be related to function. Techniques such as stable isotope probing and metagenomics offer new avenues, but research on isolated gut microorganisms and their interactions continues to be of key importance.

ACKNOWLEDGEMENTS

The Rowett Institute is supported by SEERAD (Scottish Executive Environment and Rural Affairs Department). MTR is supported by the EU framework project GEMINI (QLK3-2002-02056). AJP is supported by the EU framework V project ARTRADI (QLK2-CT-2002-00843). CL and AWW are supported by a BBSRC-SEERAD research grant.

REFERENCES

Accetto, T., Peterka, M. and Avgustin, G. (2005) Type II restriction modification systems of *Prevotella bryantii* TC1–1 and *Prevotella ruminicola* 23 strains and their effect on the efficiency of DNA introduction via electroporation. *FEMS Microbiology Letters* 247, 177–183.

Aminov, R.I., Garrigues-Jeanjean, N. and Mackie, R.I. (2001) Molecular ecology of tetracycline resistance: Development and validation of primers for detection of tetracycline resistance genes encoding ribosomal protection proteins. *Applied and Environmental Microbiology* 67, 22–32.

Amann, R.I., Binder, B.J., Olson, R.J., Chisholm, S.W., Devereux, A. and Stahl, D.A. (1990) Combination of 16S rRNA-targeted oligonucleotide probes with flow cytometry for analyzing mixed microbial populations. *Applied and Environmental Microbiology* 56, 1919–1925.

Apajalahti, J.H.A., Kettunen, A., Bedford, M.R. and Holben, W.E. (2001) Percent G+C profiling accurately reveals diet-related differences in the gastrointestinal microbial community of broiler chickens. *Applied and Environmental Microbiology* 67, 5656–5667.

Barbosa, T.M., Scott, K.P. and Flint, H.J. (1999) Evidence for recent intergeneric transfer of a new tetracycline resistance gene, *tetW*, isolated from *Butyrivibrio fibrisolvens* and the occurrence of *tetO*, in ruminal bacteria. *Environmental Microbiology* 1, 53–64.

Barcenilla, A., Pryde, S.E., Martin, J.C., Duncan, S.H., Stewart, C.S. and Flint, H.J. (2000) Phylogenetic relationships of dominant butyrate producing bacteria from the human gut. *Applied and Environmental Microbiology* 66, 1654–1661.

Barnes, E.M. (1972) The avian intestinal flora with particular reference to the ecological significance of the cecal anaerobic bacteria. *American Journal of Clinical Nutrition* 25, 1475–1497.

Billington, S.J., Songer, J.G. and Jost, B.H. (2002) Widespread distribution of a *tet*(W) determinant among tetracycline-resistant isolates of the animal pathogen *Arcanobacterium pyogenes*. *Antimicrobial Agents and Chemotherapy* 46, 1281–1287.

Coenye, T. and Vandamme, P. (2003) Intragenomic heterogeneity between multiple 16S ribosomal RNA operons in sequenced bacterial genomes. *FEMS Microbiology Letters* 228, 45–49.

Daly, K., Stewart, C.S., Flint, H.J. and Shirazi-Beechey, S.P. (2001) Bacterial diversity within the equine large intestine as revealed by molecular analysis of cloned 16S rRNA genes. *FEMS Microbiology Ecology* 38, 141–151.

Deplancke, B., Hristova, K.R., Oakley, H.A., McCracken, V.J., Aminov, R., Mackie, R.I. and Gaskins, H.R. (2000). Molecular ecological analysis of the succession and diversity of sulphate-reducing bacteria in the mouse gastrointestinal tract. *Applied and Environmental Microbiology* 66, 2166–2174.

Diez-Gonzalez, F., Bond, D.R., Jennings, E. and Russell, J.R. (1999) Alternative schemes of butyrate production in *Butyrivibrio fibrisolvens* and their relationship to acetate utilization, lactate production, and phylogeny. *Archives of Microbiology* 171, 324–330.

Duncan, S.H., Hold, G.L., Barcenilla, A., Stewart, C.S. and Flint, H.J. (2002a) *Roseburia intestinalis* sp. nov., a novel saccharolytic, butyrate-producing bacterium from human faeces. *International Journal of Systematic Evolutionary Microbiology* 52, 1615–1620.

Duncan, S.H., Barcenilla, A., Stewart, C.S., Pryde, S.E. and Flint, H.J. (2002b) Acetate utilization and butyryl CoA:acetate CoA transferase in human colonic bacteria. *Applied and Environmental Microbiology* 68, 5186–5190.

Duncan, S.H., Scott, K.P., Ramsay, A.G., Harmsen, H.J.M., Welling, G.W., Stewart, C.S. and Flint, H.J. (2003) Effects of alternative dietary substrates on competition between human colonic bacteria in an anaerobic fermentor. *Applied and Environmental Microbiology* 69, 1136–1142.

Eckburg, P.B., Bik, E.M., Bernstein, C.N., Purdom, E., Dethlefsen, L., Sargent, M., Gill, S.R., Nelson, K.E. and Relman, D.A. (2005) Diversity of the human intestinal microbial flora. *Science* 308, 1635–1638.

Finegold, S.M., Sutter, V.L. and Mathison, G.E. (1983) Normal indigenous flora. In: Hentges, D.J. (ed.) *Human Intestinal Microflora in Health and Disease*. Academic Press, New York, pp. 3–31.

Franks, A.H., Harmsen, H.J.M., Raangs, G.C., Jansen, G.J., Schut, F. and Welling, G.W. (1998) Variations of bacterial populations in human faeces quantified by fluorescent *in situ* hybridization with group-specific 16S rRNA-targeted oligonucleotide probes. *Applied and Environmental Microbiology* 64, 3336–3345.

Gong, J., Forster, R.J., Yu, H., Chambers, J.R., Sabour, P.M., Wheatcroft, R. and Chen, S. (2002) Diversity and phylogenetic analysis of bacteria in the mucosa of chicken caeca and comparison with bacteria in the cecal lumen. *FEMS Microbiology Letters* 208, 1–7.

Handfield, M. and Levesque, R.C. (1999) Strategies for isolation of *in vivo* expressed genes from bacteria. *FEMS Microbiology Reviews* 23, 69–91.

Harmsen, H.J.M., Raangs, G.C., He, T., Degener, J.E. and Welling, G.W. (2002) Extensive set of 16S rRNA-based probes for detection of bacteria in human feces. *Applied and Environmental Microbiology* 68, 2982–2990.

Hayashi, H., Sakamoto, M. and Benno, Y. (2002) Phylogenetic analysis of the human gut microbiota using 16S rDNA clone libraries and strictly anaerobic culture-based methods. *Microbiology and Immunology* 46, 535–548.

Hold, G.L., Pryde, S.E., Russell, V.J., Furrie, E. and Flint, H.J. (2002) Assessment of microbial diversity in human colonic samples by 16S rDNA sequence analysis. *FEMS Microbiology Ecology* 39, 33–39.

Hold, G.L., Schwiertz, A., Blaut, M. and Flint, H.J. (2003) Development of oligonucleotide probes to quantify specific groups of butyrate-producing bacteria within the human fecal flora. *Applied and Environmental Microbiology* 69, 4320–4324.

Kocherginskaya, S.A., Aminov, R.I. and White, B.A. (2001) Analysis of the rumen bacterial diversity under two different diet conditions using denaturing gradient gel electrophoresis, random sequencing, and statistical ecology approaches. *Anaerobe* 7, 119–134.

Knarreborg, A., Simon, M.A., Engberg, R.M., Jensen, B.B. and Tannock, G.W. (2002) Effects of dietary fat source and subtherapeutic levels of antibiotic on the bacterial community in the ileum of broiler chickens at various ages. *Applied and Environmental Microbiology* 68, 5918–5924.

Langendijk, P.S., Schut, F., Jansen, G.J., Raangs, G.C., Kamphuis, G.R., Wilkinson, M.H.F. and Welling, G.W. (1995) Quantitative fluorescence *in situ* hybridization of *Bifidobacterium* spp. with genus-specific 16S rRNA-targeted probes and its application in faecal samples. *Applied and Environmental Microbiology* 61, 3069–3075.

Larue, R., Yu, Z., Parisi, A., Egan, A.R. and Morrison, M. (2005) Novel microbial diversity adherent to plant biomass in the herbivore gastrointestinal tract, as revealed by ribosomal intergenic spacer analysis and *rrs* gene sequencing. *Environmental Microbiology* 7, 530–543.

Leitch, E.C., Walker, A., Duncan, S.H. and Flint, H.J. (2004) Comparison of the microbiota attached to insoluble food substrates with the planktonic population from *in vitro* human colon simulations. *Reproduction Nutrition and Development* 44 (1), S16.

Leser, T.D., Amenuvor, J.Z., Jensen, T.K., Lindecrona, R.H., Boye, M. and Moller, K. (2002) Culture-independent analysis of gut bacteria: the pig gastrointestinal tract microbiota revisited. *Applied and Environmental Microbiology* 68, 673–690.

Louis, P., Duncan, S.H., McCrae, S.I., Millar, J., Jackson, M.S. and Flint, H.J. (2004) Limited distribution of butyrate kinase among human colonic bacteria. *Journal of Bacteriology* 186, 2099–2106.

Loy, A., Hom, M. and Wagner, M. (2003) ProbeBase: an online resource for rRNA-targetted oligonucloetide probes. *Nucleic Acids Research* 331, 514–516.

Lu, J., Sanchez, S., Hofacre, C., Maurer, J.J., Harmon, B.G. and Lee, M.D. (2002) Evaluation of broiler litter with reference to the microbial composition as assessed by using 16S rRNA and functional gene markers. *Applied and Environmental Microbiology* 69, 901–908.

Lu, J., Idris, U., Harmon, B., Hofacre, C., Maurer, J.J. and Lee, M.D. (2003) Diversity and succession of the intestinal bacterial community of the maturing broiler chicken. *Applied and Environmental Microbiology* 69, 6816–6824.

McIntyre, A., Gibson, P.R. and Young, G.P. (1993) Butyrate production from dietary fiber and protection against large bowel cancer in a rat model. *Gut* 34, 386–391.

Mortensen, P.B. and Clausen, M.R. (1996) Short chain fatty acids in the human colon: relation to gastrointestinal health and disease. *Scandinavian Journal of Gastroenterology* 31 (216), 132–148.

Muyzer, G. and Smalla, K. (1998) Application of denaturing gradient gel electrophoresis (DGGE) and temperature gradient gel electrophoresis (TGGE) in microbial ecology. *Antonie von Leeuwenhoek* 73, 127–141.

Nagashima, K., Hisada, T., Sato, M. and Mochizuki, J. (2003) Application of new primer–enzyme combinations to terminal restriction fragment length polymorphism profiling of bacterial populations in human feces. *Applied and Environmental Microbiology* 69, 1935–1939.

Patterson, A.J., Flint, H.J. and Scott, K.P. (2004) Investigating the distribution of antibiotic resistance across Europe by the use of a macroarray system. *International Journal of Antimicrobial Agents* 24, S150–S150.

Pryde, S.E., Richardson, A.J., Stewart, C.S. and Flint, H.J. (1999) Molecular analysis of the microbial diversity present in the colonic wall, colonic lumen, and cecal lumen of a pig. *Applied and Environmental Microbiology* 65, 5372–5377.

Pryde, S.E., Duncan, S.H., Hold, G.L., Stewart, C.S. and Flint, H.J. (2002) The microbiology of butyrate formation in the human colon. *FEMS Microbiology Letters* 217, 133–139.

Radajewski, S., Ineson, P., Parekh, N.T.R. and Murrell, J.C. (2000) Stable isotope probing as a tool in microbial ecology. *Nature* 403, 646–649.

Salanitro, J.P., Fairchilds, I.G. and Zgornicki, Y.D. (1974) Isolation, characteristics and identification of anaerobic bacteria from the chicken caecum. *Applied Microbiology* 27, 678–687.

Simmering, R., Pforte, H., Jacobasch, G. and Blaut, M. (2002) The growth of the flavonoid-degrading intestinal bacterium *Eubacterium ramulus* is stimulated by dietary flavonoids *in vivo*. *FEMS Microbiology Ecology* 40, 243–248.

Small, J., Call, D.R., Brockman, F.J., Straub, T.M. and Chandler, D.P. (2001) Direct detection of 16S rRNA in soil extracts by using oligonucleotide microarrays. *Applied and Environmental Microbiology* 67, 4708–4716.

Socransky, S.S., Haffajee, A.D., Smith, C., Martin, L., Haffajee, J.A., Uzel, N.G. and Goodson, J.M. (2004) Use of checkerboard DNA–DNA hybridization to study complex microbial ecosystems. *Oral Microbiology and Immunology* 19, 352–362.

Stahl, D.A., Flesher, B., Mansfield, H.R. and Montgomery, L. (1988) Use of phylogenetically based hybridization probes for studies of ruminal microbial ecology. *Applied and Environmental Microbiology* 54, 1079–1084.

Stanton, T.B., Humphrey, S.B., Scott, K.P. and Flint, H.J. (2005) Hybrid tet genes and tet gene nomenclature: request for opinion. *Antimicrobial Agents and Chemotherapy* 49, 1265–1266.

Steele, H.L. and Streit, W.R. (2005) Metagenomics: advances in ecology and biotechnology. *FEMS Microbiology Letters* 247, 105–111.

Suau, A., Bonnet, R., Sutren, M., Godon, J.J., Gibson, G.R., Collins, M.D. and Dore, J. (1999) Direct analysis of genes encoding 16S rRNA from complex communities reveals many novel molecular species within the human gut. *Applied and Environmental Microbiology* 24, 4799–4807.

Suau, A., Rochet, V., Sghir, A., Gramet, G., Brewaeys, S., Sutren, M., Rigottier-Gois, L. and Dore, J. (2001) *Fusobacterium prausnitzii* and related species represent a dominant group within the human fecal flora. *Systematic and Applied Microbiology* 24, 139–145.

Tajima, K., Aminov, R.I., Nagamine, T., Ogata, K., Nakamura, M., Matsui, H. and Benno, Y. (1999) Rumen bacterial diversity as determined by sequence analysis of 16S rDNA libraries. *FEMS Microbiology Ecology* 29, 159–169.

Tajima, K., Aminov, R.I., Nagamine, T., Matsui, H., Nakamura, M. and Benno, Y. (2001) Diet-dependent shifts in the bacterial population of the rumen revealed with real-time PCR. *Applied and Environmental Microbiology* 67, 2766–2774.

Topping, D.L. and Clifton, P.M. (2001) Short chain fatty acids and human colonic function: roles of resistant starch and nonstarch polysaccharides. *Physiological Reviews* 81, 1031–1064.

Tyson, G.W., Chapman, J., Hugenholtz, P., Allen, E.E., Ram, R.J., Richardson, P.M., Solovyov, V.V., Rubin, E.M., Rokhsar, D.S. and Banfield, J.F. (2004) Community structure and metabolism through reconstruction of microbial genomes from the environment. *Nature* 428, 37–43.

Villedieu, A., Diaz-Torres, M.L., Hunt, N., McNab, R., Spratt, D.A., Wilson, M. and Mullany, P. (2003) Prevalence of tetracycline resistance genes in oral bacteria. *Antimicrobial Agents and Chemotherapy* 47, 878–882.

von Wintzingerode, F., Gobel, U.B. and Stackebrandt, E. (1997) Determination of microbial diversity in environmental samples: pitfalls of PCR-based methods. *FEMS Microbiology Reviews* 21, 213–229.

Walker, A.W., Duncan, S.H., McWilliam Leitch, E.C., Child, M.W. and Flint, H.J. (2005) pH and peptide supply can radically alter bacterial populations and short chain fatty acid ratios within microbial communities from the human colon. *Applied and Environmental Microbiology* 71, 3692–3700.

Wang, M., Ahrne, S., Antonsson, M. and Molin, G. (2004) T-RFLP combined with principal component analysis and 16S rRNA gene sequencing: an effective strategy for comparison of fecal microbiota in infants of different ages. *Journal of Microbiological Methods* 59, 53–69.

Weisburg, W.G., Barns, S.M., Pelletier, D.A. and Lane, D.J. (1991) 16S ribosomal DNA amplification for phylogenetic study. *Journal of Bacteriology* 173, 697–703.

Whitford, M.F., Forster, R.J., Beard, C.E., Gong, J. and Teather, R.M. (1998) Phylogenetic analysis of rumen bacteria by comparative sequence analysis of cloned 16S rDNA libraries. *Anaerobe* 4, 153–163.

Wilson, K.H. and Blitchington, R.B. (1996) Human colonic biota studied by ribosomal DNA sequence analysis. *Applied and Environmental Microbiology* 62, 2273–2278.

Wilson, K.H., Wilson, W.J., Rodasevitch, J.L., DeSantis, T.Z., Viswanathan, V.S., Kuczmarski, T.A. and Anderson, G.L. (2002) High density microarray of small subunit ribosomal RNA probes. *Applied and Environmental Microbiology* 68, 2535–2541.

Woese, C. (1987) Bacterial evolution. *Microbiological Reviews* 51, 221–271.

Wood, J., Scott, K.P., Avgustin, G., Newbold, C.J. and Flint, H.J. (1998) Estimation of the relative abundance of different *Bacteroides* and *Prevotella* ribotypes in gut samples by restriction enzyme profiling of PCR-amplified 16S rRNA gene sequence. *Applied and Environmental Microbiology* 64, 3683–3689.

Zhu, X.Y., Zhong, T., Pandya, Y. and Joerger, R.D. (2002) 16S rRNA-based analysis of microbiota from the caecum of broiler chickens. *Applied and Environmental Microbiology* 68, 124–137.

Zoetendal, E.G., Akkermans, A.D.L. and DeVos, W.M. (1998) Temperature gradient gel electrophoresis analysis of 16S rRNA from human faecal samples reveals stable and host-specific communities of active bacteria. *Applied and Environmental Microbiology* 64, 3854–3859.

Zoetendal, E.G., Collier, C.T., Koike, S., Mackie, R.I. and Gaskins, H.R. (2004) Molecular ecological analysis of the gastrointestinal microbiota: a review. *Journal of Nutrition* 134, 465–472.

Zwirglmaier, K. (2005) Fluorescence *in situ* hybridisation (FISH) – the next generation. *FEMS Microbiology Letters* 246, 151–158.

CHAPTER *8*
Microbes of the chicken gastrointestinal tract

J. Apajalahti* and A. Kettunen

*Alimetrics Ltd., Helsinki, Finland; *e-mail: juha.apajalahti@alimetrics.com*

ABSTRACT

The gastrointestinal tract of an adult chicken is inhabited by up to 10^{13} bacteria. This microbiota has a wide metabolic potential and it affects both the nutrition and health of the host. The microbial populations in the chicken gut grow rapidly, the density plateauing within 3–4 days of hatching. The structural evolution, however, is likely to continue for the entire life of a broiler chicken.

The effect of microbes on the commercial aspects of a poultry enterprise is apparent to the farmer only in extreme cases. Known pathogens, such as *Clostridium perfringens* and *Escherichia coli*, may reduce bird growth and liveability and human pathogens, such as *Salmonella enterica*, may lead to carcass condemnation and reduced income. Most of the time, however, the effect of bacteria is less obvious to the grower, and one or many of the unidentified bacterial species present may cause non-specific enteritis.

The composition of the intestinal bacterial community has been studied by many researchers. The methods used here varied markedly and, therefore, it is often impossible to know whether the varying observations are due to geography, feeding regimen, bird management or to the methods used for microflora analysis. The microbial community of the caecum is far more diverse and less well characterized than that of the crop and small intestine. The abundance of bacteria is significantly affected by feed composition and ingredients. We have numerous means of modulating the microflora and these can be applied as soon as knowledge from well-designed studies unveils the correct bacterial targets.

INTRODUCTION

The gastrointestinal tract (GIT) of warm-blooded animals, whose diet contains complex carbohydrates, is densely populated by bacteria. The exact density and composition of the microflora is markedly affected by the bacterial

© CAB International 2006. *Avian Gut Function in Health and Disease* (ed. G.C. Perry)

composition of the inoculum received at birth or hatch, the structure of the host's intestinal epithelium and the diet. Gastrointestinal microflora differentially affect the nutrition and health of different animal species. In ruminants, the resident microflora are essential in converting the diet into the form available for the animal, whereas for the monogastric animals the nutrients in the diet are already in the form available for the host without microfloral involvement.

However, microbes in the GIT of monogastric animals also greatly affect performance of the host animal. Bacteria compete with the host for dietary compounds but, on the other hand, produce metabolites that provide energy for the host. The relative intensity of such bacterial activities determines the overall effect of bacteria on the feed conversion efficiency. The effect of microflora on energy balance is difficult to quantify and, therefore, it is one of the hidden factors that can explain occasional unexpected growth performance. Bacteria affecting the health of animals are more obvious since they cause mortality or clear growth suppression. There are few bacterial diseases in broiler chickens where the causative agents are indisputably known but there are many syndromes in which the aetiology and causative agents remain to be identified. Examples of names used for such syndromes are dysbacteriosis, unspecific enteritis and wet litter problems.

The above-mentioned effects of bacteria on animal production have significant economic consequences; impaired feed conversion efficiency, reduced liveability, growth suppression and the presence of foodborne pathogens all cause financial losses for the animal producer. These are the major reasons for the recent increased interest in understanding the development of microbial community in the GIT, its composition, the role of different bacterial species and possibilities for manipulation of the composition through dietary modulation. During the last 10 years tools available for microfloral analysis have developed and these have opened new avenues for this field of research.

Particularly important developments have opened up possibilities for determining the overall abundance of bacteria and also for tracking the types of bacteria that cannot be cultured under laboratory conditions. The research done with such tools has revealed the presence of an earlier undiscovered bacterial world which first created confusion and the feeling of ignorance but, after the first reaction, has led to research projects aimed at understanding the role of the entire bacterial community on the health and performance of production animals. This chapter provides an introduction to the incoming novel data and points out specific data items that have been widely misunderstood by the general audience.

GROWTH OF GUT MICROBES

Newborn wild birds become exposed to bacteria from the very moment of hatching. Their GIT receives the first inoculum of bacteria from the surface of the eggshells which is heavily populated by bacteria from the intestine of the mother bird and the natural environment. In a modern hatchery, however, eggshells and the hatching rooms are kept as bacteria-free as possible. This is likely to affect the development of the microflora, immune system and the intestinal physiology of the broiler chickens.

If the birds are not intentionally inoculated with, e.g. competitive exclusion cultures at hatch, the first inoculum received by the newly hatched chicks may be random and, perhaps, represent species foreign to the GIT of the chickens. The effect of the first inoculum may last over the entire life of a broiler chicken by directing the development of the immune system (immunological tolerance) and the normal intestinal microflora.

Bacteria first entering the GIT of chickens have been found to grow very rapidly. The initially sterile habitats of the ileum and caecum were inhabited by 10^8 and 10^{10} bacteria per gram of digesta, respectively, in 1 day (Apajalahti *et al.*, 2004). The maximum bacterial density was found to be reached in less than 1 week; by that time the ileum had $>10^9$ and the caecum $>10^{11}$ bacteria/g of digesta (Apajalahti *et al.*, 2004). The shared data are from the study that used a culture independent method, flow cytometry, for bacterial enumeration to ensure that unculturable bacteria were also included (Apajalahti *et al.*, 2002).

If we assumed that 100 bacteria initially inoculated the caecum, and the growth proceeded unlimited with no acceleration or retardation phase, the doubling time of 40 min would lead to a density of 10^{10} bacteria/g in 18 h. Correspondingly, with 80 min doubling time it would take 36 h to reach the same bacterial density (Fig. 8.1). In reality, exponential growth is soon followed by a retardation phase caused by nutrient depletion and accumulation of inhibitory metabolites. The rapid initial growth detected in the GIT of chicks is possible only if the growth medium is highly rich in nutrients. In newly hatched chicks the most likely initial source of growth substrate for bacteria is the residue of yolk sack contents.

Once the maximum bacterial density has been reached the subsequent growth of bacteria in the GIT occurs in synchrony with the digesta passage rate, keeping the bacterial density in each part of the GIT constant. Accordingly, the turnover time of digesta in an intestinal compartment would be equal to the bacterial doubling time in the same part of the digestive tract. Since digesta passage rate in the small intestine is higher than in the caecum or cloacae, the bacteria inhabiting the small intestine have to be able to grow faster than those residing in the lower GIT to resist washout. It is worth noting that bacteria capable of adhering to intestinal linings can resist washout even when their growth rate is lower than the rate of digesta passage.

Taking into account that small intestinal digesta are very rich in nutrients it is striking how efficiently birds can keep bacterial numbers low in this critical part of the digestive tract. Chickens have several mechanisms that restrict bacterial growth in the proximal GIT, including the following: (i) chemical inhibition (e.g. acids and bile); (ii) highly competitive rates of nutrient absorption (large absorptive surface and active transport); (iii) high passage rates of digesta (washout of slowly growing bacteria); (iv) continuous sloughing of the epithelial cells and mucous (washout of adhered bacteria); and (v) the

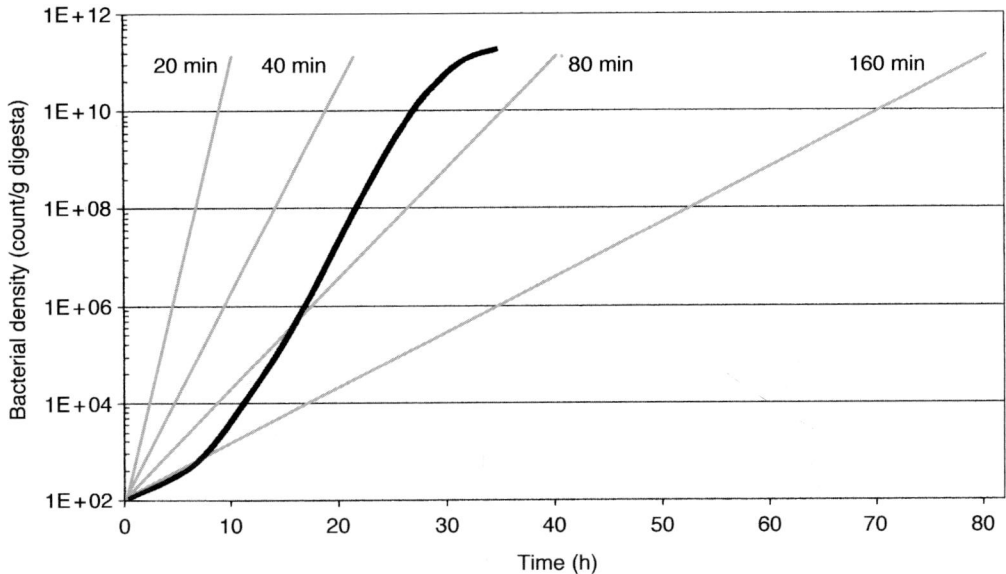

Fig 8.1. Theoretical bacterial densities reached at various doubling times. The grey lines show the time it takes for bacteria to exceed the density of 10^{11}/g of digesta assuming unlimited bacterial growth at indicated doubling times and an initial inoculum of 100 bacteria. The black line is a realistic bacterial growth curve which is in line with the bacterial densities measured in the caecum of broiler chickens (Apajalahti *et al.*, 2004).

immunological defence mechanisms (IgA and cell mediated). Failure to prevent bacterial growth in the proximal small intestine would, no doubt, lead to rapid starvation of the host.

16S RIBOSOMAL RNA-BASED BACTERIAL SURVEYS

Even in the richest culture media only a minority of bacteria can be captured and grown as pure cultures. As different bacteria may require different, mutually exclusive growth conditions it is hardly possible to find conditions that are suitable for all bacteria. Therefore, culture-based studies tend significantly to underestimate bacterial numbers and diversity in complex bacterial habitats. Furthermore, growth requirements for many bacteria remain unknown and such bacteria cannot be isolated under any known culture conditions.

Today, several culture independent methods are available for exploring total microbial communities. However, few of the DNA-based methods are totally independent of prior knowledge of the microbial community. Per cent guanine + cytosine (%G+C) profiling of the microbial community DNA is such a method, but it only pictures the total community and bacterial genera without revealing the identity of individual microbial species (Apajalahti *et al.*, 1998, 2001).

   In order to acquire more detailed information on the composition of the
microbial community, methods based on polymerase chain reaction (PCR) are
frequently used alone and in combination with %G+C profiling (Apajalahti *et
al.*, 2002; Holben *et al.*, 2002, 2004). The major principle of PCR is to have
two DNA oligomers called primers, which are complementary to the DNA
sequence of the target DNA and, therefore, selectively bind to the specific DNA.
With matching primers present, an enzyme called DNA polymerase amplifies
the piece of DNA between the two primers so many times that its concentration
reaches a detectable level. As PCR reaction is dependent on the binding of the
primers, their precise base sequence largely determines the results obtained.
   PCR reaction is a crucial element when studying the composition of
microbial communities by a method called 16S ribosomal RNA (16S rRNA)
sequencing. In this method the exact nucleotide sequence of the PCR products
amplified from the 16S rRNA gene is determined. At least one copy of 16S
rRNA gene is present in the chromosome of all prokaryotic organisms and its
nucleotide sequence has, therefore, been chosen as the basis for modern
microbial taxonomy. The identity and phylogenetic position of a microbe is
determined by its 16S rRNA sequence and the relationship to other microbial
16S rRNA sequences.
   The first step in the 16S rRNA sequencing-based survey is the
amplification of a variable region from the 16S rRNA genes of all bacteria
present. This is achieved by using primers targeted at non-variable (conserved)
regions of the gene. It would be ideal for the 16S rRNA-based surveys if
identical or highly conserved primer sequences were present in all bacteria
(universal primers), whereas the DNA sequence surrounded by the primers
would be highly variable. If this was the case and there was no bias in
amplification, the relative abundance of the 16S rRNA genes originating from
different bacteria would be exactly the same before and after the PCR reaction.
However, even though there are published lists of universal primers, no single
primer pair truly captures all microbes, but inevitably leads to a bias of varying
degree. It is worth noting that no PCR-based method locates all bacteria in an
undefined bacterial community.
   Table 8.1 shows the bacterial coverage of primer pairs that have been used
in 16S rRNA-based microbial community surveys in broiler chickens. We
compared each primer sequence to a public 16S rRNA sequence database,
Ribosomal Database Project (RDP II; Cole *et al.*, 2005). First, the percentage of
RDP-II 16S sequences with 100% similarity to primers (no mismatches) was
recorded, and then the proportion of sequences with one mismatch.
   The rationale behind the approach was that PCR with primers carrying one
mismatch were still likely to produce an amplification product but less efficiently
than with primers identical to the target DNA. For the primer pair 8F/1492R, the
coverage with no mismatches was 64% for the forward primer 8F and 84% for
the reverse primer 1492R. Assuming that the more universal primer, 1492R,
captured all the microbes that were captured by the less universal forward
primer 8F the coverage would be the 64%, as shown in Table 8.1. Based on the
analysis presented in Table 8.1, the primer pair that achieved the highest
coverage, and hence likely to lead to the smallest bias, would be 518F/928R.

Table 8.1. Coverage of primer pairs used in 16S rRNA sequencing of chicken gut microbes.

| Primer pairs | Maximum coverage (%) | | References |
	No mismatches allowed	One mismatch allowed	
518F/928R	74	94	Apajalahti and Kettunen, 2006
518F/928R	74	94	Holben *et al.*, 2004
8F/926R	39	83	Lu *et al.*, 2003
8F/1494R	62	89	Lu *et al.*, 2003
8F/1522R	44	69	Lu *et al.*, 2003
8F/1492R	64	91	Zhu *et al.*, 2002
63F/1387R	13	30	Zhu *et al.*, 2002

[a] Coverage percentages for primer pairs are based on the assumption that the primer with the higher coverage would also cover all the sequences captured by the lower coverage primer. Thus, the percentages are referred to as maximum possible coverage for a given primer pair.

Table 8.2a shows the results of two 16S-based microbial community surveys of the chicken ileum. Although one of the surveys was carried out in USA and the other one in Finland, the proportions of the microbial clusters are surprisingly similar. Based on these studies, the genus *Lactobacillus* clearly dominates ileal microbial community of the chicken (Lu *et al.*, 2003; Apajalahti and Kettunen, 2006). The presence of clostridial cluster XIVa in the ileum may reflect the presence of a caecal microbial community resulting from reverse peristalsis (Table 8.2a).

Table 8.2b shows the results of four studies that investigated the composition of the microbial community in the caecum of chickens. Geographically the research covered Georgia and Delaware in the USA, Finland in Northern Europe and a global area comprising 500 chickens from Australia, the UK, Finland, France and the USA (Zhu *et al.*, 2002; Lu *et al.*, 2003; Holben *et al.*, 2004; Apajalahti and Kettunen, 2006).

The results of the study carried out in Delaware differ significantly from those of the other studies, apparently due to low 16S sequence coverage of the primers 63F/1387R which were used in the relevant part of the study (Tables 8.1 and 8.2b). With these primers 89% of bacteria in the caecum appeared to belong to clostridial cluster IX, the abundance of which was negligible in the studies using primers with a better coverage (Table 8.2b).

Therefore, it seems likely that the results are strongly biased due to the odd primer selection (Zhu *et al.*, 2002). The surveys also referred to other introduced methodological approaches and primers. However, in the Delaware survey only the random cloning part of the study was included since it was the only one that could be considered quantitative and, therefore, comparable to other studies. The other three surveys' findings are in line with each other, indicating that the majority of the caecal microbes belonged to clostridial clusters IV and XIVa, and to the genera *Bacteroides*, *Lactobacillus* and *Bifidobacterium*.

Significant abundance of *Bifidobacterium* was found in only one of the studies. This can partly be explained by the fact that none of the primer pairs

Table 8.2. Relative abundance of bacterial groups found in the ileum (a) and caecum (b) of broiler chickens in surveys based on 16S rRNA sequencing.[a]

(a)

	Geographical location of study	
	[1]Finland	[2]Georgia, USA
Number of birds included	20	5
Proportion of bacterial groups (%)		
Lactobacillus	80	86
Clostridial cluster XIVa	10	6
Bacteroides	< 5	< 5
Enterococcus	< 5	< 5
Escherichia	< 5	< 5
Streptococcus	< 5	< 5
Others	< 5	< 5

(b)

	Geographical location of study			
	[1]Finland	[3]Global	[2]Georgia, USA	[4]Delaware, USA
Number of birds included	20	500	5	1
Proportion of bacterial groups (%)				
Clostridial cluster XIVa	42	42	45	3
Clostridial cluster IV	< 3	13	35	5
Clostridial cluster IX	< 3	< 3	< 3	89
Bacteroides	29	9	5	< 3
Lactobacillus	12	11	< 3	< 3
Bifidobacterium	< 3	8	< 3	< 3
Eubacterium desmolans	< 3	< 3	9	< 3
Escherichia	< 3	< 3	< 3	< 3
Others	13	14	5	< 3

[a] The source of data is indicated by superscripts: [1]Apajalahti and Kettunen, 2006; [2]Lu *et al.*, 2003; [3]Holben *et al.*, 2004; [4]Zhu *et al.*, 2002.

used in the three studies had a 100% match to the 16S sequences of bifidobacteria. The 16S rRNA gene of bifidobacteria shows a low degree of similarity to the genes of the other microbes and, therefore, it is hardly possible to find primers with wide general coverage among all bacteria that would also efficiently capture bifidobacteria. Consequently, the proportion of bifidobacteria is likely to be underestimated in all of the microbial surveys.

The forward primer 8F used in the Georgia study shows so much dissimilarity to bifidobacteria that it is not surprising that bifidobacteria remained undetected in that study. The primers 518F and 928R used in the global and Finnish studies show only one mismatch to bifidobacteria and should thus also capture representatives of this genus, though with reduced efficiency. In the global study, bifidobacteria comprised 8% of the caecal microbial community, whereas in the Finnish study that used identical primers the abundance of bifidobacteria remained below detection limit.

Samples from these two studies have also been analysed by %G+C profiling to capture the overall diversity of caecal bacterial communities. Results of this analysis, shown in Fig. 8.2, indicate that in the caecum of the chickens from Finland there were no microbes in the high %G+C area where bifidobacteria should show up. In contrast, the %G+C profile representing the average caecal community of the 500 birds included in the global study revealed significant abundance of bacteria in the high %G+C range consistent with the %G+C of the bifidobacteria (~ 65%G+C).

Hence, it appears that there was a true difference in the levels of bifidobacteria in the global and Finnish studies. While the level of bifidobacteria was truly low in the Finnish study, the actual levels of bifidobacteria in the US studies remain unknown.

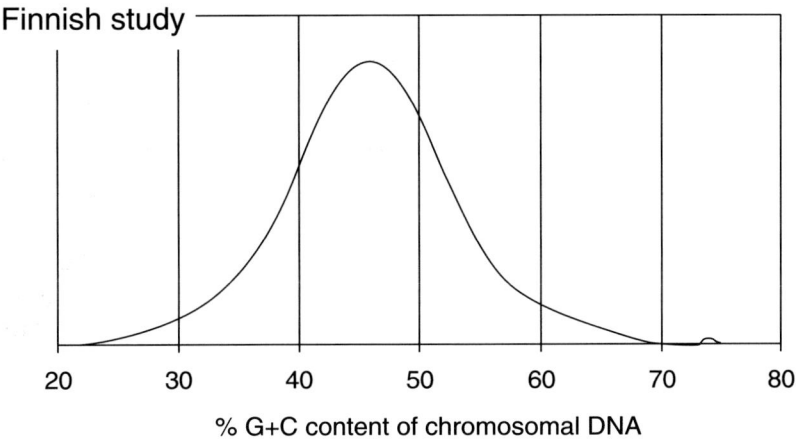

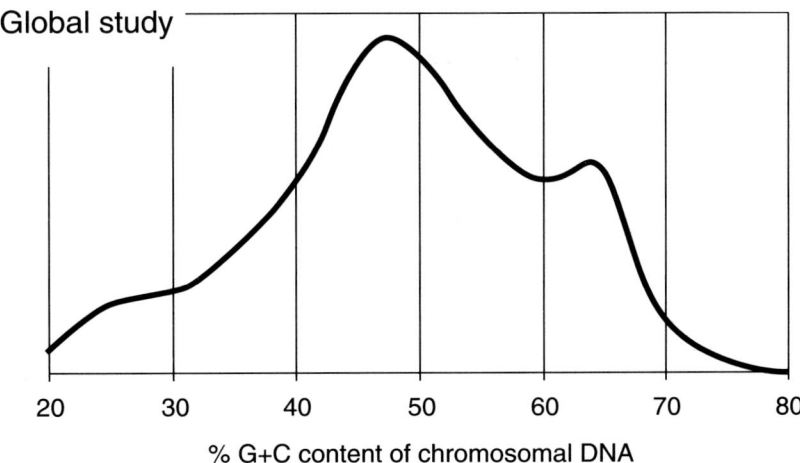

Fig. 8.2. Per cent guanine + cytosine profiles of the caecal microbial communities in the chickens included in two 16S rRNA sequencing-based surveys. The profiles presented are related to the microbial analysis presented in Table 8.2b for the Finnish and global studies.

MISINTERPRETATIONS IN BACTERIAL NOMENCLATURE

In the early days of microbiology, bacteria were classified according to their morphology, e.g. cocci (spheres) and bacilli (rods). Later on, rod-shaped bacteria were divided into those carrying endospores and those that did not. Aerobic rods with endospores formed the genus *Bacillus*, whereas the genus containing their anaerobic counterparts was named *Clostridium*. Later, chemical structure and physiological characteristics were taken into account and bacterial nomenclature advanced; bacterial species were divided into several phyla, classes, orders, families and genera.

The genus *Clostridium* is well known to people in the poultry industry, mainly because of one pathogenic species, *Cl. perfringens*. Unfortunately this species has affected the reputation of the entire bacterial class *Clostridia*, which is the most abundant bacterial class in the intestine of both broiler chickens and other homeothermic animals (Fig. 8.3). It is worth noting that the genus *Clostridium* is only one out of 15 genera in the family *Clostridiaceae*, which is only one of eight families in the order *Clostridiales* which, again, is only one of three orders in the class *Clostridia* (Cole *et al.*, 2005). Furthermore, *Cl. perfringens* is only distantly related to the majority of the more than 100 species of the genus *Clostridium*.

Recently, studies on evolutionary relationships of bacteria by 16S rRNA sequencing have led to a new nomenclature that rearranges the grouping of bacteria within the order *Clostridiales*. This classification is by no means in line with the classification presented in Fig. 8.3, but divides the bacteria in the order *Clostridiales* into clusters based on evolutionary relatedness (Collins *et al.*, 1994). *Cl. perfringens* and many other pathogenic clostridia belong to clostridial cluster I, whereas the bulk of bacteria in the chicken (and human) GIT belong to clostridial cluster XIVa (Table 8.2). Many bacterial species belonging to clostridial cluster XIVa have positive traits such as production of butyric acid, which promotes the health of the intestinal epithelium. It is obvious that a change in the mindset of laymen and preciseness in the statements of professionals is needed when dealing with the presence and role of clostridia in the gastrointestinal tract.

The use of the 16S rRNA sequencing approach for bacterial surveys has revealed the entire diversity of bacterial communities in the GIT of animals. One outcome has been the discovery of many bacterial genera and species that have never been isolated or known to exist before. This is most likely due to the fact that there are no methods of growing them under laboratory conditions. Since new bacterial taxa can be described only when their representatives have been isolated and their physiological characteristics determined in the laboratory, the unculturable bacteria discovered in 16S rRNA surveys can only be called phylotypes or operational taxonomic units (OTUs), as defined by their 16S rRNA sequence and evolutionary relationships to known bacterial species.

In our own discovery work we have found about 2000 different 16S sequences in chickens. These form, altogether, 140 genus-level ($S_{ab} < 95\%$) clusters (Cole *et al.*, 2005). The role of phylotypes in the health and

domain: Bacteria
　　phylum: *Firmicutes*
　　　　class: **Clostridia**
　　　　　　order: *Thermoanaerobacteriales*
　　　　　　order: *Haloanaerobiales*
　　　　　　order: **Clostridiales**
　　　　　　　　family: *Lachnospiraceae*
　　　　　　　　family: *Peptostreptococcaceae*
　　　　　　　　family: *Eubacteriaceae*
　　　　　　　　family: *Peptococcaceae*
　　　　　　　　family: *Heliobacteriaceae*
　　　　　　　　family: *Acidaminococcaceae*
　　　　　　　　family: *Syntrophomonadaceae*
　　　　　　　　family: **Clostridiaceae**
　　　　　　　　　genus: *Acetivibrio*
　　　　　　　　　genus: *Acidaminobacter*
　　　　　　　　　genus: *Alkaliphilus*
　　　　　　　　　genus: *Anaerobacter*
Clostridial　　　　genus: *Caloramator*
clusters I to XIX　genus: *Caloranaerobacter*
　　　　　　　　　genus: *Caminicella*
　　　　　　　　　genus: *Dorea*
　　　　　　　　　genus: *Natronincola*
　　　　　　　　　genus: *Sarcina*
　　　　　　　　　genus: *Sporobacter*
　　　　　　　　　genus: *Tepidibacter*
　　　　　　　　　genus: *Thermobrachium*
　　　　　　　　　genus: *Tindallia*
　　　　　　　　　genus: **Clostridium**

Fig. 8.3. Family tree showing taxonomic position of the bacterial genus *Clostridium* within the class *Clostridia*. Text in bold highlights taxonomic entities comprising the genus *Clostridium*. Based on evolutionary relationships bacteria belonging to the order *Clostridiales* have also been divided into clostridial clusters I to XIX. This clustering does not follow the classification shown by the family tree.

performance of the host can be revealed only through epidemiological studies and studies transferring suspect microbial communities from one host to another.

EFFECT OF ANTIMICROBIALS ON GUT MICROFLORA

The routine use of feed antibiotics during recent decades has most probably gradually shifted the microbial community of the gastrointestinal tract from microbes that are sensitive to antibiotics towards those that tolerate antibiotics. Consequently, the microflora that are currently considered normal actually represent antibiotic-tolerant microflora. This being the case, most of our knowledge and experience of poultry management is based on birds with antibiotic-tolerant GIT microflora.

Furthermore, veterinary science has grown accustomed to recognizing diseases developing from those microflora that actually represent antibiotic-tolerant microbes. The ban of routine use of growth-promoting antibiotics has changed the situation and removed the selection pressure against antibiotic-sensitive microbes. In the current situation, the selection pressure is solely based on the efficiency of microbes to grow on feed residues under the conditions of the GIT. This may lead to the outgrowth and establishment of unforeseeable microbes from the largely unknown intestinal microbial community of the animals.

Studies on the effects of antibiotics have traditionally been based on cultivation of the most commonly recognized pathogens (Butaye *et al.*, 2003). *Cl. perfringens* has been shown to be sensitive to a large variety of antibiotics, while the effects of antibiotics on coliforms, campylobacteria, enterococci and species of *Salmonella* have been more variable (Butaye *et al.*, 2003). In fact, there are few studies that have studied the effects of antibiotics on chicken GI tract microbes using modern microbial techniques.

Apajalahti and Kettunen (2006) found that monensin decreased the total number of ileal microbes, resulting in reduction of competition for nutrients. Hence, the positive effect of monensin on the body weight gain of broiler chickens may not be solely based on the prevention of coccidiosis, but also on reduction of overall bacterial growth in the small intestine. However, the composition of the microflora was unaffected by monensin in both the ileum and caecum of chickens.

When bacitracin methylene disalicylate and phenoxy methyl penicillin were tested in combination with monensin, microbial numbers in the small intestine were further reduced and body weight gain was even higher than with monensin alone (Apajalahti and Kettunen, 2006). Unlike with monensin alone, significant changes were also observed in the composition of gut microflora, including a remarkable decrease in the abundance of lactobacilli and an increase in *E. coli*.

Lee (2006) observed similar changes in the numbers of lactobacilli and *E. coli* with antibiotics. Lactobacilli appear to react to a variety of antibiotics, as their abundance was shown to decrease also with sub-therapeutic levels of avilamycin and salinomycin in a study using both DGGE gel band sequencing and lactic acid bacteria-specific PCR primers (Knarreborg *et al.*, 2002). In contrast, lactobacilli were favoured in the upper GIT of chickens when virginiamycin was added to feed (Van Kessel *et al.*, Abstracts, this volume). When an antibiotic is included in feed the overall dynamics of gastrointestinal microbial community change and, depending on the relative tolerance of different bacterial species to the antibiotic, their relative abundance may change unexpectedly.

PRACTICAL IMPLICATIONS OF THE EMERGING INFORMATION

The growth of bacteria in the GIT of chickens can be divided into at least two distinct colonization phases. First, bacteria entering the untouched environment of the chick intestine grow exponentially. During the first colonization phase, bacteria that happened to be present in the hatching environment increase rapidly, leading to a fully colonized gut in just a few days. The second maturation phase starts from the fully colonized gut and is characterized by: (i) a low growth rate equal to that of the digesta passage; and (ii) gradual selection of bacteria that most efficiently adapt to the prevailing conditions.

Kinetics of natural bacterial growth in the GIT has major implications on the use of direct-fed microbes (DFM) – also known as probiotics or competitive exclusion cultures. When given to newly hatched chicks before or at the early colonization phase, commercial cultures can be applied in numbers exceeding those of natural colonizers by several orders of magnitude, at acceptable cost. This is because of the fact that natural bacterial level in the hatcheries is low and can easily be outnumbered by DFMs. If the initial growth rate of a commercial culture was equal to that of the endogenous microflora it would comprise the bulk of the plateaued microbial community present in the ileum and caecum of birds before the bacterial maturation phase.

If the commercial DFM preparation was given to birds only after the first colonization phase (days from hatching), the cell count of the commercial culture provided would inevitably fall several orders of magnitude below that of the competing endogenous microflora and thus would meet hard competition for nutrients and the ecological niche during the bacterial maturation phase. Indeed, a commercially successful DFM culture should be provided to birds at, or very soon after, hatching and should grow at a rate comparable to or higher than authentic microflora.

At first sight, the ongoing revolution in basic GIT microbiology may appear to complicate the application of the most recent knowledge to animal farming, since new, previously unknown bacterial species are emerging and the classification of bacteria into good and bad ones may be unclear. However, it is obvious that now that we are capable of detecting and quantifying all intestinal bacteria instead of only a small fraction of them, comprehensive understanding is closer than ever.

Surveys carried out so far using the DNA sequencing approach have shown that bacteria in the gut of healthy chickens may not vary drastically in different parts of the world. The major bacterial groups in the ileum and caecum of birds in Finland and the USA appeared to be the same, even though some differences were observed in the relative abundance of the bacteria. Perhaps the common breeds of broiler chickens accommodate similar bacterial flora, but differences in feeding regimens in different countries affect the proportions of different bacteria.

Additional studies are invaluable in providing information on the effects of feed composition and environmental conditions on the microfloral composition. Well-directed bacterial surveys, even in the short term, may reveal bacterial indicators linked to performance parameters, feed conversion efficiency and apparent/true metabolizable energy.

Many new bacterial species will be found in the near future with the advanced molecular techniques available. It is most unlikely that all the new species will be neutral or beneficial for the health of the host. In fact, it is possible that some of the disease syndromes we are witnessing today on broiler farms are caused by unknown bacterial species that may have benefited from the removal of growth-promoting antibiotics. Indeed, we are getting used to microbial problems such as unspecific enteritis, dysbacteriosis and wet litter to an extent that most of us are not even expecting their aetiology to be unveiled. It is difficult to fight against undiscovered bacteria the existence of which cannot be positively proven. If the causative agents of such disorders can be discovered by these novel tools, regular monitoring of pathogen abundance would be straightforward and it would be possible to start screening for products to suppress such harmful organisms.

Relatively few bacterial poultry pathogens are known today. This may be one of the reasons for the bad reputation of clostridia. *Cl. perfringens* is a true pathogen when the conditions in the GIT support its toxin production. However, it is often found in the intestine in relatively high numbers with no symptoms of necrotic enteritis or growth suppression.

Nearly all bacterial species belonging to the bacterial class *Clostridia* are non-pathogenic and even beneficial. Recently, many novel species belonging to this class have been given fresh genus and species names with no reference to clostridia. Still, scientifically they are members of the class *Clostridia* and share significant 16S rRNA sequence homology. Strangely enough, most of the pathogenic clostridia belong to the clostridial cluster I, and are only distantly related to the bulk of clostridia. In general, one should consider the abundance of clostridia as a positive health indicator and worry only when the abundance of the cluster I exceeds the detection limit in sequencing surveys.

REFERENCES

Apajalahti, J. and Kettunen, A. (2006) Rational development of novel microbial modulators. In: Barug, D., de Jong, J., Kies, A.K. and Verstegen, M.W.A. (eds) *Antimicrobial Growth Promoters. Where do We Go From Here?* Wageningen Academic Publishers, Wageningen, Netherlands.

Apajalahti, J., Sarkilahti, L.K., Maki, B.R., Heikkinen, J.P., Nurminen, P.H. and Holben, W.E. (1998) Effective recovery of bacterial DNA and percent-guanine-plus-cytosine- based analysis of community structure in the gastrointestinal tract of broiler chickens. *Applied and Environmental Microbiology* 64, 4084–4088.

Apajalahti, J., Kettunen, A., Bedford, M.R. and Holben, W.E. (2001) Percent G+C profiling accurately reveals diet-related differences in the gastrointestinal microbial community of broiler chickens. *Applied and Environmental Microbiology* 67, 5656–5667.

Apajalahti, J., Kettunen, H., Kettunen, A., Holben, W.E., Nurminen, P.H., Rautonen, N. and Mutanen, M. (2002) Culture-independent microbial community analysis reveals that insulin in the diet primarily affects previously unknown bacteria in the mouse caecum. *Appied and Environmental Microbiology* 68, 4986–4995.

Apajalahti, J., Kettunen, A. and Graham, H. (2004) Characteristics of the gastrointestinal microbial communities, with special reference to the chicken. *World's Poultry Science Journal* 60, 223–232.

Butaye, P., Devriese, L.A. and Haesebrouck, F. (2003). Antimicrobial growth promoters used in animal feed: effects of less well known antibiotics on gram-positive bacteria. *Clinical Microbiology Reviews* 16, 175–188.

Cole, J.R., Chai, B., Farris, R.J., Wang, Q., Kulam, S.A., McGarrell, D.M., Garrity, G.M. and Tiedje, J.M. (2005) The Ribosomal Database Project (RDP-II): sequences and tools for high-throughput rRNA analysis. *Nucleic Acids Research* 33, D294–D296.

Collins, M.D., Lawson, P.A., Willems, A., Cordoba, J.J., Fernandez-Garayzabal, J., Garcia, P., Cai, J., Hippe, H. and Farrow, J.A. (1994) The phylogeny of the genus *Clostridium*: proposal of five new genera and eleven new species combinations. *International Journal of Systematic Bacteriology* 44, 812–826.

Holben, W.E., Sarkilahti, L.K., Williams, P., Saarinen, M. and Apajalahti, J. (2002) Phylogenetic analysis of intestinal microflora indicates a novel *Mycoplasma* phylotype in farmed and wild salmon. *Microbiological Ecology* 44, 175–185.

Holben, W.E., Feris, K.P., Kettunen, A. and Apajalahti, J. (2004) GC fractionation enhances microbial community diversity assessment and detection of minority populations of bacteria by denaturing gradient gel electrophoresis. *Applied and Environmental Microbiology* 70, 2263–2270.

Knarreborg, A., Simon, M.A., Engberg, R.M., Jensen, B.B. and Tannock, G.W. (2002) Effects of dietary fat source and subtherapeutic levels of antibiotic on the bacterial community in the ileum of broiler chickens at various ages. *Applied and Environmental Microbiology* 68, 5918–5924.

Lee, M.D. (2006) Molecular basis for AGP effects in animals. In: Barug, D., de Jong, J., Kies, A.K. and Verstegen, M.W.A. (eds) *Antimicrobial Growth Promoters. Where do we Go From Here?* Wageningen Academic Publishers, Wageningen, Netherlands.

Lu, J., Idris, U., Harmon, B., Hofacre, C., Maurer, J.J. and Lee, M.D. (2003) Diversity and succession of the intestinal bacterial community of the maturing broiler chicken. *Applied and Environmental Microbiology* 69, 6816–6824.

Zhu, X.Y., Zhong, T., Pandya, Y. and Joerger, R.D. (2002) 16S rRNA-based analysis of microbiota from the caecum of broiler chickens. *Applied and Environmental Microbiology* 68, 124–137.

CHAPTER 9
Mechanisms of pathogen control in the avian gastrointestinal tract

A.M. Donoghue,[1]* M.B. Farnell,[1] K. Cole[2] and D.J. Donoghue[2]

[1]Poultry Production and Product Safety Research Unit, ARS, USDA, Fayetteville, Arkansas, USA; [2]Department of Poultry Science, University of Arkansas, Fayetteville, Arkansas, USA; *e-mail: donoghue@uark.edu

ABSTRACT

The environment of the avian enteric system is complex and dynamic. With many different types of bacteria and up to 10^{12} viable bacteria/g of digesta, how does the gut maintain a commensal microflora while protecting itself from harmful organisms? Mechanisms by which pathogens are thought to be controlled in the gut include: (i) physiological barriers such as mucin and pH; (ii) competition for enteric attachment sites; (iii) competition with the host and other microbes for nutrients, such as short-chain fatty acids; (iv) production of antimicrobial compounds or agents, including bacteriocins and bacteriophage; and (v) stimulation of the immune system.

The intestinal mucosa functions as a primary host defence against the constant presence of antigens from food and the wide variety of beneficial and harmful microorganisms in the gut lumen. There is emerging evidence of a sophisticated 'cross-talk' between the enteric microflora and the host, as well as similar communication between the microbial inhabitants in the intestinal lumen. Understanding the interactions between commensal and pathogenic bacteria and the avian gastrointestinal tract will enhance the discovery of new ways to exploit the potential benefits of this complex and dynamic environment.

INTRODUCTION

The environment of the avian enteric system is complex and dynamic. With the likelihood of more than 500 different types of bacteria and the presence of up to 10^{12} viable bacteria/g of digesta, how does the gut maintain a commensal microflora while protecting itself from harmful organisms?

© CAB International 2006. *Avian Gut Function in Health and Disease* (ed. G.C. Perry)

Mechanisms controlling pathogens in the gut involve: (i) the structure, physiology and environment of the enteric system; (ii) competition for enteric attachment sites; (iii) competition with the host and other microbes for nutrients; (iv) production of antimicrobial compounds or agents such as bacteriocins and bacteriophage; and (v) stimulation of the immune system. However, some pathogenic bacteria can adapt to this hostile environment to colonize and invade the gut mucosa. Because of the dynamic state of the environment and the complexity of the interactions involved, it has been difficult to understand how different factors influence the physiology of the gut and limit pathogenic influence.

Furthermore, the characteristics that determine whether a microbe is a commensal or a pathogen are unclear; for example, certain strains of *Salmonella* can be highly infectious in poultry, whereas others act as commensals. *Campylobacter*, another foodborne pathogen, can cause illness in humans from an infective dose as low as 500 cells, whereas no sign of disease is observed in poultry colonized with 10^{10} cfu/g of caecal content. This chapter will explore the mechanisms that operate in the gut to protect the bird from pathogenic bacteria and consider briefly how some pathogens have developed strategies to survive in such a dynamic environment.

GUT MUCOSAL BARRIER

Key physico-chemical properties

The enteric system has evolved in order to balance the digestion of nutrients with protection of the host from undesirable organisms by both physical and chemical means. The mucosal layer provides a physical and a physiological barrier. Epithelial cells have an apical membrane covered by a glycocalyx and tight junctions between adjacent cells that restrict penetration by pathogens (Hecht, 1995). Segments of the gastrointestinal tract (GIT) provide an unfavourable physico-chemical environment for bacterial pathogens.

For example, the pH of the crop, proventriculus and gizzard can be highly acidic (pH range 2.5–4.8; Denbow, 2000), which is inhibitory to the colonization of acid-sensitive bacteria like *Salmonella*. The lower gut is less acidic and more favourable to many organisms, with the largest populations of anaerobic bacteria residing in the caeca. Even in these regions, however, lactic acid-producing bacteria (e.g. *Lactobacillus*, *Bifidobacterium* spp.) can reduce gut pH, thereby inhibiting growth of various pathogens (Rolf, 1991).

The role of mucins in pathogen protection

Gastrointestinal mucin is a component of the luminal barrier and an integral line of defence against invading pathogens (Moncada *et al.*, 2003). The integrity of the mucosal layer is critical in host resistance to enteric disease (Mantle and Allen-Vercoe, 1989). The epithelium of the intestinal tract is covered by a layer of mucous that is composed predominantly of mucin glycoproteins, which are synthesized and secreted by goblet cells, forming a gel that adheres to the mucosal surface (Forstner *et al.*, 1995). This layer acts as a barrier between the luminal contents and the absorptive system of the intestine, and protects the mucosal surface from exogenous or endogenous luminal irritants.

Expression of various mucins, defined by differences in either their protein backbones or glycosylation patterns, varies both between and within tissues (Gendler and Spicer, 1995). Mucin genes are regulated at the transcriptional level by cytokines (IL-4), bacterial metabolic products and growth factors (TGFα; Temann *et al.*, 1997). Mucin biosynthesis is also influenced by conditions or agents that affect the differentiation of precursor cells into mature goblet cells, as well as by those that uncouple the processes of glycosylation and protein synthesis or which influence protein synthesis generally, such as fasting or malnutrition (Sharma *et al.*, 1997; Langhout *et al.*, 1999).

Presence of this mucous layer may prevent bacterial translocation, since pathogens must first pass through the mucous layer in order to adhere to and invade the epithelial cells. Reducing mucin production in rats by starvation allowed a rapid translocation of *S. typhimurium* through the Peyer's patches (Sakamoto *et al.*, 2004). The significant atrophy of villi and reduction in mucous production that were observed provide evidence of the importance of the mucosal barrier in inhibiting bacterial translocation.

Mucin can also serve as a substrate for fermentation by commensal bacteria. Gusils *et al.* (2003) demonstrated that *Lactobacillus* spp. could adhere to purified intestinal mucin from chicken and Jonsson *et al.* (2001) found that the addition of mucin to the growth medium initiated mucin-binding properties in several strains of *Lactobacillus*. Some studies have described mucin as a site for bacterial adhesion (Vimal *et al.*, 2000), with subsequent competition between pathogenic and other bacteria (Craven and Williams, 1998; Pascual *et al.*, 1999).

The native microflora and certain pathogenic bacteria are also involved in mucin turnover by stimulating mucin gene expression (Dohrman *et al.*, 1998) and also by producing mucin-degrading enzymes (Deplancke *et al.*, 2002). For example, the human pathogen *Helicobacter pylori* uses mucous from the stomach to protect itself from the acidic environment, attaching to the mucosal cell surface by producing phospholipases, which dissolve the phospholipid layer (Bengmark and Jeppsson, 1995). In addition, this organism utilizes mucin-degrading glyocosidases and glucosulphatases to destroy glycoproteins in the mucous, leading to inflammation and ulceration (Slomiany *et al.*, 1987).

Bismuth compounds, an effective treatment for *H. pylori*-induced ulcers in humans, inhibit bacterial enzymes that *H. pylori* utilizes to obtain nutrients and to penetrate the mucosal layer. We have also used these compounds, with some success, to reduce caecal colonization of broilers with *C. jejuni* (Farnell *et al.*, 2006a), suggesting that manipulation of enzymes in the mucosal layer may be an effective strategy for reducing pathogen colonization of the GIT.

Mucin dynamics can also be altered by dietary supplements. Fernandez *et al.* (2000) fed chickens a diet containing xylanase, which reduces *C. jejuni* colonization by decreasing the viscosity of caecal mucous, altering gut transit time and possibly by 'flushing' *C. jejuni* from the GIT. Smirnov *et al.* (2004) observed that goblet cell density was greater in the ileum and jejunum and mucin glycoprotein levels were lower in the duodenum of chicks fed antibiotic growth promoters. In the same study, use of probiotics increased the goblet cell cup in the lower intestines of chicks.

In vitro studies utilizing *L. plantarum* 299v demonstrated the ability of probiotics to inhibit enteropathogenic organisms by inducing the expression of intestinal mucin genes (Mack *et al.*, 1999). Thus, mucin plays an important role in protecting the gut from pathogens.

THE ROLE OF MICROFLORA IN PATHOGENIC PROTECTION

Maturation of the enteric system and the influence of microflora on the development of pathogen resistance

The indigenous microflora is a key component in protecting the gut from invasion by pathogens. This is one of the best-known phenomena in relation to the control of enteric pathogens. The GIT of the newly hatched neonate is sterile and highly susceptible to pathogen colonization, whereas the mature bird is much more resistant to colonization (Nurmi and Rantala, 1973; Mead, 1998).

Several studies have shown that a newly hatched chick can be infected with only a few cells of *Salmonella,* whereas the older bird has a mature microflora capable of resisting pathogens (see review by Barrow, 2000). The protective influence of maternal transfer of enteric microflora is known for various warm-blooded species, including humans. Unfortunately, in many poultry operations, transfer of microflora from the hen to her offspring no longer occurs, because chicks are raised separately from parent flocks.

The concept of accelerating development of the normal enteric microflora, thereby increasing the resistance of young poultry to infection, was first described by Nurmi and Rantala (1973). These researchers collected microflora from the alimentary tract of mature chickens and used it to inoculate newly hatched chicks, thereby reducing considerably *Salmonella* colonization. This strategy has been called 'competitive exclusion', 'the Nurmi effect' or 'probiotic supplementation' and, subsequently, numerous studies have demonstrated reductions in *Salmonella* colonization of poultry using mixed, undefined enteric cultures (Pivnick and Nurmi, 1982; Mead and Impey, 1986; Bailey, 1987; Stavric and D'Aoust, 1993; Mead, 2000, 2002; Nisbet, 2002).

Several mechanisms have been proposed for the protection provided by the enteric microflora and therefore by effective competitive exclusion cultures (Nurmi *et al.*, 1992; Corrier and Nisbet, 1999), including competition for binding sites, competition for nutrients, production of antibacterial substances and immuno-stimulation. Native microflora may competitively exclude pathogenic bacteria by blocking potential adherence sites on gastrointestinal epithelia, thus increasing resistance to *Salmonella*. Protection by this mechanism is thought to be primarily physical, rather than involving any immediate change in bacterial metabolism, because the effect is so rapid.

Evidence of competition for unspecified receptors is demonstrated by the mat of microbial cells and interconnecting fibres of the glycocalyx that form a physical barrier in the GIT of older birds. This has also been observed in chicks after the oral administration of microbial preparations (Mead, 2000).

Several investigators have attempted to exploit and improve the competitive exclusion phenomenon by mimicking properties of efficacious bacteria, using defined cultures or by measuring beneficial effects within the GIT. For example, Schoeni and Doyle (1992) isolated caecum-colonizing bacteria that produced anti-*Campylobacter* metabolites from *C. jejuni*-free hens and demonstrated that these isolates could protect chicks against a subsequent challenge with *C. jejuni*. Attempts were made to focus on mechanisms of *Campylobacter* colonization in birds, especially the chemo-attraction of *Campylobacter* to mucin and the ability of the organism to use mucin as a growth substrate. Bacteria were isolated that could grow on mucin as a sole substrate and were presumed to occupy the same niche in the caecum as *Campylobacter*.

In other studies, bacterial strains isolated from washed caeca were shown to possess hydrophobic properties and their use improved the efficacy of competitive exclusion cultures in chickens (Stavric and D'Aoust, 1993). Hydrophobicity of the outer bacterial surface has also been implicated in attachment of the organisms to host tissue (Kiely and Olson, 2000).

A competitive exclusion culture was developed from the microflora occurring in the same niche as that occupied by *Campylobacter*, using scrapings of intestinal mucosa (Stern, 1994). Koenen *et al.* (2004b) developed a method for *in vitro* selection of lactic acid bacteria with immuno-modulating properties in chickens.

Together with others, we have developed methods of selecting candidate microbes on their ability to out-compete foodborne pathogens grown *in vitro* (Hargis *et al.*, 2003; Donoghue *et al.*, 2004a, b). We identified 137 isolates capable of competing successfully *in vitro* against *S. enteritidis*, from a pool of more than 8 million isolates. These cultures have demonstrated prophylactic and therapeutic activity against *Salmonella* (Bielke *et al.*, 2003; Tellez *et al.*, 2005) and, to some extent, *Campylobacter* (Donoghue *et al.*, 2004b) in chickens and turkeys.

Competition for nutrients

Competition for essential nutrients between native microflora and pathogenic microbes may be a limiting factor in colonization of the gut by invading pathogens (Lan *et al.*, 2005). Iron is an essential nutrient for all living organisms and the acquisition of iron is an important aspect of bacterial colonization (Ratledge and Dover, 2000; van Vliet *et al.*, 2002; Ho *et al.*, 2004). Limited availability of iron delays the growth of bacteria and triggers their expression of virulence factors (Litwin and Calderwood, 1993). In contrast, excessive iron uptake leads to iron toxicity and oxidative stress, which halts bacterial growth (Touati, 2000).

In host tissues, most of the iron is stored intracellularly, bound by transferrin in serum or bound at mucosal surfaces by lactoferrin (Woolridge and Williams, 1993). Therefore, the resulting availability of free iron in the host is extremely limited. In response to this limited availability of iron, many Gram-negative bacteria – such as *Salmonella* – produce small, iron-chelating molecules called siderophores that have a high affinity for ferric iron (Kingsley *et al.*, 1995; Neilands, 1995). Although *Campylobacter* spp. possess transport systems for siderophores, these organisms are unable to synthesize iron chelators (Field *et al.*, 1986). It has been reported, however, that *Campylobacter* spp. can acquire iron by using siderophores produced by other bacteria (Pickett *et al.*, 1992).

Many *in vitro* studies have shown that limiting the availability of iron in culture media can inhibit growth of certain strains of *E. coli*, *Salmonella* spp. and *Campylobacter* spp. (Field *et al.*, 1986; Palyada *et al.*, 2004; Holmes *et al.*, 2005). The iron content can be limited by adding natural or synthetic iron chelators, such as ovotransferrin, lactoferrin or EDTA, to the medium (Chart and Rowe, 1993; Lisiecki *et al.*, 2000; Ho *et al.*, 2004).

Competition for nutrients other than iron – such as the amino acids serine, threonine, aspartic acid and arginine – has also been suggested as a means by which the intestinal microflora could inhibit invading pathogens (Ushijima and Seto, 1991). For example, limiting the serine concentration *in vitro* slowed the growth of *S. typhimurium*, but not that of selected organisms from the chicken caecal microflora (Ha *et al.*, 1995).

ANTIBACTERIAL PROPERTIES OF SHORT-CHAIN FATTY ACIDS

Short-chain fatty acids (SCFA) are antibacterial organic acids produced by organisms such as *Lactobacillus* and *Bifidobacterium* spp. during fermentation of indigestible carbohydrates in the GIT (Topping and Clifton, 2001; Ricke, 2003; Miller, 2004). The predominant SCFA present in the GIT are acetic, propionic and butyric acids, which are all weak acids with pKa values of approximately 4.8 (Topping and Clifton, 2001).

SCFA increase from undetectable levels in the caeca of day-old broilers to the highest concentrations at 15 days of age (van der Wielen *et al.*, 2000). In the early stages of chick growth, concentrations of SCFA in the caeca are too low and the pH too high to prevent multiplication of *Salmonella*. As the microflora becomes more complex, concentrations of SCFA increase and conditions are then less favourable for *Salmonella* (Barnes *et al.*, 1979; Nisbet *et al.*, 1994, 1996; Corrier *et al.*, 1995).

SCFA have been reported to inhibit growth or reduce levels of *S. enteritidis*, *S. typhimurium*, *S. pullorum*, *E. coli*, *C. jejuni* and *C. coli* (van der Wielen *et al.*, 2000; Topping and Clifton, 2001; Chaveerach *et al.*, 2002; Ricke, 2003). Although not fully understood, the antibacterial mechanisms include bacteriostatic and bactericidal properties, depending on the physiological status of the bacteria and the physico-chemical characteristics of the external environment (see review, Ricke, 2003). Undissociated SCFA diffuse across bacterial lipid membranes, decreasing intracellular pH and causing cellular damage or death of those microbes that are sensitive to such conditions (van der Wielen *et al.*, 2000).

Furthermore, SCFA can also depress bacterial growth, since additional energy is required to return the internal pH of the cells to homeostatic levels (van Immerseel *et al.*, 2004a). Organic acids are also thought to interfere with cytoplasmic membrane structure and proteins, uncoupling electron transport systems and reducing subsequent ATP production (see review, Ricke, 2003). Other mechanisms attributed to organic acids include interference with nutrient transport, cytoplasmic leakage due to membrane damage, disruption of outer membrane permeability and anion accumulation (Ricke, 2003).

Many factors affect the efficacy of SCFA against pathogens, including diet, gut transit time, microflora and gut pH. SCFA are reported to be more bactericidal at a lower pH, probably due to the increased availability of undissociated acid in a low-pH environment (Corrier *et al.*, 1990; van Immerseel *et al.*, 2004a).

Although the addition of SCFA has been shown to decrease bacterial growth *in vitro*, researchers have had limited success in administering SCFA *in vivo*. The antibacterial effects of SCFA dissipate after reaching the crop (Thompson and Hinton, 1997), due in part to the fact that some organic acids are absorbed and never reach all locations in the GIT. For example, radio-labelled propionate added to feed did not reach the intestines or caeca in high enough concentrations to be effective, and was largely absorbed in the crop, gizzard and proventriculus (Hume *et al.*, 1993).

Van Immerseel *et al.* (2004b) hypothesized that the encapsulation of SCFA would allow these compounds to remain undissociated and reach the lower GIT. They found that chicks challenged with *S. enteritidis* and treated orally with encapsulated butyric acid had significantly less caecal colonization than had positive control birds. Al-Tarazi and Alshawabkeh (2003) reported that the administration of formic and propionic acids via feed reduced *S. pullorum*-related mortality in broilers by 58% and caecal colonization by 75%.

While the direct, oral administration of SCFA has had limited success, utilizing probiotics and/or prebiotics (non-digestible food ingredients that feed the enteric microflora, e.g. oligosaccharides) to increase SCFA is more effective. The effectiveness of butyric acid, the optimal acid for preventing bacterial invasion/colonization (van Immerseel *et al.*, 2004a), was increased by supplementing the diet with fructo-oligosaccharide and *Bifidobacterium* spp. (Le Blay *et al.*, 1999; Kaplan and Hutkins, 2000).

Corrier *et al.* (1990) fed dietary lactose to chickens in combination with a competitive exclusion preparation and demonstrated a significant increase in lactic acid, a decrease in caecal pH and a 3.5–4.0 log reduction in *S.* Typhimurium. The combined treatment also increased concentrations of acetic, propionic and butyric acids in the caeca of these birds. Kubena *et al.* (2001a, b) treated chicks with a competitive exclusion preparation and found that concentrations of propionic acid were increased and that treated birds had greater protection from caecal colonization by *S. typhimurium*.

BACTERIOCINS

Bacteriocins are peptides or proteins produced by bacteria that kill or inhibit the growth of other bacteria (Cleveland *et al.*, 2001). Bacteriocins have activity against a number of pathogenic, Gram-negative bacteria (Mota-Meira *et al.*, 2000; Arques *et al.*, 2004) and are one of the proposed mechanisms of action of competitive exclusion preparations (Nurmi and Rantala, 1973; Mead, 2000; Patterson and Burkholder, 2003). However, evidence for the role of bacteriocins in the gastrointestinal tract has been limited (Rolf, 1991).

Nevertheless, the administration of bacteriocins isolated from *L. salivarius* and *Paenibacillus polymyxa* reduced *Campylobacter* colonization to undetectable levels in the caeca of chickens (Stern *et al.*, 2005) and turkeys (Cole *et al.*, 2006), whereas 10^6 cfu/g of *Campylobacter* were detected in the caeca of positive control birds. In addition, we observed that administration of these bacteriocins significantly reduced crypt depth and goblet cell density in the duodenum of turkey poults (Cole *et al.*, 2006). Since *Campylobacter* preferentially colonizes mucin in the caecal crypts of poultry (Beery *et al.*, 1988; Meinersmann *et al.*, 1991), this may provide clues as to how bacteriocins alter the colonization sites and render them less suitable for *Campylobacter*.

INTERACTION OF GASTROINTESTINAL MICROFLORA WITH THE MUCOSAL IMMUNE SYSTEM

The intestine is a major part of the immune system, the impact of which on the avian GIT is discussed in detail by Davison *et al.* and Klasing (Chapters 6 and 14, respectively, this volume). The GIT can prevent enteric infections by secreting defensins and secretory IgA, while phagocytes and M cells continually survey the microflora of the lumen and lamina propria for potential pathogens (Sansonetti, 2004; Sugiarto and Yu, 2004). Microbial components of the native microflora and probiotics have both been shown to stimulate the avian immune system (Farnell *et al.*, 2003; Galdeano and Perdigon, 2004; Koenen *et al.*, 2004a) and to reduce organ invasion by *Salmonella in vivo* (Lowry *et al.*, 2005).

Heterophils, the avian equivalent of mammalian neutrophils, kill pathogens by the release of toxic oxygen metabolites (oxidative burst), lytic enzymes and antimicrobial peptides (degranulation). We have observed increases in oxidative burst and degranulation *in vitro* when neonatal chicken heterophils were stimulated with killed probiotic bacteria, suggesting that the avian immune response may be potentiated by appropriate manipulation of the gut microflora (Farnell *et al.*, 2006b).

VIRUSES AS ENTERIC DEFENCE AGENTS: BACTERIOPHAGES

Bacteriophages are small viruses that are found in the enteric microflora and are highly specific for target bacterial species (Gorski and Weber-Dabrowska,

2005). When lytic phages encounter an appropriate host bacterium, they attach to and penetrate the cell wall, replicate inside the cell and kill it. Although the Western world has largely ignored bacteriophages, human bacteriophage therapy is said to have been practised in Eastern Bloc countries for more than 50 years, with considerable success (Kutter, 1997; Alisky *et al.*, 1998).

The potential benefits of bacteriophage use in the poultry industry were first explored in 1925, when phages specific for *S. pullorum* were isolated from the faeces of hens that had recovered from a *pullorum* infection. Bacteriophages that were active against *S. enteritidis* and *C. jejuni* significantly reduced contamination of chicken and turkey skin (Goode *et al.*, 2003) and carcasses (Higgins *et al.*, 2005). Bacteriophages lytic for *S. typhimurium* and *S. enteritidis* and administered orally became established in the caeca, controlling colonization and reducing bird mortality (Berchieri *et al.*, 1991; Sklar and Joerger, 2001).

Similarly, caeca colonized following a natural challenge by *Campylobacter* had reduced levels of colonization in the presence of endemic bacteriophages (Connerton *et al.*, 2004; Attebury *et al.*, 2005). Bacteriophages may be important in the population dynamics of enteric bacteria and there is recent evidence of immuno-suppressive activity (Gorski and Weber-Dabrowska, 2005).

DYNAMICS OF HUMAN PATHOGENIC BACTERIA IN THE AVIAN GIT

Understanding the strategies by which zoonotic bacteria survive and adapt in the avian gut is important, because a major mode of carcass contamination occurs during processing, when edible meat is exposed to intestinal contents. *Campylobacter* and *Salmonella* are the most prevalent pathogens derived from poultry that infect humans through foodborne illness (CDC, 2004). These pathogens present an interesting challenge to intervention strategies, because only infrequently does colonization cause clinical disease in poultry (Hargis *et al.*, 2001).

Other foodborne pathogens, including *Listeria monocytogenes* and *Clostridium perfringens,* can also colonize the avian gut and are potentially pathogenic to humans. Enteric *E. coli* isolates from avian species tend to be non-pathogenic to humans; however, there is some evidence that chickens can be colonized by *E. coli* 0157:H7, a highly pathogenic organism (Stavric *et al.*, 1993; Best *et al.*, 2003).

Enteric bacteria must cross the adherent mucin layer to attach properly to the gut mucosa (Sharma *et al.*, 1997). The ability of bacteria to bind to the mucosal epithelium is influenced by flagellar function and other bacterial adherence factors. Bacteria have developed several strategies to penetrate the defensive barriers of the gut. Modifications of the fimbriae and flagella are associated with pathogenicity in certain strains of *Salmonella*. Type 1 fimbriae mediate mannose-sensitive haemagglutination and are thought to play a role in the initiation of enteric colonization (Duguid and Campbell, 1969). Flagella also play a key role in *S. typhimurium* pathogenesis in chickens, since reduced motility diminishes the pathogenicity of this organism (Allen-Vercoe and Woodward, 1999).

In addition to adherence factors, *S. typhimurium* secretes exopolysaccharide to form a biofilm that facilitates attachment to chicken gut epithelium (Ledeboer and Jones, 2005). Mutations in the *wcaM* gene (partly responsible for colonic acid biosynthesis) and the *yhjN* gene (possibly involved in cellulose biosynthesis) were found to disrupt biofilm formation (Ledeboer and Jones, 2005) and may provide clues for reducing colonization.

Campylobacter colonization occurs primarily in the lower intestines, where the organism does not adhere directly to intestinal epithelial cells but locates mainly in the mucous layer of the caecal crypts (Beery *et al.*, 1988; Meinersmann *et al.*, 1991). Adherence of bacteria to mucosal surfaces is mediated by chemotactic factors (Freter and O'Brien, 1981) and bacterial adhesins, which can include fimbriae, flagella, capsules, glycocalyces, lipopolysaccharides and cell-associated lectins (Beachey, 1981). A polar flagellum confers motility on *Campylobacter* and the spiral shape of the cell allows the organism to move about in the viscous mucous layer (Newell *et al.*, 1985; Lee *et al.*, 1986).

Ziprin *et al.* (2001) demonstrated that the genes involved in adherence of *Campylobacter* to mucin in the GIT: (i) *cadF* (mutations prevent the production of fibronectin-binding protein); (ii) *dnaJ* (mutations reduce the organism's ability to respond well to thermal stressors); (iii) *ciaB* (mutants do not invade cells *in vitro*); and (iv) *pldA* (mutants do not produce proteins with phospholipase activity, thereby reducing their ability to break down the mucin layer) are important to caecal colonization in chickens.

These mutants were avirulent in chickens, even at large challenge doses. Recent evidence supports the pathogenicity of *Campylobacter* for humans, but not for poultry. *Campylobacter*-infected humans produce antibodies against cytolethal distending toxin, the best-characterized *Campylobacter* toxin, whereas no immunoglobulins were detected in chicks colonized asymptomatically (AbuOun *et al.*, 2005).

CONCLUSIONS

The mechanisms involved in excluding pathogens are diverse, complex and dynamic in the intestinal environment (Table 9.1). Synergistic interactions between the microflora and the avian host are likely to be important in controlling colonization by pathogenic bacteria. Future research to elucidate these complex interactions may lead to the discovery of new strategies in promoting intestinal integrity and the enhancement of avian health.

Table 9.1. Defence mechanisms of the avian gastrointestinal tract.

Mechanism	Mode of action	References
Physical barriers		
Mucin	Mucin secretion and type affect microflora	Deplanke and Gaskins, 2001
pH	Low pH of upper GIT inhibits growth of some enteric bacteria	Fooks and Gibson, 2002
Nutrient competition	Bacteria must compete with the GIT for nutrients	Bedford, 2000
Peristalsis	Movement of digesta and mucin prevents bacterial adherence	Montagne *et al.*, 2004
Oxygen tension	The anaerobic environment of the GIT inhibits some microbes	Rychlik and Barrow, 2005
Gut microflora		
Competition for adhesion sites	Bacteria compete for adhesion sites	Bailey, 1987
Nutrient competition	Bacteria compete for nutrients	Bedford, 2000
Bacteriocins	Antimicrobial compounds produced by other bacteria to inhibit competitors	Cole *et al.*, 2005; Stern *et al.*, 2005
Bacteriophages	Viruses that replicate within and lyse specific bacteria	Berchieri *et al.*, 1999
Short-chain fatty acids	Antimicrobial compounds that can inhibit the growth of some bacteria	Ricke, 2003
Competitive exclusion		
Mature microflora	Microflora from healthy adults protects neonates	Nurmi and Rantala, 1973
Continuous flow culture	Defined probiotic culture grown in a continuous-flow fermenter	Nisbet *et al.*, 1994, 1996
Mucosal scrapings	Microflora collected from mucosal scrapings that reduce *Campylobacter*	Stern, 1994
Immuno-modulation	Probiotic bacteria that stimulate an immune response	Koenen *et al.*, 2004a, b
Bactericidal compounds	Caecal bacteria that secrete metabolites bactericidal to *C. jejuni*	Schoeni and Doyle, 1992
In vitro competition	Enteric bacteria that outcompete pathogens *in vitro*	Bielke *et al.*, 2003; Donoghue *et al.*, 2004b
Mucosal immunity		
Immune surveillance	M cells and phagocytes constantly monitor the GIT for pathogens	MacPherson and Harris, 2004
Defensins	Antimicrobial peptides expressed in the villus crypts	Sugiarto and Yu, 2004
Secretory IgA	Secreted by B cells to bind to bacteria and prevent bacterial attachment	MacPherson *et al.*, 2000
Mucin secretion	Regulated by pattern-recognition receptors; flow and type affect microflora	Moncada *et al.*, 2003

REFERENCES

AbuOun, M., Manning, G. Cathraw, S.A., Ridley, A., Ahmed, I.H., Wassenaar, T.M. and Newell, D.G. (2005) Cytolethal distending toxin (CDT)-negative *Campylobacter jejuni* strains and anti-CDT neutralizing antibodies are induced during human infection but not during colonization in chickens. *Infection and Immunity* 73, 3053–3062.

Alisky, J., Iczkowski, K., Rapoport, A. and Troitsky, N. (1998) Bacteriophages show promise as antimicrobial agents. *Journal of Infection* 36, 5–15.

Allen-Vercoe, E. and Woodward, M.J. (1999) The role of flagella, but not fimbriae, in the adherence of *Salmonella enterica* serotype Enteritidis to chick gut explant. *Journal of Medical Microbiology* 48, 771–780.

Al-Tarazi, Y.H. and Alshawabkeh, K. (2003) Effect of dietary formic and propionic acids on *Salmonella* Pullorum shedding and mortality in layer chicks after experimental infection. *Journal of Veterinary Medicine* 50, 112–117.

Arques, J.L., Fernandez, J., Gaya, P., Nunez, M., Rodriguez, E. and Medina, M. (2004) Antimicrobial activity of reuterin in combination with nisin against foodborne pathogens. *International Journal of Food Microbiology* 95, 225–229.

Attebury, R.J., Dillon, E., Swift, C., Connerton, P.L., Frost, J.A., Dodd, C.E.R., Rees, C.E.D. and Connerton, I.F. (2005) Correlation of *Campylobacter* bacteriophage with reduced presence of hosts in broiler chicken caeca. *Applied and Environmental Microbiology* 71, 4885–4887.

Bailey, J.S. (1987) Factors affecting microbial competitive exclusion in poultry. *Food Technology* 41, 88–92.

Barnes, E.M., Impey, C.S. and Stevens, B.J. (1979) Factors affecting the incidence and anti-salmonella activity of the anaerobic caecal flora of the young chick. *Journal of Hygiene* 82, 263–283.

Barrow, P.A. (2000) The paratyphoid salmonellae. *Revue Scientifique et Technique* 19, 351–375.

Beachey, E.H. (1981) Bacterial adherence: adhesin–receptor interactions mediating the attachment of bacteria to mucosal surfaces. *Journal of Infectious Diseases* 143, 325–345.

Bedford, M.R. (2000) Exogenous enzymes in monogastric nutrition: their current value and future benefits. *Animal Feed Science and Technology* 86, 1–13.

Beery, J.T., Hughdahl, M.B. and Doyle, M.P. (1988) Colonization of the gastrointestinal tracts of chicks by *Campylobacter jejuni*. *Applied and Environmental Microbiology* 54, 2365–2370.

Bengmark, S. and Jeppsson, B. (1995) Gastrointestinal surface protection and mucosa reconditioning. *Journal of Parenteral and Enteric Nutrition* 19, 410–415.

Berchieri, A., Lovell, M.A. and Barrow, P.A. (1991) The activity in the chicken alimentary tract of bacteriophages lytic for *Salmonella* Typhimurium. *Research in Microbiology* 142, 541–549.

Best, A., La Ragione, R.M., Cooley, W.A., O'Connor, C.D., Velge, P. and Woodward, M.J. (2003) Interaction with avian cells and colonization of specific pathogen-free chicks by Shiga-toxin negative *Escherichia coli* O157:H7 (NCTC 12900). *Veterinary Microbiology* 93, 207–222.

Bielke, L.R., Elwood, A.L., Donoghue, D.J., Donoghue, A.M., Newberry, L.A., Neighbor, N.K. and Hargis, B.M. (2003) Approach for selection of individual enteric bacteria for competitive exclusion in turkey poults. *Poultry Science* 82, 1378–1382.

Centers for Disease Control (CDC) (2004) Preliminary FoodNet data on the incidence of infection with pathogens commonly transmitted through food – selected sites, USA, 2003. *Morbidity and Mortality Weekly Report, 30 April*, 53, 338–343.

Chart, H. and Rowe, B. (1993) Iron restriction and the growth of *Salmonella* Enteritidis. *Epidemiology and Infection* 110, 41–47.

Chaveerach, P., Keuzenkamp, D.A., Urlings, H.A.P., Lipman, L.J.A. and van Knapen, F. (2002) *In vitro* study on the effects of organic acids on *Campylobacter jejuni/coli* populations in mixtures of water and feed. *Poultry Science* 81, 621–628.

Cleveland, J., Montville, T.J., Nes, I.F. and Chikindas, M.L. (2001) Bacteriocins: safe, natural antimicrobials for food preservation. *International Journal of Food Microbiology* 71, 1–20.

Cole, K., Farnell, M.B., Donoghue, A.M., Stern, N.J., Svetoch, E.A., Eruslanov, B.N., Kovalev, Y.N., Perelygin, V.V., Levchuck, V.P., Reyes-Herrera, I., Blore, P.J. and Donoghue, D.J. (2006) Bacteriocins reduce *Campylobacter* colonization and alter gut architecture in turkey poults. *Poultry Science* 85, 1570–1575.

Connerton, P.L., Loc Carrillo, C.M., Swift, C., Dillon, E., Scott, A., Rees, C.E., Dodd, C.E.R., Frost, J. and Connerton, I.F. (2004) Longitudinal study of *Campylobacter jejuni* bacterio-phages and their hosts from broiler chickens. *Applied and Environmental Microbiology* 70, 3877–3883.

Corrier, D.E. and Nisbet, D.J. (1999) Competitive exclusion in control of *Salmonella enterica* serovar Enteritidis in laying poultry in: *Salmonella enterica* Enteritidis. In: Saeed, A.M., Gast, R.K., Potter, M.E. and Wall, P.G. (eds) *Humans and Animals: Epidemiology, Pathogenesis, and Control.* Iowa State University Press, Ames, Iowa, pp. 391–396.

Corrier, D.E., Hinton, A., Ziprin, R.L., Beier, R.C. and DeLoach, J.R. (1990) Effect of dietary lactose on caecal pH, bacteriostatic volatile fatty acids, and *Salmonella* Typhimurium colo-nization of broiler chicks. *Avian Diseases* 34, 617–625.

Corrier, D.E., Nisbet, D.J., Scanlan, C.M., Hollister, A.G. and DeLoach, J.R. (1995) Control of *Salmonella* Typhimurium colonization in broiler chicks with a continuous-flow characterized mixed culture of bacteria. *Poultry Science* 74, 916–924.

Craven, S.E. and Williams, D.D. (1998) *In vitro* attachment of *Salmonella* Typhimurium to chicken caecal mucus: effect of cations and pretreatment with *Lactobacillus* spp. isolated from the intestinal tracts of chickens. *Journal of Food of Protection* 61, 265–271.

Denbow, D.M. (2000) Gastrointestinal anatomy and physiology. In: Whittow G.C. (ed.) *Sturkies Avian Physiology.* Academic Press, San Diego, California, pp. 299–326.

Deplancke, B. and Gaskins, H.R. (2001) Microbial modulation of innate defense: goblet cells and the intestinal mucus layer. *American Journal of Clinical Nutrition* 73, 1131S–1141S.

Deplancke, B., Vidal, O., Ganessunker, D., Donovan, S.M., Mackie, R.I. and Gaskins, H.R. (2002) Selective growth of mucolytic bacteria including *Clostridium perfringens* in a neonatal piglet model of total parenteral nutrition. *American Journal of Clinical Nutrition* 76, 1117–1125.

Dohrman, A., Miyata, S., Gallup, M., Li, J.D., Chapelin, C.A., Coste, A.E., Escudier, E., Nadel, J. and Basbaum, C. (1998) Mucin gene (MUC 2 and MUC 5AC) up-regulation by Gram-positive and Gram-negative bacteria. *Biochimica et Biophysica Acta* 1406, 251–259.

Donoghue, A.M., Huff, W.E., Hargis, B.M., Tellez, G. and Donoghue, D.J. (2004a). Bacteriophage and probiotics: their role in the control of *Salmonella* in poultry. *Fifth Asian Pacific Poultry Health Proceedings*, Gold Coast, Australia, pp. 10–15.

Donoghue, D.J., Hargis, B.M., Tellez, G. and Donoghue, A.M. (2004b). Competitive exclusion as a means of controlling *Campylobacter* in poultry. *Fifth Asian Pacific Poultry Health Proceedings*, Gold Coast Australia, pp. 16–20.

Duguid, J.P. and Campbell, I. (1969) Antigens of the type-1 fimbriae of *Salmonellae* and other enterobacteria. *Journal of Medical Microbiology* 2, 535–553.

Farnell, M.B., He, H. and Kogut, M.H. (2003) Differential activation of signal transduction path-ways mediating oxidative burst by chicken heterophils in response to stimulation with lipopolysaccharide and lipoteichoic acid. *Inflammation* 27, 225–231.

Farnell, M.B., Donoghue, A.M., Cole, K., Reyes-Herrera, I., Solis de los Santos, F., Dirain, M.L.S., Blore, P.J., Pandya, K. and Donoghue, D.J. (2006a) Effect of oral administration of bismuth compounds on *Campylobacter* colonization in broilers. *Poultry Science* 85, 2009–2011.

Farnell, M.B., Donoghue, A.M., de los Santos, F.S., Blore, P.J., Tellez, G., Hargis, B.M. and Donoghue, D.J. (2006b) Upregulation of oxidative burst and degranulation in chicken heterophils stimulated with probiotic bacteria. *Poultry Science* 85, 1900–1906.

Fernandez, F., Sharma, R., Hinton, M. and Bedford, M.R. (2000) Diet influences the colonization of *Campylobacter jejuni* and distribution of mucin carbohydrates in the chick intestinal tract. *Cellular and Molecular Life Sciences* 57, 1793–1801.

Field, L.H., Headley, V.L., Payne, S.M. and Berry, L.J. (1986) Influence of iron on growth, morphology, outer membrane protein composition, and synthesis of siderophores in *Campylobacter jejuni*. *Infection and Immunity* 54, 126–132.

Fooks, L.J. and Gibson, G.R. (2002) Probiotics as modulators of the gut flora. *British Journal of Nutrition* 88, S39-S49.

Forstner, J.F., Oliver, M.G. and Sylvester, F.A. (1995) Production, structure and biologic relevance of gastrointestinal mucins. In: Blaser, J., Smith, P.D., Ravdin, J.I., Greenberg, H.B. and Guerrant, R.L. (eds) *Infections of the Gastrointestinal Tract*. Raven Press, New York, pp. 71–88.

Freter, R. and O'Brien, P.C.M. (1981) Role of chemotaxis in the association of motile bacteria with intestinal mucosa: fitness and virulence of nonchemotactic *Vibrio cholerae* mutants in infant mice. *Infection and Immunity* 34, 222–233.

Galdeano, C.M. and Perdigon, G. (2004) Role of viability of probiotic strains in their persistence in the gut and in mucosal immune stimulation. *Journal of Applied Microbiology* 97, 673–681.

Gendler, S.J. and Spicer, A.P. (1995) Epithelial mucin genes. *Annual Review of Physiology* 57, 607–634.

Goode, D., Allen, V.M. and Barrow, P.A. (2003) Reduction of experimental *Salmonella* and *Campylobacter* contamination of chicken skin by application of lytic bacteriophages. *Applied and Environmental Microbiology* 69, 5032–5036.

Gorski, A. and Weber-Dabrowska, B. (2005) The potential role of endogenous bacteriophages in controlling invading pathogens. *Cellular and Molecular Life Sciences* 62, 511–519.

Gusils, C., Oppezzo, O., Pizarro, R. and Gonzalez, S. (2003) Adhesion of probiotic *Lactobacilli* to chick intestinal mucus. *Canadian Journal of Microbiology* 49, 472–478.

Ha, S.A., Ricke, S.C., Nisbet, D.J., Corrier, D.E. and DeLoach, J.R. (1995) Serine utilization as a potential competition mechanism between *Salmonella* Typhimurium and a chicken caecal bacterium. *Journal of Food Protection* 57, 1074–1079.

Hargis, B.M., Caldwell, D.J. and Byrd, J.A. (2001) Microbiological pathogens: live poultry considerations. In: Sams, A.R. (ed.) *Poultry Meat Processing*. CRC Press, Boca Raton, Florida, pp. 121–135.

Hargis, B.M., Tellez, G.I., Nava, G., Donoghue, A.M., Vicente, J.L., Higgins, S.E., Donoghue, D.J. and Wolfenden, A.D. (2003) The role of beneficial microflora in controlling enteric bacterial diseases: probiotics, prebiotics, and competitive exclusion. *Cornell Nutrition Conference for Feed Manufactures*, East Syracuse, New York, 21–23 October, pp. 109–118.

Hecht, G. (1995) Bugs and barriers: enteric pathogens exploit yet another epithelial function. *Journal American Physiology* 10, 160–166.

Higgins, J.P., Higgins, S.E., Guenther, K.L., Huff, W., Donoghue, A.M., Donoghue, D.J. and Hargis, B.M. (2005) Use of a specific bacteriophage treatment to reduce *Salmonella* in poultry products. *Poultry Science* 84, 1141–1145.

Ho, W.L., Yu, R.C. and Chou, C.C. (2004) Effect of iron limitation on the growth and cytotoxin production of *Salmonella choleraesuis* SC-5. *International Journal of Food Microbiology* 90, 295–302.

Holmes, K., Mulholland, F., Pearson, B.M., Pin, C., McNicholl-Kennedy, J., Ketley, J.M. and Wells, J.M. (2005) *Campylobacter jejuni* gene expression in response to iron limitation and the role of fur. *Microbiology* 151, 243–257.

Hume, M.E., Corrier, D.E., Ambrus, S., Hinton, A. and DeLoach, J.R. (1993) Effectiveness of dietary propionic acid in controlling *Salmonella* Typhimurium colonization in broiler chicks. *Avian Diseases* 37, 1051–1056.

Jonsson, H., Strom, E. and Roos, S. (2001) Addition of mucin to the growth medium triggers mucus-binding activity in different strains of *Lactobacillus reuteri in vitro*. *FEMS Microbiology Letters* 204, 19–22.

Kaplan, H. and Hutkins, R.W. (2000) Fermentation of fructooligosaccharides by lactic acid bacteria and bifidobacteria. *Applied and Environmental Microbiology* 66, 2682–2684.

Kiely, L.J. and Olson, N.F. (2000) The physiochemical surface characteristics of *Lactobacillus casei*. *Food Microbiology* 17, 277–291.

Kingsley, R., Rabsch, W., Stephens, P., Roberts, M., Reissbrodt, R. and Williams, P.H. (1995) Iron-supplying systems of *Salmonella* in diagnostics, epidemiology and infection. *FEMS Immunology and Medical Microbiology* 11, 257–267.

Koenen, M., van der Hulst, E., Leering, R.M., Jeurissen, S.H.M. and Boersma, W.J.A. (2004a) Development and validation of a new *in vitro* assay for selection of probiotic bacteria that express immune-stimulating properties in chickens *in vivo*. *FEMS Immunology and Medical Microbiology* 40, 119–127.

Koenen, M.E., Kramer, J., van der Hulst, R., Heres, L., Jeurissen, S.H.M. and Boersma, W.J.A. (2004b) Immunomodulation by probiotic lactobacilli in layer- and meat-type chickens. *British Poultry Science* 45, 355–366.

Kubena, L.F., Bailey, R.H., Byrd, J.A., Young, C.R., Corrier, D.E., Stanker, L.H. and Rottinghaus, G.E. (2001a) Caecal volatile fatty acids and broiler chick susceptibility to *Salmonella* Typhimurium colonization as affected by aflatoxins and T-2 toxin. *Poultry Science* 80, 411–417.

Kubena, L.F., Byrd, J.A., Young, C.R. and Corrier, D.E. (2001b) Effects of tannic acid on caecal volatile fatty acids and susceptibility to *Salmonella* Typhimurium colonization in broiler chicks. *Poultry Science* 80, 1293–1298.

Kutter, E. (1997) Phage therapy: bacteriophages as antibiotics. http://www.evergreen.edu/phage/phagetherapy/phagetherapy.htm (accessed 12 August 2005).

Lan, Y., Verstegen, M.W.A., Tamminga, S. and Williams, B.A. (2005) The role of the commensal gut microbial community in broiler chickens. *World's Poultry Science Journal* 61, 95–104.

Langhout, D.J., Schutte, J.B., Van Leeuwen, P.V., Wiebenga, J. and Tamminga, S. (1999) Effect of dietary high- and low-methylated citrus pectin on the activity of the ileal microflora and morphology of the small intestinal wall of broiler chicks. *British Poultry Science* 40, 340–347.

Le Blay, G., Michel, C., Blottiere, H.M. and Cherbut, C. (1999) Prolonged intake of fructo-oligosaccharides induces a short-term elevation of lactic acid-producing bacteria and persistent increase in caecal butyrate in rats. *Journal of Nutrition* 129, 2231–2235.

Ledeboer, N.A. and Jones, B.D. (2005) Exopolysaccharide sugars contribute to biofilm formation by *Salmonella enterica* serovar Typhimurium on HEp-2 cells and chicken intestinal epithelium. *Journal of Bacteriology* 187, 3214–3226.

Lee, A., O'Rourke, J.L., Barrington, P.J. and Trust, T.J. (1986) Mucus colonization as a determinant of pathogenicity in intestinal infection by *Campylobacter jejuni*: a mouse caecal model. *Infection and Immunity* 51, 536–546.

Lisiecki, P., Wysocki, P. and Mikucki, J. (2000) Susceptibility of *Staphylococci* to natural and synthetic iron chelators. *Medycyna Doswiadczalna i Mikrobiologia* 52, 103–110.

Litwin, C.M. and Calderwood, S.B. (1993) Role of iron in regulation of virulence genes. *Clinical Microbiology Reviews* 6, 137–149.

Lowry, V.K., Farnell, M.B., Ferro, P.J., Swaggerty, C.L., Bahl, A. and Kogut, M.H. (2005) Purified beta-glucan as an abiotic feed additive up-regulates the innate immune response in immature chickens against *Salmonella enterica* serovar Enteritidis. *International Journal of Food Microbiology* 98, 309–318.

Mack, D.R., Michail, S., Wei, S., McDougall, L. and Hollingsworth, M.S. (1999) Probiotics inhibit enteropathogenic *E. coli* adherence *in vitro* by inducing intestinal mucin gene expression. *American Journal of Physiology* 274, G941–G950.

MacPherson, A.J. and Harris, N.L. (2004) Interactions between commensal intestinal bacteria and the immune system. *Nature Reviews in Immunology* 4, 478–485.

MacPherson, A.J., Dominique, G., Sainsbury, E., Harriman, G.R., Hensgartner, H. and Zinkernagel, R.M. (2000) A primitive T cell-independent mechanism of intestinal mucosal IgA responses to commensal bacteria. *Science* 288, 2222–2226.

Mantle, M. and Allen-Vercoe, A. (1989) Gastrointestinal mucus. In: Davison, T.F. (ed.) *Gastrointestinal Secretion.* J.S. University Press, London, pp. 202–229.

Mead, G.C. (1998) Critical control of major pathogens in poultry. In: Hakkinen, M., Nuotio, L., Nurmi, E. and Mead, G.C. (eds) *Pathogenic Micro-organisms in Poultry and Eggs: Safe Chicken for the Next Century.* COST Action 97, Luxembourg, pp. 6–12.

Mead, G.C. (2000) Prospects for 'competitive exclusion' treatment to control salmonellas and other foodborne pathogens in poultry. *Veterinary Journal* 159, 111–123.

Mead, G.C. (2002) Factors affecting intestinal colonization of poultry by *Campylobacter* and role of microflora in control. *World's Poultry Science Journal* 58, 169–178.

Mead, G.C. and Impey, C.S. (1986) Current progress in reducing *Salmonella* colonization of poultry by 'competitive exclusion'. *Society for Applied Bacteriology Symposium Series* 15, 67S–75S.

Meinersmann, R.J., Rigsby, W.E., Stern, N.J., Kelley, L.C., Hill, J.E. and Doyle, M.P. (1991) Comparative study of colonizing and noncolonizing *Campylobacter jejuni. American Journal of Veterinary Research* 52, 1518–1522.

Miller, S.J. (2004) Cellular and physiological effects of short-chain fatty acids. *Mini Reviews in Medicinal Chemistry* 4, 839–845.

Moncada, D.M., Kammanadiminti, S.J. and Chadee, K. (2003) Mucin and Toll-like receptors in host defence against intestinal parasites. *Trends in Parasitology* 19, 305–311.

Montagne, L., Piel, M.S. and Lallès, J.P. (2004) Effect of diet on mucin kinetics and composition: nutrition and health implications. *Nutrition Reviews* 62, 105–144.

Mota-Meira, M., LaPointe, G., Lacroix, C. and Lavoie, M.C. (2000) MICs of mutacin B-Ny266, nisin A, vancomycin, and oxacillin against bacterial pathogens. *Antimicrobial Agents and Chemotherapy* 44, 24–29.

Neilands, J.B. (1995) Siderophores: structure and function of microbial iron transport compounds. *Journal of Biological Chemistry* 270, 26723–26726.

Newell, D.G., McBride, H. and Dolby, J.M. (1985) Investigations on the role of flagella in the colonization of infant mice with *Campylobacter jejuni* and attachment of *Campylobacter jejuni* to human epithelial cell lines. *Journal of Hygiene* 95, 217–227.

Nisbet, D. (2002) Defined competitive exclusion cultures in the prevention of enteropathogen colonization in poultry and swine. *Antonie van Leeuwenhoek* 81, 481–486.

Nisbet, D.J., Corrier, D.E., Scanlan, C.M., Hollister, A.G., Beier, R.C. and DeLoach, J.R. (1994) Effect of dietary lactose and cell concentration on the ability of a continuous flow-derived bacterial culture to control *Salmonella* caecal colonization in broiler chickens. *Poultry Science* 73, 56–62.

Nisbet, D.J., Corrier, D.E., Ricke, S., Hume, M.E., Byrd, J.A. and DeLoach, J.R. (1996) Maintenance of the biological efficacy in chicks of a cecal competitive-exclusion culture against *Salmonella* by continuous-flow fermentation. *Journal of Food Protection* 59, 1279–1283.

Nurmi, E. and Rantala, M. (1973) New aspects of *Salmonella* infection in broiler production. *Nature* 241, 210–211.

Nurmi, E., Nuotio, L. and Schneitz, C. (1992) The competitive exclusion concept: development and future. *International Journal of Food Microbiology* 15, 237–240.

Palyada, K., Threadgill, D. and Stintzi, A. (2004) Iron acquisition and regulation in *Campylobacter jejuni. Journal of Bacteriology* 186, 4714–4729.

Pascual, M., Hugas, M., Badiola, J.I., Monfort, J.M. and Garriga, M. (1999) *Lactobacillus salivarius* CTC2197 prevents *Salmonella* Enteritidis colonization in chickens. *Applied and Environmental Microbiology* 65, 4981–4986.

Patterson, J.A. and Burkholder, K.M. (2003) Application of prebiotics and probiotics in poultry production. *Poultry Science* 82, 627–631.

Pickett, C.J., Auffenberg, T., Pesci, E.C., Sheen, V.L. and Jusuf, S.S. (1992) Iron acquisition and hemolysin production by *Campylobacter jejuni*. *Infectious Immunity* 60, 3872–3877.

Pivnick H. and Nurmi, E. (1982) The Nurmi concept and its role in the control of *Salmonellae* in poultry. In: Davies R. (ed.) *Developments in Food Microbiology 1.* Applied Sciences Publishers Ltd., Essex, England, pp. 41–70.

Ratledge, C. and Dover, L.G. (2000) Iron metabolism in pathogenic bacteria. *Annual Review of Microbiology* 54, 881–941.

Ricke, S.C. (2003) Perspectives on the use of organic acids and short chain fatty acids as antimicrobials. *Poultry Science* 82, 632–639.

Rolf, R.D. (1991) Population dynamics of the intestinal tract. In: Blankenship, L.C. (ed.) *Colonization Control of Human Enteropathogens in Poultry.* Academic Press, San Diego, California, pp. 59–76.

Rychlik, I. and Barrow, P.A. (2005) *Salmonella* stress management and its relevance to behaviour during intestinal colonization and infection. *FEMS Microbiology Reviews* 29, 1021–1040.

Sakamoto, K., Mori, Y., Takagi, H., Iwata, H., Yamada, T., Futamura, N., Sago, T., Ezaki, T., Kawamura, Y. and Hirose, H. (2004) Translocation of *Salmonella* Typhimurium in rats on total parenteral nutrition correlates with changes in intestinal morphology and mucus gel. *Nutrition* 20, 372–376.

Sansonetti, P.J. (2004) War and peace at mucosal surfaces. *Nature Reviews in Immunology* 4, 953–964.

Schoeni, J.L. and Doyle, M.P. (1992) Reduction of *Campylobacter jejuni* colonization of chicks by caecum-colonizing bacteria producing anti-*C. jejuni* metabolites. *Applied and Environmental Microbiology* 58, 664–670.

Sharma, R., Fernandez, F., Hinton, M. and Schumacher, U. (1997) The influence of diet on the mucin carbohydrates in the chick intestinal tract. *Cellular and Molecular Life Sciences* 53, 935–942.

Sklar, I.B. and Joerger, R.D. (2001) Attempts to utilize bacteriophage to combat *Salmonella enterica* serovar Enteritidis infection in chickens. *Journal of Food Safety* 21, 15–29.

Slomiany, B., Bilski, J., Sarosiek, J., Murty, V.L., Dworkin, B., Van Horn, K., Zielenski, J. and Slomiany, A. (1987) *Campylobacter pylordis* degrades mucin and undermines gastric mucosal integrity. *Biochemical and Biophysical Research Communications* 144, 307–314.

Smirnov, A., Sklan, D. and Uni, Z. (2004) Mucin dynamics in the chick small intestine are altered by starvation. *Journal of Nutrition* 134, 736–742.

Stavric, S., Buchanan, B. and Gleeson, T.M. (1993) Intestinal colonization of young chicks with *Escherichia coli* O157:H7 and other verotoxin-producing serotypes. *Journal of Applied Bacteriology* 74, 557–563.

Stavric, S. and D'Aoust, J.Y. (1993) Undefined and defined bacterial preparations for the competitive exclusion of *Salmonella* in poultry – a review. *Journal of Food Protection* 56, 173–180.

Stern, N.J. (1994) Mucosal competitive exclusion to diminish colonization of chickens by *Campylobacter jejuni*. *Poultry Science* 73, 402–407.

Stern, N.J., Svetoch, E.A., Eruslanov, B.V., Kovalev, Y.N., Volodina, L.I., Perelygin, V.V., Mitsevich, E.V., Mitsevich, I.P. and Levchuk, V.P. (2005) *Paenibacillus polyxma* purified bacteriocin to control *Campylobacter jejuni* in chickens. *Journal of Food Protection* 68, 1450–1453.

Sugiarto, H. and Yu, P.-L. (2004) Avian antimicrobial peptides: the defence role of beta defensins. *Biochemical and Biophysical Research Communications* 323, 721–727.

Tellez, G., Higgins, S.E., Donoghue, A.M. and Hargis, H.A. (2005) Digestive physiology and the role of the micro-organism. *Journal of Applied Poultry Research* 15, 136–144.

Temann, U.A., Prasad, B., Gallup, M.W., Basbaum, C., Ho, S.B., Flavell, R.A. and Rankin, J.A. (1997) A novel role for murine IL-4 *in vivo*: induction of MUC5AC gene expression and mucin hypersecretion. *American Journal of Respiratory Cell and Molecular Biology* 16, 471–478.

Thompson, J.L. and Hinton, M. (1997) Antibacterial activity of formic and propionic acids in the diet of hens on *Salmonellas* in the crop. *British Poultry Science* 38, 59–65.

Topping, D.L. and Clifton, P.M. (2001) Short-chain fatty acids and human colonic function: roles of resistant starch and nonstarch polysaccharides. *Physiological Reviews* 81, 1031–1064.

Touati, D. (2000) Iron and oxidative stress in bacteria. *Archives of Biochemistry and Biophysics* 373, 1–6.

Ushijima, T. and Seto, A. (1991) Selected faecal bacteria and nutrients essential for antagonism of *Salmonella* Typhimurium in anaerobic continuous flow cultures. *Journal of Medical Microbiology* 35, 111–117.

van der Wielen, P.W.J.J., Biesterveld, S., Notermans, S., Hofstra, H., Urlings, B.A.P. and van Knapen, F. (2000) Role of volatile fatty acids in development of the caecal microflora in broiler chickens during growth. *Applied and Environmental Microbiology* 66, 2536–2540.

van Immerseel, F., De Buck, J., De Smet, I., Pasmans, F., Haesebrouck, F. and Ducatelle, R. (2004a) Interactions of butyric acid- and acetic acid-treated *Salmonella* with chicken primary caecal epithelial cells *in vitro*. *Avian Diseases* 48, 384–391.

van Immerseel, F., Fievez, V., De Buck, J., Pasmans, F., Martel, A., Haesebrouck, F. and Ducatelle, R. (2004b) Microencapsulated short-chain fatty acids in feed modify colonization and invasion early after infection with *Salmonella* Enteritidis in young chickens. *Poultry Science* 83, 69–74.

van Vliet, A.H.M., Ketley, J.M., Park, S.F. and Penn, C.W. (2002) The role of iron in *Campylobacter* gene regulation, metabolism, and oxidative stress defense. *FEMS Microbiology Reviews* 26, 173–186.

Vimal, D.B., Khullar, M., Gupta, S. and Ganguly, N.K. (2000) Intestinal mucins: the binding sites for *Salmonella* Typhimurium. *Molecular and Cellular Biochemistry* 204, 107–117.

Woolridge, K.G. and Williams, P.H. (1993) Iron uptake mechanisms of pathogenic bacteria. *FEMS Microbiology Reviews* 12, 325–348.

Ziprin, R.L., Young, C.R., Byrd, J.A., Stanker, L.H., Hume, M.E., Gray, S.A., Kim, B.J. and Konkel, M.E. (2001) Role of *Campylobacter jejuni* potential virulence genes in caecal colonization. *Avian Diseases* 45, 549–557.

PART IV
Nutritional effects

Chapter 10
Effect of non-starch polysaccharidases on avian gastrointestinal function

M.R. Bedford

Zymetrics, Marlborough, UK; e-mail: Michael.Bedford@onetel.net

ABSTRACT

Non-starch polysaccharidases, which include cellulases and xylanases, are routinely used in poultry diets containing wheat, barley, oats, triticale and rye and, more recently, maize. The beneficial responses observed may be based on one or more of three proposed mechanisms of action, namely: (i) cereal endosperm cell wall hydrolysis; (ii) reduction in intestinal viscosity; and (iii) provision of fermentable oligomeric substrates as a result of cell wall hydrolysis. Since each mechanism elicits a specific series of responses at the intestinal level, and the extent of involvement of each will vary depending upon circumstances, there are a variety of potential responses observed. Changes in intestinal function may be driven by changes in nutrient supply and/or changes in the physico-chemical conditions (the viscosity of the intestinal contents, the nutrient-binding ability of the fibre fraction and the thickness of the unstirred water layer) of the intestine.

Nutrient supply encompasses not only improved nutrient extraction by the host, and thus reduced availability for the intestinal flora, but also the provision of oligomeric substrates through cell wall hydrolysis. As a result of such changes there may be subsequent, interrelated changes in the structure and size of the intestinal tract, in pancreatic enzyme secretion, in the quantity and chemistry of mucin secreted and in the quantity and species distribution of resident intestinal flora. The qualitative and quantitative changes in intestinal function observed are very much dependent upon the status in the absence of enzyme, which highlights the fact that responses upon use of enzymes are variable, being greatest in poor-quality diets.

INTRODUCTION

Enzymes which target the non-starch polysaccharides (NSP) of the cereals commonly used in poultry diets, be it wheat, barley, oats, triticale, rye or even maize, have traditionally been viewed as tools for improving the digestibility of a ration. The use of cellulase or xylanase-rich preparations in oats/barley- or wheat/rye/triticale-based diets, respectively, has been shown to improve growth rate and feed conversion ratio of birds on a relatively consistent basis (Elwinger and Saterby, 1987; Pettersson *et al.*, 1990; Friesen *et al.*, 1991; Campbell and Bedford, 1992).

Such a response is principally a result of the fact that NSPs seem to be anti-nutritive and poultry do not express enzymes capable of degrading them. Significant debate exists with regards to why, exactly, the NSPs are anti-nutritive and, moreover, how these enzymes elicit such a response. The most commonly discussed mechanisms include cereal cell wall dissolution and subsequent exposure of the contents to more efficient digestion, and reduction in intestinal aqueous phase viscosity by depolymerization of high-molecular weight, soluble, viscous cell wall polymers (Choct, 1992; Cowan, 1992; Bedford, 1993). A third mechanism is based on the knowledge that the activity of NSP enzymes in the intestine produces growth-promoting prebiotics (Morgan *et al.*, 1992; Choct *et al.*, 1996).

In truth it is likely that all three mechanisms play a role in any response observed, the contribution of each mechanism to the final result varying with each particular situation. Since each mechanism results in quite different interactions with gut function, the impact that the NSPs have and thus the benefit of use of a non-starch polysaccharidase (Nsp'ase) on avian gut function will vary from test to test. Nevertheless, Nsp'ases tend to reduce gut size as a proportion of body weight (although in absolute terms the intestine may be larger since enzyme-fed birds are often larger than equivalent controls) and alter function due to their facilitation of digestion, although some data with older birds suggest the opposite (Jaroni *et al.*, 1999; Brenes *et al.*, 2002, 2003).

In general, if digestion is consistently suboptimal, whether due to ingredient quality, microbial interaction or anti-nutritive factors, the gastrointestinal tract (GIT) responds by increasing in both size (surface area) and digestive enzyme output. Even though diet digestibility may be maintained as a result of such adaptation, the energy cost of digestion is greater and thus feed efficiency falls. Use of a Nsp'ase may improve animal performance through improved diet digestibility, reduced endogenous inputs or, most probably, through both mechanisms.

This chapter will review some of these interactions and attempt to provide guidance on what should be expected in different sections of the GIT under a variety of conditions, although it should be noted that such arbitrary segmentation of the GIT does not take sufficient note of the interactive and integrative processes between such segments which bind the whole process of digestion together.

THE ROLE OF NSP'ASES IN AVIAN NUTRITION: RELEVANCE IN DIFFERENT SEGMENTS OF THE GASTROINTESTINAL TRACT

Crop

The crop is a diverticulum of the oesophagus into which a proportion of the feed enters for storage if the proventriculus–gizzard complex is already occupied with feed (Moran, 1982). In the modern day broiler, which is fed *ad libitum*, it would be easy to assume that the use of the crop would be limited, with the bird simply topping up its proventriculus–gizzard as and when required. In actual fact it is probable that, under commercial conditions, the crop is extensively used by many broilers. Individual bird behaviour will influence the use of the crop.

Marini (2003) reported that 18–21-week-old turkey toms varied considerably in their feeding patterns, with some eating a little and often, yet others tending to have fewer and more intensive visits to the feeder. This latter strategy would probably result in extensive use of the crop compared with the former, and suggests that the flock behaves as a group of individuals rather than as a homogenous population, with some clearly making more use of the crop than others. In addition, the introduction of lighting programmes throughout much of the commercial world has resulted in periods of darkness during which the bird does not feed. As a result the flock as a whole adapts its feeding behaviour so that it can remain fully fed during the dark period.

Birds tend to fill their crop before the lights go out and 'gorge' when the lights come on (Moran, 1982). As a result, it is likely that the crop plays a significant role in food storage in commercially produced broilers, a practice which may not be replicated in much of the reported literature due to the preponderance of 24- or 23-h lighting in such work. Use of the crop is significant since the conditions therein are favourable for the activity of many Nsp'ases – and phytase for that matter.

It is a well-hydrated environment, and depending upon the population of resident lactobacilli and thus the concentration of lactic acid, the pH (ranges from 4 to 7.8; Moran, 1982) can be quite favourable for many of the fungal cellulases and xylanases in commercial use today (i.e. pH 3.5–5.5). Since feed can reside in the crop for several hours (Moran, 1982), this organ may play a significant role in determining the extent to which diets will respond to the use of Nsp'ases.

Not only will these enzymes initiate attack on the fibre of the grains being fed – thereby reducing viscosity and potentially increasing access to fermentable nutrient contents – but they will also produce fermentable oligomeric carbohydrates that may influence bacterial populations and their metabolic processes. They may also effectively increase the surface area available for bacterial attachment by 'roughening' the surface of endosperm cell walls, a mechanism that has been proposed in ruminants (Nsereko *et al.*, 2000).

Evidence that such events occur is provided by Jozefiak *et al.*, who observed significant increments in lactic acid content of the crop when the relevant enzymes were added to barley- or wheat-based diets (Jozefiak *et al.*, Abstract, this volume). Furthermore, the use of Nsp'ases has been shown to influence crop microfloral populations (Danicke *et al.*, 1999c; Murphy *et al.*, 2004), although differences were noted between enzyme sources, suggesting that not all enzymes bring about the same benefits. As a result, it is likely that the species and quantities of organisms seeding the proventriculus–gizzard and points lower down the digestive tract will be influenced by activity in the crop. Thus, although the effects may appear regional and minimal, it is possible that they are more far-reaching than may be envisaged at first sight.

From a structural viewpoint, Nsp'ase use clearly results in reduced crop size when highly viscous diets (Lazaro *et al.*, 2004) or diets rich in lupins (Brenes *et al.*, 2002) are used, but effects in wheat-based diets were found to be minimal (Brenes *et al.*, 2002; Wu and Ravindran, 2004). It is clear also that effects vary. Differences in microbial populations were observed in the crop on feeding different xylanases, suggesting that there would have been effects on nutrient and NSP digestibility, but in related work no such effects were observed (Murphy *et al.*, 2005a, b). Perhaps the differences were due to the endpoints being measured or to the conditions of the study in question. Regardless, it is important that the effects of Nsp'ase enzymes on crop physiology/microbiology are viewed in the context of the management conditions of the trial concerned. Conditions of use should mimic commercial conditions as closely as possible if the interpretation desired is for commercial production.

Proventriculus–gizzard

The fact that Nsp'ases elicit a response in poultry diets may seem surprising, since the chicken is an omnivore and has evolved to digest diets in which cereal grains play a significant part. However, under commercial conditions the bird is exposed to a diet which is dominated by one or two ingredients, thus increasing 'exposure' to NSPs and, moreover, the form in which the diet is fed may exacerbate their detrimental effects. Fine grinding of diets increases surface area and subsequent pelleting increases the dissolution rate of soluble NSPs (Silversides and Bedford, 1999), leading to increased intestinal viscosity (Teitge *et al.*, 1991).

Although many argue that fine grinding improves access of the contents to digestion (Carrte, 2004), it seems that fine grinding can limit development of the gizzard and, as a result, ironically, lead to larger average particle sizes being found in the small intestine than would be the case if the ingredient was coarsely ground (Svihus *et al.*, 1997; Hetland *et al.*, 2004). Activation of the gizzard through the use of whole grains or cellulose has been shown to dramatically improve starch digestion in wheat-rich diets, further suggesting that the capacity of the digestive tract to deal with incoming starch is reduced if gizzard activity is compromised.

Thus the practice of feeding of cereal-rich, finely ground, pelleted diets – as is the norm in commercial operations – has elevated the problems associated with cereal NSPs beyond that which would be experienced in the wild. Use of

Nsp'ases seems to enhance the ability of the gizzard to grind and break open cereal endosperm cell walls (Morgan *et al.*, 1995; Bedford and Autio, 1996), and in fact the benefits of xylanase use on NSP and DM digestibility have been observed to be most significant at the level of the gizzard (Murphy *et al.*, 2005a, b), suggesting this is a major site of Nsp'ase activity. Whether this is due to softening of the cell walls directly or to increased gizzard activity is not clear.

Gizzard activity is controlled at several levels and may, for example, be responding to changes in fermentation patterns further down the digestive tract which will be discussed later. Regardless, feed form and enzyme use does interact, with some authors noting that the use of whole grains to stimulate the activity of the proventriculus–gizzard can reduce the need for added exogenous enzymes (Jones and Taylor, 2001). Use of Nsp'ases tends to reduce the size of most digestive organs, including the gizzard, probably due to their facilitation of digestion resulting in feedback mechanisms down-regulating intestinal growth (Brenes *et al.*, 2002, 2003; Lazaro *et al.*, 2004). As will be discussed later, some of the benefits of NSP'ase use will inevitably be due to savings in endogenous inputs into the digestive process.

Small intestine

The small intestine is where most of the benefits from use of Nsp'ases and thus most of the changes in intestinal function have been observed. Many factors interrelate to determine the scale of response observed, with some of these factors being poorly understood. Increased viscosity can independently reduce digestibility of all nutrients through direct effects on diffusion of solutes in the luminal aqueous phase (the effects being magnified for the larger solutes such as micelles, more so for the long-chain fatty acid micelles (Danicke *et al.*, 1997, 2000)) and through increasing the effective thickness of the unstirred water layer.

Also, with increased intestinal viscosity, the bird must adapt in order to maintain thorough mixing of digesta, which is particularly important for emulsification of fats. Fat-soluble vitamins are particularly prone to the interaction between fat type and intestinal viscosity since they co-migrate with the larger, less digestible micelles. This is clearly underlined by the fact that marked effects on digestibility and liver storage of Vitamin E have been observed in rye-based diets where tallow was the principal source of fat (Danicke *et al.*, 1999a).

The scale and even the direction of response observed depends upon, amongst other variables, the viscosity produced at the level of the small intestine. Viscosity of the aqueous phase of the digesta increases dramatically as material leaves the gizzard and enters the small intestine (Bedford and Classen, 1993; Choct *et al.*, 1999; Lazaro *et al.*, 2004). Re-analysis of the data from Bedford and Classen (1992) indicates that intake, and thus intestinal transit rate, tends to increase with increasing small intestinal viscosity to a maximum of approximately 20–30 mPas. This seems to be an attempt to maintain nutrient intake in response to recognition of reduced ration density. As a result, gain tends to be maintained and thus FCR deteriorates as intestinal viscosity increases to approximately 20–30 mPas.

Therefore, when an Nsp'ase is employed in diets which are < 30 mPas (which constitutes the vast majority of those featured in the literature on wheat, and much of the literature on barley), the observed response tends to be a reduction in feed intake, no improvement in gain but a benefit with respect to FCR. Intestinal changes tend to be muted but often are seen as small reductions in intestinal size and/or endogenous enzyme output (Maisonnier *et al.*, 2003; Engberg *et al.*, 2004). Feeding an Nsp'ase in such situations will not only result in reduced intestinal size – as discussed previously (Simon, 1998; Brenes *et al.*, 2002; Wu *et al.*, 2004) – but also result in improved villus structure, even in mature laying hens (Jaroni *et al.*, 1999).

As intestinal viscosity increases above this threshold, the intestines struggle to maintain flow rate by peristalsis, particularly in the younger bird (Almirall *et al.*, 1995), with the result that the bird is no longer able to move the material through the intestine rapidly enough to maintain daily nutrient requirements. Gain, intake and FCR deteriorate in unison and intestinal size and enzyme output continue to increase. Use of enzymes in such diets (mostly relating to studies where rye is the dominant grain, with some barley studies and a small number of those on wheat) often leads to dramatic improvements in all performance parameters: digestibility of nutrients as well as a marked reduction in digestive inputs as measured by size of digestive tract and input of endogenous enzymes (Friesen *et al.*, 1991; Silva and Smithard, 1997; Silva *et al.*, 1997).

The rate at which enterocytes are replaced in the crypts is also significantly reduced, and individual villus height increases (with concomitant increase in villus height/crypt depth ratio) when Nsp'ases are used in particularly challenging diets, which results in considerable savings in endogenous losses (Silva and Smithard, 1996; Mathlouthi *et al.*, 2002).

Addition of Nsp'ases clearly increases the availability of nutrients for the host at the expense of the microflora in the small intestine and this, coupled with the fact that their use also increases the rate at which digesta pass through the small intestine (i.e. residence time is reduced) as a result of more effective peristalsis (Danicke *et al.*, 1999c), means that the luminal microfloral populations often decrease. Since the signals that encourage incremental intestinal outputs are either indirectly related to the presence of undigested nutrients or to the presence of certain microfloral metabolites (Moran, 1982), it is not surprising that Nsp'ase use will often lead to savings in endogenous losses. Such savings can be very significant when highly viscous diets are employed, which partially explains why the performance enhancements observed are often not fully described or valued in digestibility studies (Danicke, 2001).

Where microbial populations have been investigated, the effects of added Nsp'ases have been to reduce fermentation in the small intestine (Danicke *et al.*, 1999b), whilst concomitantly increasing caecal fermentation (Choct *et al.*, 1999). Others have noted that xylanases can reduce total CFU in the small intestine of young broilers fed wheat-based diets (Vahjen *et al.*, 1998). It is also clear that the microflora is intimately involved in the small intestinal response to Nsp'ases, since its absence mutes or even removes the problems associated with feeding viscous diets (Langhout et al., 2000; Maisonnier et al., 2003).

Bacteria are particularly involved in the effects on fat digestion since they produce bile acid hydrolases to protect themselves from this anti-microbial and, in doing so, compromise fat micelle formation. Feeding of highly viscous diets has been shown to reduce the content and/or conjugation status of bile acids and thus addition of the relevant enzyme (or additional bile salts) leads to improved fat digestion (Smits et al., 1998; Hubener et al., 2002; Mathlouthi et al., 2002), which confirms the interrelationship between microbial status, viscosity and fat emulsification and digestion.

Caecum

The caecum is a source of Nsp'ase activity in its own right as a result of the resident microflora (Hubener et al., 2002). When feeding very poor-quality cereals it is clear that performance is poor but mostly due to extreme individual bird variability (Choct et al., 1999). The 'better performers' on barley- and rye-based diets are probably those individuals supporting caecal flora which produce Nsp'ases that have some benefit if refluxed (Choct et al., 2001). Not only does the addition of Nsp'ases reduce the bird-to-bird variability, but it has also been suggested that in extremely viscous diets they can remove plugs of soluble material that block the entrance to the caecum, with concomitant dramatic effects on fermentation profiles (Choct et al., 1995, 1999). In general, the effects of Nsp'ases on caecal fermentation are dependent upon the age of the bird, the enzyme used, its dosage and the dietary quality into which it is administered.

The older the bird, the greater its ability to deal with poorly digested diets and thus the smaller the response to added enzymes.

The type and dose of enzyme influences the identity and size of the molecules concerned, some of which will stimulate caecal fermentation in both rats and chickens, with others having no effect (Choct et al., 1999). If a large excess of enzyme is utilized it is possible that significant quantities of free sugars will be generated, which may lead to osmotic problems and wet litter (Schutte, 1990).

The response in caecal fermentation patterns have been shown to be dependent upon the principal ingredients in the ration (Jozefiak et al., 2004) and on the enzyme type used (Murphy et al., 2004). In some cases addition of the exogenous enzyme has resulted in either: (i) elevated fermentation patterns (Choct et al., 1999; Jozefiak et al., 2004); (ii) no effect at all (Jozefiak et al., 2004); or (iii) in some cases a reduction in microbial populations (Murphy et al., 2004).

It is likely that the variation in response is a result of the balance between: (i) the effect the enzyme has on removing fermentable starch and protein from the caeca (which would probably be the case in very poorly digested diets); and (ii) its ability to deliver fermentable oligomers from NSP degradation into the caecum, both of which processes depend upon diet quality, enzyme type/dose and bird age. Such age-related changes in mucosally attached bacteria populations in response to xylanase use in wheat/rye-based diets have been reported.

At 7 days of age, addition of a xylanase to a wheat/rye diet tended to increase populations of enterobacteria and decrease those of lactic acid bacteria, whereas at 28 days of age enterobacterial numbers fell, there was no effect on lactic acid bacteria and Gram-positive cocci increased (Hubener *et al.*, 2002). The authors of this study suggested that the effect of adding an Nsp'ase was to shift bacterial fermentation to more proximal sites of the intestinal tract, and it is likely that in some circumstances the oligosaccharides released by NSP'ase activity are utilized almost completely in the small intestine and thus have no influence on caecal fermentation, particularly in older birds where the ileum is a sizeable fermentation vessel in its own right.

It has been suggested that the microfloral responses to xylanase and to fermentable oligosaccharides may be partially explained by changes in mucin chemistry (Fernandez *et al.*, 2000, 2002). The mechanism by which these changes are brought about is likely to be a result of interactions between the microflora and the intestines. On the use of a xylanase in a wheat-based diet it was noted that there was an increase in glycosylation of crypt-surface sialic acid residues and, as a result, colonization by *Campylobacter jejuni* was reduced.

Microbial populations and fermentation patterns also react to the relative balance between fermentable carbohydrate and nitrogen entering the caecum. A recent review showed that feeding higher-protein diets where the protein was not rapidly digested led to delivery of protein to the hindgut where putrefaction (which yields ammonia and amines in addition to VFAs) takes precedence over CHO fermentation (Apajalahti, 2005) and, as a result, pH increased and intestinal health could suffer. Since Nsp'ases produce fermentable oligosaccharides from largely unfermentable substrates its seems that enzymes may play a pivotal role in describing the structure of the populations which reside in the small and large intestine through provision of carbohydrates to the caecum.

CONCLUSIONS

Clearly, cereal NSPs differ between sources and, with this, so does their influence on intestinal function and microbial populations. It is clear, for example, that in extreme circumstances highly viscous rye-based diets are far more likely to create the environment for colonization by *Clostridium perfringens* (Riddell and Kong, 1992), and that such colonization is seen in all parts of the intestinal tract with the exception of the gizzard (Craven, 2000).

Most commercial avian production is not exposed to such extremes however, and as a result the responses to added Nsp'ases in scientific reports must be considered in the context of the conditions described. Cereal type, quality and quantity in the ration, the age of bird, feed processing conditions, fat type and quantity, presence of antibiotics, cleanliness and many other parameters influence the response observed. Such interactions have been demonstrated through mechanistic, but also through empirical, analysis of animal performances (Rosen, 2000). The changes in intestinal function that can be expected are probably the result of a multifactorial suite of interacting intestinal, dietary and microbial factors that underlie the complexity of the response observed.

REFERENCES

Almirall, M., Francesch, M., Perez-Vendrell, A.M., Brufau, J. and Esteve-Garcia, E. (1995) The differences in intestinal viscosity produced by barley and beta-glucanase alter digesta enzyme activities and ileal nutrient digestibilities more in broiler chicks than in cocks. *Journal of Nutrition* 125, 947–955.

Apajalahti, J. (2005) Comparative gut microflora, metabolic challenges, and potential opportunities. *Journal of Applied Poultry Research* 14, 1–10.

Bedford, M.R. (1993) Mode of action of feed enzymes. *Journal of Applied Poultry Research* 2, 85–92.

Bedford, M.R. and Autio, K. (1996) Microscopic Examination of feed and digesta from wheat-fed broiler chickens and its relation to bird performance. *Poultry Science* 75, 14.

Bedford, M.R. and Classen, H.L. (1992) Reduction of intestinal viscosity through manipulation of dietary rye and pentosanase concentration is effected through changes in the carbohydrate composition of the intestinal aqueous phase and results in improved growth rate and food conversion efficiency of broiler chicks. *Journal of Nutrition* 122, 560–569.

Bedford, M.R. and Classen, H.L. (1993) An *in vitro* assay for prediction of broiler intestinal viscosity and growth when fed rye-based diets in the presence of exogenous enzymes. *Poultry Science* 72, 137–143.

Brenes, A., Marquardt, R.R., Guenter, W. and Viveros, A. (2002) Effect of enzyme addition on the performance and gastrointestinal tract size of chicks fed lupin seed and their fractions. *Poultry Science* 81, 670–678.

Brenes, A., Slominski, B.A., Marquardt, R.R., Guenter, W. and Viveros, A. (2003) Effect of enzyme addition on the digestibilities of cell wall polysaccharides and oligosaccharides from whole, dehulled, and ethanol-extracted white lupins in chickens. *Poultry Science* 82 1716–1725.

Campbell, G.L. and Bedford, M.R. (1992) Enzyme applications for monogastric feeds: a review. *Canadian Journal of Animal Science* 72, 449–466.

Carre, B. (2004) Causes for variation in digestibility of starch among feedstuffs. *World's Poultry Science Journal* 60, 76–89.

Choct, M. (1992) Relationship between soluble arabinoxylan and the nutritive value of wheat for broiler chickens. *Proceedings of the 13th Western Nutrition Conference*, 16–17 September, Saskatoon, Canada, pp. 41–54.

Choct, M., Hughes, R.J., Wang, J., Bedford, M.R., Morgan, A.J. and Annison, G. (1995) Feed enzymes eliminate the anti-nutritive effect of non-starch polysaccharides and modify fermentation in broilers. *Australian Poultry Science Symposium Proceedings* 7, 121–125.

Choct, M., Hughes, R.J., Wang, J., Bedford, M.R., Morgan, A.J. and Annison, G. (1996) Increased small intestinal fermentation is partly responsible for the anti-nutritive activity of non-starch polysaccharides in chickens. *British Poultry Science* 37, 609–621.

Choct, M., Hughes, R.J. and Bedford, M.R. (1999) Effects of a xylanase on individual bird variation, starch digestion throughout the intestine, and ileal and caecal volatile fatty acid production in chickens fed wheat. *British Poultry Science* 40, 419–422.

Choct, M., Hughes, R.J., Perez-Maldonado, R. and van Barneveld, R.J. (2001) The metabolisable energy value of sorghum and barley for broilers and layers. *Australian Poultry Science Symposium Proceedings* 13, 39–42.

Cowan, W.D. (1992) Advances in feed enzyme technology. *Agri-Food Industry Hi-Tech*, May/June, 34–40..

Craven, S.E. (2000) Colonization of the intestinal tract by *Clostridium perfringens* and fecal shedding in diet-stressed and unstressed broiler chickens. *Poultry Science* 79, 843–849.

Danicke, S. (2001) Estimation of endogenous N-losses from the digestive tract of broilers by N-balance technique and linear regression analysis: effects of xylanase addition to a rye-based diet. *Landbauforschung Volkenrode* 51, 33–40.

Danicke, S., Simon, O., Jeroch, H. and Bedford, M.R. (1997) Interactions between dietary fat type and xylanase supplementation when rye-based diets are fed to broiler chickens. 2: performance, nutrient digestibility and the fat-soluble vitamin status of livers. *British Poultry Science* 38, 546–556.

Danicke, S., Jeroch, H., Bottcher, W., Bedford, M.R. and Simon, O. (1999a) Effect of dietary fat type, pentosan level and xylanases on digestibility of fatty acids, liver lipids and vitamin E in broilers. *Fett/Lipid* 101, S90–S100.

Danicke, S., Vahjen, W., Simon, O. and Jeroch, H. (1999b) Effects of dietary fat type and xylanase supplementation to rye-based broiler diets on selected bacterial groups adhering to the intestinal epithelium, on transit time of feed, and on nutrient digestibility. *Poultry Science*, 78, 1292–1299.

Danicke, S., Vahjen, W., Simon, O. and Jeroch, H. (1999c) Effects of dietary fat type and xylanase supplementation to rye-based broiler diets on selected bacterial groups adhering to the intestinal epithelium, on transit time of feed, and on nutrient digestibility. *Poultry Science*, 78, 1292–1299.

Danicke, S., Jeroch, H., Bottcher, W. and Simon, O. (2000). Interactions between dietary fat type and enzyme supplementation in broiler diets with high pentosan contents: effects on precaecal and total tract digestibility of fatty acids, metabolizability of gross energy, digesta viscosity and weights of small intestine. *Animal Feed Science Technology* 84, 279–294.

Elwinger, K. and Saterby, B. (1987) The use of B-glucanase in practical broiler diets containing barley or oats. Effect of enzyme level, type and quality of grain. *Swedish Journal of Agricultural Research* 17, 133–140.

Engberg, R.M., Hedemann, M.S., Steenfeldt, S. and Jensen, B.B. (2004) Influence of whole wheat and xylanase on broiler performance and microbial composition and activity in the digestive tract. *Poultry Science* 83, 925–938.

Fernandez, F., Sharma, R., Hinton, M. and Bedford, M.R. (2000) Diet influences the colonization of *Campylobacter jejuni* and distribution of mucin carbohydrates in the chick intestinal tract. *Cellular and Molecular Life Sciences* 57, 1793–1801.

Fernandez, F., Hinton, M. and Van Gils, B. (2002) Dietary mannan-oligosaccharides and their effect on chicken caecal microflora in relation to *Salmonella* Enteritidis colonization. *Avian Pathology* 31, 49–58.

Friesen, O.D., Guenter, W., Rotter, B.A. and Marquardt, R.R. (1991) The effects of enzyme supplementation on the nutritive value of rye grain (*Secale cereale*) for the young broiler chick. *Poultry Science* 70, 2501–2508.

Hetland, H., Choct, M. and Svihus, B. (2004) Role of insoluble non-starch polysaccharides in poultry nutrition. *World's Poultry Science Journal* 60, 413–420.

Hubener, K., Vahjen, W. and Simon, O. (2002) Bacterial responses to different dietary cereal types and xylanase supplementation in the intestine of broiler chicken. *Archives of Animal Nutrition* 56, 167–187.

Jaroni, D., Scheideler, S.E., Beck, M.M. and Wyatt, C. (1999) The effect of dietary wheat middlings and enzyme supplementation – II: apparent nutrient digestibility, digestive tract size, gut viscosity, and gut morphology in two strains of leghorn hens. *Poultry Science* 78, 1664–1674.

Jones, G.P.D. and Taylor, R.D. (2001) The incorporation of whole grain into pelleted broiler chicken diets: production and physiological responses. *British Poultry Science* 42, 477–483.

Jozefiak, D., Rutkowski, A., Fratczak, M. and Boros, D. (2004) The effect of dietary fibre fractions from different cereals and microbial enzyme supplementation on performance, ileal viscosity and short-chain fatty acid concentrations in the caeca of broiler chickens. *Journal of Animal Feed Science* 13, 487–496.

Langhout, D.J., Schutte, J.B., de Jong, J., Sloetjes, H., Verstegen, M.W.A. and Tamminga, S. (2000) Effect of viscosity on digestion of nutrients in conventional and germ-free chicks. *British Journal of Nutrition* 83, 533–540.

Lazaro, R., Latorre, M.A., Medel, P., Gracia, M. and Mateos, G.G. (2004) Feeding regimen and enzyme supplementation to rye-based diets for broilers. *Poultry Science* 83, 152–160.

Maisonnier, S., Gomez, J., Bree, A., Berri, C., Baeza, E. and Carre, B. (2003) Effects of microflora status, dietary bile salts and guar gum on lipid digestibility, intestinal bile salts, and histomorphology in broiler chickens. *Poultry Science* 82, 805–814.

Marini, P. (2003) Feeding behaviour of large turkey toms from 18–21 weeks of age. *Proceedings of the 26th Technical Turkeys Conference,* Shrigley Hall Hotel, Macclesfield, UK, 23–25 April, 51, 9–10.

Mathlouthi, N., Lalles, J.P., Lepercq, P., Juste, C. and Larbier, M. (2002) Xylanase and beta-glucanase supplementation improve conjugated bile acid fraction in intestinal contents and increase villus size of small intestine wall in broiler chickens fed a rye-based diet. *Journal of Animal Science* 80, 2773–2779.

Moran, E.T. (1982) In: *Comparative Nutrition of Fowl and Swine: the Gastrointestinal Systems.* University of Guelph, Ontario, Canada.

Morgan, A.J., Mul, A.J., Beldman, G. and Voragen, A.G.J. (1992) Dietary oligosaccharides – new insights. *Agro-Food Industry Hi-Tech* November/December, 35–38.

Morgan, A.J., Bedford, M.R., Tervila-Wilo, A., Autio, K., Hopeakoski-Nurminen, M., Poutanen, K. and Parkkonen, T. (1995) How enzymes improve the nutritional value of wheat. *Zootecnica International* April, 44–48.

Murphy, T.C., Bedford, M.R. and McCracken, K.J. (2004) The effect of a range of xylanases added to wheat-based diet on broiler performance and microfloral populations. *British Poultry Science* 45, S61–S62.

Murphy, T.C., Bedford, M.R. and McCracken, K.J. (2005a) Effect of a range of xylanases on apparent nutrient digestibility along the digestive tract of broilers upon supplementation of a wheat-based diet. *British Poultry Abstracts* 1, 40.

Murphy, T.C., Bedford, M.R. and McCracken, K.J. (2005b) Xylanase action on non-starch poly-saccharides along the digestive tract of broilers. *British Poultry Abstracts* 1, 38.

Nsereko, V.L., Morgavi, D.P., Rode, L.L.M., Beauchemin, K.A. and McAllister, T.A. (2000) Effects of fungal enzyme preparations on hydrolysis and subsequent degradation of alfalfa hay fiber by mixed rumen microorganisms *in vitro. Animal Feed Science Technology* 88, 153–170.

Pettersson, D., Graham, H. and Aman, P. (1990) Enzyme supplementation of low or high crude protein diets for broiler chickens. *Animal Production* 51, 399–404.

Riddell, C. and Kong, X.-M. (1992) The influence of diet on necrotic enteritis in broiler chickens. *Avian Diseases* 36, 499–503.

Rosen, G.D. (2000) Multifactorial assessment of exogenous enzymes in broiler pronutrition. Targets and problems. Initial results. *European Symposium on Feed Enzymes,* 3. (In press)

Schutte, J.B. (1990) Nutritional implications and metabolizable energy value of D-Xylose and L-Arabinose in chicks. *Poultry Science* 69, 1724–1730.

Silva, S.S.P. and Smithard, R.R. (1996) Exogenous enzymes in broiler diets: crypt cell pro-liferation, digesta viscosity, short chain fatty acids and xylanase in the jejunum. *British Poultry Science* 37 (Suppl.), S77–S79.

Silva, S.S.P. and Smithard, R.R. (1997) Digestion of protein, fat and energy in rye-based broiler diets is improved by the addition of exogenous xylanase and protease. *British Poultry Science* 38 (Suppl.), S38–S39.

Silva, S.S.P., Gilbert, H.J. and Smithard, R.R. (1997) Exogenous polysaccharides do not improve digestion of fat and protein by increasing trypsin or lipase activities in the small intestine. *British Poultry Science* 38 (Suppl.), S39–S40.

Silversides, F.G. and Bedford, M.R. (1999) Effect of pelleting temperature on the recovery and efficacy of a xylanase enzyme in wheat-based diets. *Poultry Science* 78, 1184–1190.

Simon, O. (1998) The mode of action of NSP hydrolysing enzymes in the gastrointestinal tract. *Journal of Animal Feed Science* 7, 115–123.

Smits, C.H.M., Veldman, A., Verkade, H.J. and Beynen, A.C. (1998) The inhibitory effect of carboxymethylcellulose with high viscosity on lipid absorption in broiler chickens coincides with reduced bile salt concentration and raised microbial numbers in the small intestine. *Poultry Science* 77, 1534–1539.

Svihus, B., Herstad, O., Newman, C.W. and Newman, R.K. (1997) Comparison of performance and intestinal characteristics of broiler chickens fed on diets containing whole, rolled or ground barley. *British Poultry Science* 38, 524–529.

Teitge, D.A., Campbell, G.L., Classen, H.L. and Thacker, P.A. (1991) Heat pretreatment as a means of improving the response to dietary pentosanase in chicks fed rye. *Canadian Journal of Animal Science* 71, 507–513.

Vahjen, W., Glaser, K., Schafer, K. and Simon, O. (1998) Influence of xylanase-supplemented feed on the development of selected bacterial groups in the intestinal tract of broiler chicks. *Journal of Agricultural Science (Cambridge)* 130, 489–500.

Wu, Y.B. and Ravindran, V. (2004) Influence of whole wheat inclusion and xylanase supplementation on the performance, digestive tract measurements and carcass characteristics of broiler chickens. *Animal Feed Science Technology* 116, 129–139.

Wu, Y.B., Ravindran, V., Thomas, D.G., Birtles, M.J. and Hendriks, W.H. (2004) Influence of phytase and xylanase, individually or in combination, on performance, apparent metabolisable energy, digestive tract measurements and gut morphology in broilers fed wheat-based diets containing adequate level of phosphorus. *British Poultry Science* 45, 76–84.

CHAPTER 11
Effects of amino acid and protein supply on nutrition and health

M.T. Kidd* and A. Corzo

Department of Poultry Science, Mississippi State University, Mississippi, USA;
*e-mail: mkidd@poultry.msstate.edu

ABSTRACT

Modern commercial broiler strains accrete more breast meat and are marketed earlier than were previous strains, in addition to typically having different feed intake patterns. Much recent research has addressed increasing dietary amino acid density to optimize growth and tissue accretion rates of modern commercial broilers. Increasing amino acid density, especially early in the bird's life, results in improved live production and carcass traits, resulting in increased profitability. However, supplying amino acids in excess can be costly and increase nitrogen excretion, whereas limiting their supply can result in suboptimal yields and economic returns.

Essential amino acid (i.e. threonine, isoleucine, valine, arginine and tryptophan) experimentation with respect to their effect on broiler live performance, carcass traits and health is discussed. Although feeding deficient levels of isoleucine, valine and arginine suppressed some aspects of the immune system, fortifying these amino acids to levels above surfeit was without effect on immune system measurements.

Our findings suggest that the threonine, valine and isoleucine requirements for functionality of a normal immune system do not exceed needs for growth. Arginine, however, has vast metabolic functions related to immunity and further testing is warranted to determine if its need for health is higher than that for growth in commercial broilers. Although the dietary supply of amino acids is critical for good growth and tissue accretion, marginal limitations or excesses do not seem to impact bird health. Future amino acid research via early feeding and *in ovo* may result in early growth and health benefits.

INTRODUCTION

Supplying broilers diets that are adequate in all essential nutrients via least-cost formulation is of utmost importance in optimizing productive efficiency. Consumer-driven trends (i.e. animal welfare concerns, the removal of antimicrobials from diets, feeding environmentally friendly diets and heightened meat quality of broiler chicken meat) are changing our perspectives in dietary formulation. Hence, in addition to least-cost diets that increase efficiency and carcass yields, improved broiler health and disease resistance are also targets achievable by diet.

Literature reviews have shown that some ingredients and nutrients in broiler diets impact immunity (Koutsos and Klasing, 2001; Humphrey and Klasing, 2004; Kidd, 2004) and animal welfare (Whitehead, 2002). Recent findings in the UK indicate that the welfare of broilers is impacted more from housing conditions than from stocking density (Dawkins *et al.*, 2004). These authors concluded that nutrition should be one of the key researched areas for improving animal welfare regulations.

Therefore, findings point to the importance of nutrition on health and immunity of broilers and should be further researched so that least-cost diets can be formulated for health as well as for productive efficiency. Although beyond the scope of this review, future research should further elucidate specific mechanisms of innate and adaptive immunity so that optimizing health efficiency does not compromise productive efficiency.

Genetic selection for increased growth rate has been shown to be negatively correlated with immunity (Siegel and Gross, 1980). Much of this chapter is concerned with optimizing amino acid nutrition of modern, high-yielding broiler strains. Unfortunately, experimentation directed at evaluating amino acid density does not typically address immunity. As such, health implications of broilers in the former studies are addressed through mortality data. However, recent research concerning the typical amino acids limiting after threonine (i.e. isoleucine and valine) in least-cost broiler diets on immune system functions will be addressed, in addition to threonine needs for intestinal function.

AMINO ACID DENSITY

When attempting to reduce production costs, increasing some dietary essential amino acids over levels that would be considered sufficient is sometimes viewed with a certain degree of scepticism. There are, however, recent studies that address the question of whether reducing dietary cost by decreasing amino acid density would necessarily translate into increased profitability. Another point of concern for nutritionists centres on whether the increase in amino acid density results in any deleterious effects on bird health or an incidence of mortality. While all of the aforementioned items are valid points of concern, they may not necessarily have a common answer if one takes into consideration the different responses observed across broiler strains, gender and market type.

The first parameters that can be used to evaluate the response of broilers to various dietary amino acid density regimes would be those related to body weight and feed conversion. Illustrated are the responses observed by three different broiler strain crosses (two multi-purpose versus one high-yield) to body weight gain and feed conversion (see Table 11.1).

It will be observed how amino acid density impacted live performance of broilers regardless of the broiler strain type used. However, the magnitude of the amino acid density effect may vary among strain crosses. Evidently, it has previously been reported how dietary amino acid density may have interactive effects within strain crosses. While the Ross × Ross 508 broiler has shown an improvement in various parameters, including breast meat yield when fed higher-amino acid-density diets (Kidd *et al.*, 2004), the Arbor Acres Plus has shown diminishing improvements in these same parameters with increased amino acid density (Corzo *et al.*, 2004).

Other studies have also reported differences in dietary amino acid density impact on different broiler strain crosses (Smith *et al.*, 1998; Corzo *et al.*, 2005a). Thus, it becomes necessary to make a distinction between the nutrient needs for optimum growth between different broiler strain crosses. Recently, a high-yield strain cross (Ross × Ross 708) broiler was evaluated for the effect of amino acid density on performance, carcass traits and economic returns (Kidd *et al.*, 2005). It was clear how feeding high-amino acid-density regimes throughout life resulted in broilers with improved live performance and carcass and breast meat yields.

Furthermore, an increase in profitability was noted in birds fed on high-amino acid-density diets through the grower period, in contradiction to the old belief of 'lower diet cost represents higher economic returns', based on the contribution of diet to the total scheme of production costs. However, future research is necessary to validate any increase in profitability associated with

Table 11.1. Strain cross body weight and feed conversion in response to various dietary amino acid density levels.[a]

Parameter[b]	0–28 days			0–42 days			0–56 days		
	A	B	C	A	B	C	A	B	C
Body weight (kg)									
H	1.20[c]	1.23[c]	1.14[d]	2.19[ab]	2.24[c]	2.07[e]	3.11[c]	3.11[c]	2.93[d]
L	1.16[d]	1.16[d]	1.04[e]	2.14[d]	2.15[d]	1.94[f]	3.06[cd]	3.06[cd]	2.79[e]
Feed conversion (g/g)									
H	1.43[e]	1.44[de]	1.43[e]	1.68[de]	1.66[e]	1.69[de]	1.86[d]	1.85[d]	1.88[d]
L	1.47[cd]	1.47[cd]	1.49[c]	1.72[d]	1.71[d]	1.78[c]	1.89[d]	1.88[d]	1.96[c]

[a] Strain A = multipurpose; strain B = multipurpose; strain C = high-yield; adapted from Corzo *et al.*, 2005a.
[b] H = high amino acid density; L = low amino acid density. Diets were isocaloric. The average percentage of digestible amino acid increase for TSAA, Lys, Thr and Ile for the H versus the L diets was 7, 8.5, 7.4 and 6%, respectively.
[c–f] Values within comparisons with different superscripts differ ($p < 0.05$).

dietary amino acid density, especially if strain cross and highly fluctuating feed ingredient costs are taken into consideration.

Currently, human obesity is a topic of rising concern. There is a demand for leaner foods and, with it, a reduction in animal meat fat composition. Various nutritional strategies have been reported over the years as the means to reduce compositional fat of poultry meat. Among these is the increase in amino acid density of the diets fed to broilers. Abdominal fat is typically a good indicator of overall body fat of broilers. Illustrated is the cumulative or phase-feeding effect that differing dietary amino acid-density regimes had on abdominal fat percentage of broilers.

Figure 11.1a represents the values of all three strain crosses (two multi-purpose and one high-yield) averaged for amino acid density, considering no observed interactive effects. A 10 and 8% reduction in relative abdominal fat values of broilers was observed when feeding high-amino acid-density diets throughout life when compared to low-density diets, at 42 and 56 days, respectively, regardless of broiler strain.

Figure 11.1b represents the effect of amino acid phase feeding on abdominal fat percentage values. For example, the HMML abbreviation represents the abdominal fat percentage of broilers fed an amino acid dietary regime consisting of high, medium, medium and low diets during the starter, grower, finisher and withdrawal feeding phases, respectively.

A progressive increase in abdominal fat percentage was observed as broilers were chronologically exposed to diets with lower amino acid density values throughout growth. In this particular case, a 24% increase in relative abdominal fat was observed when comparing the highest with the lowest amino acid-density regimes (HHHH versus LLLL). In agreement with the previously discussed results, Wijtten *et al.* (2004) reported a decrease in both absolute and relative values of lipid content of the carcasses of broilers fed increasing levels of amino acid density. It could be suggested that both examples illustrate the potential for the reduction in lipid tissue deposition in the broiler carcass.

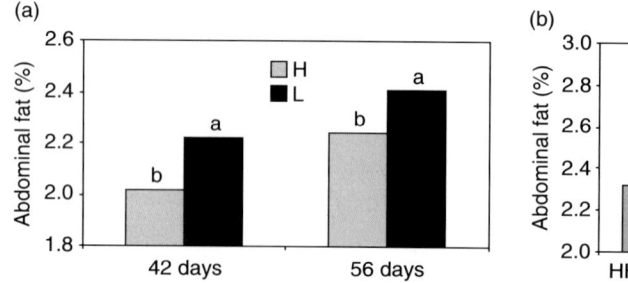
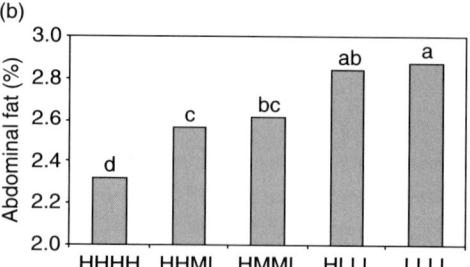

Fig. 11.1. Dietary amino acid density effects (high density versus low density (a) and varying amino acid phase density (b)) on abdominal fat relative values (g fat/g carcass $\times$ 100). H, high amino acid density; M, medium amino acid density; L, low amino acid density. Values within comparisons with different letters differ ($p > 0.05$). Adapted from Corzo *et al.*, 2005 (a) and Kidd *et al.*, 2004 (b).

In addition to the previously discussed benefits, there was a report of increased bird uniformity – or rather a reduction in flock body weight variability – with the use of high-amino acid-density diets (see Fig. 11.2; Corzo *et al.*, 2004).

A previously discussed effect on body weight as mediated by amino acid density is illustrated in Fig. 11.2a, with a concomitant reduction in bird-to-bird body weight variability (coefficient of variation, %). This increase in flock uniformity, while somewhat unexpected, is perhaps the result of satisfying the essential amino acid needs of the majority of a population of birds, considering variations in their cumulative nutrient consumption. While this represents an increase in the similarity of the whole weights of the different carcass cut-up parts, it may also be seen favourably by processing plants as an aid in obtaining an end product with more consistent mass and yield characteristics.

There have been reports of an increase in the incidence of mortality with further supplementation of amino acids, particularly lysine (Latshaw, 1993; Kidd *et al.*, 1998). However, there currently appears to be general agreement that moderate increases in amino acid density above recommended levels (NRC, 1994) have no implications in the incidence of mortality (Bartov and Plavnik, 1998; Smith *et al.*, 1998; Corzo *et al.*, 2004, 2005a; Wijtten *et al.*, 2004; Kidd *et al.*, 2005). Furthermore, under heat stress conditions, any increase in dietary amino acid density has not resulted in associated increases in mortality (Temim *et al.*, 2000; Dozier and Moran, 2001).

Apart from the potential for satisfying the quantitative needs for amino acids in a broiler population, there may be some additional growth stimulation mechanisms associated with an increase in dietary amino acid density. One possible explanation may be an increase in IGF-I concentrations, leading to protein anabolism (Rosebrough *et al.*, 1996; Tesseraud *et al.*, 2003). But perhaps an unrecognized plethora of explanations may rely on up- or down-regulating effects that individual or combined amino acids may have on metabolic actions regarding tissue synthesis and deposition (Fafournoux *et al.*, 2000; German *et al.*, 2003). However, we are just beginning to understand the possible role amino acids may play in these regulatory activities.

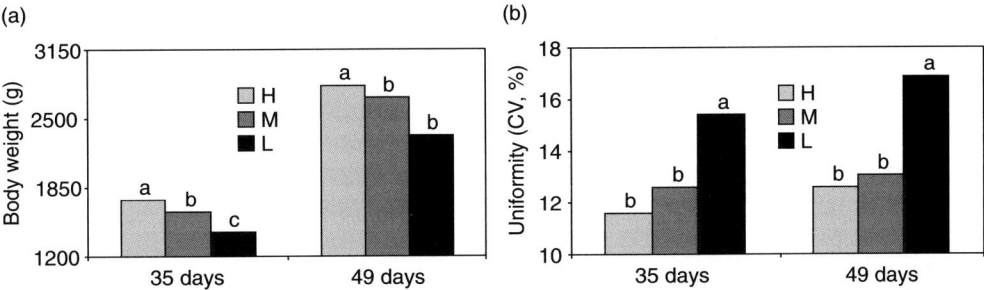

Fig. 11.2. Dietary amino acid density effects on flock body weight (a) and uniformity (b). H, high amino acid density; M, medium amino acid density; L, low amino acid density. Values within comparisons with different letters differ ($p > 0.05$). Adapted from Corzo *et al.*, 2004.

AMMONIA

Broiler companies are taking measures to reduce dietary ammonia. The goal for ammonia concentrations in broiler houses in the USA is to be < 25 ppm. Hence, decreased productive efficiency (Miles *et al.*, 2004) and health (Kristensen and Wathes, 2000) are observed in broilers exposed to ammonia. As the former section denotes improvements in broiler growth and carcass traits with increased amino acid density, it must be pointed out that surplus protein increased nitrogen excretion and ammonia (Ferguson *et al.*, 1998).

Furthermore, each 1% decrease in surplus protein yields around 8% decrease in nitrogen excretion (Ferguson *et al.*, 1998; Kidd *et al.*, 2001a). Measures such as amino acid balance and minimizing amino acid excesses, especially towards the end of flock life, will allow for increased productive efficiency without increasing surplus protein – and hence ammonia. Ammonia regulations in company animal welfare guidelines are tied to footpad and eye lesions. Further, means to reduce ammonia are warranted because levels as low as 25–50 ppm have been shown to increase the incidence of respiratory disease in chickens (Kling and Quarles, 1974).

SPECIFIC EFFECTS FOR CRITICAL AMINO ACIDS IN DIETARY FORMULATION

As interest mounts concerning nutritional modulation that boosts immune systems, knowledge of amino acid minima that could be adjusted in diets to optimize immunity is warranted. Can some amino acids in least-cost formulation be increased over levels thought to optimize growth performance resulting in benefits in immunity? A recent review on nutrition and immunity (Kidd, 2004) demonstrated that numerous amino acids (i.e. betaine as an amino acid derivative, and arginine) have the potential to offer immune requirements that differ from those associated with growth performance. Additionally, will diets formulated on a digestible amino acid basis, that use supplemental amino acids in order to minimize surplus protein, limit the functional capacity of the immune system? The former question will be addressed from recent research in our laboratory on isoleucine and valine on immunity, and tryptophan on physiological stress.

Arginine

Arginine's role in immune system function has been studied extensively in poultry (as reviewed by Kidd, 2004). Although numerous effects of arginine on immunity and disease have been observed, published reports do not make mention of practical levels of arginine that could be employed in dietary formulation to boost immunity. Thus, most experimentation with arginine in poultry elucidates effects and mechanisms on immunity, but the levels tested are typically too deficient or excessive to be considered in practice. In an attempt to determine a practical level of arginine that could be used in dietary formulation, Kidd *et al.* (2001b) fed varying levels of arginine near published guidelines (NRC, 1994) and noted increased plasma arginine with increased dietary arginine, but no effects on immunity.

However, research evaluating dietary arginine near levels observed in practice has resulted in reduced mortality during an *Eimeria* challenge (Kidd *et al.*, 2002) and reduced skin-infected scratches at slaughter (Corzo *et al.*, 2003). Increasing levels of dietary arginine in dietary formulation are usually achieved via increasing soybean meal. Arginine requirements for optimal immunity and health in dietary formulation remain elusive.

Isoleucine

Low-protein diets containing commercially available amino acid supplements (i.e. methionine, lysine and threonine) can be limiting in isoleucine as it is typically the fourth most limiting amino acid. Hence, care must be taken in establishing an isoleucine minimum in dietary formulation so that it does not become limiting.

Konashi *et al.* (2000) studied groups of amino acids and their effect on immunity in chickens and found that immunodepression for some parameters was more pronounced with the branch-chain amino acids than with other groups of amino acids.

Hale *et al.* (2004) conducted dietary dose-response research with isoleucine in anticipation of predicting quadratic trends for various immune parameters and, thus, an isoleucine recommendation for immunity. Published dose–response research to predict immunity needs, as opposed to only growth performance needs, is sparse. It is warranted as it can provide definitive estimations of differences in nutrient needs for immunity versus more typically measured parameters. However, parameters tested by Hale *et al.* (2004) in a dietary isoleucine dose–response test failed to yield recommendations (see Table 11.2).

Hence, quadratic trends for immune organ weights, T cell receptor expression, cutaneous basophil hypersensitivity and primary titres to sheep erythrocytes were not found. Although an acute isoleucine deficiency (58% of the recommended level) depressed some aspects of immunity, feeding diets marginally deficient in isoleucine did not appear to alter broiler chicken immune responses.

Valine

Depending on the type of broiler diet and ingredients used, valine may be co-limiting or more limiting than isoleucine. Preliminary research with valine on immunity of broilers was conducted by Thornton *et al.* (2004). Although a deficiency of valine (88% of the recommended level) depressed spleen development, quadratic trends for valine and immune parameters were not found. Taken as a whole with the data on isoleucine, above, it appears that the

Table 11.2. Impact of dietary isoleucine dose–response on immune parameters in broilers.[a]

Parameter	Dietary isoleucine (%)					
	0.42	0.50	0.58	0.66	0.74	0.82
Bursa (% of BW)	0.19	0.17	0.15	0.13	0.15	0.13
Spleen (% of BW)	0.08	0.08	0.09	0.07	0.10	0.09
Thymus (% of BW)	0.12	0.13	0.15	0.12	0.15	0.11
Toe web swelling (%)[b]	15.3	18.9	27.7	24.8	24.9	28.0
Titre (Log_2)[c]	8.43	6.93	9.93	8.33	8.14	8.36
CD8 + α[d]	447.8	460.3	518.3	428.5	587.5	483.8
CD8 + B[d]	257.0	398.5	306.5	335.0	315.8	461.8
TCR1[d]	182.8	217.3	227.5	194.3	215.8	252.3

[a] Data did not fit quadratic trends.
[b] Toe web swelling represents change in thickness after birds received an intradermal injection of 100 mg of phytohaemagglutinin-P.
[c] Titres represent log_2 readings from a primary antibody response to sheep red blood cells.
[d] Relative fluorescence units of T cell receptors.

branched-chain amino acid minimum constraints in dietary formulation should be optimized for growth performance and carcass yield, rather than for immunity.

Threonine

Threonine is important for proper intestinal function. Its role is heavily associated with its content in mucins. Mucins represent glycosylated proteins that are differentially expressed in the intestinal epithelium. Threonine represents more than 40% of their amino acid residues (Bengmark and Jeppsson, 1995). A significant portion of mucin also contains serine (Bengmark and Jeppsson, 1995). Threonine can be transaminated to glycine and serine, whereas the reverse pathway has been shown to be negligible (Baker *et al.*, 1972). Although threonine has been shown not to impact immune system development (Kidd *et al.*, 2001a) and immune responsiveness (Kidd *et al.*, 1997) in broilers, its ability to promote health via heightened intestinal functionality should be further addressed.

Because threonine is typically the third limiting amino acid it would be advantageous to know what minimum to use in least-cost formulation depending on dietary ingredients and anticipated gut challenges. Differences in threonine minima have been shown in mature Cobb broilers reared in clean versus dirty environments (Kidd *et al.*, 2003a). Birds in the dirty environment responded to higher levels of threonine than those documented in the literature, whereas birds in the clean environment responded to threonine levels in agreement with published findings (Kidd *et al.*, 2003a). However, it has been demonstrated that broiler chicks do not have heightened threonine needs during a mild *Eimeria acervulina* infection (Kidd *et al.*, 2003b).

Although nutritional experimentation conducted in differing environments requires somewhat large and time-consuming experiments, more research should be conducted with threonine in this manner to better understand if its minimum need varies as environment changes. Further, threonine requirement should be measured in diets containing marginally deficient lysine so that ratios to lysine can be predicted rather than optimal threonine levels.

Tryptophan

Diets fed to commercial broiler chickens are seldom limiting in dietary tryptophan. However, the inclusion of certain grains in feed formulation may result in this amino acid becoming the fourth limiting one. Consequently, dietary deficiency of this amino acid could have deleterious effects considering this amino acid's functionalities. Apart from being an obvious constituent of body proteins, it has precursor roles for neurotransmitters such as melatonin and serotonin, as well as for the vitamin niacin.

Additionally, this amino acid has shown to have ketogenic activity in the chicken. It's deficiency in diets fed to broilers has led to increasing nervousness (Corzo *et al.*, 2005b). However, no degree of physiological stress was found when tryptophan was fed in deficient amounts to broilers at 20 days of age, as observed in the lack of alteration in those blood chemistry parameters commonly associated with the presence of stress (see Table 11.3).

CONCLUSIONS

Although manipulation of the broiler growth curve via environmental and nutritional means has been shown to produce advantages (i.e. bone quality), feeding higher amino acid levels to optimize growth and meat yield in nutrient density studies does not seem to have any adverse effects on health. Some amino acids have been shown to heighten immune system parameters at

Table 11.3. Blood plasma cholesterol, heterophil/lymphocyte ratio (H/L), glucose, high/low-density lipoprotein ratio (HD/LD) and corticosterone level of broiler chicks fed diets containing graduations of tryptophan (adapted from Corzo *et al.*, 2005c).

Dietary tryptophan (%)	Cholesterol (mg/dl)	H/L[a]	Glucose (mg/dl)	HD/LD[b]	Corticosterone (ng/ml)
0.13	138.1	36.6	238.5	2.85	1.01
0.15	125.4	35.8	239.4	3.09	1.20
0.17	123.9	38.7	244.3	2.47	1.09
0.19	130.3	23.0	249.0	2.78	1.03
0.21	130.4	29.9	248.7	4.39	1.11
0.25	125.4	27.1	249.4	2.76	1.17
0.25	128.7	32.1	258.1	2.48	1.12

[a] Values are expressed as heterophil/lymphocyte cell count × 100.
[b] Values are expressed as high/low-density lipoprotein.

dietary levels in excess of growth needs. However, until further information is gained regarding amino acid needs for both innate and adaptive immunity, dietary formulation of amino acids should be concerned with optimizing growth and carcass responses. Amino acid minima in least-cost formulated broiler diets that support growth seem to support good health as measured by mortality and immune response data.

ACKNOWLEDGEMENT

This is Journal Article Number BC10788 from the Mississippi Agricultural and Forestry Experiment Station supported by MIS-322140. Use of trade names in this publication does not imply endorsement by the Mississippi Agricultural and Forestry Experiment Station of the products, nor similar ones not mentioned.

REFERENCES

Baker, D.H., Hill, T.M. and Kleiss, A.J. (1972) Nutritional evidence concerning formation of glycine from threonine in the chick. *Journal of Animal Science* 34, 582–586.

Bartov, I. and Plavnik, I. (1998) Moderate excess of dietary protein increases breast meat yield of broiler chicks. *Poultry Science* 77, 680–688.

Bengmark, S. and Jeppsson, B. (1995) Gastrointestinal surface protection and mucosa reconditioning. *Journal of Parenteral and Enteral Nutrition* 19, 410–415.

Corzo, A., Moran, Jr., E.T. and Hoehler, D. (2003) Arginine need of heavy broiler males: Applying the ideal protein concept. *Poultry Science* 82, 402–407.

Corzo, A., McDaniel, C.D., Kidd, M.T., Miller, E.R., Boren, B.B. and Fancher, B.I. (2004) Impact of dietary amino acid concentration on growth, carcass yield, and uniformity of broilers. *Australian Journal of Agricultural Research* 55, 1133–1138.

Corzo, A., Kidd, M.T., Burnham, D.J., Miller, E.R., Branton, S.L. and Gonzalez-Esquerra, R. (2005a) Dietary amino acid density effects on growth and carcass of broilers differing in strain cross and sex. *Journal of Applied Poultry Research* 14, 1–9.

Corzo, A., Moran, Jr. E.T., Hoehler, D. and Lemme, A. (2005b) Tryptophan need of broiler males from forty-two to fifty-six days of age. *Poultry Science* 84, 226–231.

Corzo, A., Kidd, M.T., Thaxton, J.P. and Kerr, B.J. (2005c) Dietary tryptophan effects on growth and stress of male broiler chicks. *British Poultry Science* 46, 478–484.

Dawkins, M.S., Donnelly, C.A. and Jones, T.A. (2004) Chicken welfare is influenced more by housing conditions than by stocking density. *Nature* 427, 342–344.

Dozier, W.A., III and Moran, Jr. E.T. (2001) Response of early- and late-developing broilers to nutritionally adequate and restrictive feeding regimens during the summer. *Journal of Applied Poultry Research* 10, 92–98.

Fafournoux, P., Bruhat, A. and Jousse, C. (2000) Amino acid regulation of gene expression. *Biochemical Journal* 351, 1–12.

Ferguson, N.S., Gates, R.S., Taraba, J.L., Cantor, A.H., Pescatore, A.J., Ford, M.J. and Burnham, D.J. (1998) The effect of dietary crude protein on growth, ammonia concentration, and litter composition in broiler. *Poultry Science* 77, 1481–1487.

German, J.B., Roberts, M.A. and Watkins, S.M. (2003) Genomics and metabolomics as markers for the interaction of diet and health: lessons from lipids. *Journal of Nutrition* 133, 2078S–2083S.

Hale, L.L., Pharr, G.T., Burgess, S.C., Corzo, A. and Kidd, M.T. (2004) Isoleucine needs of thirty-to-forty-day-old female chickens: immunity. *Poultry Science* 83, 1979–1985.

Humphrey, B.D. and Klasing, K.C. (2004) Modulation of nutrient metabolism and homeostasis by the immune system. *World's Poultry Science Journal* 60, 90–100.

Kidd, M.T. (2004) Nutritional modulation of immune function in broilers. *Poultry Science* 83, 650–657.

Kidd, M.T., Kerr, B.J. and Anthony, N.B. (1997) Dietary interactions between lysine and threonine in broilers. *Poultry Science* 76, 608–614.

Kidd, M.T., Kerr, B.J., Halpin, K.M., McWard, G.W. and Quarles, C.L. (1998) Lysine levels in starter and grower-finisher diets affect broiler performance and carcass traits. *Journal of Applied Poultry Research* 7, 351–358.

Kidd, M.T., Gerard, P.D., Heger, J., Kerr, B.J., Rowe, D., Sistani, K. and Burnham, D.J. (2001a) Threonine and crude protein responses in broiler chicks. *Animal Feed Science and Technology* 94, 57–64.

Kidd, M.T., Peebles, E.D., Whitmarsh, S.K., Yeatman, J.B. and Wideman, Jr, R.F. (2001b) Growth and immunity of broiler chicks as affected by dietary arginine. *Poultry Science* 80, 1535–1542.

Kidd, M.T., Thaxton, J.P., Yeatman, J.B., Barber, S.J. and Virden, W.S. (2002) Arginine responses in broilers: live performance. *Poultry Science* 80 (1), 114.

Kidd, M.T., Barber, S.J., Virden, W.S., Dozier, W.A. III, Chamblee, D.W. and Wiernusz, C. (2003a) Threonine responses of Cobb male finishing broilers in differing environmental conditions. *Journal of Applied Poultry Research* 12, 115–123.

Kidd, M.T., Pote, L.M. and Keirs, R.W. (2003b) Lack of interaction between dietary threonine and *Eimeria acervulina* in chicks. *Journal of Applied Poultry Research* 12, 124–129.

Kidd, M.T., McDaniel, C.D., Branton, S.L., Miller, E.R., Boren, B.B. and Fancher, B.I. (2004) Increasing amino acid density improves live performance and carcass yields of commercial broilers. *Journal of Applied Poultry Research* 13, 593–604.

Kidd, M.T., Corzo, A., Hoehler, D., Miller, E.R. and Dozier, W.A. III (2005) Broiler responsiveness (Ross x 708) to diets varying in amino acid density. *Poultry Science* 84, 1389–1396.

Kling, H.F. and Quarles, C.L. (1974) Effect of atmospheric ammonia and the stress of infectious bronchitis vaccination on leghorn males. *Poultry Science* 53, 1161–1167.

Konashi, S., Takahashi, K. and Akiba, Y. (2000) Effects of dietary essential amino acid deficiencies on immunological variables in broiler chickens. *British Journal of Nutrition* 83, 449–456.

Koutsos, E.A. and Klasing, K.C. (2001) Interactions between the immune system, nutrition, and productivity of animals. In: Garnsworthy, P.C. and Wiseman, J. (eds) *Recent Advances in Animal Nutrition*. Nottingham University Press, Nottingham, UK, pp. 173–190.

Kristensen, H.H. and Wathes, C.M. (2000) Ammonia and poultry welfare: a review. *World's Poultry Science Journal* 56, 235–245.

Latshaw, J.D. (1993) Dietary lysine concentrations from deficient to excessive and the effects on broiler chicks. *British Poultry Science* 34, 951–958.

Miles, D.M., Branton, S.L. and Lott, B.D. (2004) Atmospheric ammonia is detrimental to the performance of modern commercial broilers. *Poultry Science* 83, 1650–1654.

National Research Council (NRC) (1994) *Nutrient Requirements of Poultry*, 9th revision edition. National Academy Press, Washington DC.

Rosebrough, R.W., Mitchell, A.D. and McMurty, P.J. (1996) Dietary crude protein changes rapidly after metabolism and plasma insulin-like growth factor I concentrations in broiler chickens. *Journal of Nutrition* 126, 2888–2898.

Siegel, P.B. and Gross, W.B. (1980) Production and persistency of antibodies in chickens to sheep erythrocytes. 1: directional selection. *Poultry Science* 59, 1–5.

Smith, E.R., Pesti, G.M., Bakalli, R.I., Ware, G.O. and Menten, J.F.M. (1998) Further studies on the influence of genotype and dietary protein on the performance of broilers. *Poultry Science* 77, 1678–1687.

Temim, S., Chagneau, M., Guillaumin, S., Michel, J., Peresson, R. and Tesseraud, S. (2000) Does excess dietary protein improve growth performance and carcass characteristics in heat-exposed chickens? *Poultry Science* 79, 312–317.

Tesseraud, S., Pym, R.A., Bihan-Duval, E.L. and Duclos, M. (2003) Response of broilers selected on carcass quality to dietary protein supply: live performance, muscle development, and circulating insulin-like growth factors (IGF-I and -II). *Poultry Science* 82, 1011–1016.

Thornton, S.A., Pharr, G.T., Corzo, A., Branton, S.L. and Kidd, M.T. (2004) Valine needs for immune responses in male broilers from day 21 to 42. *Poultry Science* 83 (1), 186.

Whitehead, C.C. (2002) Nutrition and poultry welfare. *World's Poultry Science Journal* 58, 349–356.

Wijtten, P.J.A., Prak, R., Lemme, A. and Langhout, D.J. (2004) Effect of different dietary ideal protein concentrations on broiler performance. *British Poultry Science* 45, 504–511.

CHAPTER 12
The role of feed processing on gastrointestinal function and health in poultry

B. Svihus

Department of Animal and Aquacultural Sciences, Norwegian University of Life Sciences, Ås, Norway; e-mail: birger.svihus@umb.no

ABSTRACT

Nearly all feeds used in commercial poultry production are subjected to some form of feed processing. A majority of the feed is ground through a hammer mill and formed into pellets by pressing the heated feed through a pellet press. The two major effects of processing are thus: (i) changes in the micro- and macrostructure of the feed; and (ii) heat-induced chemical changes of some components in the feed. It has been shown that the microstructure of the feed, i.e. the particle distribution of the dissolved feed particles, has a significant impact on gut function. Gizzard activity increases significantly, resulting in a very finely ground feed material entering the small intestine. An increased feed volume in the gizzard and a decrease in gizzard pH when coarse material is fed results in a lower jejunal microflora number and changes in microfloral composition. In addition, secretion of bile acids and pancreatic enzymes is enhanced.

The macrostructure determines to a large extent the feed intake pattern. It has been shown that a feed structure that allows for a high feed intake, such as pellets, may affect starch digestibility negatively. This may be associated with overconsumption of feed, and may be linked to a suboptimal feed flow regulation due to reduced gizzard function. Although poultry feeds are subjected to only moderate heat treatment, heat-induced chemical changes may take place. A small fraction of the starch will gelatinize, solubility of fibres may increase and proteins may form indigestible new bonds. In addition, vitamins and enzymes may lose their effect.

INTRODUCTION

More than 600 million t of processed concentrates are produced annually on a worldwide basis (Gill, 2003). Poultry feeds are the largest product group, with

37% of the total feed production. A typical poultry feed production process consists of grinding the feed ingredients in a hammer mill through a 3–4 mm sieve. Feed for layers is commonly fed as a mash feed, while other types of feed are passed through a conditioner followed by pelleting. The temperature will normally reach a maximum of 80–90°C during the pelleting process.

Although there has been surprisingly little research on the interactions between feed processing and nutrition, it is evident that the physical and chemical changes that take place during processing may have a large impact on the performance of birds, both directly through effects on feed digestibility and indirectly through effects on feed intake pattern and gut function. In this chapter, some effects of processing will be covered. Since any effect on nutrient absorption efficiency will potentially affect gut function and gut health, nutritional aspects will also be addressed.

EFFECTS OF STRUCTURE

One of the functions of feed processing is to affect the structure of the feed. Grinding is performed to reduce the particle size of feed ingredients, while the pelleting process is performed to increase the particle size of the feed. Thus, in a pelleted feed, structure concerns both the particle distribution of the ingredients that the pellet consists of and the structure of the pellets themselves. It is thus pertinent to describe the structure of the particles inside the pellets as the 'microstructure' and the structure of the pellets themselves as the 'macrostructure'. As pellets normally disintegrate quite rapidly once they are moistened in the upper digestive tract, the macrostructure will affect only feed intake pattern and total feed intake. The microstructure, on the other hand, will affect gut function through the effects of particle size and distribution.

Effects of microstructure

Although grinding represents a considerable cost in terms of energy consumption and feed mill capacity, cereals used for poultry are usually finely ground in a hammer mill fitted with a screen between 3–4.5 mm in size. One major reason for this fine grinding is to ensure a good pellet quality. Fine grinding of cereals has also been considered to be beneficial for a proper utilization of the feed, but this view is not supported by most scientific literature (Farrell *et al.*, 1983; Reece *et al.*, 1985; Reece *et al.*, 1986a, b; Deaton *et al.*, 1989; Proudfoot and Hulan, 1989; Lott *et al.*, 1992; Nir *et al.*, 1994a, b; Hamilton and Proudfoot, 1995; Nir *et al.*, 1995; Hamilton and Kennie, 1997; Svihus *et al.*, 2004b).

Some studies have even shown an improved performance with the use of coarsely ground cereals in mash form (Reece *et al.* 1985; Nir *et al.*, 1990, 1995). Birds prefer coarse particles and will eat more of a coarsely ground than of a finely ground mash diet (Reece *et al.*, 1985; Portella *et al.*, 1988 a, b; Nir *et al.*, 1990, 1994b).

It is thus necessary to use pelleted diets to separate the effect of feed structure on feed intake stimulation from the effect of feed structure on feed utilization. Experiments conducted with pelleted diets containing cereals ground to different sizes usually demonstrate that coarse grinding gives similar weight gain and feed utilization to those of fine grinding (Reece et al., 1985, 1986a, b; Proudfoot and Hulan, 1989; Nir et al., 1995; Hamilton and Kennie, 1997; Svihus et al., 2004b). Considering the numerous recent results showing that the young broiler chicken is able efficiently to utilize diets with up to 400 g unground wheat/kg (see Svihus et al., 2004a, for references), this is not surprising.

It has been shown that coarse particles are very finely ground in the gizzard (Hetland et al., 2002) and that this grinding does not affect passage rate through the intestinal tract (Svihus et al., 2002). This indicates that from a nutritional point of view the cereals used for poultry feed can be ground more coarsely than is current practice. A coarser grinding would save considerable energy and time in the grinding process. Reece et al. (1986a) found that the energy cost for hammer grinding of maize could be reduced from 2.9 kwh/t to 2.2 kwh/t by increasing the sieve size from 4.76 to 6.35 mm. In addition, coarser grinding will reduce the amount of fines in the pelleted product and thus, potentially, reduce transport loss and dust problems in the feed mill and on the farm.

One concern with using coarsely ground cereals in the feed has been the pellet quality. As the proportion of coarse particles increases in the pellet, this may result in more fractures due to reduced particle surface area and weak spots in the pellet. Although Svihus et al. (2004b) observed a moderate reduction in pellet durability with a coarser grinding, other data have shown improved pellet quality with a coarser grinding (Reece et al., 1986a, b).

Roller mills are not commonly used in feed production, despite the fact that they grind with a lower energy consumption per ton of feed (Nir et al., 1990), give less dust during grinding and are less noisy (Heimann, 2002). It has also been shown that roller mills grind cereals more uniformly and produce a lower amount of fines than do hammer mills (Nir et al., 1990, 1995). As roller mills are considered to be particularly effective for coarse grinding, the advantages of using a roller mill will increase when a coarse grinding is used. A major disadvantage of roller mills is the higher cost for purchase and maintenance (Heimann, 2002).

Birds eat and swallow food without prior chewing. Although secretion of saliva does occur, in the chicken and turkey the saliva does not contain amylase and the time exposed to saliva in the mouth is short (Duke, 1986). Thus food enters the crop or, if the gizzard is empty, passes directly to the proventriculus and gizzard largely as intact food particles. The crop may store large amounts of food for a considerable amount of time. Svihus et al. (2002) found up to 30 g feed in the crop 30 min after feeding starved broiler chickens weighing 1100–1400 g, with half this feed still remaining in the crop after 3 h.

Although no digestive enzymes are secreted and no grinding occurs in the crop, feed may be softened and endogenous and microbial enzymes may be activated. The main body of the gizzard comprises two thick, opposed lateral muscles and two thin anterior and posterior muscles. A thick layer of glycoprotein that is hardened by the low pH covers the inside of the gizzard. This koilin layer is composed of a combination of rod-like and granular secretions from the gizzard's glands, and is continuously renewed by these glands as it is utilized. Some birds, however, are known to shed the whole koilin layer at intervals (McLelland, 1979). None of the domesticated bird species have, as yet, been reported to exhibit such behaviour, but unpublished observations of gizzards lacking the koilin layer in our laboratory indicate that this may happen in turkeys. The rod-like, hardened glycoprotein will stand out from the remaining glycoprotein matrix and will give the surface its characteristic sandpaper-like appearance (Hill, 1971). The thick muscles grind material by contraction that rubs the material against the koilin layer on the inside of the gizzard. The small muscles move material between contractions of the large muscles.

A prerequisite for utilizing coarsely ground or un-ground cereals is a well-functioning gizzard. Feeding studies with broilers have shown that the gizzard becomes stimulated when the feed's microstructure is improved, e.g. when whole cereals replace ground cereals. It has also been shown that replacing 50% ground wheat with the same quantity of ground oats gives the same increase in gizzard weight as that achieved by replacing 50% ground wheat with whole wheat (Hetland et al., 2002). This indicates that a relatively small amount of insoluble fibre (e.g. 10% oat hulls coming from 50% oats in the diet) stimulates the gizzard to the same extent as 50% whole wheat.

Current data indicate that the gizzard weight is less affected by cereal structure in layers. However, insoluble fibre seems to play an even more profound role for gizzard activity in the more mature digestive tract of the layer. This corresponds with an earlier experiment with layers, where consumption of 4% of feed as wood shavings resulted in a 50% heavier gizzard, whereas including 40% whole wheat in wheat-based diets increased the gizzard weight by only 10% (Hetland et al., 2003a).

Coarse feed particles need to be ground to a certain critical size before they can leave the gizzard (Clemens et al., 1975; Moore, 1999), causing the volume of gizzard contents to increase when diets with whole cereals or insoluble fibre are fed (Hetland et al., 2003a, b). Hetland et al. (2002) thus found that 80–90% of particles in the duodenum of broilers were smaller than 300 μm and that this was independent of dietary particle size. This suggests that coarse particles accumulate in the gizzard and are retained longer than are other nutrients. Hetland et al. (2005) showed that finely ground oat hulls passed through the gizzard much faster than coarse oat hulls. Thus, an increased volume of gizzard contents in birds fed diets with whole grains or high levels of insoluble fibre has been observed (Hetland et al., 2003a, b). Indeed, the gizzard has a remarkable ability to grind all organic constituents of feed to a very consistent particle size range regardless of the original particle size of the feed, with some ground as finely as 0.05 mm.

It has also been shown that stimulation of the gizzard function by structural components results in a reduction in gizzard pH (Engberg et al., 2004; Bjerrum et al., 2005). This, combined with an increased retention time, improves the potential for the gizzard to function as a barrier to foodborne microorganisms. The same authors also observed a signifcant reduction in pathogenic microflora when ground wheat was replaced by whole wheat. Field evidence that more microstructure in the feed is beneficial to gut health has also been provided by Branton *et al.* (1987), where mortality due to necrotic enteritis was much lower in birds given coarsely ground diets than in broilers given finely ground diets.

A significant increase in amylase activity and bile acid concentration has been observed when diets contain more structural components (Svihus *et al.*, 2004a). This may indicate that an increased secretory activity is the cause of the improvements in nutritive value sometimes observed with whole wheat or other structural components (Kiiskinen, 1996; Plavnik *et al.*, 2002; Hetland *et al.*, 2003a; Svihus *et al.*, 2004a). This hypothesis is also supported by the fact that amylase activity in the jejunum and starch digestibility in the anterior, median and posterior ileum for individual birds had a correlation of 0.56, 0.54 and 0.47, respectively (Svihus *et al.*, 2004a).

The cause of this increased secretory activity remains unclear, but may be associated with a stimulation of pancreatic secretion caused by an increase in gizzard activity. Hetland *et al.* (2003a) found a significant increase in amylase activity and bile acid secretion when gizzard activity was stimulated by oat hulls. The same tendency, although not significant for amylase activity, was seen when gizzard activity was stimulated by whole wheat.

The major stimuli causing increased exogenous enzyme secretion by the pancreas are the vagus nerve and cholecystokinin (CCK). Cholecystokinin is mainly produced in the pyloric region of birds (Denbow, 2000) and acts via the vagus nerve to stimulate pancreatic enzyme secretion (Li and Owyang, 1993). Thus, a stimulating effect of increased gizzard activity on enzyme secretions may be mediated through increased CCK release in the pyloric region of the gizzard. Gabriel *et al.* (2003) found tendencies for reduced levels of brush border enzyme activity by the inclusion of whole wheat. It is thus possible that a stimulatory effect of coarse feed structure is limited to pancreatic enzyme secretions.

Cholecystokinin has also been shown to stimulate gastrointestinal refluxes (Duke, 1992). Svihus *et al.* (2004a) observed a significant increase in the content of bile acids in the gizzard when large wheat particles were fed. This indicates increased gastroduodenal refluxes, which may be caused by increased CCK secretion. As indicated by the performance data, the increased gastroduodenal reflux does not reduce feed consumption. The results discussed above clearly suggest that from a nutritional point of view the microstructure of feeds for poultry can be very coarse, for example by using coarsely ground or even un-ground cereals. The results also indicate that a coarse microstructure may be beneficial from a gastrointestinal health point of view.

Effects of macrostructure

It is well known that feed intake will improve with pelleted diets compared to that with mash diets. In broilers an increase in feed intake of 25% is often observed (Svihus *et al.*, 2004b). As opposed to the microstructure, improvements in macrostructure do not result in increased gizzard weight or gizzard volume. In fact, a reduction in relative gizzard weight and gizzard feed content is observed (Engberg *et al.*, 2002; Svihus *et al.*, 2004b). This is due to the fact that pellets disintegrate quite rapidly when they are moistened in the upper digestive tract.

It has been shown that the pelleting process results in a significantly finer microstructure (Svihus *et al.*, 2004b). This is caused by the grinding action of the rolls in the pellet press, and may be one contributing factor to the smaller relative size of the gizzard. As relative size of the gizzard is reduced with increasing weight of the bird (Hetland *et al.*, 2002), this may also contribute to a smaller relative size of the gizzard in birds fed pellets. Pelleting results in significant improvements in the feed/gain ratio (Engberg *et al.*, 2002; Svihus *et al.*, 2004b), but this is not often associated with improvements in digestibility or metabolizable energy content.

In fact, a reduction in starch digestibility has been observed when pelleted diets were compared to mash diets (Svihus and Hetland, 2001), and based on these observations it has been hypothesized that some birds are overeaters when presented with pelleted diets. Engberg *et al.* (2002) also showed a significant reduction in enzyme activities when pelleted diets were used instead of mash diets.

The hypothesis of some birds being overconsumers is supported by further studies in our laboratory indicating a slight negative correlation between feed intake of individual birds on identical diets and the apparent metabolizable energy (AME) value determined (see Figs 12.1 and 12.2). In these studies, male Ross 208 broilers were raised in individual cages from 13 days of age and pelleted diets were provided *ad libitum*. AME content and correlation between

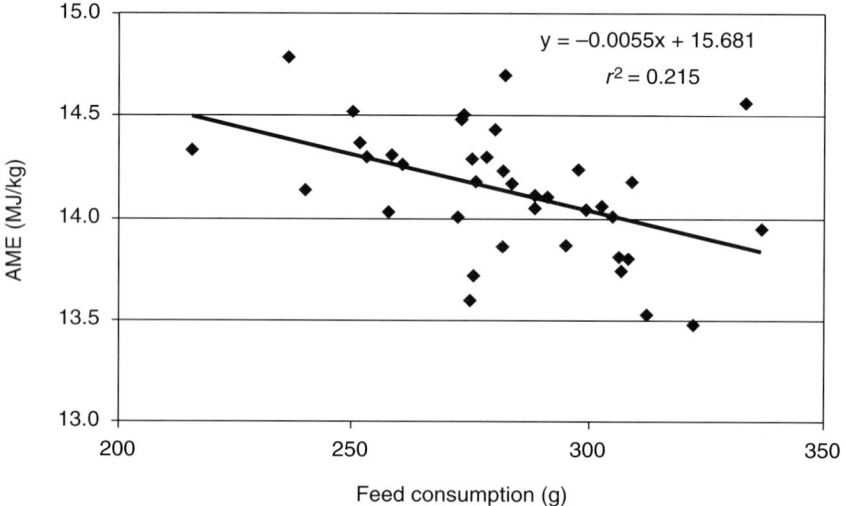

Fig. 12.1. Relationship between feed consumption and apparent metabolizable energy (AME) determined on individual birds fed on a pelleted, maize-based diet.

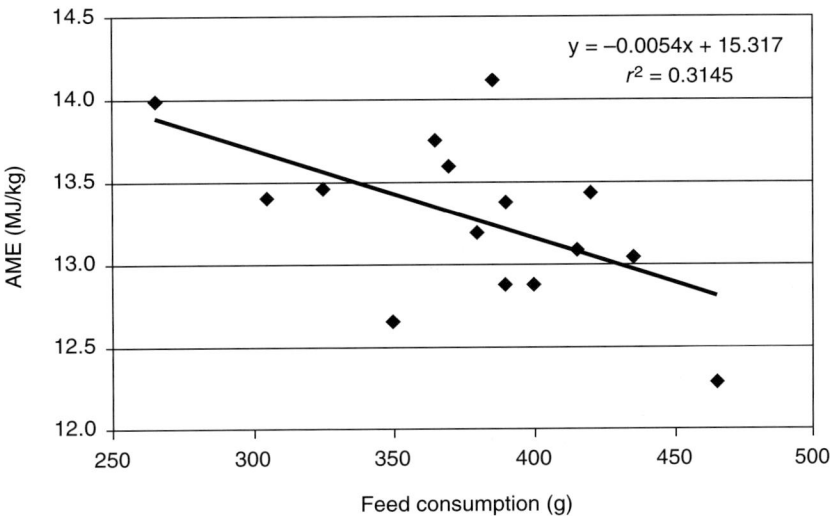

Fig. 12.2. Relationship between feed consumption and apparent metabolizable energy (AME) determined on individual birds fed on a pelleted, wheat-based diet.

feed intake and AME was determined by monitoring feed and faeces quantitatively from days 19–22, combined with determination of energy content using a calorimeter.

These data also show that feed intake varies between birds. It is thus possible that overconsumption among individual birds in a flock results in impaired starch digestion due to an increased passage rate. This is consistent with many observations of reduced variation among individuals concurrent with an increased starch digestibility (Svihus and Hetland, 2001). The question may then arise on the causes of overconsumption among individual birds.

Peter Siegel was among the first to indicate that intensive breeding for weight gain had resulted in broilers that were capable of overconsuming feed due to disturbance of the appetite control centre in the brain (Siegel and Dunnington, 1987). Lacy *et al.* (1985) showed that broilers, as opposed to layers, did not respond to intra-hepatic glucose infusions with reduced feed intake. This indicates that peripheral appetite control centres have also been disturbed in broilers.

It can thus be postulated that modern breeds of broilers may overconsume feeds, with a resulting impairment in digestibility. Our results indicate that there may be an interaction between gizzard function and over-consumption of feeds. When feeds have more structure, either through the use of whole cereals or through the addition of large fibre particles (oat hulls or wood shavings) into the diet, an improvement in starch digestibility and a reduced variation between birds is commonly observed (Rogel *et al.*, 1987; Svihus and Hetland, 2001; Hetland *et al.*, 2002).

Our hypothesis is that an active and well-developed gizzard will function as a food intake regulator that will ensure that the bird does not overconsume feeds. This is due to both: (i) the active retention of large particles until they are broken down, which results in a larger filling of the gizzard and thus less room for more feed to be consumed; and (ii) a stimulative effect of gizzard activity on gastroduodenal refluxes.

Effects of heat on nutritional value of pelleted feeds

For poultry, the majority of heat treatment occurs during conditioning and pelleting. During conditoning, the temperature is usually raised to 70–80°C by the addition of steam. This temperature may increase to 85–90°C during the pelleting process. Although these temperatures are moderate compared to those achieved during extrusion and expansion of feeds, they may result in the destruction of the three-dimensional structure of proteins and starch, and also in new covalent bond formations such as disulphide bridges, iso-peptide bonds and Maillard-reactions.

There is a scarcity of published results on changes in protein availability during pelleting. Carré *et al.* (1991) did not find any increase in protein digestibiliy in broiler chickens as a result of pelleting, while Vande Ginste and De Schrijver (1998) found a small but significant increase in protein digestibility in finishing pigs when pelleting was compared to mash feeding. In extruded diets, a small decrease in protein digestibility is often observed (Ljøkjel *et al.*, 2004), but since extrusion is a much more severe heating process – with temperatures usually exceeding 110°C – it is logical to conclude that protein digestibility is probably not affected negatively by the pelleting process.

Starch is gelatinized to some extent during the pelleting process. The extent of starch gelatinization is usually found to vary between 5 and 20% (Svihus *et al.*, 2004b), but no improvements in starch digestibility have been established as a consequence of the pelleting process, except for a significant improvement from pelleting of peas (Carré *et al.*, 1991). This latter improvement has been associated with large, starch-containing particles in the non-pelleted diets.

The antinutritive effect of soluble fibres in the diet is well known, the major mechanism for which is an increased viscosity of the gut contents, and thus a reduced digestibility and nutrient absorption. These changes may also affect gut development and chyme mobility, and they have been shown to have an impact on gut microflora composition. Thus, the viscosity effect of soluble fibres is important for gut health and function. It has been shown that processing may increase the solubility of fibres, thus resulting in an increased viscosity of gut contents and reduced nutrient digestibility. Graham *et al.* (1989) observed increased solubility of barley β-glucans after pelleting, and Cowieson *et al.* (2005) observed an increased viscosity of wheat diets with increasing conditioning temperature.

In an experiment carried out at the Norwegian University of Life Sciences, complete diets containing 77% barley, oats or wheat were fed to broiler chickens from 7 days of age. The cereal used in each diet was either added untreated, or it was heat-treated by conditioning at 90°C followed by pelleting and regrinding. At 21 days of age the birds were killed and digesta contents

Table 12.1. Properties of gut contents in broilers fed complete cold-pelleted diets based on 77% cereals either non-heated or pelleted at 90°C.

| | Cereal source | | | | | |
| | Oats | | Barley | | Wheat | |
	Non-heated	Heat-treated	Non-heated	Heat-treated	Non-heated	Heat-treated
Jejunal viscosity (cP)	25.9[b]	111.4[a]	26.6[b]	80.8[a]	15.4[b]	30.2[a]
DM in chyme (%)	15.7[a]	12.7[b]	18.0[a]	16.6[b]	25.0[a]	19.8[b]
Dirtiness score[c]	2.9[b]	4.2[a]	2.9[b]	3.8[a]	2.9[b]	3.2[a]

[ab] Within each cereal category, numbers within a row which do not share common superscripts are significantly different ($p < 0.05$) in a Ryan-Einot-Gabriel-Welsh test.
[c] Subjective score of cage dirtiness, with the highest score of 5 for the dirtiest cages.

collected. The results presented in Table 12.1 show a tremendous increase in the jejunal viscosity of broilers when the cereals were heat-treated, followed by a typical increase in water content in the chyme and higher stickiness of the faeces.

It is well known that many vitamins and carotenoids are susceptible to heat denaturation. Andersen and Sunderland (2002) found that 67% of vitamin E was recovered after an extrusion process, and that recovery decreased with increasing water content. This indicates that vitamin E is heat-sensitive, although it is reasonable to conclude that much less would be destroyed during a pelleting process, due to less water and lower temperatures. Marchetti *et al.* (1999) observed significant reductions in the recovery of crystalline vitamins after pelleting. For vitamins K and C recovery was 50%, while the B-vitamins generally had recoveries between 82 and 96%. Exceptions were pyridoxine and folic acid, with recoveries of 75 and 65%, respectively. Thus, pelleting can have a large impact on the content of some vitamins and this must be taken into account during feed processing.

Enzymes are now routinely used in most poultry feeds. Since enzymes are proteins which act as catalysts due to their specific three-dimensional structures, they are susceptible to heat destruction through denaturation. Inborr and Bedford (1994) demonstrated that β-glucanase activity was reduced to 66% of its initial value after conditioning at 75°C followed by pelleting. After conditioning at 95°C, this was reduced to 16% of the initial activity. Even though more heat-stable enzyme products have been developed using coating technologies or by selection of less heat-sensitive proteins, reduced activity due to processing can still be expected.

Based on the above it can be concluded that the temperature should be kept as low as possible during feed processing to avoid reduction in the nutritional value of the feed. This also calls for caution in using processes such as expansion and extrusion for the production of poultry feeds, as these processes involve high temperatures and high pressure. McCracken (2002), while reviewing the literature on feed processing, reported data showing a reduced performance by expander treatment of poultry feeds.

CONCLUSION

Micro- and macrostructure of the feed as determined by processing has a large impact on gut function and digestive processes. Gut health may also be affected through effects on available substrate in the lower digestive tract, and through changes in pH and retention time in the upper digestive tract. Heat treatment does not seem to have any significant beneficial effect on nutrient digestion, and may have a negative effect on nutrient digestibility through solubilization of fibres and through destruction of vitamins and enzymes added to the feed. Thus, temperature during processing should be kept as low as possible.

REFERENCES

Andersen, J.S. and Sunderland, R. (2002) Effect of extruder moisture and dryer processing temperature on vitamin C and E and astaxanthin stability. *Aquaculture* 207, 137–149.

Bjerrum, L., Pedersen, K. and Engberg, R.M. (2005) The influence of whole wheat feeding on salmonella infection and gut flora composition in broilers. *Avian Diseases* 49, 9–15.

Branton, S.L., Reece, F.N. and Hagler, W.M. (1987) Influence of a wheat diet on mortality of broiler chickens associated with necrotic enteritis. *Poultry Science* 66, 1326–1330.

Carré, B., Beaufils, E. and Melcion, J.P. (1991) Evaluation of protein and starch digestibilities and energy value of pelleted or unpelleted pea-seeds from winter or spring cultivars in adult and young chickens. *Journal of Agricultural and Food Chemistry* 39, 468–472.

Clemens, E.T., Stevens, C.E. and Southworth, M. (1975) Sites of organic acid production and pattern of digesta movement in the gastrointestinal tract of geese. *Journal of Nutrition* 105, 1341–1350.

Cowieson, A.J., Hruby, M. and Isaksen, M.F. (2005) The effect of conditioning temperature and exogenous xylanase addition on the viscosity of wheat-based diets and the performance of broiler chickens. *Proceedings of the 15th European Symposium on Poultry Nutrition*, WPSA, Balatonfüred, Hungary, pp. 256–259.

Deaton, J.W., Lott, B.D. and Simmons, J.D. (1989) Hammer mill *versus* roller mill grinding of corn for commercial egg layers. *Poultry Science* 68, 1342–1344.

Denbow, D.M. (2000) Gastrointestinal anatomy and physiology, In: Whittow, G.C. (ed.) *Sturkie's Avian Physiology*, Academic Press, New York, pp. 299–325.

Duke, G.E. (1986) Alimentary canal: anatomy, regulation of feeding, and motility. In: Sturkie, P.D. (ed.) *Avian Physiology* (4th edn), Springer-Verlag, New York, pp. 269–288.

Duke, G.E. (1992) Recent studies on regulation of gastric motility in turkeys. *Poultry Science* 71, 1–8.

Engberg, R.M., Hedemann, M.S. and Jensen, B.B. (2002) The influence of grinding and pelleting of feed on the microbial composition and activity in the digestive tract of broiler chickens. *British Poultry Science* 43, 569–579.

Engberg, R.M., Hedemann, M.S., Steenfeldt, S. and Jensen, B.B. (2004) Influence of whole wheat and xylanase on broiler performance and microbial composition and activity in the digestive tract. *Poultry Science* 83, 925–938.

Farrell, D.J., Thomson, E. and Choice, A. (1983) Effects of milling and pelleting of maize, barley and wheat on their metabolizable energy value for cockerels and chicks. *Animal Feed Science and Technology* 9, 99–105.

Gabriel, I., Mallet, S. and Leconte, M. (2003) Differences in the digestive tract characteristics of broiler chickens fed on complete pelleted diet or on whole wheat added to pelleted protein concentrate. *British Poultry Science* 44, 283–290.

Gill, C. (2003) Back to basics of growth. *Feed International* 23 (January), 6–9.

Graham, H., Fadel, J.G., Newman, C.W. and Newman, R.K. (1989) Effect of pelleting and β-glucanase supplementation on the ileal and faecal digestibility of a barley-based diet in the pig. *Journal of Animal Science* 67, 1293–1298.

Hamilton, R.M.G. and Kennie, J. (1997) The effects of lighting program, ingredient particle size and feed form on the performance of broiler turkeys. *Canadian Journal of Animal Science* 77, 503–508.

Hamilton, R.M.G. and Proudfoot, F.G. (1995) Ingredient particle size and feed texture: effects on the performance of broiler chickens. *Animal Feed Science and Technology* 51, 203–210.

Heimann, M. (2002) The 'Bottom line' of grinding. *Feed International* 23 (May), 32–34.

Hetland, H., Svihus, B. and Olaisen, V. (2002) Effect of feeding whole cereals on performance, starch digestibility and duodenal particle size distribution in broiler chickens. *British Poultry Science* 43, 416–423.

Hetland, H., Svihus, B. and Krogdahl, Å. (2003a) Effects of oat hulls and wood shavings on digestion in broilers and layers fed diets based on whole or ground wheat. *British Poultry Science* 44, 275–282.

Hetland, H., Svihus, B., Lervik, S. and Moe, R. (2003b) Effect of feed structure on performance and welfare in laying hens housed in conventional and furnished cages. *Acta Agriculturae Scandinavica* 53, 92–100.

Hetland, H., Svihus, B. and Choct, M. (2005) Role of insoluble fibre on gizzard activity in layers. *Journal of Applied Poultry Research* 14, 38–46.

Hill, K.J. (1971) The structure of the alimentary tract. In: Bell, D.J. and Freeman, B.M. (eds), *Physiology and Biochemistry of the Domestic Fowl*, Academic Press, London, pp. 1–23.

Inborr, J. and Bedford, M.R. (1994) Stability of feed enzymes to steam pelleting during feed processing. *Animal Feed Science and Technology* 46, 179–196.

Kiiskinen, T. (1996) Feeding whole grain with pelleted diets to growing broiler chickens. *Agricultural and Food Science in Finland* 5, 167–175.

Lacy, M.P., Van Krey, H.P., Skewes, P.A. and Denbow, D.M. (1985). Effect of intrahepatic glucose infusions on feeding in heavy and light breed chicks. *Poultry Science* 64, 751–756.

Li, Y. and Owyang, C. (1993) Vagal afferent pathway mediates physiological action of cholecystokinin on pancreatic-enzyme secretion. *Journal of Clinical Investigations* 92, 418–424.

Ljøkjel, K., Sørensen, M., Storebakken, T. and Skrede, A. (2004) Digestibility of protein, amino acids and starch in mink (*Mustela vison*) fed diets processed by different extrusion conditions. *Canadian Journal of Animal Science* 84, 673–680.

Lott, B.D., Day, E.J., Deaton, J.W. and May, J.D. (1992). The effect of temperature, dietary energy level, and corn particle size on broiler performance. *Poultry Science* 71, 618–624.

Marchetti, M., Tossani, N., Marchetti, S. and Bauce, G. (1999) Stability of crystalline and coated vitamins during manufacture and storage of fish feeds. *Aquaculture Nutrition* 5, 115–120.

McCracken, K.J. (2002) Effects of physical processing on the nutritive value of poultry diets. In: McNab, J.M. and Boorman, K.N. (eds) *Poultry Feedstuffs. Supply, Composition and Nutritive Value*, CABI Publishing, Wallingford, UK, pp. 301–316.

McLelland, J. (1979) Digestive system. In: King, A.S. and McLelland, J. (eds) *Form and Function in Birds*, Academic Press, London, pp. 69–181.

Moore, S.J. (1999) Food breakdown in an avian herbivore; who needs teeth? *Australian Journal of Zoology* 47, 625–632.

Nir, I., Melcion, J.P. and Picard, M. (1990) Effect of particle size of sorghum grains on feed intake and performance of young broilers. *Poultry Science* 69, 2177–2184.

Nir, I., Shefet, G. and Aaroni, Y. (1994a) Effect of particle size on performance. 1. Corn. *Poultry Science* 73, 45–49.

Nir, I., Hillel, R., Shefet, G. and Nitsan, Z. (1994b) Effect of grain particle size on performance. 2. Grain texture interactions. *Poultry Science* 73, 781–791.

Nir, I., Hillel, R., Ptichi, I. and Shefet, G. (1995) Effect of particle size on performance. 3. Grinding pelleting interactions. *Poultry Science* 74, 771–783.

Plavnik, I., Macovsky, B. and Sklan, D. (2002) Effect of feeding whole wheat on performance of broiler chickens. *Animal Feed Science and Technology* 96, 229–236.

Portella, F.J., Caston, L.J. and Leeson, S. (1988a) Apparent feed particle size preference by laying hens. *Canadian Journal of Animal Science* 68, 915–922.

Portella, F.J., Caston, L.J. and Leeson, S. (1988b) Apparent feed particle size preference by broilers. *Canadian Journal of Animal Science* 68, 923–930.

Proudfoot, F.G. and Hulan, H.W. (1989) Feed texture effects on the performance of roaster chickens. *Canadian Journal of Animal Science* 69, 801–807.

Reece, F.N., Lott, B.D. and Deaton, J.W. (1985) The effects of feed form, grinding method, energy level, and gender on broiler performance in a moderate (21°C) environment. *Poultry Science* 64, 1834–1839.

Reece, F.N., Lott, B.D. and Deaton, J.W. (1986a) Effects of environmental temperature and corn particle size on response of broilers to pelleted feed. *Poultry Science* 65, 636–641.

Reece, F.N., Lott, B.D. and Deaton, J.W. (1986b) The effects of hammer mill screen size on ground corn particles size, pellet durability, and broiler performance. *Poultry Science* 65, 1257–1261.

Rogel, A.M., Balnave, D., Bryden, W.L. and Annison, E.F. (1987) Improvement of raw potato starch digestion in chickens by feeding oat hulls and other fibrous feedstuffs. *Australian Journal of Agricultural Research* 38, 629–637.

Siegel, P.B. and Dunnington, E.A. (1987). Selection for growth in chickens. *CRC Critical Reviews in Poultry Biology* 1, 1–24.

Svihus, B. and Hetland, H. (2001) Ileal starch digestibility in growing broiler chickens fed on a wheat-based diet is improved by mash feeding, dilution with cellulose or whole wheat inclusion. *British Poultry Science* 42, 633–637.

Svihus, B., Hetland, H., Choct, M. and Sundby, F. (2002) Passage rate through the anterior digestive tract of growing broiler chickens fed on diets with ground and whole wheat. *British Poultry Science* 43, 662–668.

Svihus, B., Juvik, E., Hetland, H. and Krogdahl, Å. (2004a) Causes for improvement in nutritive value of broiler chicken diets with whole wheat instead of ground wheat. *British Poultry Science* 45, 1–6.

Svihus, B., Kløvstad, K.H., Perez, V., Zimonja, O., Sahlström, S., Schüller, R.B., Jeksrud, W.K. and Prestløkken, E. (2004b) Nutritional effects of pelleting of broiler chicken diets made from wheat ground to different coarsenesses by the use of roller mill and hammer mill. *Animal Feed Science and Technology* 117, 281–293.

Vande Ginste, J. and De Schrijver, R. (1998) Expansion and pelleting of starter, grower and finisher diets for pigs: effects on nitrogen retention, ileal and total tract digestibility of protein, phosphorus and calcium and *in vitro* protein quality. *Animal Feed Science and Technology* 72, 303–314.

CHAPTER *13*
Wet litter: its causes and prevention and the role of nutrition

S.R. Collett

The University of Georgia, College of Veterinary Medicine, Poultry Diagnostic and Research Center, Athens, Georgia, USA; e-mail: colletts@uga.edu

ABSTRACT

Poultry litter becomes wet when the rate of water addition (urine/faeces/spillage) exceeds the rate of removal (evaporation). Anti-nutritional factors, toxins, pathogens and nutrient imbalances may cause wet litter directly by altering normal digestive physiology or indirectly by disturbing normal gut ecology.

Poor-quality ingredients and those with excess oligosaccharides or minerals can cause a nutritionally induced polydypsia, polyuria and diarrhoea which may increase water output sufficiently to cause wet litter. If the situation persists for long enough the ensuing inflammatory response causes mild to severe gastroenteritis which further increases water output. The damage so caused to the cytoskeleton of the gastrointestinal tract reduces the surface area for nutrient absorption and allows the opportunity for pathogen proliferation, gut colonization and even invasion and systemic disease.

Historically the solution to the downstream effect of compromised digestion and absorption on gut ecology has been to supplement birds with antibiotic growth promoters, a practice that has not been permitted within Europe from January 2006. The only alternative is to use a holistic approach in managing the gut ecology by manipulating the microbial populations and host inflammatory response through non-antibiotic nutritional techniques.

Enzymes, probiotics, prebiotics, immune modulators and mycotoxin binders have all shown promise in this regard. Seeding the gastrointestinal tract with bacteria at hatch and subsequently managing the gastrointestinal environment, whilst at the same time reducing anti-nutritional factors and the use of precision diet formulation, has proved advantageous in minimizing the incidence of wet litter in birds whilst maintaining and improving bird performance.

INTRODUCTION

It is important to the wellbeing, welfare and productivity of commercial poultry that housing conditions are maintained within *the comfort zone* and feed and water are available *ad libitum*. Under these controlled conditions water excretion (urine and faeces) closely parallels intake since evaporative water loss is limited. In pursuance of homeostasis, water balance is maintained by the bird as either positive (growth) or neutral.

Years of breeding and selection to improve growth rate, feed efficiency and egg production have increased appetite and hence daily feed and water intake. At standard temperatures broilers will consume approximately 1.8–2.0 times more water than feed (by weight), and water deprivation can severely affect feed intake and growth rate (Kellerup *et al.*, 1965; Leeson and Summers, 2005). During the log phase of growth, commercial poultry excrete exponentially increasing quantities of faecal and urinary water into the bedding/litter. It is crucial that the poultry house ventilation system design and operation is efficient enough to prevent the litter moisture content from exceeding an optimal 25%.

WATER BALANCE

Avian osmoregulation involves balancing water and electrolyte intake with excretion via the kidney (urine), gastrointestinal tract (faeces), skin and respiratory tract (evaporation). Since feed is generally low in moisture (10%) and metabolic water production is limited by diet formulation ($\sim$ 0.14g/kcal of dietary energy), moisture intake is primarily controlled by drinking ($\sim$ 80%) (Leeson and Summers, 2005).

Water consumption is requirement-driven and the satiety centre is stimulated by cellular dehydration (osmoreceptors), extracellular dehydration (mechanoreceptors) and angiotensin II secretion (renin–angiotensin axis) (Kaufman *et al.*, 1980; Kanosue *et al.*, 1990; Goldstein and Skadhauge, 2000). All three mechanisms of thirst stimulation are activated when birds are dehydrated and over-consumption frequently occurs because intestinal absorption is sluggish and cellular rehydration takes time. In order to maintain water balance under such conditions, excess water is excreted as dilute urine (Takei *et al.*, 1988).

Water intake is ambient temperature-dependent because evaporative cooling mechanisms account for 50–80% of moisture loss, depending on the state of hydration (Goldstein and Skadhauge, 2000). Consumed water also acts as a heat-sink in heat-stressed birds, providing an important means of thermoregulation. Insensible moisture loss during normal respiration increases markedly when panting is initiated in response to heat stress and, interestingly, 50–80% of total evaporative loss occurs through the skin, despite the lack of sweat glands (Marder and Ben-Asher, 1983; Goldstein and Skadhauge, 2000).

The control of this through cutaneous evaporation is not completely understood but is partly regulated by adrenergic-induced vasodilatation in response to elevated ambient temperatures (Marder and Ben-Asher, 1983; Marder and Raber, 1989).

The kidneys rid the body of excess water and solutes by a combination of glomerular filtration, tubular re-absorption and secretion. Excretory function is controlled by several hormones, including: (i) arginine vasotocin (antidiuretic); (ii) renin/angiotensin (diuretic/antidiuretic and natiuretic/antinatiuretic); (iii) aldosterone (antinatiuretic); (iv) atrial natriuretic peptide (diuretic and naturietic); and (v) parathyroid hormone (calcium mobilization and excretion of phosphorus) (Morild *et al.*, 1985; Clark and Mok, 1986; Gray *et al.*, 1988; Gray, 1993; Goldstein and Skadhauge, 2000).

Urinary excretion is somewhat unique in the avian species, since first the ureters open into the coprodeum and, secondly, the urine passes retrograde up the colon to the caeca before being evacuated via the cloaca with the faeces (Goldstein and Skadhauge, 2000). The content of the urine is significantly altered during its passage through the coprodeum, colon and caeca (Rice and Skadhauge, 1982).

Approximately 95% of the daily glomerular filtrate water (11 times total body water) is normally re-absorbed, but both filtration rate and reabsorption rate are altered by the avian antidiuretic hormone, arginine vasotocin (AVT), which is released by the neurohypophysis in response to an elevation in extracellular osmolality or a drop in blood pressure (Gray *et al.*, 1988; Goldstein and Skadhauge, 2000).

AVT reduces filtration rate and increases the rate of water reabsorption (70–99%), thus increasing urine concentration from typically iso-osmotic to approximately $3 \times$ the osmolality of plasma in the dehydrated state. Despite this limited concentrating ability in the avian kidney, the efficiency of the avian species in conserving water is similar to that of mammals. This is in part due to their being uricotelic (excretion of insoluble urates limits water loss), but also to the reprocessing of urine that takes place in the distal gastrointestinal tract.

Retrograde flow of urine from the urethra to the caeca via the coprodeum and colon, before excretion via the cloaca, facilitates reclamation of otherwise wasted water, electrolytes, nitrogen and energy. Sodium chloride, water, glucose, fatty acids, amino acids, phosphate and ammonia are actively absorbed from – while potassium is secreted into – the lumen of the lower gastrointestinal tract (Goldstein and Skadhauge, 2000). The relative efficiency of these mechanisms is heavily influenced by the luminal concentration of the candidate concerned (Arnason *et al.*, 1986; Arnason and Skadhauge, 1991; Goldstein and Skadhauge, 2000).

Chickens on grain-based diets (approximately 30 meq Na^+/kg) excrete a lot of NH_4^+, phosphate and K^+, while Na^+ and Cl^- together only account for just over 10% of the total osmolality. When the Na^+ concentration in the ration is raised threefold there is a corresponding increase in urine Na^+ and Cl^- ($<$ 50% of osmotic space) at the expense of the NH_4^+, while other solute concentrations change very little (Goldstein and Skadhauge, 2000). Provided the drinking water solute concentration does not exceed the regulatory capacity of the kidney, the bird is able to cope with relatively high dietary salt concentrations (Roberts and Hughes, 1983).

Although an increase in urinary output can cause wet litter, the condition is usually, and often incorrectly, interpreted as an increase in true faecal water consequent to diarrhoea or enteritis. Several feed ingredient characteristics can alter faecal water content directly by increasing ingesta osmolarity or by reducing transit time and absorptive surface area/function, thus compromising water absorption and stimulating intake.

Although feed passage rate has increased with appetite and intake (breeding and selection), this has been at the expense of time spent in the proventriculus and gizzard, and retention time in the small intestine has thus remained fairly constant (Washburn, 1991; Denbow, 2000).

Feed passage studies using insoluble (Branch and Cummings, 1978; Uden *et al.*, 1980; Ferrando *et al.*, 1987) or soluble markers (Vergara *et al.*, 1989) indicate that average retention time is around 5–9 h. Although feed markers first appear 1.6–2.6 h after ingestion, the time taken for complete clearance appears to have been poorly documented. This is unfortunate, since although feed transit time is inversely proportional to intake in broilers, retro-peristalsis has been shown to be an important part of avian digestive physiology, particularly after withholding feed (Clench *et al.*, 1989; Almirall and Esteve-Garcia, 1994).

Even normal dark periods could cause sufficient feed withdrawal to stimulate retro-peristalsis, so experimental feed passage studies may be confounded and may not accurately reflect reality. Any stimulus that alters feed and water intake may affect passage time, digestion and absorption efficiency and nutrient throughflow.

Apart from the direct feed efficiency implication of reduced digestion and absorption, the throughflow of undigested nutrients impacts downstream gut ecology (Gidenne, 1997). The morphology of the ileocaecal junction is such that only fluid or very small (dissolved or suspended) particles enter the caecal pouch when intra-luminal pressure increases during convergence of rectal retro-peristaltic and ileal peristaltic contractions (Duke, 1982; Denbow, 2000). Since the retro-peristaltic contractions of the colon/rectum are almost continuous, 87–97% of the caecal fluid originates from the urine (Akester *et al.*, 1967; Bjornhag and Sperber, 1977; Duke, 1982; Denbow, 2000).

The caeca have the capacity to absorb amino acids, and degradation of microbial protein by caecal proteases is thought to provide an additional source of amino acid nutrition (Bjornhag and Sperber, 1977). Urine-derived nitrogen (urates, ammonia, urea, creatinine, amino acids, etc.) reaching the caeca by retro-peristalsis and undigested (faecal) protein provide a source of nitrogen for microbial amino acid synthesis (Goldstein and Skadhauge, 2000). Unfortunately potentially toxic compounds such as ammonia, amines, phenols and indoles are also generated by the proteolytic and ureolytic activity of the caecal flora on non-digested nutrients that make their way through to the caecal pouches. (Gidenne, 1997). These toxic compounds affect floral ecology in the rabbit and the same is probably true for the broiler.

Extended feed withdrawal – as would occur with fever response or feed outages – would increase the amount of retro-peristalsis and alter the gastrointestinal environment sufficiently to impact the microbial ecology of the upper gastrointestinal tract (Clench *et al.*, 1989; Klasing, 1998). If liver nutrient reserve was exhausted by feed withdrawal, urine nitrogen (negative nitrogen balance) would also increase and require further caecal microbiotic adaptation (Denbow, 2000).

The flora of the lower gastrointestinal tract spend the majority of their existence in *stationary phase*, because intense competition for a limited source of nutrients causes near-starvation. Stationary phase evolution occurs very rapidly and continuously through mutation, selection and takeover (Zambrano *et al.*, 1993; Finkel and Kolter, 1999).

Selection favours gene expression over function and, in the case of *E. coli*, enhanced amino acid scavenging capacity emerges (Zinser and Kolter, 2004). Rapid expression of new and advantageous metabolic pathways may redefine the organism's niche and mean the loss of functionally beneficial genes (Zinser and Kolter, 2004). Any factor that reduces digestion efficiency in the upper gastrointestinal tract changes the nutrient supply to the lower tract and will probably favour specific stationary-phase mutants.

NUTRITIONAL FACTORS

The following intrinsic dietary characteristics: (i) nutrient digestibility; (ii) nutrient solubility; (iii) protein/energy ratio; (iv) diet specification; (v) exogenous enzyme inclusion; and (vi) anti-nutrient content and extrinsic characteristics: (1) rate of feed passage; (2) endogenous enzyme secretion; and (3) pH all have a profound effect on feed efficiency. In pursuance of gut health the nutritionist needs to consider ingredient blend in addition to nutrient specification (Bedford, 1996).

Several feed ingredient characteristics will affect passage time and faecal moisture content, including particle size, viscosity in solution, digestibility (starch) and lipid/protein/mineral content (Sibbald, 1979; Sell *et al.*, 1983; Classen, 1996; Refstie *et al.*, 1999; Weurding *et al.*, 2001; Leeson and Summers, 2005). Regulations add further complications, since without meat-and-bone meal as an optional ingredient for balancing protein amino acid composition the nutritionist is forced to rely more heavily than perhaps desired on soy as a source of protein.

Any physiological perturbation that negatively affects nutrient assimilation (intake, digestion and absorption) will increase nutrient throughflow and could alter the microbial population of the distal gut sufficiently to cause wet litter. The growth-enhancing effect of dietary enzymes is comparable to that of antibiotics when tested in controlled experimental conditions, suggesting a common mechanism of action, i.e. manipulation of the gut ecology (Rosen, 2000a, b, 2001). This may be one of the explanations why there is individual and ingredient variation in response to enzyme supplementation (Bedford, 1996; Kocher *et al.*, 1997).

Protein

Soy is the largest source of protein in avian diets, and heat inactivation of the naturally occurring trypsin inhibitor (also present in flax seed and barley) is crucial to the bird's ability to efficiently digest the protein component of this ingredient.

High-protein diets, essential to attain broiler muscle tissue accretion rates, increase the risk of downstream gut health challenges by increasing the chance of protein throughflow. Peptic digestion is already marginal, because selection for growth rate has reduced feed retention time in the crop and gizzard, thus reducing enzyme/nutrient contact time (Sklan *et al.*, 1975; Denbow, 2000; Klipper *et al.*, 2004). Apart from the obvious inefficiencies of nutrient wastage arising from poor digestibility or rapid feed passage, undigested proteins (high molecular weight) reaching the caeca are strongly inflammatory and thus further reduce feed efficiency (Lillehoj and Trout, 1996; Klipper *et al.*, 2001, 2004). This is especially so with soluble protein, because liquids pass through the digestive tract 15% faster than do solids (Sklan *et al.*, 1975; Sklan and Hurwitz, 1980; Klipper *et al.*, 2004).

During the first 3 days post-hatch, the chick immune system learns to tolerate innocuous gut antigens because intact antigens are able to transgress the immature gut cytoskeleton, enter the circulation and make contact with developing lymphocytes in the thymus, bursa and spleen (Brandtzaeg, 1989; Lillehoj and Trout, 1996; Klipper *et al.*, 2001, 2004; Ueda *et al.*, 2001; Brandtzaeg, 2002). The degree of immune tolerance induced during this time is proportional to the dose and frequency of antigen exposure and inversely proportional to antigen solubility (Lillehoj and Trout, 1996; Klipper *et al.*, 2001).

Gut barrier function matures by day 3 and oral exposure to 'new' innocuous antigens after this will lead to an inappropriate immune response which is likely to reduce feed efficiency (Uni *et al.*, 1999, 2000; Geyra *et al.*, 2001; Klipper *et al.*, 2001, 2004). Pre-starter rations must contain the correct balance of feed ingredient antigens to prepare the gut to tolerate feed antigen exposure after maturation of gut barrier function (day 3).

Maternally derived antibodies play a relatively minor protective role at the enteric interface but they are crucial in guiding the development of the hatchling immune system along the fine line between tolerance and response. Maternal antibodies to hazardous/pathogen antigens block development of tolerance in the same way that they interfere with early vaccination (Klipper *et al.*, 2004). This enables the developing immune system to differentiate potentially hazardous/pathogenic antigens from innocuous feed/commensal antigens and thus to initiate response or tolerance (Klipper *et al.*, 2004).

The common practice of hyper-immunizing parent flocks to protect offspring against early exposure to harmful antigens probably generates additional gut health benefits. By manipulating the level and extent of maternal antibody transfer it is possible to shape the balance between tolerance and response in the broiler for life (Klipper *et al.*, 2004). Avian gut health management needs to start at the parent flock level.

The amount of nitrogen reaching the caeca is influenced by the following: (i) the amount of protein in the diet; (ii) the efficiency of protein digestion/absorption in the upper gastrointestinal tract; (iii) the amino acid profile of the protein; and (iv) the state of nitrogen balance. Compromised protein digestion increases caecal nitrogen either by increasing nutrient throughflow or body protein turnover rate (nitrogen excretion via the urine) (Akester *et al.*, 1967; Duke, 1982; Denbow, 2000).

Volatile fatty acids (VFA) are by-products of uric acid degradation by caecal flora and, despite passive absorption, caecal VFA concentrations (acetate > propionate > butyrate) are very high (125 nM) (Annison *et al.*, 1968; Sudo and Duke, 1980). Since these weak organic acids have antibacterial activity they probably play an important role in balancing the caecal ecology (Cherrington *et al.*, 1990, 1991; Davidson, 1997). Inadequate heat treatment of soybean meal or high levels of inclusion can lead to enteritis and wet litter (Leeson and Summers, 2005).

Exogenous enzymes added to the diet to promote protein digestion affect caecal floral communities by reducing the amount of protein nitrogen reaching the caeca, especially if nutrient credits are allocated to the enzyme during formulation.

Carbohydrate

Water-soluble, non-starch polysaccharides (NSP) adversely affect digestibility by stimulating mucous production and increasing ingestal viscosity, frequently resulting in wet litter (Choct and Annison, 1992; Iji, 1999; Collier *et al.*, 2003). Grains such as wheat (5–8% pentosans and α-amylase inhibitors), rye (3–7% β-glucans) and barley (3–15% β-glucans and trypsin inhibitor) are rich in water-soluble NSP, and there is ample research to demonstrate that the use of exogenous enzymes improves digestibility and reduces the wet litter problems associated with these ingredients (Choct and Annison, 1992; Rosen, 2000a, 2001; Leeson and Summers, 2005).

This problem is not, however, unique to the traditional carbohydrate-rich ingredients. The metabolizable energy value of soybean meal is compromised by the lack of α1:6 galactosidase and, furthermore, the non-digestible stachyose (5%), raffinose (1%) and sucrose (6%) accelerate feed passage, reduce fibre digestion and alter caecal floral composition, thus contributing to wet litter problems (Leeson and Summers, 2005). Other compounds such as the lectins and isoflavones present in soybean meal are thought to account for the balance (50%) of the anti-nutritional effect of inadequate heat treatment of this ingredient (Leeson and Summers, 2005).

The negative effect of the NSP tends to dissipate as birds mature, which suggests that microbiotic adaptation probably occurs with time. An increase in the mucolytic component of the small intestinal microbiota and/or an increase in NSP-digesting organisms in the distal part of the gastrointestinal tract is most likely.

Lipid

All fats and oils have the potential to become oxidized, and the resulting rancid fats compromise digestibility and can cause gastrointestinal disturbance and wet litter directly (steatorrhoea) or indirectly by affecting gut flora (oxidative). Unprotected fatty acids released by oil seed processing (grinding or chemical extraction) are very susceptible to oxidative rancidity (Leeson and Summers, 2005).

The type and quality of dietary lipids affects their digestion and absorption and – indirectly – the nature and extent of the bird's inflammatory response, and consequently influences the extent of tissue damage following antigenic stimulation of the gut lining (Leeson and Summers, 2005). Dietary polyunsaturated fatty acids (PUFA), for example, provide the building blocks for cell membrane synthesis and indirectly determine the type of immune response that follows cell damage, since the cell membrane lipids provide the substrate for immune system communication molecule synthesis (Korver and Klasing, 1995, 1997; Klasing, 1998).

Cereal grains are high in linoleic acid (n-6 PUFA precursor for arachidonic acid), which generates prostaglandins, leukotreins and thromboxanes, while fish oil is high in n-3 PUFA, which generates Interleukin-1 and prostaglandin-E (Fritsche *et al.*, 1991; Korver and Klasing, 1997). Of all the vegetable oils flax seed oil contains the highest levels of omega-3 fatty acids (50% linolenic acid), thus providing a very suitable substitute for fish meal as a source of omega-3 fatty acids (Leeson and Summers, 2005).

Mineral balance

Renal mineral excretion is a process of ion exchange across the renal tubule membrane, so the anion–cation balance of the diet affects both water intake and urinary output (Mongin, 1982). The relatively high levels of potassium in soybean (and molasses) can, for example, be sufficient to induce polydipsia, polyuria and wet litter.

Most diets will have added salt and, since sodium is the primary extracellular cation, maintenance of sodium balance by the kidney is crucial in the control of extracellular fluid volume and blood pressure (Goldstein and Skadhauge, 2000; Leeson and Summers, 2005). Sodium excretion is regulated by arginine vasotocin (antidiuretic), rennin/angiotensin (naturesis and diuresis), aldosterone (sodium conservation) and atrial naturetic peptide (naturesis and diuresis) (Arnason *et al.*, 1986; Gray, 1993; Goldstein and Skadhauge, 2000). Increased salt intake requires elevated sodium excretion which, by necessity, means the concomitant loss of an equivalent anion (usually Cl⁻) and water. The polyuria induced by increased sodium excretion is exacerbated by elevated chlorine, since this ion contributes to the osmolarity of the urine (Goldstein and Skadhauge, 2000).

Minor sodium excess is controlled by reducing intestinal uptake but, as the concentration in the diet increases, renal naturesis follows (Goldstein and Skadhauge, 2000). Arginine vasotocin release in response to increased extracellular fluid osmolarity initially reduces urinary output but, as the extracellular fluid volume expands, AVT levels decline and elevated blood ANP increases water and sodium excretion – potentially causing wet litter (Koike *et al.*, 1979; Gray, 1993). The polydypsia/polyuria induced by elevated dietary salt can, to a degree, be countered by partial replacement of salt-derived sodium by sodium bicarbonate, thereby reducing chlorine intake, but this does potentially compromise heat stress coping mechanisms (Leeson and Summers, 2005).

The nutritionist regulates dietary calcium and phosphorus levels by imposing stringent maximum and minimum specification constraints, because both the amount and ratio of these minerals is important to productivity (Leeson and Summers, 2005). Calcium is reabsorbed from the glomerular filtrate very efficiently (>98%) by a mechanism that operates close to maximum, while phosphorus is, in contrast, reabsorbed fairly inefficiently (40%) (Wideman, 1987).

Parathyroid hormone (PTH) secretion in response to low blood Ca enhances intestinal absorption and reduces renal excretion (Wideman, 1987). Elevated blood Ca proportionally increases the calcium concentration of the glomerular filtrate, which easily exceeds reabsorption capacity and consequently increases Ca excretion (Clark and Mok, 1986; Wideman, 1987). PTH increases phosphorus excretion by inhibiting renal tubular reabsorption and secretion (Clark and Mok, 1986; Wideman, 1987).

Excess calcium excretion can cause renal pathology (calcinosis/urolithiasis), resulting in compromised water retention and diuresis/wet litter (Shane *et al.*, 1969; Wideman *et al.*, 1985). Dolomitic limestone contains relatively high levels of magnesium (8–10%) and, apart from competing with calcium for absorption, Mg excretion can cause diuresis and wet litter (Leeson and Summers, 2005).

TOXICITY

Nephrotoxins such as ochratoxin (especially type A), citrinin and oosporin can compromise renal function, causing polyuria/polydypsia and wet litter (Leeson and Summers, 2005). Mycotoxicoses are difficult to quantify as there appears to be significant interaction and potentiation.

NUTRITIONAL INTERVENTION

Some ingredient characteristics such as water content, viscosity and non-digestible nutrient composition can be enhanced with concomitant enzyme and or osmolyte usage, while others such as mineral content require a more fundamental approach (diet formulation) (Bedford, 2000; Rosen, 2001; Leeson and Summers, 2005).

The task of gut environment management is almost neglected because the internalized external environment of the gastrointestinal lumen is complex and poorly understood. This is rather ironic, since the efficiency of the assimilation process – and hence feed efficiency – is dependent on the integrity of the absorptive membrane, which is in turn dependent on the status of the gut environment. Dietary ingredient mix and nutrient composition is fundamental to the establishment and maintenance of the complex microbial ecology of the gastrointestinal tract (Collett, 2005).

Modern molecular techniques have helped elucidate details of the gastrointestinal environment that are revolutionizing our understanding. 16S ribosomal RNA-based techniques have enhanced the accuracy of gut floral profiling (highlighting the inadequacy of culture techniques) and are making the detailed study of gut ecology feasible (Barnes and Impey, 1970; Barnes *et al.*, 1972; Salanitro *et al.*, 1974, 1978; Barnes, 1977; Cherrington *et al.*, 1991; Amann *et al.*, 1995; Anadon *et al.*, 1995; Vaughan *et al.*, 2000; Gong *et al.*, 2002; Stappenbeck *et al.*, 2002; Lu *et al.*, 2003).

Gene sequencing of purported commensals and host cascade reaction modulation studies (such as pro- and anti-inflammatory pathways) have already opened the door on the vast host/flora cellular communication process that maintains immune–microbe homeostasis (Elson and Cong, 2002; McFall-Ngai, 2002; Neish, 2002; Lu *et al.*, 2003; Xu and Gordon, 2003; Xu *et al.*, 2003; Kelly *et al.*, 2004).

Gut microbial imbalance is a fundamental cause of wet litter and there are several opportunities for intervention to enhance gut health and productivity by managing this ecosystem (Collett, 2005):

1. Seeding of the hatchling gut begins with vertical transmission of parent gut flora, but is effectively modified with early administration of effective probiotics or competitive exclusion products. To be successful these must initiate the development of a primary flora that will rapidly evolve into a stable and favourable climax flora by creating suitable gut conditions and excluding unfavourable organisms.

2. Preparing the gut environment (pH, redox potential) for early transition from primary to climax flora through water/feed acidification. Candidates need to be weak acids that are buffered to withstand the neutralizing effect of minerals dissolved in the drinking water and have dissociation characteristics that make them active in the small intestine.

3. Excluding pathogens from colonizing the gut by competitive and selective exclusion. It is important that the selective exclusion product is compatible with (does not exclude) the organisms used for competitive exclusion or as a probiotic.

4. Enhancing resilience by stimulating the protective immune response while suppressing the acute phase or fever response.

5. Decreasing nutrient throughflow by enhancing nutrient digestion and absorption (exogenous enzyme addition and nutrient modification, feeding and lighting programmes, careful use of antibiotics) to avoid caecal floral upset.

On-farm risk analysis is necessary to determine which interventions are most appropriate and to establish the most profitable strategy for improving gut health.

CONCLUSIONS

Nutritionally induced wet litter may result from increased urine production (polyuria) or compromised water absorption (diarrhoea/enteritis). The stage at which aberration reflects as a clinical problem is governed by: (i) the capacity and operation of the house ventilation system; (ii) bird stocking density; (iii) litter material quantity and quality; and (iv) prevailing climatic conditions.

Urinary output increases in response to feeding diets that require increased excretion of nitrogen (governed by protein quantity and quality or amino acid profile) and minerals in response to dietary excess or imbalance. Faecal water excretion increases as a consequence of reduced water absorption and can be the result of ingredient characteristics that affect:

1. The process of absorption and feed passage time, including particle size, viscosity in solution, digestibility (starch), mineral content and lipid/protein quantity and quality.
2. The integrity of the gastrointestinal lining directly (toxic) or indirectly (perturbation of the microbial ecology of the gastrointestinal tract) by causing an inflammatory process and cell damage.

In pursuance of gut health the nutritionist needs to consider ingredient blend in addition to nutrient specification. Seeding the gastrointestinal tract with bacteria at hatch and subsequently managing the gastrointestinal environment, whilst at the same time reducing anti-nutritional factors and the use of precision diet formulation, has proved advantageous in minimizing the incidence of wet litter in birds whilst maintaining and improving bird performance.

REFERENCES

Akester, A.R., Anderson, R.S., Hill, K.J. and Osbaldiston, G.W. (1967) A radiographic study of urine flow in the domestic fowl. *British Poultry Science* 8, 209–212.

Almirall, M. and Esteve-Garcia, E. (1994) Rate of passage of barley diets with chromium oxide: influence of age and poultry strain and effect of beta-glucanase supplementation. *Poultry Science* 73, 1433–1440.

Amann, R.I., Ludwig, W. and Schleifer, K.H. (1995) Phylogenetic identification and *in situ* detection of individual microbial cells without cultivation. *Microbiology Review* 59, 143–169.

Anadon, A., Martinez-Larranaga, M.R., Diaz, M.J., Bringas, P., Martinez, M.A., Fernanadez-Cruz, M.L., Fernandez, M.C. and Fernandez, R. (1995) Pharmacokinetics and residues of enrofloxacin in chickens. *American Journal of Veterinary Research* 56, 501–506.

Annison, E.F., Hill, K.J. and Kenworthy, R. (1968) Volatile fatty acids in the digestive tract of the fowl. *British Journal of Nutrition* 22, 207–216.

Arnason, S.S. and Skadhauge, E. (1991) Steady-state sodium absorption and chloride secretion of colon and coprodeum, and plasma levels of osmoregulatory hormones in hens in relation to sodium intake. *Journal of Comparative Physiology* (B) 161, 1–14.

Arnason, S.S., Rice, G.E., Chadwick, J. and Skadhauge, E. (1986) Plasma levels of arginine vasotocin, prolactin, aldosterone and corticosterone during prolonged dehydration in the domestic fowl: effect of dietary NaCl. *Journal of Comparative Physiology* (B) 156, 383–397.

Barnes, E.M. (1977) Ecological concepts of the anaerobic flora in the avian intestine. *American Journal of Clinical Nutrition* 30, 1793–1798.

Barnes, E.M. and Impey, C.S. (1970) The isolation and properties of the predominant anaerobic bacteria in the caeca of chickens and turkeys. *British Poultry Science* 11, 467–481.

Barnes, E.M., Mead, G.C., Barnum, D.A. and Harry, E.G. (1972) The intestinal flora of the chicken in the period 2 to 6 weeks of age, with particular reference to the anaerobic bacteria. *British Poultry Science* 13, 311–326.

Bedford, M. (1996) Interaction between ingested feed and the digestive system in poultry. *Journal of Applied Poultry Research* 5, 86–95.

Bedford, M. (2000) Removal of antibiotic growth promoters from poultry diets: implications and strategies to minimise subsequent problems. *World's Poultry Science Journal* 56, 347–365.

Bjornhag, D. and Sperber, I. (1977) Transport of various food components through the digestive tract of turkeys, geese and guinea fowl. *Swedish Journal of Agricultural Science* 7, 57–66.

Branch, W.J. and Cummings, J.H. (1978) Comparison of radio-opaque pellets and chromium sesquioxide as inert markers in studies requiring accurate faecal collections. *Gut* 19, 371–376.

Brandtzaeg, P. (1989) Overview of the mucosal immune system. *Current Topics in Microbiology and Immunology* 146, 13–25.

Brandtzaeg, P.E. (2002) Current understanding of gastrointestinal immunoregulation and its relation to food allergy. *Annals of the New York Academy of Sciences* 964, 13–45.

Cherrington, C.A., Hinton, M. and Chopra, I. (1990) Effect of short-chain organic acids on macromolecular synthesis in *Escherichia coli*. *Journal of Applied Bacteriology* 68, 69–74.

Cherrington, C.A., Hinton, M., Mead, G.C. and Chopra, I. (1991) Organic acids: chemistry, antibacterial activity and practical applications. *Advances in Microbial Physiology* 32, 87–108.

Choct, M. and Annison, G. (1992) The inhibition of nutrient digestion by wheat pentosans. *British Journal of Nutrition* 67, 123–132.

Clark, N.B. and Mok, L.L. (1986) Renal excretion in gull chicks: effect of parathyroid hormone and calcium loading. *American Journal of Physiology* 250, R41–R50.

Classen, H.L. (1996) Cereal grain starch and exogenous enzymes in poultry diets. *Animal Feed Science Technology* 62, 21–27.

Clench, M.H., Pineiro-Carrero, V.M. and Mathias, J.R. (1989) Migrating myoelectric complex demonstrated in four avian species. *American Journal of Physiology* 256, G598–G603.

Collett, S. (2005) Strategies for improving gut health in commercial operations. *3rd International Poultry Broiler Nutritionists Conference – Poultry Beyond 2010*, Auckland, New Zealand.

Collier, C.T., van der Klis, J.D., Deplancke, B., Anderson, D.B. and Gaskins, H.R. (2003) Effects of tylosin on bacterial mucolysis, *Clostridium perfringens* colonization, and intestinal barrier function in a chick model of necrotic enteritis. *Antimicrobial Agents Chemotherapy* 47, 3311–3317.

Davidson, P. (1997) Chemical preservatives and natural antimicrobial compounds. In: Doyle, M., Beuchat, L. and Montville, T. (eds) *Food Microbiology – Fundamentals and Frontier*, American Society for Microbiology, Washington DC, pp. 520–556.

Denbow, D. (2000) Gastrointestinal anatomy and physiology. In: Wihittow, G. (ed.) *Sturkie's Avian Physiology*. Academic Press, New York, pp. 299–325.

Duke, G.E. (1982) Gastrointestinal motility and its regulation. *Poultry Science* 61, 1245–1256.

Elson, C.O. and Cong, Y. (2002) Understanding immune-microbial homeostasis in intestine. *Immunologic Research* 26, 87–94.

Ferrando, C., Vergara, P., Jimenez, M. and Gonalons, E. (1987) Study of the rate of passage of food with chromium-mordanted plant cells in chickens (*Gallus gallus*). *Quarterly Journal of Experimental Physiology* 72, 251–259.

Fritsche, K.L., Cassity, N.A. and Huang, S.C. (1991) Effect of dietary fat source on antibody production and lymphocyte proliferation in chickens. *Poultry Science* 70, 611–617.

Geyra, A., Uni, Z. and Sklan, D. (2001) Enterocyte dynamics and mucosal development in the posthatch chick. *Poultry Science* 80, 776–782.

Gidenne, T. (1997) Caeco-colic digestion in the growing rabbit: impact of nutritional factors and related disturbances. *Livestock Production Science* 51, 73–78.

Goldstein, D. and Skadhauge, E. (2000) Renal and extrarenal regulation of body fluid compartments. In: Wihittow, G. (ed.) *Sturkie's Avian Physiology*. Academic Press, New York.

Gong, J., Forster, R.J., Yu, H., Chambers, J.R., Sabour, P.M., Wheatcroft, R. and Chen, S. (2002) Diversity and phylogenetic analysis of bacteria in the mucosa of chicken caeca and comparison with bacteria in the caecal lumen. *FEMS Microbiology Letters* 208, 1–7.

Gray, D.A. (1993) Plasma atrial natriuretic factor concentrations and renal actions in the domestic fowl. *Journal of Comparative Physiology* (B) 163, 519–523.

Gray, D.A., Naude, R.J. and Eramus, T. (1988) Plasma arginine vasotocin and angitensin II in the water-deprived ostrich (*Struthio camelus*). *Comparative Biochemistry and Physiology* 89, 251–256.

Iji, P. (1999) The impact of cereal non-starch polysaccharides on intestinal development and function in broiler chickens. *World's Poultry Science Journal* 55, 375–387.

Kanosue, K., Schmid, H. and Simon, E. (1990) Differential osmoresponsiveness of periventricular neurons in duck hypothalamus. *American Journal of Physiology* 258, R973–R981.

Kaufman, S., Kaesermann, H.P. and Peters, G. (1980) The mechanism of drinking induced by parenteral hyperonocotic solutions in the pigeon and in the rat. *Journal of Physiology* 301, 91–99.

Kellerup, S.U., Parker, J.E. and Arscott, G.H. (1965) Effects of restricted water consumption on broiler chicks. *Poultry Science* 44, 78–83.

Kelly, D., Campbell, J.I., King, T.P., Grant, G., Jannson, E.A., Coutts, E.A., Pettersson, S. and Conway, S. (2004) Commensal anaerobic gut bacteria attentuate inflammation by regulating nuclear-cytoplasmic shuttling of PPAR-gamma and RelA. *Nature Immunology* 5, 104–112.

Klasing, K.C. (1998) Nutritional modulation of resistance to infectious diseases. *Poultry Science* 77, 1119–1125.

Klipper, E., Sklan, D. and Freidman, A. (2001) Response, tolerance and ignorance following oral exposure to a single dietary protein antigen in *Gallus domesticus*. *Vaccine* 19, 2890–2897.

Klipper, E., Sklan, D. and Freidman, A. (2004) Maternal antibodies block induction of oral tolerance in newly hatched chicks. *Vaccine* 22, 493–502.

Kocher, A., Hughes, R. and Barr, A. (1997) Beta-gluconase reduces but does not eliminate variation in AME of barley varieties. *Australian Poultry Science Symposium*.

Koike, T.I., Pryor, L.R. and Neldon, H.L. (1979) Effects of salt infusion on plasma immuno-reactive vasotocin in conscious chickens. *General and Comparative Endocrinology* 37, 451–458.

Korver, D. and Klasing, K. (1995) n-3 polyunsaturated fatty acids improve growth rate of broiler chickens and decrease interleukin-1 production. *Poultry Science* 74 (Suppl. 15).

Korver, D.R. and Klasing, K.C. (1997) Dietary fish oil alters specific and inflammatory immune responses in chicks. *Journal of Nutrition* 127, 2039–2046.

Leeson, S. and Summers, J.D. (2005) *Commercial Poultry Nutrition*. University Books, Canada.

Lillehoj, H.S. and Trout, J.M. (1996) Avian gut-associated lymphoid tissues and intestinal immune responses to *Eimeria* parasites. *Clinical Microbiology Reviews* 9, 349–360.

Lu, J., Idris, U., Harmon, B., Hofacre, C., Maurer, J.J. and Lee, M.D. (2003) Diversity and succession of the intestinal bacterial community of the maturing broiler chicken. *Applied Environmental Microbiology* 69, 6816–6824.

Marder, J. and Ben-Asher, J. (1983) Cutaneous water evaporation, I: Its significance in heat-stressed birds. *Comparative Biochemistry and Physiology* A 75, 425–431.

Marder, J. and Raber, P. (1989) Beta-adrenergic control of trans-cutaneous evaporative cooling mechanisms in birds. *Journal of Comparative Physiology* 159, 97–103.

McFall-Ngai, M.J. (2002) Unseen forces: the influence of bacteria on animal development. *Developmental Biology* 242, 1–14.

Mongin, P. (1982) Transport of electrolytes and water in the upper jejunum of the fowl *in vivo* perfusion. *Pflugers Archiv: European Journal of Physiology* 392, 251–256.

Morild, I., Monwinckel, R., Bohle, A. and Ja, C. (1985) The juxtaglomerular apparatus in the avian kidney. *Cell Tissue Research* 240, 209–214.

Neish, A.S. (2002) The gut microflora and intestinal epithelial cells: a continuing dialogue. *Microbes and Infection/Institut Pasteur* 4, 309–317.

Refstie, S., Svihus, B., Shearer, K. and Storebakken, T. (1999) Nutrient digestibility in Atlantic salmon and broiler chickens related to viscosity and non-starch polysaccharide content in different soybean products. *Animal Feed Science Technology* 79, 331–345.

Rice, G.E. and Skadhauge, E. (1982) Caecal water and electrolyte absorption and the effects of acetate and glusose, in dehydrated, low-NaCl-diet hens. *Journal of Comparative Physiology* 147, 61–64.

Roberts, J.R. and Hughes, M.R. (1983) Glomerular filtration rate and drinking rate in Japanese quail, *Coturnix coturnix japonica*, in response to acclimation to saline drinking water. *Canadian Journal of Zoology* 61, 2394.

Rosen, G. (2000a) Enzymes for broilers: A multi-factorial assessment. *Feed International* 21, 14–18.

Rosen, G. (2000b) Multi-factorial assessment of exogenous enzymes in broiler pronutrition: Target and problems. *Proceedings of the 3rd European Symposium on Feed Enzymes*, Noordwijkerhout, Netherlands.

Rosen, G.D. (2001) Multi-factorial efficacy evaluation of alternatives to antimicrobials in pronutrition. *British Poultry Science* 42, S104–S105.

Salanitro, J.P., Blake, I.G. and Muirehead, P.A. (1974) Studies on the cecal microflora of commercial broiler chickens. *Applied Microbiology* 28, 439–447.

Salanitro, J.P., Blake, I.G., Muirehead, P.A., Malio, M. and Goodman, J.R. (1978) Bacteria isolated from the duodenum, ileum, and caecum of young chicks. *Applied Environmental Microbiology* 35, 782–790.

Sell, J., Eastwood, J. and Mateos, G. (1983) Influence of supplemental fat on metabolizable energy and ingesta transit time in laying hens. *Nutrition Report International* 28, 487–495.

Shane, S.M., Young, R.G. and Krook, L. (1969) Renal and parathyroid changes produced by high calcium intake in growing pullets. *Avian Diseases* 13, 558–567.

Sibbald, I.R. (1979) Passage of feed through the adult rooster. *Poultry Science* 58, 446–459.

Sklan, D. and Hurwitz, S. (1980) Protein digestion and absorption in young chicks and turkeys. *Journal of Nutrition* 110, 139–144.

Sklan, D., Dubrov, D., Eisner, U. and Hurwitz, S. (1975) 51Cr-EDTA, 91Y and 141Ce as non-absorbed reference substances in the gastrointestinal tract of the chicken. *Journal of Nutrition* 105, 1549–1552.

Stappenbeck, T.S., Hooper, L.V. and Gordon, J.I. (2002) Developmental regulation of intestinal angiogenesis by indigenous microbes via Paneth cells. *Proceedings of the National Academy of Sciences of the USA of America* 99, 15451–15455.

Sudo, S. and Duke, G. (1980) Kinetics of absorption of volatile fatty acids from the caeca of domestic turkeys. *Comparative Biochemistry and Physiology* 67, 231–237.

Takei, Y., Okawara, Y. and Kobayashi, H. (1988) Water intake induced by water deprivation in the quail, *Coturnix coturnix japonica*. *Journal of Comparative Physiology* (B) 158, 519–525.

Uden, P., Colucci, P.E. and Van Soest, P.J. (1980) Investigation of chromium, cerium and cobalt as markers in digesta. Rate of passage studies. *Journal of the Science of Food and Agriculture* 31, 625–632.

Ueda, Y., Hachimura, S., Somaya, T., Hisatsune, T. and Kaminogawa, S. (2001) Apoptosis of antigen-specific T cells induced by oral administration of antigen: comparison of intestinal and non-intestinal immune organs. *Bioscience, Biotechnology and Biochemistry* 65, 1170–1174.

Uni, Z., Noy, Y. and Sklan, D. (1999) Posthatch development of small intestinal function in the poult. *Poultry Science* 78, 215–222.

Uni, Z., Geyra, A., Ben-Hur, H. and Sklan, D. (2000) Small intestinal development in the young chick: crypt formation and enterocyte proliferation and migration. *British Poultry Science* 41, 544–551.

Vaughan, E.E., Schut, F., Heilig, H.G., Zoetendal, E.G., de Vos, W.M. and Akkermans, A.D. (2000) A molecular view of the intestinal ecosystem. *Current Issues in Intestinal Microbiology* 1, 1–12.

Vergara, P., Ferrando, C., Jiminez, M., Fernandez, E. and Gonalons, E. (1989) Factors determining gastrointestinal transit time of several markers in the domestic fowl. *Quarterly Journal of Experimental Physiology* 74, 867–874.

Washburn, K.W. (1991) Efficiency of feed utilization and rate of feed passage through the digestive system. *Poultry Science* 70, 447–452.

Weurding, R.E., Veldman, A., Veen, W.A., van der Aar, P.J. and Verstegen, M.W. (2001) Starch digestion rate in the small intestine of broiler chickens differs among feedstuffs. *Journal of Nutrition* 131, 2329–2335.

Wideman, R.F.J. (1987) Renal regulation of avian calcium and phosphorus metabolism. *Journal of Nutrition* 117, 808–814.

Wideman, R.F.J., Closser, J.A., Roush, W.B. and Cowen, B.S. (1985) Urolithiasis in pullets and laying hens: role of dietary calcium and phosphorus. *Poultry Science* 64, 2300–2307.

Xu, J. and Gordon, J.I. (2003) Inaugural article: honor thy symbionts. *Proceedings of the National Academy of Sciences of the USA of America* 100, 10452–10459.

Xu, J., Bjursell, M.K., Himrod, J., Deng, S., Carmichael, L.K., Chiang, H.C., Hooper, L.V. and Gordon, J.I. (2003) A genomic view of the human–*Bacteroides thetaiotaomicron* symbiosis. *Science* 299, 2074–2076.

Zambrano, M.M., Siegele, D.A., Almiron, M., Tormo, A. and Kolter, R. (1993) Microbial competition: *Escherichia coli* mutants that take over stationary phase cultures. *Science* 259, 1757–1760.

Zinser, E.R. and Kolter, R. (2004) *Escherichia coli* evolution during stationary phase. *Research in Microbiology* 155, 328–336.

CHAPTER 14
Micronutrient supply: influence on gut health and immunity

K.C. Klasing

Department of Animal Science, University of California Davis, Davis, California, USA; e-mail: kcklasing@ucdavis.edu

ABSTRACT

The intestine is sensitive to the deficiency or excess of several vitamins and minerals. A deficiency of vitamin A interferes with the development and differentiation events necessary for maintaining intestinal architecture. A deficiency of vitamin B_6, pantothenic acid, thiamin, niacin, or riboflavin interferes with cellular proliferation and diminishes the size of the villi. However, the gut is not the most sensitive indicator of a deficiency. Some nutrients, including long-chain fatty acids, carotenoids and vitamins A, D and E, modulate regulatory pathways of systems involved in intestinal homeostasis and health, including the immune system.

By shifting the balance of regulatory processes, the dietary levels of these nutrients change gut health, sometimes for the good and sometimes to the detriment of the bird, depending on the pathogen milieu. Enteric infections markedly diminish the absorption of many micronutrients, especially the fat-soluble vitamins. The decrease in absorption may be related to: (i) decreased intestinal integrity and function due to pathology inflicted by the pathogen; (ii) increased rate of passage of digesta through the intestines; or (iii) as an integral part of the inflammatory response to the pathogens. Thus, micronutrient nutrition can affect gut health, but gut health can also affect micronutrient nutrition.

INTRODUCTION

The diet is an important determinant of the morphology, defences and microflora of the gut. Dietary fibre, starch, protein and other macronutrients dominate much of the research on gut health because they have considerable effects on gut morphology and microbial ecology (Klasing, 2005). The supply of micronutrients, including vitamins and minerals, often has more subtle

© CAB International 2006. *Avian Gut Function in Health and Disease* (ed. G.C. Perry)

effects on gut health by influencing developmental and defensive processes of the gastrointestinal tract. This chapter reviews the priority of the gut for micronutrients, the effect of deficient or excess levels of micronutrients on gut health and the emerging concept of immunomodulation by micronutrients (see Table 14.1).

The gut is sensitive to deficiency of several nutrients (Klasing and Austic, 2003), and providing adequate levels of these nutrients is critical for gut health (more is good). However, several nutrients, including iron and biotin, are critical nutrients for the growth of microbes, so supplying excess levels may, in some situations, negatively impact health (more is bad). Some nutrients modulate regulatory pathways of systems involved in intestinal homeostasis and health, including the immune system. By shifting the balance of regulatory processes, the dietary levels of these nutrients change gut health, sometimes for the good and sometimes to the detriment of the bird (more is different). Nutritional approaches relevant to one micronutrient may be inappropriate for others; consequently, an understanding of the influences of micronutrients on the integrity of the gut and on the regulation of gut defences is needed for optimal health.

REQUIREMENT AND PRIORITY OF THE GUT FOR NUTRIENTS

The gastrointestinal tract (GIT), especially the intestine, is among the most metabolically active tissues of the body and its epithelia have a very high rate of cell turnover and loss of protein and cells into the lumen. Consequently its requirement (g/kg tissue/d) for most nutrients is unmatched by any other tissue. The intestines are buffered from many nutrient deficiencies because the nutrients in food pass through its epithelium, giving this tissue first refusal at utilizing them. Absorption of some nutrients occurs mostly after the mid-duodenum, so tissues proximal to this region do not have first refusal and must receive these nutrients via the blood – and consequently must compete with other tissues.

Table 14.1. Examples of micronutrients that impact gut health and the mechanisms involved.

Mechanism	Micronutrients
Sensitivity of proximal gut to low levels of essential nutrients	Vitamins A, B_6, pantothenic acid, thiamin, niacin, riboflavin and vitamin E–selenium
Novel nutrient needs during stress or disease	Betaine, vitamin C, glutamine and nucleotides
Sensitivity of gut to toxicities	Copper, salt, heavy metals, vitamins A and E
Deprivation of micronutrients limits microbial proliferation	Iron, biotin
Antimicrobial actions of micronutrients	Copper, zinc
Modulation of the immune system by micronutrients	Long-chain polyunsaturated fatty acids, carotenoids and vitamins A, C, D and E

Severe deficiencies of several nutrients have been shown to cause pathology in the gizzard or proximal duodenum, including: (i) vitamin A (Klasing and Austic, 2003); (ii) vitamin B_6 (Daghir and Haddad, 1981); (iii) pantothenic acid (Ringrose *et al.*, 1931; Gries and Scott, 1972); (iv) thiamin (Gries and Scott, 1972); (v) niacin (Gries and Scott, 1972); (vi) riboflavin (Gries and Scott, 1972); and (vii) vitamin E–selenium (Cantor *et al.*, 1982; Van Vleet, 1982).

However, other organs, as well as growth and efficiency of feed use, are more sensitive to low dietary levels of these nutrients, so the gut is not normally used as the sentinal tissue for diagnosis of specific nutrient deficiencies.

MICRONUTRIENT DEFICIENCIES OR EXCESSES AND GUT HEALTH

The impact of micronutrient deficiencies on gut health may be due to either direct consequences on the gut tissue or indirect complications of the deficiency on various systemic systems, especially the immune system. Because of the high priority of the gut for most nutrients, this latter mechanism often dominates aetiology. The gut is protected from the massive population of microbes found in its lumen by leukocytes that mature in the thymus, bursa and bone marrow. These primary immune tissues are metabolically highly active and are sensitive to micronutrient deficiencies.

The thymus is especially sensitive, and its size and integrity are often used as indicators of marginal nutrient deficiencies. Lymphocytes produced in the thymus and bursa are critical for defence against enteric pathogens and for tolerance to commensal microflora and the macromolecules in food. Thus, impairments in the development of leukocytes in the periphery can result in diminished intestinal defences and increased incidence of pathogen-induced pathology and food-induced intolerances.

A vicious cycle often ensues, with enteric infections and hypersensitivities resulting in malabsorption, exacerbating nutritional deficiencies, which further compromise immunity. Thus, while the intestines – by virtue of their first refusal priority for nutrients – are not usually the primary site of pathology due to micronutrient malnutrition, enteric diseases secondary to compromised immunity are a major contributor to the morbidity. In field cases of micronutrient deficiencies, the stunting, runting and unthrifty syndromes that accompany micronutrient deficiencies are often the result of these secondary effects on intestinal immunity. A deficiency of most of the micronutrients impacts on intestinal immune defences and the effects of vitamin A are reviewed in detail below.

The GIT is the first organ system to encounter excessive levels of nutrients and food contaminants and is the site where toxicosis is often expressed. For example, copper is often added to feed because of its growth-promoting actions and the gizzard is very sensitive to high copper levels (Fisher *et al.*, 1973; Poupoulis and Jensen, 1976; Jensen *et al.*, 1991). Gizzards from chicks fed toxic levels of copper have linings that are markedly thickened and folded and have erosions, fissures and a warty appearance. Histologically, there are haemorrhages under the thickened koilin layer. Salt toxicity causes haemorrhages and severe congestion in the gastrointestinal tract (Klasing and Austic, 2003). Excess vitamin E causes cornification of the oesophageal mucous membrane (Wang, 1993). The gut is also sensitive to acute toxicity due to heavy metals, such as cadmium and lead (Fox, 1975; NRC, 2005).

Vitamin A

The active form of vitamin A, retinoic acid, serves as a regulatory signal for Hox expression, which determines the morphological patterning of the avian gut during embryonic development (Huang *et al.*, 1998). After hatch, the continued differentiation of a number of cell lineages is controlled by retinoic acid. Deficiency of vitamin A results in the de-differentiation of simple epithelial membranes into squamous epithelia (squamous metaplasia). Marginal deficiencies of vitamin A cause epithelial changes in the oropharynx and oesophagus in chicks with normal growth performance (Aye *et al.*, 2000). In more severe deficiencies, the intestinal epithelia also undergo structural and functional changes that cause maldigestion and malabsorption.

Uni and co-workers (Uni *et al.*, 1998) found that the jejunal villi of broiler chickens grown on a vitamin A-deficient diet were shorter, thicker and had a much higher enterocyte density and smaller enterocyte size. The abrupt withdrawal of vitamin A from the diet resulted in a rapid decline in the number of goblet cells in the epithelium (Rojanapo *et al.*, 1980).

High enterocyte density is due to increased proliferation and decreased protein accretion. The enterocytes of vitamin A-deficient chicks expressed lower levels of intestinal brush border enzymes (Uni *et al.*, 2000), which probably contributes to malabsorption. Impaired growth rate of deficient broiler chicks appears to be secondary to diminished digestive function (Klasing and Austic, 2003). These changes are reversible upon supplementation of vitamin A (Uni *et al.*, 1998).

Vitamin A, again as retinoic acid, regulates leukocyte function, especially lymphocytes (Halevy *et al.*, 1994; Semba, 1998). In chickens, low vitamin A depressed CD4/CD8 T-lymphocyte ratios (Lessard *et al.*, 1997), specific antibody production to protein antigens (Friedman and Sklan, 1989a; Sklan *et al.*, 1994) and *in vitro* proliferation responses of T-lymphocytes to both antigens (Friedman and Sklan, 1989b; Sklan *et al.*, 1994) and mitogens (Halevy *et al.*, 1994; Dalloul *et al.*, 2002). Duodenal and jejunal intra-epithelial lymphocytes from chicks fed a low-vitamin A diet expressed diminished levels of CD3, CD4, CD8, $\alpha\beta$ TCR and $\gamma\delta$ TCR (Dalloul *et al.*, 2002).

In the small intestinal mucosa of mice, vitamin A deficiency resulted in diminished IL-4, IL-5 and IL-6, leading to decreased total and antigen-specific IgA production (Nikawa *et al.*, 1999). Interestingly, in broilers the proportion of intra-epithelial lymphocytes that were IgA positive increased during a vitamin A deficiency (Dalloul *et al.*, 2002), probably because of greater levels of challenges from microbes residing in the intestines. In vitamin A-deficient chickens, changes in the response to Newcastle Disease Virus or coccidia indicated a shift from Th1- towards Th2-type responses (Lessard *et al.*, 1997; Dalloul *et al.*, 2002).

In cockatiels, antibody production is more sensitive to vitamin A malnutrition than are classic indicators of vitamin A deficiency, such as histological changes in mucosal membranes (Koutsos *et al.*, 2003). Friedman and Sklan (1989a) and Sklan *et al.*(1994) titrated the response to increasing dietary vitamin A from deficient to very high levels and found a robust increase in antigen-specific B and T lymphocyte responses when a deficiency was corrected. However, they also saw immunomodulatory effects at very high levels, illustrating both the required nutritional effects of vitamin A and the immunomodulatory actions of this vitamin, which are mediated by a separate mechanism (see the section on interactions between micronutrients and gut microflora, below).

Low dietary vitamin A impairs intestinal barrier functions by impairing mucin production of goblet cells (De Luca *et al.*, 1969; Uni *et al.*, 2000) and the integrity of tight junctions between epithelial cells (Basova *et al.*, 2002). Foci of vitamin A-deficient epithelia may be the sites of penetration of bacteria and viruses contributing to an increased severity of infections by pathogens and an increase in incidence of secondary infections (Ross and Stephensen, 1996).

In the intact animal, the effects of vitamin A nutrition on the intestines are probably due to a combination of the following: (i) effects on the differentiation of the epithelium; (ii) diminished barrier functions; (iii) malabsorption resulting in greater substrate for luminal bacterial overgrowth; and (iv) impaired function of resident leukocytes. This combination results in decreased resistance to a variety of infectious diseases (Cook, 1991; Semba, 1998; Klasing and Austic, 2003).

Erasmus and Scott (1960) found that chicks fed a low-vitamin A diet had higher mortality due to either an *Eimeria acervulina* or *E. tenella* infection. In vitamin A-replete broilers, CD8-positive T lymphocytic proportions increased in the intestinal epithelium following an *E. acervulina* challenge, but this change was blocked by vitamin A deficiency (Dalloul *et al.*, 2002).

The increase in interferon-γ and IL-2 due to a coccidial challenge was also blunted (Dalloul *et al.*, 2002, 2003). In this study the vitamin A-deficient birds had a higher burden of infection, as indicated by higher oocyst shedding, indicating that the diminished immune response, coupled with changes in epithelial integrity, resulted in lower resistance. Vitamin A deficiency also increased the burden of *Ascaridia galli* in challenge experiments (Zoltowska *et al.*, 1995).

Vitamin A is the most toxic vitamin and the immune system is very sensitive to excess (Sklan *et al.*, 1994; Koutsos *et al.*, 2003). Furthermore, vitamin A has immunoregulatory properties that are distinct from its required functions in the differentiation of epithelia and leukocytes. Thus, deficiencies must be avoided but supplementation must be done judiciously.

Selenium and vitamin E

Deficiency of vitamin E and/or selenium (Se) can produce myopathic changes in the smooth muscle that lines the gut (Yarrington and Whiehair, 1975; Cantor *et al.*, 1982). Lesions are especially extensive in the gizzard, which becomes pale and infiltrated by heterophils. Smooth muscle cells show degeneration of the sarcoplasmic reticulum and mitochondria. Smooth muscle in the proximal duodenum is similarly affected. Oxidized fat in the diet exacerbates the impact of a vitamin deficiency on the integrity of the gut (Dam, 1970).

As with vitamin A, a deficiency of vitamin E and/or Se impairs most measures of immune function (Marsh *et al.*, 1986; Tengerdy, 1990; Erf *et al.*, 1998; Leshchinsky and Klasing, 2001). The primary lymphoid organs are major targets of Se and vitamin E dietary deficiencies (Marsh *et al.*, 1986). Supplementation of marginally deficient diets with Se or vitamin E reduced mortality and morbidity of chickens infected with *E. tenella* (Colnago *et al.*, 1984). These same investigators found that immunization of chickens against coccidiosis was enhanced by Se or vitamin E supplementation. Supplementation of diets that were already sufficient in vitamin E (25 mg/kg) with additional vitamin E did not improve protection (Allen and Fetterer, 2002).

Increasing vitamin E to levels well above the requirement has immunomodulating properties that are mediated by mechanisms different from its roles that prevent nutritional deficiencies (see section on interactions between micronutrients and gut microflora). Thus, nutritional vitamin E should be viewed in two ways: (i) preventing deficiency is necessary for normal immunity and is likely to be beneficial; however, (ii) adding high levels is immunomodulatory and the benefits are dependent upon the type of pathogen challenge. Furthermore, high levels of vitamin E antagonize vitamin A and can cause symptoms of vitamin A deficiency (Wang, 1993).

Other micronutrients

Several micronutrients that are not nutritionally essential in healthy birds become important during challenges to gut health, including betaine, vitamin C, glutamine and nucleotides. Betaine is not considered to be dietarily essential because it can be synthesized from choline or *de novo* from serine and methionine (Klasing, 1998). However, betaine has osmoprotective functions in the chick intestine and this property appears to be important in protecting the epithelium of the gut, especially during challenges by coccidia (Kettunen *et al.*, 2001a, b).

In addition to protecting enterocytes, betaine is osmoprotective for macrophages, increasing their chemotaxis and NO release in response to coccidia (Klasing *et al.*, 2002). Thus dietary betaine mitigates the extent of pathology that accompanies some types of coccidial infection (Augustine *et al.*, 1997; Matthews *et al.*, 1997; Augustine and Danforth, 1999; Klasing *et al.*, 2002; Fetterer *et al.*, 2003).

Similarly, vitamin C, glutamine and nucleotides are also synthesized by poultry, though some avian species are unable to synthesize vitamin C (Klasing, 1998). Social, physical and infectious stresses deplete tissue ascorbate levels due to a mismatch between synthesis and utilization rates (Pardue and Thaxton, 1986). Glutamine is depleted by inflammatory stresses (Klasing and Austic, 1984; Newsholme, 2001). Rapidly proliferating cells –

especially enterocytes and activated leukocytes – require large amounts of glutamine and nucleotides, making these nutrients conditionally essential during infection or injury (Newsholme, 2001). Thus, supplementation to purified diets that have low levels decreases morbidity following a variety of enteric infections.

INTERACTIONS BETWEEN MICRONUTRIENTS AND GUT MICROFLORA

Nutritional need of microflora for micronutrients

The developing embryo and hatchling have several systems that influence the colonization of the gut by commensal microflora. Microflora requires specific micronutrients for proliferation and pathogenicity. The sequestration of micronutrients by binding proteins in egg albumin is an important determinant of microbial growth (Romanoff and Romanoff, 1949).

Iron and biotin are tightly chelated to ovotransferrin and avidin, respectively. Although some species of bacteria can synthesize biotin, many pathogens have lost this capability and require biotin. Among avian pathogens, biotin is required for the growth of *Candida*, *Listeria*, *Staphylococcus*, *Clostridia*, *Streptococcus*, *Salmonella*, *E. coli* and eukaryotic parasites, including coccidia. Iron is required for all bacteria, though the requirement is lower for some taxa such as lactobacilli (Weinberg, 1997).

Avidin is a basic glycoprotein with an extraordinarily high dissociation constant for biotin of 10^{-15}. This is the highest non-covalent binding found in nature and confers remarkable stability to the biotin–avidin complex. In addition to being found in egg white, avidin is produced by the chicken intestine following tissue trauma or infection (Elo and Korpela, 1984; Kunnas *et al.*, 1993). Although little research has examined the roles of avidin or ovotransferrin in the protection of the avian gut, they are important at limiting pathogen growth *in vitro* (Klasing and Peng, 2004).

Given the analogous role of lactoferrin in mammals (Chierici, 2001; Lonnerdal, 2003), these two nutrient-binding proteins of birds are likely to be important in directing the development of a healthy population of commensal microflora and in diminishing the growth of pathogens. Evidence supporting this contention includes the following: (i) addition of iron to eggs inoculated with *Camplyobacter jejuni* enhances the infection rate of hatchlings (Clark and Bueschkens, 1985); (ii) iron stimulates the growth of enteric *S*. Enteritidis (Clay and Board, 1991; Chen *et al.*, 2001); and (iii) iron acquisition systems are important virulence factors for several avian pathogens (Dozois *et al.*, 2003; Ewers *et al.*, 2005).

Antimicrobial effects of micronutrients

Copper (Cu) salts are antimicrobial, anti-parasitic (Gabrashanska *et al.*, 1993; Aarestrup and Hasman, 2004) and change gut microflora and caecal morphology (Jensen and Maurice, 1978). When fed at levels (125–250 mg/kg) that are greatly above nutritional requirements (10 mg/kg), Cu increases growth rates of broiler chickens (Pesti and Bakalli, 1996; Ewing *et al.*, 1998).

In pigs, the growth-promoting activity of Cu is associated with changes in gut microbial populations that are similar to those induced by antibiotics (Hojberg *et al.*, 2005).

Zinc (Zn) is also bacteriocidal for a variety of avian pathogens (Aarestrup and Hasman, 2004), improves growth rates in chicks orally challenged by *S. Typhimurium* (Hegazy and Adachi, 2000), modifies the ecology of intestinal bacteria (Hojberg *et al.*, 2005) and is sometimes used as a growth promoter. However, high levels of Zn increase the frequency of lesions caused by *Clostridium perfringens*, possibly by increasing alpha toxin production and protecting it from the neutralizing actions of trypsin (Baba *et al.*, 1992).

MICRONUTRIENT IMMUNOMODULATION

Some essential nutrients, when fed at dietary levels that are clearly above the nutritional requirement, modulate the immune system (see Fig. 14.1). Nutrients that are not normally considered as being nutritionally essential may also modulate immunity. This nutritional modulation may then impact on a bird's resistance to enteric pathogens. A relationship between immunomodulatory

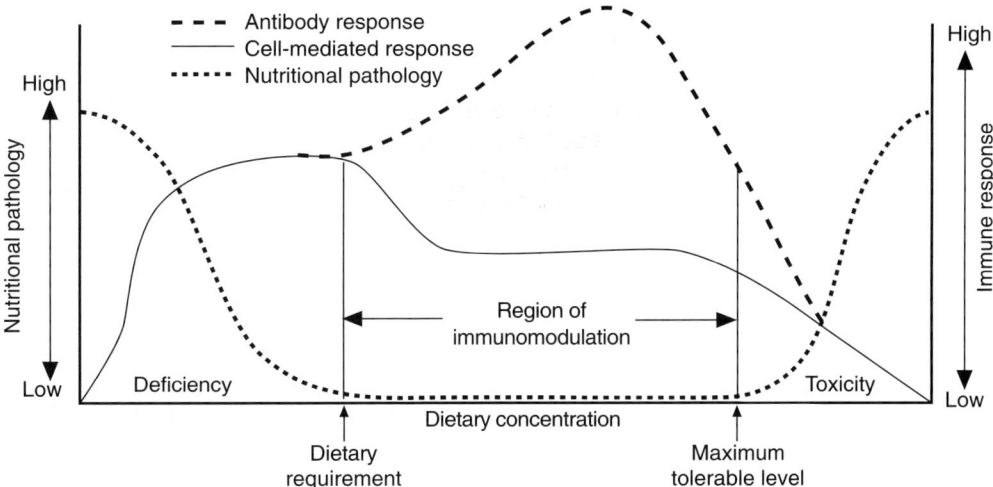

Fig. 14.1. The dietary concentration of a nutrient influences immunocompetence and the severity of nutritional pathologies. At low dietary levels (deficiency) and at high levels (toxicity), tissue pathology occurs. At levels between the dietary requirement and the maximum tolerable level, nutritional pathologies are absent. Typically, all components of immunity increase with increasing dietary nutrient concentrations at levels below the requirement. At levels above the requirement, the different components of the immune system are affected differently (immunomodulation). In the case of n-3 PUFA, antibody responses are increased at nutrient levels in excess of the requirement while cell-mediated immunity and inflammation decrease.

nutrients and pathological consequences of infectious challenges has been observed for most enteric pathogens, but coccidiosis has received the most attention (Allen *et al.*, 1998).

Nutrients that have strong immunomodulating activities include metabolizable energy, long-chain polyunsaturated fatty acids (PUFA), carotenoids and vitamins A, C, D and E (Klasing, 2001). For most of these nutrients, receptors have been identified that bind the nutrient and modify signal transduction cascades important in the regulation and response of leukocytes. For example, some PUFA bind to peroxisome proliferators activated receptor (PPAR) nuclear hormone receptors, which then change the activity of transcription factors such as NFκB and AP1.

Similarly, vitamin A metabolites bind to retinoic acid X receptor (RXR) nuclear hormone receptors and regulate NFκB and AP1, which are important for the initiation of pro-inflammatory responses and the balance between Th1 and Th2 lymphocyte responses. Friedman and Sklan (1989a) and Sklan *et al.* (1994) titrated the amount of vitamin A needed to maximize lymphocyte immunomodulation and found that more than 10-fold higher dietary levels were needed for immunomodulation than for maximal weight gain or prevention of deficiency symptoms (i.e. the nutritional requirement). Similarly, Leshchinsky *et al.* (2001) determined that immunomodulation by vitamin E was maximal at dietary levels of between five and ten times the requirement.

The immunomodulatory effect of nutrients is now well appreciated and used extensively in human and veterinary medicine. Both experimental and clinical results clearly indicate that the specific immunomodulatory actions of a nutrient must be understood before its application in human or animal populations because its efficacy is context dependent. Unlike increases in nutrients from deficient to sufficient levels where most indices of immunocompetence are elevated, increases from sufficient to immunomodulatory levels cause some components of immunity to be elevated and others to be diminished; in other words, the type and intensity of responses have been changed or modulated.

The resulting changes may be protective for some pathogens, i.e. those that are controlled by the elevated responses, but contraindicated for other pathogens, i.e. those that are controlled by diminished responses. This is illustrated by n-3 PUFA, which diminish pro-inflammatory and Th1-mediated lymphocyte responses, but augment Th2 responses. The general result is improved resistance to challenges controlled by Th2 responses but increased susceptibility to challenges controlled by Th1 responses and inflammation (Anderson and Fritsche, 2002).

INTESTINAL HEALTH AND MICRONUTRIENT STATUS

Enteric infections markedly diminish the absorption of many nutrients. In general, the absorption of micronutrients is impacted to a greater degree than that of macronutrients such as carbohydrates, protein or fat. Three general mechanisms mediate the malabsorption. First, diminished absorption can be a direct result of pathological changes in the integrity and function of the intestinal epithelium. This is especially the case with infections that inflict pathology to the absorptive regions of the small intestines.

Secondly, malabsorption can be due to an increase in the rate of passage of digesta through the intestines, because high rates of transit diminish the time available for absorption. Thirdly, the reduction in absorption of some nutrients is orchestrated by the immune response to the pathogen. The changes in nutrient absorption that are mediated by the immune system may have protective value for the host by depriving microbes of nutrients that they need for proliferation (e.g. iron) or for defences against the immune system's effector mechanisms (e.g. antioxidants).

The absorption of vitamin A and carotenoids is particularly decreased during enteric infections (Marusich *et al.*, 1973; Allen, 1992; West *et al.*, 1992). The magnitude of the decrease can not be accounted for by a generalized decrease in lipid digestion and absorption. Absorption of vitamin A and carotenoids is also diminished during systemic infections, so at least part of the malabsorption is probably a component of the immune response. The absorption of iron is also markedly decreased during enteric infections. Some of this decrease is correlated with damage to iron-absorbing regions of the intestines (Turk, 1981), but absorption is also diminished during systemic infections (Hill *et al.*, 1977), and research in rodents indicates that pro-inflammatory cytokines mediate reduced iron absorption (Steele *et al.*, 2005).

CONCLUSIONS

In general, micronutrient malnutrition does not affect gut health as greatly as do macronutrients. Much of the gut receives first refusal at nutrients arriving from the diet, endowing it with priority over other tissues. Thus, gut pathology is not commonly a sentinel for malnutrition. However, a variety of micronutrients can affect the ecology of microflora inhabiting the gut, either by serving as nutrients for the microbes or by affecting the immune system, which controls population dynamics in the gut. Consequently, micronutrient impacts on gut health are subtle but important.

REFERENCES

Aarestrup, F.M. and Hasman, H. (2004) Susceptibility of different bacterial species isolated from food animals to copper sulphate, zinc chloride and antimicrobial substances used for disinfection. *Veterinary Microbiology* 100, 83–89.

Allen, P.C. (1992) Effect of coccidiosis on the distribution of dietary lutein in the chick. *Poultry Science* 71, 1457–1463.

Allen, P.C., and Fetterer, R.H. (2002) Effects of dietary vitamin E on chickens infected with *Eimeria maxima*: observations over time of primary infection. *Avian Diseases* 46, 839–846.

Allen, P.C., Danforth, H.D. and Augustine, P.C. (1998) Dietary modulation of avian coccidiosis. *International Journal of Parasitology* 28, 1131–1140.

Anderson, M. and Fritsche, K.L. (2002) (n-3) Fatty acids and infectious disease resistance. *Journal of Nutrition* 132, 3566–3576.

Augustine, P.C. and Danforth, H.D. (1999) Influence of betaine and salinomycin on intestinal absorption of methionine and glucose and on the ultrastructure of intestinal cells and parasite developmental stages in chicks infected with *Eimeria acervulina*. *Avian Diseases* 43, 89–97.

Augustine, P.C., McNaughton, J.L., Virtanen, E. and Rosi, L. (1997) Effect of betaine on the growth performance of chicks inoculated with mixed cultures of avian *Eimeria* species and on invasion and development of *Eimeria tenella* and *Eimeria acervulina in vitro* and *in vivo*. *Poultry Science* 76, 802–809.

Aye, P.P., Morishita, T.Y., Saif, Y.M., Latshaw, J.D., Harr, B.S. and Cihla, F.B. (2000) Induction of vitamin A deficiency in turkeys. *Avian Diseases* 44, 809–817.

Baba, E., Fuller, A.L., Gilbert, J.M., Thayer, S.G. and McDougald, L.R. (1992) Effects of *Eimeria brunetti* infection and dietary zinc on experimental induction of necrotic enteritis in broiler chickens. *Avian Diseases* 36, 59–62.

Basova, N.A., Markov, I.U.G. and Berzin, N.I. (2002) Effects of natural (zinc, vitamin E) and synthetic (diludin) antioxidants on the intestinal permeability in chicks with vitamin A deficiency. *Rossiiskii Fiziologicheskii Zhurnal Imeni I. M. Sechenova* 88, 650–657.

Cantor, A.H., Moorhead, P.D. and Musser, M.A. (1982) Comparative effects of sodium selenite and selenomethionine upon nutritional muscular dystrophy, selenium-dependent glutathione peroxidase and tissue selenium concentrations of turkey poults. *Poultry Science* 61, 478–484.

Chen, H., Anantheswaran, R.C. and Knabel, S.J. (2001) Optimization of iron supplementation for enhanced detection of *Salmonella* Enteritidis in eggs. *Journal of Food Protection* 6, 1279–1285.

Chierici, R. (2001) Antimicrobial actions of lactoferrin. *Advances in Nutritional Research* 10, 247–269.

Clark, A.G. and Bueschkens, D.H. (1985) Laboratory infection of chicken eggs with *Campylobacter jejuni* by using temperature or pressure differentials. *Applied and Environmental Microbiology* 49, 1467–1471.

Clay, C.E. and Board, R.G. (1991) Growth of *Salmonella* Enteritidis in artificially contaminated hens' shell eggs. *Epidemiology and Infection* 106, 271–281.

Colnago, G.L., Jensen, L.S. and Long, P.L. (1984) Effect of selenium and vitamin E on the development of immunity to coccidiosis in chickens. *Poultry Science* 63, 1136–1143.

Cook, M.E. (1991) Nutrition and the immune response of the domestic fowl. *Critical Reviews in Poultry Biology* 3, 167–190.

Daghir, N.J. and Haddad, K.S. (1981) Vitamin B_6 in the etiology of gizzard erosion in growing chickens. *Poultry Science* 60, 988–992.

Dalloul, R.A., Lillehoj, H.S., Shellem, T.A. and Doerr, J.A. (2002) Effect of vitamin A deficiency on host intestinal immune response to *Eimeria acervulina* in broiler chickens. *Poultry Science* 81, 1509–1515.

Dalloul, R.A., Lillehoj, H.S., Shellem, T.A. and Doerr, J.A. (2003) Intestinal immunomodulation by vitamin A deficiency and lactobacillus-based probiotic in *Eimeria acervulina*-infected broiler chickens. *Avian Diseases* 47, 1313–1320.

Dam, H. (1970) Interrelations between vitamin E and polyunsaturated fatty acids. *Bibliotheca Nutritio et Dieta* 15, 114–128.

De Luca, L., Little, E.P. and Wolf, G. (1969) Vitamin A and protein synthesis by rat intestinal mucosa. *Journal of Biological Chemistry* 244, 701–708.

Dozois, C.M., Daigle, F. and Curtiss, R. 3rd, (2003) Identification of pathogen-specific and conserved genes expressed *in vivo* by an avian pathogenic *Escherichia coli* strain. *Proceedings of the National Academy of Sciences of the USA* 100, 247–252.

Elo, H.A. and Korpela, J. (1984) The occurrence and production of avidin: a new conception of the high-affinity biotin-binding protein. *Comparative Biochemistry and Physiology B: Biochemistry and Molecular Biology* 78, 15–20.

Erasmus, J. and Scott, M.L. (1960) A relationship between coccidiosis and vitamin A nutrition in chickens. *Poultry Science* 39, 565–572.

Erf, G.F., Bottje, W.G., Bersi, T.K., Headrick, M.D. and Fritts, C.A. (1998) Effects of dietary vitamin E on the immune system in broilers: altered proportions of CD4 T cells in the thymus and spleen. *Poultry Science* 77, 529–537.

Ewers, C., Janssen, T., Kiessling, S., Philipp, H.C. and Wieler, L.H. (2005) Rapid detection of virulence-associated genes in avian pathogenic *Escherichia coli* by multiplex polymerase chain reaction. *Avian Diseases* 49, 269–273.

Ewing, H.P., Pesti, G.M., Bakalli, R.I. and Menten, J.F. (1998) Studies on the feeding of cupric sulfate pentahydrate, cupric citrate and copper oxychloride to broiler chickens. *Poultry Science* 77, 445–448.

Fetterer, R.H., Augustine, P.C., Allen, P.C. and Barfield, R.C. (2003). The effect of dietary betaine on intestinal and plasma levels of betaine in uninfected and coccidia-infected broiler chicks. *Parasitology Research* 90, 343–348.

Fisher, C., Laursen-Jones, A.P., Hill, K.J. and Hardy, W.S. (1973) The effect of copper sulphate on performance and the structure of the gizzard in broilers. *British Poultry Science* 14, 55–68.

Fox, M.R. (1975) Protective effects of ascorbic acid against toxicity of heavy metals. *Annals of the New York Academy of Sciences* 258, 144–150.

Friedman, A. and Sklan, D. (1989a) Antigen-specific immune response impairment in the chick as influenced by dietary vitamin A. *Journal of Nutrition* 119, 790–795.

Friedman, A. and Sklan, D. (1989b) Impaired T lymphocyte immune response in vitamin A-depleted rats and chicks. *British Journal of Nutrition* 62, 439–449.

Gabrashanska, M., Galvez-Morros, M.M. and Garcia-Martinez, O. (1993) Application of small doses of copper salts (basic and neutral) to *Ascaridia galli*-infected chicks. *Journal of Helminthology* 67, 287–290.

Gries, C.L. and Scott, M.L. (1972) The pathology of thiamin, riboflavin, pantothenic acid and niacin deficiencies in the chick. *Journal of Nutrition* 102, 1269–1285.

Halevy, O., Arazi, Y., Melamed, D., Friedman, A. and Sklan, D. (1994) Retinoic acid receptor-alpha gene expression is modulated by dietary vitamin A and by retinoic acid in chicken T lymphocytes. *Journal of Nutrition* 124, 2139–2146.

Hegazy, S.M. and Adachi, Y. (2000) Comparison of the effects of dietary selenium, zinc and selenium and zinc supplementation on growth and immune response between chick groups that were inoculated with *Salmonella* and aflatoxin or *Salmonella*. *Poultry Science* 79, 331–335.

Hill, R., Smith, I.M., Mohammadi, H. and Licence, S.T. (1977) Altered absorption and regulation of iron in chicks with acute *Salmonella gallinarum* infection. *Research in Veterinary Science* 22, 371–375.

Hojberg, O., Canibe, N., Poulsen, H.D., Hedemann, M.S. and Jensen, B.B. (2005) Influence of dietary zinc oxide and copper sulfate on the gastrointestinal ecosystem in newly weaned piglets. *Appied Environmental Microbiology* 71, 2267–2277.

Huang, D., Chen, S.W., Langston, A.W. and Gudas, L.J. (1998) A conserved retinoic acid-responsive element in the murine Hoxb-1 gene is required for expression in the developing gut. *Development* 125, 3235–46.

Jensen, L.S., Dunn, P.A. and Dobson, K.N. (1991) Induction of oral lesions in broiler chicks by supplementing the diet with copper. *Avian Diseases* 35, 969–973.

Jensen, L.S. and Maurice, D.V. (1978) Effect of high dietary copper on the caeca of chicks. *Poultry Science* 57, 166–170.

Kettunen, H., Peuranen, S. and Tiihonen, K. (2001a) Betaine aids in the osmoregulation of duodenal epithelium of broiler chicks and affects the movement of water across the small intestinal epithelium *in vitro*. *Comparative Biochemistry and Physiology A: Molecular and Integrative Physiology* 129, 595–603.

Kettunen, H., Tiihonen, K., Peuranen, S., Saarinen, M.T. and Remus, J.C. (2001b) Dietary betaine accumulates in the liver and intestinal tissue and stabilizes the intestinal epithelial structure in healthy and coccidia-infected broiler chicks. *Comparative Biochemistry and Physiology A: Molecular and Integrative Physiology* 130, 759–769.

Klasing, K.C. (1998) *Comparative Avian Nutrition*. CAB International, Wallingford, UK.

Klasing, K.C. (2001) Protecting animal health and well being: Nutrition and immune function. In: National Research Council (ed.) *Scientific Advances in Animal Nutrition: Promises for the New Century*. National Academy Press, Washington DC, pp. 13–20.

Klasing, K.C. (2005) Interplay between diet, microbes and immune defences of the gastrointestinal tract. In: Stark, J.M. and Wang, T. (eds) *Consequences of Feeding in Vertebrates*. Oxford University Press, Oxford, UK.

Klasing, K.C. and Austic, R.E. (1984) Changes in plasma, tissue and urinary nitrogen metabolites due to an inflammatory challenge. *Proceedings of the Society of Experimental Biology and Medicine* 176, 276–284.

Klasing, K.C. and Austic, R.E. (2003) Nutritional diseases. In: Saif, Y.M. (ed.) *Diseases of Poultry*. Blackwell Publishing, Ames, Iowa, pp. 1027–1053.

Klasing, K.C. and Peng, R. (2004) Biotin is the first limiting for growth of *Salmonella*. in chickens and iron is the second limiting. *Proceedings of the Comparative Nutrition Society* 4, 54–58.

Klasing, K.C., Adler, K.L., Remus, J.C. and Calvert, C.C. (2002) Dietary betaine increases intraepithelial lymphocytes in the duodenum of coccidia-infected chicks and increases functional properties of phagocytes. *Journal of Nutrition* 132, 2274–2282.

Koutsos, E.A., Tell, L.A., Woods, L.W. and Klasing, K.C. (2003) Adult cockatiels (*Nymphicus hollandicus*) at maintenance are more sensitive to diets containing excess vitamin A than to vitamin A-deficient diets. *Journal of Nutrition* 133, 1898–1902.

Kunnas, T.A., Wallen, M.J. and Kulomaa, M.S. (1993) Induction of chicken avidin and related mRNAs after bacterial infection. *Biochimica et Biophysica Acta* 1216 (3), 441–445.

Leshchinsky, T.V. and Klasing, K.C. (2001) Relationship between the level of dietary vitamin E and the immune response of broiler chickens. *Poultry Science* 80, 159–159.

Lessard, M., Hutchings, D. and Cave, N.A. (1997) Cell-mediated and humoral immune responses in broiler chickens maintained on diets containing different levels of vitamin A. *Poultry Science* 76, 1368–1378.

Lonnerdal, B. (2003) Nutritional and physiologic significance of human milk proteins. *American Journal of Clinical Nutrition* 77, 1537S–1543S.

Marsh, J.A., Combs, G.F., Jr., Whitacre, M.E. and Dietert, R.R. (1986) Effect of selenium and vitamin E dietary deficiencies on chick lymphoid organ development. *Proceedings of the Society of Experimental Biology and Medicine* 182, 425–436.

Marusich, W.L., Ogrins, E.F., Schildknecht, E., Brown, P.R. and Mitrovic, M. (1973). The effect of subclinical infections with *Eimeria praecox* and *Eimeria tenella* on pigmentation and vitamin A absorption in broilers. *British Poultry Science* 14, 541–546.

Matthews, J.O., Ward, T.L. and Southern, L.L. (1997) Interactive effects of betaine and monensin in uninfected and *Eimeria acervulina*-infected chicks. *Poultry Science* 76, 1014–1019.

Newsholme, P. (2001) Why is L-glutamine metabolism important to cells of the immune system in health, postinjury, surgery or infection? *Journal of Nutrition* 131, 2515S–2522S; discussion 2523S–2524S.

Nikawa, T., Odahara, K., Koizumi, H., Kido, Y., Teshima, S., Rokutan, K. and Kishi, K. (1999) Vitamin A prevents the decline in immunoglobulin A and Th2 cytokine levels in small intestinal mucosa of protein-malnourished mice. *Journal of Nutrition* 129, 934–941.

NRC (2005) *Mineral Tolerance of Animals; Second Revised Edition*. National Academy Press, Washington DC.

Pardue, S.L. and Thaxton, J.P. (1986) Ascorbic acid in poultry: a review. *World's Poultry Science Journal* 42, 107–123.

Pesti, G.M. and Bakalli, R.I. (1996) Studies on the feeding of cupric sulfate pentahydrate and cupric citrate to broiler chickens. *Poultry Science* 75, 1086–1091.

Poupoulis, C. and Jensen, L.S. (1976) Effect of high dietary copper on gizzard integrity of the chick. *Poultry Science* 55, 113–121.

Ringrose, A.T., Norris, L.C. and Heuser, G.F. (1931) The occurrence of a pellagra-like syndrome in chicks. *Poultry Science* 10, 166–177.

Rojanapo, W., Lamb, A.J. and Olson, J.A. (1980) The prevalence, metabolism and migration of goblet cells in rat intestine following the induction of rapid, synchronous vitamin A deficiency. *Journal of Nutrition* 110, 178–188.

Romanoff, A.L. and Romanoff, A.J. (1949) *The Avian Egg*. J. Wiley, New York.

Ross, A.C. and Stephensen, C.B. (1996) Vitamin A and retinoids in antiviral responses. *The Faseb Journal: Official Publication of the Federation of American Societies for Experimental Biology* 10, 979–985.

Semba, R.D. (1998) The role of vitamin A and related retinoids in immune function. *Nutrition Reviews* 56, S38–S48.

Sklan, D., Melamed, D. and Friedman, A. (1994) The effect of varying levels of dietary vitamin A on immune response in the chick. *Poultry Science* 73, 843–847.

Steele, T.M., Frazer, D.M. and Anderson, G.J. (2005) Systemic regulation of intestinal iron absorption. *International Union of Biochemistry and Molecular Biology* 57, 499–503.

Tengerdy, R.P. (1990) Immunity and disease resistance in farm animals fed vitamin E supplement. *Advances in Experimental Medicine and Biology* 262, 103–110.

Turk, D.E. (1981) Coccidial infections and iron absorption. *Poultry Science* 60, 323–326.

Uni, Z., Zaiger, G. and Reifen, R. (1998) Vitamin A deficiency induces morphometric changes and decreased functionality in chicken small intestine. *British Journal of Nutrition* 80, 401–407.

Uni, Z., Zaiger, G., Gal-Garber, O., Pines, M., Rozenboim, I. and Reifen, R. (2000) Vitamin A deficiency interferes with proliferation and maturation of cells in the chicken small intestine. *British Poultry Science* 41, 410–415.

Van Vleet, J.F. (1982) Amounts of eight combined elements required to induce selenium–vitamin E deficiency in ducklings and protection by supplements of selenium and vitamin E. *American Journal of Veterinary Research* 43, 1049–1055.

Wang, C.H. (1993) Cornification of esophagus induced by excessive vitamin E. *Nutrition* 9, 225–228.

Weinberg, E.D. (1997). The *Lactobacillus* anomaly: total iron abstinence. *Perspectives in Biology and Medicine* 40, 578–583.

West, C.E., Sijtsma, S.R., Kouwenhoven, B., Rombout, J.H. and van der Zijpp, A.J. (1992) Epithelia-damaging virus infections affect vitamin A status in chickens. *Journal of Nutrition* 122, 333–339.

Yarrington, J.T. and Whiehair, C.K. (1975) Ultrastructure of gastointestinal smooth muscle in ducks with a vitamin E–selenium deficiency. *Journal of Nutrition* 105, 782–790.

Zoltowska, K., Dziekonska-Rynko, J., Olender, H. and Jablonowski, Z. (1995) Effect of vitamin A and protein levels in the diet of chickens infected with *Ascaridia galli* on activity of digestive enzymes in pancreas and small intestine. *Wiadomosci Parazytologiczne* 41, 421–428.

PART V
Pathology

CHAPTER 15
Virally induced gastrointestinal diseases of chickens and turkeys

J.S. Guy

College of Veterinary Medicine, North Carolina State University, Raleigh, North Carolina; e-mail: jim_guy@ncsu.edu

ABSTRACT

Virally induced gastrointestinal diseases are common causes of production losses in chickens and turkeys. The mechanisms by which viruses produce gastrointestinal diseases differ among the viruses that replicate in this system, with the outcome of infection determined by variables that include site of virus replication and interaction with other infectious agents, nutrition and management. Depending on these variables, viral infection results in either inapparent infection or gastrointestinal disease characterized by proventriculitis or enteritis.

Proventriculitis results from viral replication in proventricular glandular epithelium, with consequent impairment of pepsinogen and hydrochloric acid synthesis and impaired nutrient absorption. Pathogenesis of virally induced enteric disease is more complex, in that disease may occur by multiple pathways. For viruses that replicate in differentiated villous epithelium, enteric disease may occur by a perturbation of normal absorption and digestion, potentiation of pathogenesis of other infectious agents or by a combination of these effects; the role of nutrition as a complicating factor requires additional research.

An alternative mechanism for inducing enteritis is exhibited by the avian adenovirus, haemorrhagic enteritis virus; this virus perturbs intestinal function by damaging cells within the *lamina propria*. Regardless of the pathogenic mechanism involved, these viral diseases result in decreased efficiency of feed utilization and this, by itself, is an important source of economic loss. However, these infections often have other adverse effects on productivity, including decreased growth rates, decreased flock uniformity, increased susceptibility to other infectious agents and increased mortality. Control of these virally induced gastrointestinal diseases ultimately is dependent on improved understanding of the viruses involved.

INTRODUCTION

Several different viruses have been identified as causes of gastrointestinal tract diseases in chickens and turkeys, and several others have been labelled as causes based on electron microscopic identification in tissues and/or intestinal contents of affected poultry. Virally induced gastrointestinal infections occur in birds of all age groups but tend to predominate in young birds. Clinically, these gastrointestinal diseases result in a broad range of outcomes ranging from inapparent, economically insignificant effects to severe and economically devastating disease. In field situations, virally induced gastrointestinal infections almost always are complicated by other infectious agents as well as by management, nutritional and environmental factors; thus, the true role of viruses in naturally occurring gastrointestinal diseases is often difficult to assess.

The gastrointestinal tract of chickens and turkeys is a large organ comprised of many different cell types that potentially serve as targets for infection by many different viruses. However, most virally induced gastrointestinal diseases in chickens and turkeys are due to virally induced damage to only two cell types: (i) glandular epithelium in the proventriculus; and (ii) differentiated enterocytes that cover the intestinal villi.

The proventricular glandular epithelium is responsible for the production of digestive secretions (HCl and pepsinogen), and villous enterocytes are responsible for both digestion (production of disaccharidases, peptidases, etc.) and absorption of nutrients and water. Haemorrhagic enteritis virus (HEV) produces enteric disease in a unique manner by damaging cells within the intestinal *lamina propria*. Not surprisingly, virally induced damage of proventricular glandular epithelium, villous enterocytes and *lamina propria* cells potentially leads to disease, either in the form of proventriculitis or enteritis.

In today's modern poultry production systems where feed represents the single largest monetary investment, these diseases result in decreased efficiency of feed utilization and this, by itself, may result in economic loss. However, these viral infections often have other adverse effects on productivity, including increased mortality, decreased growth rates, decreased flock uniformity and increased susceptibility to other infectious agents. In addition, viral infections of the gastrointestinal tract are likely to be responsible for the development of a number of other, extra-gastrointestinal, diseases. Virally induced damage may provide a portal of entry for other infectious agents or it may promote their attachment and/or proliferation in the gastrointestinal tract.

Viral infection may result in nutritional deficiencies, especially those related to fat-soluble vitamins and minerals. Abnormal feathering, rickets, osteoporosis and other skeletal abnormalities are frequently seen in young, meat-type birds that experience gastrointestinal disease. Nutritional deficiency also may impair the growth and development of lymphoid organs, and this may result in immunological deficiency and increased susceptibility to other infectious diseases.

The gastrointestinal tract is a hostile environment for viruses. Viruses that reach susceptible gastrointestinal cells by ingestion must evade destruction by acids, bile and digestive enzymes that are secreted into the gastrointestinal lumen. Additionally, for the virus to initiate infection it must penetrate the glycocalyx/mucous coating of epithelium in proventriculus and/or intestines and find appropriate receptors for attachment. Viruses may circumvent these barriers by viraemic spread to susceptible cells, as is the case with adenoviruses such as haemorrhagic enteritis virus.

The mechanisms by which viruses cause gastrointestinal disease vary among the different viruses that replicate in this organ system. The outcome of these infections is determined by several variables, including: (i) host factors (age, immune status, genetics); (ii) site of virus replication (cell tropism); (iii) degree of virally induced damage; and (iv) interaction with non-viral cofactors such as other infectious agents. Depending on these variables, the outcome may be either: (i) inapparent infection; or (ii) gastrointestinal disease characterized by proventriculitis or enteritis.

Virally induced proventriculitis, referred to as transmissible viral proventriculitis (TVP), is a poorly understood disease of broiler chickens that is characterized by proventricular enlargement, necrosis of proventricular glandular epithelium and lymphocytic inflammation (Goodwin *et al.*, 1996; Goodwin and Hafner, 2003). The disease has been reported in broiler chickens in the USA, Netherlands and Australia. It is associated with proventricular fragility, impaired feed digestion ('feed passage'), impaired growth ('runting') and poor feed conversion (Goodwin *et al.*, 1996).

In a study involving a large broiler-producing company in the USA, feed conversion in TVP-affected broiler flocks was estimated to be ten points less than in unaffected flocks (Goodwin and Hafner, 2003). Additionally, proventricular fragility associated with proventriculitis-affected chickens may result in increased costs at processing due to rupture of the proventriculus and spillage of gut contents into the abdominal cavity. Rupture of the proventriculus at processing results in increased numbers of downgrades, condemnations and reprocessed carcasses, with estimated losses to the US broiler industry of millions of dollars each year (Leonard and Schmittle, 1995). TVP also has food safety implications, as rupture of the proventriculus may result in faecal contamination of carcasses (Thayer and Walsh, 1993).

Rotavirus, turkey coronavirus and haemorrhagic enteritis virus (HEV) are known causes of enteric disease in chickens and turkeys (Guy, 2003; McNulty, 2003; Pierson and Fitzgerald, 2003). In addition, these and other viruses (astrovirus, reovirus, torovirus) probably play key aetiological roles in several poorly understood enteric diseases of chickens and turkeys, including malabsorption syndrome, runting/stunting syndrome and pale bird syndrome in chickens, and poult enteritis and poult enteritis-mortality syndrome in turkeys.

Based on experimental studies, it is likely that these diseases have complex, multifactorial aetiologies with viruses being important initiating factors. These infections, with the exception of haemorrhagic enteritis, have similar clinical features in chickens and turkeys. They generally affect birds in the first 3 weeks of life and generally are characterized by diarrhoea, growth retardation and poor feed utilization. Sequelae to these infections often include abnormal feather development ('helicopter' chicks or poults) and rickets. Mortality may be increased in affected flocks.

A disease characterized by poor growth, retarded feather development, diarrhoea and various other clinical abnormalities was first identified in the late 1970s in broiler chickens (McNulty and McFerran, 1993). The disease has been referred to variously as malabsorption syndrome, pale bird syndrome, infectious stunting syndrome, runting stunting syndrome (RSS), pale bird syndrome and helicopter disease. Poor growth and retarded feather development were consistently observed, along with a variety of inconsistently occurring signs including diarrhoea, increased mortality, pancreatic atrophy, proventriculitis, rickets and lymphoid atrophy.

The term RSS appears to be the most acceptable for this disease as it most accurately reflects the most consistent clinical findings (McNulty and McFerran, 1993). Viruses associated with or proposed as causes of RSS include reoviruses, rotaviruses, parvoviruses, enterovirus-like viruses and adenoviruses. However, experimental attempts to reproduce this disease with these agents have been inconclusive, thus the aetiology of this disease remains undetermined.

Poult enteritis is a term used to encompass the enteric diseases of young turkeys of unknown aetiology. Clinical features include diarrhoea, impaired growth and poor feed utilization; in severe cases, runting, immune dysfunction and increased mortality may be observed. Viruses associated with poult enteritis include rotavirus types A and D, reovirus, astrovirus and enterovirus-like viruses.

In 1991, a particularly severe form of poult enteritis characterized by high mortality was recognized in North Carolina, USA. This disease, referred to as poult enteritis mortality syndrome (PEMS) is a transmissible, infectious disease that generally affects turkeys between the ages of 7 and 28 days (Barnes and Guy, 2003). PEMS-affected birds exhibit depression, diarrhoea, anorexia, growth depression and increased mortality; total mortality in severely affected flocks may be as high as 60%.

Birds examined at necropsy have pale, thin-walled and distended intestines, thymic atrophy and bursal atrophy. Microscopic lesions in affected birds generally include villous atrophy and moderate to marked lymphoid depletion in spleen, bursa of Fabricius and thymus. Flocks that have recovered from PEMS generally exhibit stunting, lack of size uniformity, increased susceptibility to other diseases, increased time-to-market and increased feed conversion. Several infectious agents have been associated with PEMS including reovirus, rotavirus types A and D, astroviruses, turkey coronavirus, *Salmonella* spp., *E. coli*, *Campylobacter* spp. and *Cryptosporidium* spp.

The avian adenovirus, haemorrhagic enteritis virus (HEV), is a well-established cause of enteritis in turkeys at 4–8 weeks of age (Pierson and Fitzgerald, 2003). Birds exhibit depression, bloody droppings and sudden death. Mortality in field outbreaks ranges from 1–60%, but mortality of approximately 80% may be observed in experimentally inoculated birds.

PROVENTRICULITIS

Transmissible viral proventriculitis (TVP) is presently considered to be a disease of unknown aetiology. Several different viruses have been associated as the cause of TVP, including reovirus, infectious bronchitis virus (IBV), infectious bursal disease virus, group I avian adenovirus and a recently described adenovirus-like virus (Goodwin *et al.*, 1996; Goodwin and Hafner, 2003; Guy *et al.*, 2005).

Additionally, other infectious, nutritional and toxic factors have been associated with this disease, including *Cryptosporidium* spp., a recently described *Clostridium* sp., low-fibre diets and high levels of dietary copper sulphate, biogenic amines and mycotoxins (Goodwin and Hafner, 2003). A viral aetiology is suggested by: (i) experimental reproduction with filtrates of diseased proventriculi; (ii) an inflammatory response comprised predominantly of CD8$^+$ T cells; and (iii) enhancement of disease by suppression of T-cell, but not B-cell, immunity (Pantin-Jackwood *et al.*, 2004 a, b; Guy *et al.*, 2005).

Viruses associated with proventriculitis

Reoviruses

Reoviruses are non-enveloped, spherical and have a diameter of approximately 75 nm (McNulty, 1993). Replication of the virus occurs exclusively in the cytoplasm, sometimes forming paracrystalline arrays. They are resistant to inactivation by heat, ether and pH 3. Reoviruses are ubiquitous in poultry populations and are commonly isolated from tissues (respiratory, intestines, proventriculi) of diseased and clinically normal birds. They have been associated with being causal agents of TVP based on their frequent isolation from TVP-affected chickens (Brugh and Wilson, 1986).

Infectious bronchitis virus

Infectious bronchitis virus (IBV) is a member of the genus *Coronavirus* within the Coronaviridae family. Coronaviruses are RNA-containing viruses that infect a wide variety of avian and mammalian species (Wege *et al.*, 1982). They are characterized on the basis of their distinctive morphology: pleomorphic, enveloped particles with diameter of 60–220 nm and having long (12–24 nm), widely spaced, petal-shaped surface projections (Wege *et al.*, 1982). Coronavirus replication occurs exclusively in the cytoplasm without the formation of inclusion bodies.

The principal site of IBV replication is in respiratory epithelium, but virus replication also may occur in other tissues, including kidney, intestine and oviduct (Wege *et al.*, 1982). Additionally, IBV replication has been detected in mucosal epithelium of proventriculus based on immunohistochemistry (Guy *et al.*, 2005). IBV has been associated with being a causal agent of TVP based on recovery from proventriculi of TVP-affected chickens. Recent studies in China have asserted that IBV is the aetiological agent of TVP based on experimental reproduction studies (Yu *et al.*, 2001); however, these studies have not been substantiated by other investigators (Guy *et al.*, 2005).

Infectious bursal disease virus

Infectious bursal disease virus (IBDV), a member of the Birnaviridae family, is a non-enveloped, icosahedral virus with a diameter of 55–65 nm (Lukert and Saif, 2003). IBDV is a common infection of commercial chickens and is present worldwide. The virus is the cause of an acute, highly contagious disease of chickens characterized by lymphoid necrosis in the bursa of Fabricius and immunosuppression. IBDV has been associated as a cause of TVP based on recovery of the virus from diseased proventriculi (Huff *et al.*, 2001).

IBDV has been shown to produce lesions in the mucosal epithelium of proventriculi of SPF chickens; however, these lesions are transient and inconsistent with those observed in chickens with naturally occurring disease (Skeeles *et al.*, 1998). In addition, recent studies failed to reproduce TVP with various IBDV strains, and the virus was inconsistently detected in proventriculi of inoculated chickens using immunohistochemistry or reverse-transcriptase polymerase chain reaction procedures (Pantin-Jackwood and Brown, 2003).

Adenoviruses

Adenoviruses are non-enveloped, icosahedral viruses that vary in size from 70–100 nm in diameter; they possess a DNA genome and morphogenesis occurs in cell nuclei (McFerran, 2003). The family is made up of two genera: *Mastadenovirus* (mammalian adenoviruses) and *Aviadenovirus* (avian adenoviruses). The avian adenoviruses (AAVs) comprise three distinct subgroups. Group I AAVs comprise the *Aviadenovirus* genus within the Adenoviridae family; groups II and III AAVs currently are unassigned members, but classification has been proposed for these viruses within two new genera, *Siadenovirus* and *Atadenovirus*, respectively (McFerran, 2003).

While all AAVs are quite similar based on morphological and physiochemical criteria, they may be distinguished and further classified based on antigenic and genomic differences. Groups I, II and III AAVs may be distinguished by antigenic differences within group-specific antigens or by genetic sequence differences identifiable by polymerase chain reaction procedures.

Group I AAVs have been associated as causal agents of TVP in broiler chickens based on their detection in proventriculi of diseased chickens, but their role in this disease remains unproven (Kouwenhoven *et al.*, 1978). Haemorrhagic enteritis virus (HEV), a group II AAV, is a well-established cause of enteric disease in turkeys (Pierson and Fitzgerald, 2003); Group III AAVs have not been associated as causal agents of gastrointestinal disease in poultry.

In 1996, Goodwin *et al.* identified a 60–70 nm, adenovirus-like virus (AdLV) in glandular epithelium of proventriculi collected from TVP-affected broiler chickens. The consistent finding of this virus in lesional sites suggested it as the likely aetiology, and the term 'transmissible viral proventriculitis' was designated to describe this disease. In a subsequent study, Huff *et al.* (2001) identified morphologically similar, intranuclear viruses in proventricular lesions of TVP-affected broiler chickens. The virus detected in these studies was not identified; however, virion size and the site of viral morphogenesis are consistent with adenoviruses.

Recent studies support the findings of Goodwin *et al.* (1996) and indicate an aetiological role for AdLV in this disease (Guy *et al.*, 2005). Intranuclear, approximately 70 nm, AdLVs were identified by thin-section EM in epithelia of proventriculi collected from TVP-affected chickens. The AdLV, designated AdLV (R11/3), subsequently was adapted to growth in embryonated SPF chicken eggs. Embryonated chicken eggs were amniotically (*in ovo*) inoculated, allowed to hatch and examined for evidence of AdLV (R11/3) infection in chicks at 2 days of age (8 days post-inoculation).

Virus propagation was evident in *in ovo*-inoculated chicks by the following means: (i) gross and microscopic lesions in proventriculi consistent with TVP; (ii) immunohistochemical localization of AdLV (R11/3) antigens in proventricular glandular epithelium; (iii) thin-section electron microscopic detection of intranuclear, approximately 70 nm AdLVs within proventricular epithelium; and (iv) negative-stain electron microscopic detection of extracellular, approximately 70 nm, AdLVs in intestinal contents. Microscopic lesions in proventriculi of chicks inoculated *in ovo* with AdLV (R11/3) were consistent with those described in both birds with naturally occurring TVP and chickens experimentally inoculated orally with proventricular homogenates collected from TVP-affected birds (Huff *et al.*, 2001; Pantin-Jackwood *et al.*, 2004b).

Antigenic and genomic analyses using indirect immunofluorescence and PCR procedures, respectively, indicated that AdLV (R11/3) is distinct from group I, II and III AAVs (Guy *et al.*, 2005). Additional studies, particularly gene sequencing studies, are needed in order to definitively identify this virus.

Pathogenesis of virally induced proventriculitis

The pathogenesis of TVP is poorly understood; however, histopathologic and electron microscopic studies suggest a likely explanation. Microscopic lesions primarily involve the glandular epithelium within the proventriculus (Goodwin and Hafner, 2003; Guy *et al.*, 2005). Glandular epithelium undergoes degeneration and necrosis, and this is accompanied by lymphocytic inflammation, ductal epithelial hyperplasia and replacement of glandular epithelium with ductal epithelium.

Based on damage observed in glandular epithelium, Goodwin *et al.* (1996) suggested that clinical effects observed in this disease were probably attributable to destruction of glandular epithelial cells that produce and secrete pepsinogen and hydrochloric acid. Conceivably, the loss of these digestive secretions would impair digestion leading to impaired feed digestion ('feed passage'), poor feed conversion and impaired growth. Increased friability of the organ and tendency to rupture at processing could be explained by a lymphocytic inflammatory response that replaces normal tissue and results in weakening of the proventricular wall.

ENTERITIS

Several different viruses have been definitively identified as causal agents of enteritis in chickens and/or turkeys. These include haemorrhagic enteritis virus (a group II AAV), astroviruses, rotaviruses, torovirus and turkey coronavirus. In addition, a number of other viruses have been associated with being causal agents of enteritis based on their detection in tissues and/or faeces/intestinal contents by virus isolation, immunohistochemistry or electron microscopy; these include group I AAVs, reoviruses and enterovirus-like viruses. The importance of this latter group of viruses as causal agents of enteric disease in chickens and turkeys awaits further investigation.

Viruses associated with enteritis

Adenoviruses

Haemorrhagic enteritis virus (HEV) is a well-established cause of enteric disease in turkeys (Pierson and Fitzgerald, 2003). HEV is present in most turkey-producing areas of the world including the USA, Canada, the UK, Germany, Australia, India, Israel and Japan. Serologic evidence indicates a high incidence of HEV infection in turkeys, but the incidence of clinical disease is low. This is believed to be due to the presence of avirulent or low-virulence HEV strains.

HEV replication occurs primarily in cells of the reticuloendothelial system (RES) – mainly in the nucleus of infected cells with the development of intranuclear inclusion bodies. HEV has not been successfully cultivated in conventional avian cell cultures or embryonated eggs; however, the virus has been propagated in turkey lymphoblastoid cells and turkey leukocyte cultures (Nazerian and Fadly, 1982). Virus infectivity is readily destroyed by treatment with a variety of disinfectants; however, the virus is resistant to inactivation by lipid solvents (chloroform and ether) and long-term storage (6 months at 4°C, 4 weeks at 37°C). HEV may survive in carcasses for several weeks at 37°C.

Gross lesions are observed primarily in spleen and intestines. Spleens are enlarged and mottled; intestines are distended, congested and filled with bloody exudate. Microscopic lesions that characterize the disease are present in intestines and cells of the RES. Microscopic lesions in intestines include congestion, degeneration of enterocytes at villous tips, sloughing of villous tips and haemorrhage into the intestinal lumen. Intranuclear inclusion bodies are sometimes observed in RES cells within the *lamina propria*. In the spleen, lesions include hyperplasia of white pulp, necrosis of lymphocytes and the presence of intranuclear inclusion bodies within RES cells.

Group 1 AAVs have been associated with being causal agents of enteric disease in chickens and turkeys but their role in this disease remains unproven. These viruses are infrequently identified in cases of enteritis in chickens and turkeys. They may be detected in infected birds by virus isolation, immunohistochemistry or histopathology. Microscopically, these viruses are recognized by the presence of characteristic, basophilic intranuclear inclusion bodies within differentiated villous enterocytes.

Rotaviruses

Rotaviruses are non-enveloped, spherical and have a diameter of approximately 70 nm (McNulty, 2003). Intact viruses consist of two icosahedral capsid shells (approximately 50 and 70 nm in diameter); they have a distinctive 'wheel-like' appearance under negative-stain electron microscopy owing to a smooth outer rim and capsomers of the inner capsid that radiate toward the rim. Replication and assembly occur in the cytoplasm with virus particles usually being found within vacuoles. Rotaviruses are environmentally stable; they are relatively heat-stable and infectivity is not affected by ether or a pH level of 3.

Classification of avian rotaviruses has been based primarily on cross-immunofluorescence studies and polyacrylamide gel electrophoresis (PAGE) analyses of ds RNA segments. Avian rotaviruses that cross-react by FA with antisera prepared against group A mammalian rotaviruses are classified as group A avian rotaviruses (McNulty, 2003). Rotaviruses which lack the group A antigen are referred to as 'atypical' rotaviruses. Three antigenically distinct, 'atypical' avian rotaviruses have been identified in avian species; one of these has been classified as group D, but the other two remain unclassified. Group D rotaviruses have been found only in avian species. PAGE analysis of ds RNA also is useful for classification of avian rotaviruses; electrophoretic migration of RNA segments is a useful indicator of serogroup classification and RNA profiles may be useful in epidemiological studies.

Rotavirus infections of avian species vary from subclinical to severe (McNulty, 2003). Diarrhoea is the principal manifestation of the disease in clinically affected birds; decreased weight gain, dehydration and increased mortality may also be observed. In general, experimental inoculation of chickens and turkeys with avian rotaviruses results in mild-to-inapparent infection. Variation in severity of rotavirus infections are probably due to differences in virulence of rotavirus strains or to the interaction of other infectious, environmental or management factors (McNulty, 2003).

Astroviruses

Astroviruses are small, roughly spherical viruses, 28–31 nm in diameter (Reynolds and Schultz-Cherry, 2003). They possess a characteristic morphological feature: a five- or six-pointed star that covers the surface of approximately 10% of virus particles.

Astrovirus infections have been identified primarily in young turkeys, 1-to-4 weeks of age (Reynolds and Schultz-Cherry, 2003). Clinical signs are variable but include diarrhoea, nervousness, litter eating and growth depression. Morbidity generally is high but mortality is low. Decreased weight gain and impaired absorption of D-xylose were observed in experimentally infected SPF turkey poults (Reynolds and Saif, 1986).

Turkey coronavirus

Turkey coronavirus (TCV) is a member of the Coronaviridae family. It is morphologically, antigenically and genetically similar to IBV, but biologically different (Guy, 2003). Unlike IBV, replication of TCV is restricted to intestines (differentiated villous enterocytes) and epithelium of the bursa of Fabricius (Guy, 2003). The virus has been identified in turkeys in the USA, Canada and the UK. Between 1951 and 1971, considerable economic losses in turkey flocks in the USA and Canada were ascribed to TCV infection; economic losses were attributed to decreased weight gain and increased mortality (Patel *et al.*, 1977).

Turkeys of all ages are susceptible to TCV. The infection is characterized by a short incubation period (1–3 d), depression, watery diarrhoea, weight loss and dehydration. In field cases, clinical signs occur suddenly, usually with high morbidity. Birds exhibit depression, anorexia, decreased water consumption, watery diarrhoea, dehydration, hypothermia and weight loss. Gross lesions consist of distended, watery intestines and villous atrophy.

TCV-infected flocks generally experience increased mortality, growth depression and poor feed conversion compared with uninfected flocks (Rives and Crumpler, 1998); however, turkeys experimentally infected using egg-adapted strains experience only mild disease, moderate growth depression and negligible mortality (Guy *et al.*, 2000).

Reoviruses

Reoviruses commonly are isolated from droppings/intestinal contents of enteritis-affected and clinically normal birds. Reovirus pathogenicity studies indicate that intestines are important sites of infection, regardless of the route of inoculation (Kibenge *et al.*, 1985; Jones *et al.*, 1989). Reovirus replication initially occurs in villous epithelium of the small intestines, with subsequent viraemic spread to other tissues, including lymphoid organs (Jones *et al.*, 1989).

While reoviruses are known to replicate in intestines of poultry and are frequently isolated from cases of enteritis, a causative role in these diseases remains unproven. Reoviruses have been shown to enhance the pathogenicity of a variety of other infectious agents including coccidia (Ruff and Rosenberger, 1985), *Cryptosporidium* spp. (Guy *et al.*, 1987), *Escherichia coli* (Rosenberger *et al.*, 1985) and chicken anaemia virus (Engstrom *et al.*, 1988). Enhanced pathogenicity of these and other infectious agents may occur as a result of virally induced immunosuppression (Rinehart and Rosenberger, 1983; Spackman *et al.*, 2005). It is interesting to speculate that this may be the mechanism by which reoviruses contribute to the pathogenesis of enteric diseases in chickens and turkeys.

Toroviruses

A turkey torovirus recently was identified as a potential aetiological agent of enteritis in turkeys (Ali and Reynolds, 2003). Toroviruses are classified in the genus *Torovirus* within the Coronaviridae family. Turkey toroviruses are RNA-containing, enveloped viruses with pleomorphic morphology, being 60–95 nm in diameter. The principal site of turkey torovirus replication is in villous epithelium along the mid-portion of the villus. Experimental, oral exposure of young turkeys produced diarrhoea, decreased weight gain, decreased intestinal disaccharidase activity and decreased D-xylose absorption.

Parvoviruses

Parvoviruses have not been conclusively identified in chickens and turkeys. Trampel *et al.* (1982) reported the finding of parvovirus-like viruses in enteritis-affected turkeys, but this finding has not been verified by other investigators. Similarly, Kisary *et al.* (1984) reported the identification of a parvovirus in chickens with malabsorption syndrome and experimental reproduction of disease with this virus (Kisary, 1985); however, these findings have not been substantiated by other investigators.

Pathogenesis of virally induced enteritis

Virally induced enteric disease in chickens and turkeys is produced most commonly by viruses that target differentiated enterocytes that cover intestinal villi. Viruses that cause disease as a result of replication in enterocytes comprising the crypts of Lieberkuhn (e.g. parvoviruses) have not been conclusively identified in chickens and turkeys. The avian adenovirus, haemorrhagic enteritis virus (HEV) produces enteritis by an alternative method, by damaging cells within the *lamina propria* of intestinal villi.

Viruses that replicate in differentiated villous enterocytes include rotaviruses, reovirus, turkey coronavirus, astrovirus, torovirus, group 1 AAVs and enterovirus-like viruses. Based on experimental studies these viruses, by themselves, typically produce only mild disease; however, naturally occurring enteric diseases are commonly complicated by other cofactors including other infectious agents, nutrition and management.

Experimental challenge studies conducted in laboratory settings often fail to provide a clear understanding of the role of these viruses in naturally occurring disease, as these studies do not replicate field conditions (i.e. environmental, management, nutrition and microbial flora). Thus, diseases such as malabsorption syndrome, runting/stunting syndrome, poult enteritis and poult enteritis-mortality syndrome, which are probably multifactorial in nature, generally cannot be reproduced solely by viruses and, thus, continue to be referred to as diseases of 'unknown' aetiology.

Virus replication in differentiated villous enterocytes potentially may produce enteric disease via several different pathways, including: (i) damage to villous enterocyte cells leading to impaired digestion and absorption; (ii) potentiation of pathogenicity of other infectious agents; or (iii) a combination of these effects. The interaction of virally induced damage with nutritional and management factors is poorly understood.

Differentiated villous enterocytes are non-proliferating cells that have both absorptive and digestive functions (Moon, 1978). In contrast, crypt enterocytes are undifferentiated cells that are the progenitors of villous enterocytes; these cells actively secrete fluid into the intestinal lumen. In health, the absorptive capacity of the villous enterocytes exceeds the secretory capacity of the crypt enterocytes, thus net absorption occurs along the villous surface.

Viral infection and consequent damage to villous enterocytes disrupt this balance by reducing the surface area of the gut mucosa (villous atrophy) and by altering the function of individual villous enterocytes. These effects lead to a generalized reduction in digestive and absorptive capacity, functional deficits that are referred to as malabsorption and maldigestion. These functional deficits are complicated by osmotic effects in which undigested and unabsorbed feed remains in the intestinal lumen and acts to hold water.

Fermentation in the large intestines results in production of additional, osmotically active, low-molecular weight molecules that further increase the retention of water in the lumen of the intestinal tract. Unabsorbed fluids are passed down the intestinal tract to the large intestine; diarrhoea results when the absorptive capacity of the large intestine is exceeded. These effects, by themselves, are probably common and economically important sequelae of intestinal infections caused by rotaviruses, turkey coronavirus and astroviruses, and possibly by others such as reovirus and group I avian adenoviruses.

Malabsorption and maldigestion that occur consequent to virally induced damage of villous enterocytes may result in nutritional deficiencies, especially those related to fat-soluble vitamins and minerals. These nutritional deficiencies may be observed clinically as abnormal feathering, rickets and osteoporosis. Additionally, these nutritional deficiencies may impair growth and development of lymphoid organs, leading to immunological deficiency and increased susceptibility to other infectious agents.

Virally induced damage to villous enterocytes may also result in enteritis by potentiating the pathogenesis of other infectious agents. While this is a probable pathogenic mechanism of virally induced gastrointestinal diseases of chickens and turkeys, it has received little attention. Virally induced damage of enterocytes may potentiate other infectious agents by impairing normal defence mechanisms (i.e. glycocalyx, mucin, microvilli), thereby promoting intestinal attachment and/or entry of infectious agents.

Virally induced enterocyte damage may provide a portal of entry for other enteric pathogens such as *E. coli* and *Salmonella* spp. Alternatively, virally induced enterocyte damage may potentiate the proliferation of other infectious agents by altering the intestinal luminal environment, a consequence of inflammatory products secreted into the lumen and presence of undigested feedstuffs.

Previous investigations have demonstrated synergistic interactions between rotavirus and enterotoxigenic *E. coli* in virally induced enteric diseases of mammalian species (Marshall, 2002); however, little work has been done to evaluate interactions of viruses and other infectious agents in the pathogenesis of enteritis in poultry. Recently, turkey coronavirus was shown to enhance intestinal colonization of enteropathogenic *E. coli* (EPEC) in young turkeys, and this combination of pathogens has been proposed as an aetiological explanation for PEMS (Guy *et al.*, 2000).

TCV and EPEC were consistently identified in PEMS-affected turkey flocks; however, experimental challenge studies failed to establish an aetiological role for either agent in this disease. No clinical signs, no weight gain disturbance and only mild intestinal lesions were observed in turkeys infected with only EPEC; only mild weight gain depression and villous atrophy were observed in turkeys infected with only TCV. However, experimental co-infection of young turkeys with both TCV and EPEC produced severe intestinal disease clinically indistinguishable from PEMS. High mortality, marked weight gain depression and extensive intestinal lesions were observed in turkeys coinfected with EPEC + TCV.

Additional experimental studies indicated that TCV infection promoted EPEC colonization of intestinal epithelium. Enhanced colonization of turkey intestines by EPEC in TCV + EPEC-infected turkeys was evident by increased frequency and distribution of AE lesions and by increased EPEC shedding in intestinal contents; however, the mechanism by which TCV promotes EPEC pathogenicity remains undetermined. TCV infection of intestinal enterocytes possibly resulted in enterocyte damage that favoured EPEC adherence and colonization. Alternatively, TCV infection results in changes in the intestinal luminal environment that potentiated EPEC proliferation, or TCV infection altered immune responses to EPEC.

Additional studies aimed at determining the mechanism by which TCV infection potentiates EPEC infection are needed as these studies may provide insights into the pathogenesis of other enteric diseases of poultry. It is interesting to speculate that other enteric diseases of poultry of unknown or poorly understood aetiology (e.g. poult enteritis, runting/stunting syndrome of chickens) may ultimately be explained by the interaction of viruses and other infectious or by non-infectious factors.

The role of nutrition in the pathogenesis of virally induced enteric diseases of poultry has received little attention. However, based on human and mammalian animal studies, it is likely that poor nutrition and poor feed quality contribute to the severity of virally induced enteric diseases. Studies of rotavirus infection in malnourished humans and laboratory animal models have demonstrated that malnutrition exacerbates the severity and prolongs the duration of clinical disease by impairing rotavirus immune responses (Offor *et al.*, 1985; Zijlstra *et al.*, 1999). The influence of particular components of poultry diets, such as fat content and the presence of noxious compounds (e.g. rancid fat, mycotoxins) on viral enteric infections in poultry has not been examined.

The mechanism by which HEV causes intestinal disease and haemorrhage has not been conclusively determined. Unlike the enteric viruses discussed above, HEV does not replicate in villous enterocytes. HEV replication primarily occurs outside the gastrointestinal tract in lymphoid tissues and macrophages; virus replication is detectable, but only to a limited extent, within cells of the *lamina propria*. It has been suggested that HEV replication occurs in intestinal villous endothelial cells and that this may result in vascular damage and ischaemic necrosis of intestinal villi.

However, more recently, it has been suggested that release of large quantities of pro-inflammatory cytokines during HEV infection, most importantly tumour necrosis factor, initiates systemic shock and leads to development of vascular lesions in the intestines, the proposed shock organ of turkeys (Rautenschlein and Sharma, 2000). Such cytokine-induced vascular damage in the *lamina propria* leads to necrosis of villous tips and intestinal haemorrhage, the characteristic intestinal lesion seen in this disease. This theory is supported by experimental studies in which treatment of HEV-infected turkeys with thalidomide, a potent TNF down-regulatory drug, prevented development of intestinal lesions.

CONTROL OF VIRALLY INDUCED GASTROINTESTINAL DISEASES

Control of viruses associated as causes of gastrointestinal diseases of chickens and turkeys is best accomplished by maintaining premises free of these agents, when possible, by implementing biosecurity measures that prevent entrance and interrupt transmission. Transmission of these viruses generally occurs by the faecal–oral route; flock-to-flock spread most commonly occurs mechanically via movement of people and equipment. Contaminated litter is the most likely source of infection for susceptible flocks, as well as for maintaining infection on contaminated premises.

At present, no control methods are available for preventing or ameliorating TVP in commercial broiler chickens. Future control procedures will be dependent upon additional studies aimed at evaluating the epidemiology and immunology of AdLV (R11/3), the presumptive aetiology of this disease. The presumptive identification of this virus as an adenovirus bodes well for potential vaccine development. Most adenovirus infections are preventable by immunization, requiring only development of humoral (IgG) immunity, not mucosal (IgA) immunity.

HEV is controlled in endemic areas by immunization. Vaccination via drinking water may be achieved using naturally occurring, attenuated strains of HEV, or a closely related virus called marble spleen disease virus. These virus vaccines are prepared by propagation in turkeys, with infected spleens being used as the source of vaccinal virus (Thorsen *et al.*, 1982). HEV has also been propagated in lymphoblastoid cell culture and utilized successfully as a live, drinking water vaccine (Fadly *et al.*, 1986).

Coronaviruses and toroviruses are readily inactivated by most common disinfectants. Successful eradication of these viruses requires specific diagnostic procedures to identify infected flocks. Elimination from contaminated premises may then be accomplished via depopulation followed by thorough cleaning and disinfection (Guy, 2003). Following cleaning and disinfection procedures, it is best to allow premises to remain free of birds for a minimum of 3–4 weeks.

Astroviruses, reoviruses and rotaviruses are relatively resistant to inactivation and are excreted in faeces in large numbers. They may survive in litter and on contaminated equipment for prolonged periods of time and this may be the source of infection for subsequent poultry flocks on all-in/all-out sites. Alternatively, these viruses may be transmitted to subsequent flocks as a result of either: (i) egg transmission, a result of virus contamination of the outer surface of the egg; or (ii) transmission after hatching. Specific control procedures for these viruses have not been developed.

The ubiquity of astroviruses, reoviruses and rotaviruses and their resistance to inactivation probably preclude the rearing of commercial poultry flocks free of these viruses. Thus, control is aimed at ensuring thorough cleaning and disinfection of facilities between flocks in order to reduce environmental contamination and degree of exposure of young poultry.

A greater understanding of the viruses that cause gastrointestinal tract disease in poultry, including immunology, pathogenesis and epidemiology of these viruses, will be necessary for development of improved control procedures. Further progress in understanding these viruses and their role in gastrointestinal diseases of poultry undoubtedly will be aided by future improvements in *in vitro* cell culture techniques and by increased application of new technologies in diagnosis such as virus-specific monoclonal antibodies and polymerase chain reaction procedures.

Additionally, new technologies will be applied to understanding the mechanisms of immunological protection and to devising methods of effectively inducing local immune responses to these viruses. Such studies may, in the future, lead to the development of vaccines for control of those viruses that are refractory to eradication efforts, such as rotaviruses, astroviruses and reoviruses.

REFERENCES

Ali, A. and Reynolds, D.L. (2003) Turkey torovirus infection. In: Barnes, H.J., Glisson, J.R., Fadly, A.M., McDougald, L.R., Saif, Y.M. and Swayne, D.E. (eds) *Diseases of Poultry* (11th edn). Iowa State University Press, Ames, Iowa, pp. 332–337.

Barnes, H.J. and Guy, J.S. (2003) Poult enteritis-mortality syndrome. In: Barnes, H.J., Glisson, J.R., Fadly, A.M., McDougald, L.R., Saif, Y.M. and Swayne, D.E. (eds) *Diseases of Poultry* (11th edn). Iowa State University Press, Ames, Iowa, pp. 1171–1180.

Brugh, M. and Wilson, R.L. (1986) Effect of dietary histamine on broiler chickens infected with avian reovirus S1133. *Avian Diseases* 30, 199–203.

Engstrom, B.E., Fossum, O. and Luthman, M. (1988) Blue wing disease: experimental infection with a Swedish isolate of chicken anemia agent and an avian reovirus. *Avian Pathology* 17, 33–50.

Fadly, A.M., Nazerian, K., Nagaraja, K. and Below, G. (1986) Field vaccination against haemorrhagic enteritis of turkeys by a cell-culture live-virus vaccine. *Avian Diseases* 29, 768–777.

Goodwin, M.A. and Hafner, S. (2003) Viral proventriculitis. In: Barnes, H.J., Glisson, J.R., Fadly, A.M., McDougald, L.R., Saif, Y.M. and Swayne, D.E. (eds) *Diseases of Poultry* (11th edn). Iowa State University Press, Ames, Iowa, pp. 383–388.

Goodwin, M.A., Hafner, S., Bounous, D.I., Latimer, K.S., Player, E.C., Niagro, F.D., Campagnoli, R.P. and Brown, J. (1996) Viral proventriculitis in chickens. *Avian Pathology* 25, 369–379.

Guy, J.S. (2003) Turkey coronavirus enteritis. In: Barnes, H.J., Glisson, J.R., Fadly, A.M., McDougald, L.R., Saif, Y.M. and Swayne, D.E. (eds) *Diseases of Poultry* (11th edn). Iowa State University Press, Ames, Iowa, USA, pp. 300–307.

Guy, J.S., Levy, M.G., Ley, D.H., Barnes, H.J. and Gerig, T.M. (1987) Experimental reproduction of enteritis in bobwhite quail (*Colinus virginianus*) with *Cryptosporidium* and reovirus. *Avian Diseases* 31, 713–722.

Guy, J.S., Smith, L.G., Breslin, J.J., Vaillancourt, J.P. and Barnes, H.J. (2000) High mortality and growth depression experimentally produced by dual infection with enteropathogenic *Escherichia coli* and turkey coronavirus. *Avian Diseases* 44, 105–113.

Guy, J.S., Barnes, H.J., Smith, L.G., Fuller, F.J. and Owen, R. (2005) Partial characterization of an adenovirus-like virus isolated from broiler chickens with transmissible viral proventriculitis. *Avian Diseases* 49, 344–351.

Huff, G.R., Zheng, Q., Newberry, L.A., Huff, W.E., Balog, J.M., Rath, N.C., Kim, K.S., Martin, E.M., Goeke, S.C. and Skeeles, J.K. (2001) Viral and bacterial agents associated with experimental transmission of infectious proventriculitis of broiler chickens. *Avian Diseases* 45, 828–843.

Jones, R.C., Islam, M.R. and Kelly, D.F. (1989) Early pathogenesis of experimental reovirus infection in chickens. *Avian Pathology* 18, 239–253.

Kibenge, F.S.B., Gwaze, G.E., Jones, R.C., Chapman, A.F. and Savage, C.E. (1985) Experimental reovirus infection in chickens: observations on early viremia and virus distribution in bone marrow, liver, and enteric tissues. *Avian Pathology* 14, 87–98.

Kisary, J. (1985) Experimental infection of chicken embryos and day-old chickens with parvovirus of chicken origin. *Avian Pathology* 14, 1–7.

Kisary, J., Nagy, B. and Bitay, Z. (1984) Presence of parvoviruses in the intestine of chickens showing stunting syndrome. *Avian Pathology* 13, 339–343.

Kouwenhoven, B., Davelaar, F.G. and Van Walsum, J. (1978) Infectious proventriculitis causing runting in broilers. *Avian Pathology* 7, 183–187.

Leonard, J. and Schmittle, S. (1995) Proventriculitis, proventriculosis distinctions are important. *Feedstuffs*, 2 January 1995.

Lukert, P.D. and Saif, Y.M. (2003) Infectious bursal disease. In: Barnes, H.J., Glisson, J.R., Fadly, A.M., McDougald, L.R., Saif, Y.M. and Swayne, D.E. (eds) *Diseases of Poultry* (11th edn). Iowa State University Press, Ames, Iowa, pp. 161–179.

Marshall, J.A. (2002) Mixed infections of intestinal viruses and bacteria in humans. In: Brogden, K. and Guthmiller, J. (eds) *Polymicrobial Diseases*. ASM Press, Washington DC, pp. 299–316.

McFerran, J.B. (2003) Avian adenovirus infections. In: Barnes, H.J., Glisson, J.R., Fadly, A.M., McDougald, L.R., Saif, Y.M. and Swayne, D.E. (eds) *Diseases of Poultry* (11th edn). Iowa State University Press, Ames, Iowa, pp. 213–214.

McNulty, M.S. (1993) Reovirus. In: *Virus Infections of Vertebrates*. Volume 4, Elsevier Science Publishers, New York, pp. 181–193.

McNulty, M.S. (2003) Rotaviruses. In: Barnes, H.J., Glisson, J.R., Fadly, A.M., McDougald, L.R., Saif, Y.M. and Swayne, D.E. (eds) *Diseases of Poultry* (11th edn). Iowa State University Press, Ames, Iowa, pp. 308–320.

McNulty, M.S. and McFerran, J.B. (1993) The runting stunting syndrome – general assessment. In: *Virus Infections of Vertebrates*. Volume 4. Elsevier Science Publishers, New York, pp. 519–529.

Moon, H.W. (1978) Mechanisms in the pathogenesis of diarrhoea: a review. *Journal of the American Veterinary Medical Association* 172, 443–448.

Nazerian, K. and Fadly, A.M. (1982) Propagation of virulent and avirulent turkey haemorrhagic enteritis virus in cell culture. *Avian Diseases* 26, 816–827.

Offor, E., Riepenhoff-Talty, M. and Ogra, P.L. (1985) Effect of malnutrition on rotavirus infection in suckling mice: kinetics of early infection. *Proceedings of the Society of Experimental Biology and Medicine* 178, 85–90.

Pantin-Jackwood, M.J. and Brown, T.P. (2003) Infectious bursal disease virus and proventriculitis in broiler chickens. *Avian Diseases* 47, 681–690.

Pantin-Jackwood, M.J., Brown, T.P and Huff, G.R. (2004a) Proventriculitis in broiler chickens: immunohistochemical characterization of the lymphocytes infiltrating the proventricular glands. *Veterinary Pathology* 41, 641–648.

Pantin-Jackwood, M.J., Brown, T.P., Kim, Y. and Huff G.R. (2004b) Proventriculitis in broiler chickens: effect of immunosuppression. *Avian Diseases* 48, 300–316.

Patel, B., Gonder, E. and Pomeroy, B.S. (1977) Detection of turkey coronavirus enteritis (bluecomb) in field epornitics using direct and indirect fluorescent antibody tests. American *Journal of Veterinary Research* 38, 1407–1411.

Pierson, F.W. and Fitzgerald, S.D. (2003) Haemorrhagic enteritis and related infections. In: Barnes, H.J., Glisson, J.R., Fadly, A.M., McDougald, L.R., Saif, Y.M. and Swayne, D.E. (eds) *Diseases of Poultry* (11th edn). Iowa State University Press, Ames, Iowa, pp. 237–247.

Rautenschlein, S. and Sharma, J.M. (2000) Immunopathogenesis of haemorrhagic enteritis virus (HEV) in turkeys. *Developmental and Comparative Immunology* 24, 237–246.

Reynolds, D.L. and Saif, Y.M. (1986) Astrovirus: a cause of an enteric disease in turkey poults. *Avian Diseases* 30, 89–98.

Reynolds, D.L. and Schultz-Cherry, S.L. (2003) Astrovirus infections. In: Barnes, H.J., Glisson, J.R., Fadly, A.M., McDougald, L.R., Saif, Y.M. and Swayne, D.E. (eds) *Diseases of Poultry* (11th edn). Iowa State University Press, Ames, Iowa, pp. 237–247.

Rinehart, C.L. and Rosenberger, J.K. (1983) Effects of avian reoviruses on the immune responses of chickens. *Poultry Science* 62, 1488–1489.

Rives, D.V. and Crumpler, D.B. (1998) Effect of turkey coronavirus infection on commercial turkey flock performance. In: *Proceedings of the American Veterinary Medical Association*, Baltimore, Maryland, p. 189.

Rosenberger, J.K., Fries, P.A., Cloud, S.S. and Wilson, R.A. (1985) *In vitro* and *in vivo* characterization of avian *Escherichia coli*. II. Factors associated with pathogenicity. *Avian Diseases* 29, 1094–1107.

Ruff, M.D. and Rosenberger, J.K. (1985) Concurrent infections with reoviruses and coccidia in broilers. *Avian Diseases* 33, 535–544.

Skeeles, J.K., Newberry, L.A., Beasley, J.N. and Hopkins, B.A. (1998) Histologic comparison of lesions induced in the proventriculus and other areas of the intestinal tract of chickens experimentally infected with both classic and variant strains of infectious bursal disease virus. *Poultry Science* 77 (1), 133.

Spackman, E., Pantin-Jackwood, M., Day, J.M. and Sellers, H. (2005) The pathogenesis of turkey origin reoviruses in turkeys and chickens. *Avian Pathology* 34, 291–296.

Thayer, S.G. and Walsh, J.L. (1993) Evaluation of cross-contamination on automatic viscera removal equipment. *Poultry Science* 72, 741–746.

Thorsen, J., Weninger, N., Weber, L. and Van Dijk, C. (1982) Field trials of an immunization procedure against haemorrhagic enteritis of turkeys. *Avian Diseases* 26, 473–477.

Trampel, D.W., Kinden, D.A., Solorzano, R.F. and Stogsdill, P.L. (1982) Parvovirus-like enteropathy in Missouri turkeys. *Avian Diseases* 27, 49–54.

Wege, H., Siddel, S. and ter Meulen, V. (1982) The biology and pathogenesis of coronaviruses. *Current Topics in Microbiology and Immunology* 99, 165–200.

Yu, L., Low, S., Wang, Z., Nam, S.J., Liu, W. and Kwang, J. (2001) Characterization of three infectious bronchitis virus isolates from China associated with proventriculitis in vaccinated chickens. *Avian Diseases* 45, 416–424.

Zijlstra, R.T., McCracken, B.A., Odle, J., Donovan, S.M., Gelberg, H.B., Petschow, B.W., Zuckerman, F.A. and Gaskins, H.R. (1999) Malnutrition modifies pig small intestinal inflammatory responses. *American Society of Nutritional Sciences* 129, 838–843.

Chapter 16
The gastrointestinal tract as a port of entry for bacterial infections in poultry

J.P. Christensen,* M.S. Chadfield, J.E. Olsen and M. Bisgaard

*Department of Veterinary Pathobiology, The Royal Veterinary and Agricultural University, Frederiksberg, Denmark; *e-mail: jpc@kvl.dk*

ABSTRACT

A substantial amount of information is available concerning gastrointestinal colonization and invasion of the chicken gut by Gram-negative microorganisms such as *Salmonella enterica* and *Campylobacter jejuni*. Both bacterial and host factors of importance have been documented in relation to invasion of the chicken gastrointestinal tract. Both groups of microorganisms are associated with the gastrointestinal tract and have received major attention due to the zoonotic aspect of these infections.

In contrast, other important poultry pathogens without this potential have been poorly characterized concerning sites of infection and pathogenesis. After the ban on antibiotic growth promoters an increasing number of infections due to *Enterococcus* and *Streptococcus* spp. have been observed in poultry in Denmark.

This chapter describes the invasive potential of these organisms in the chicken gut compared to the same ability among different serotypes of *Salmonella enterica*. In addition, the invasive potential of *Pasteurella multocida* is described. The methods used included an *in vivo* intestinal loop model in chickens and several cell culture lines. Surprisingly, some of our most recent investigations have demonstrated significant differences in the invasive properties in the gut of *Enterococcus hirae* and different *Streptococcus* species using both the loop model and avian macrophage and intestinal epithelial cell cultures.

The Gram-positive cocci have mainly been associated with septicaemic conditions in young chickens, endocarditis and septicaemia in adult birds, and also with amyloid arthropatia during rearing. Similar disease conditions in humans – in particular, endocarditis – may be observed. The significance of these findings, including comparative aspects, will be discussed.

© CAB International 2006. *Avian Gut Function in Health and Disease* (ed. G.C. Perry)

INTRODUCTION

Bacterial invasion, defined as pathogen-induced entry into eukaryotic cells, is a widespread trait associated with disease-causing microbes. A number of bacterial pathogens, including *Salmonella* and *Escherichia coli*, can utilize the intestine as a portal for entry during infection (Barnes *et al.*, 2003; Amy *et al.*, 2004). The molecular mechanisms used by these bacterial pathogens to invade and circumvent the natural defences of the host are complex, involving a large number of bacterial genes. In general, most investigations elucidating the invasion processes have been performed with mammalian hosts *in vivo* and corresponding cell-line culture assays *in vitro*.

Scant information is available concerning bacterial intestinal invasion processes in the chicken and what little there is focuses mainly upon aspects of colonization and invasion by salmonella. This has been primarily due to the lack of an appropriate avian model for invasion and avian epithelial cell lines for *in vitro* studies.

The avian intestinal lymphoid tissue consists of the bursa of Fabricius, diffuse mucosal lymphoid infiltration, caecal tonsils, Meckel's diverticulum and the Peyer's patches (Lillehoj and Trout, 1996). Apart from descriptions of the intestinal lymphoid tissue as being a target for invasion, little information is available concerning possible specific anatomical sites of the avian gut involved in the invasive process.

The caecal tonsils and the Peyer's patches have consistently been reported as being the major sites of invasion for at least *Salmonella* (Barrow *et al.*, 1994). The caecal tonsils are the most concentrated lymphoid tissue in the intestine: the lymphoid cells of these organs are observed in the epithelium, in the subepithelial zone and in the *lamina propria*. The epithelial cells associated with the tonsils are capable of processing antigens.

The Peyer's patch structures present in chickens undergo development from two Peyer's patch structures in day-old chicks to up to five structures diffusely scattered in the intestine of 12-week-old chickens, with a subsequent decline to a single Peyer's patch structure in the ileocaecal junction at the age of 56 weeks. The lymphoid tissue in the patches consists of germinal centres and diffuse lymphoid tissue. Intestinal antigens are absorbed and processed by germinal centre macrophages, macrophages in the diffuse lymphoid tissue and by the epithelial cells (Lillehoj and Trout, 1996).

Specialized M cells covering the lymphoid structures mentioned above have been reported as one of the main cell types involved in invasion; however, accumulating evidence demonstrates that invasion also takes place through the normal enterocytes of the chicken (Aabo *et al.*, 2000, 2002). This also seems to be the case in other species, e.g. pigs and cattle (Bolton *et al.*, 1999a).

An attempt is made in this chapter to provide an overview of different aspects of intestinal invasion by selected important bacterial pathogens related to the chicken, in addition to providing information concerning some zoonotic infections. Emphasis will be on current knowledge of host–pathogen interactions with focus primarily on the host and techniques used to investigate the invasion process, and less on bacterial virulence factors.

SALMONELLA

Salmonella is generally recognized as infecting animals, including chickens, by the faecal–oral route (Barrow *et al.*, 1994). Other routes such as the airway and skin infections have been reported (Cox *et al.*, 1996). The *Salmonella* genus, on the basis of pathogenesis, can be divided into two major groups: (i) host-adapted serovars; and (ii) serovars with a broad host range.

The former group typically produce systemic disease in their respective hosts via invasion of the intestine, but are rarely involved in food poisoning (Barrow, 2005). The host-adapted serovar Gallinarum, with the biovars gallinarum and pullorum, are typical of this group and are strictly associated with avian species in which they produce systemic disease, often with high mortality.

The latter group comprises the remaining serovars (more than 2400), which have a broad host range and are often responsible for cases of human food poisoning, but only produce systemic disease in chickens under certain circumstances, including very young or old animals, stressful conditions or in connection with concurrent infections (Barrow, 2005; Beal *et al.*, 2005). *Salmonella* Typhimurium and *S.* Enteritidis are the most prominent members of the latter group and among the most virulent serovars of this broad, host-range group.

Distinct differences between the two groups of *Salmonella* concerning the pathogenesis of infections in chickens have been demonstrated. However, these differences are not linked to different invasive properties between members of the two groups. Both groups invade readily at all sites in the small intestine but apparently more so in the proximal than the distal part of the small intestine (Aabo *et al.*, 2002; Chadfield *et al.*, 2003). The prime site for invasion, however, seems to be the caecal tonsils, as significantly higher counts have been observed here following the inoculation of bacteria into caecal loops (Chadfield *et. al.*, 2003) (see Figs 16.1 and 16.2).

However, there has been no indication that the host-adapted serovar *S.* Gallinarum has a predilection for this route. Similar lack of correlation between intestinal invasion and host specificity in *Salmonella* has also been demonstrated in the bovine, porcine (Bolton *et al.*, 1999b) and ovine host (Uzzau *et al.*, 2001).

Recently, it has been demonstrated *in vivo* that the broad host range serovars such as *S.* Enteritidis and *S.* Typhimurium cause rapid inflammation of the intestinal mucosa during or following the invasive process, with infiltration of large numbers of heterophilic granulocytes limiting the infection to the gut, whereas the host-adapted serovar *S.* Pullorum does not produce inflammation to the same extent and this occurs, apparently, later in the infectious process (Henderson *et al.*, 1999). In this way the host-adapted serovars may evade the local immune response and subsequently go on to cause severe systemic disease.

Results indicate that the underlying mechanism is related to differences in cytokine responses during invasion (Kaiser *et al.*, 2000). Invasion of primary chick kidney cells (CKC) by *S.* Enteritidis and *S.* Typhimurium cause an eight-

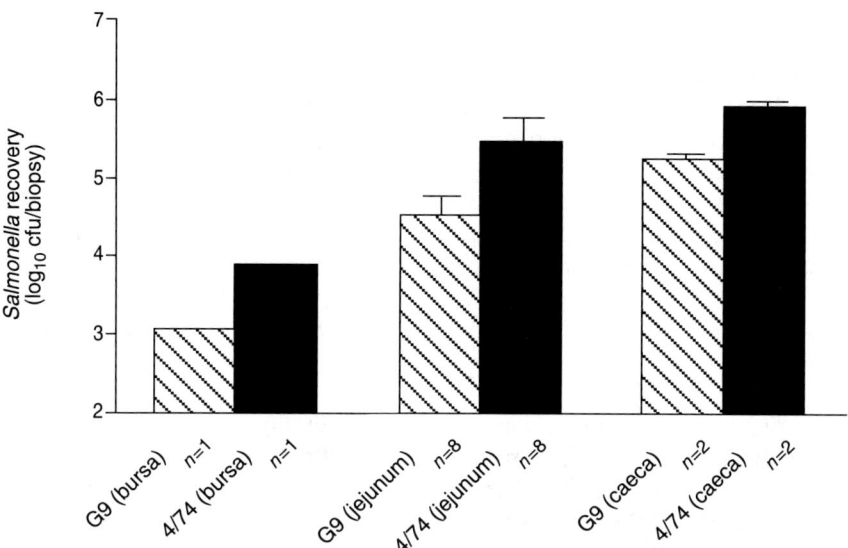

Fig. 16.1. Different levels of invasion of *Salmonella* Gallinarium (G9) and *S.* Typhimurium (4/74) at different sites of the intestine and at the bursa of Fabricius (from Chadfield *et al.*, 2003). Although the highest level of invasion is observed in the caecal tonsils, the interstrain variability is the same at all sites of invasion.

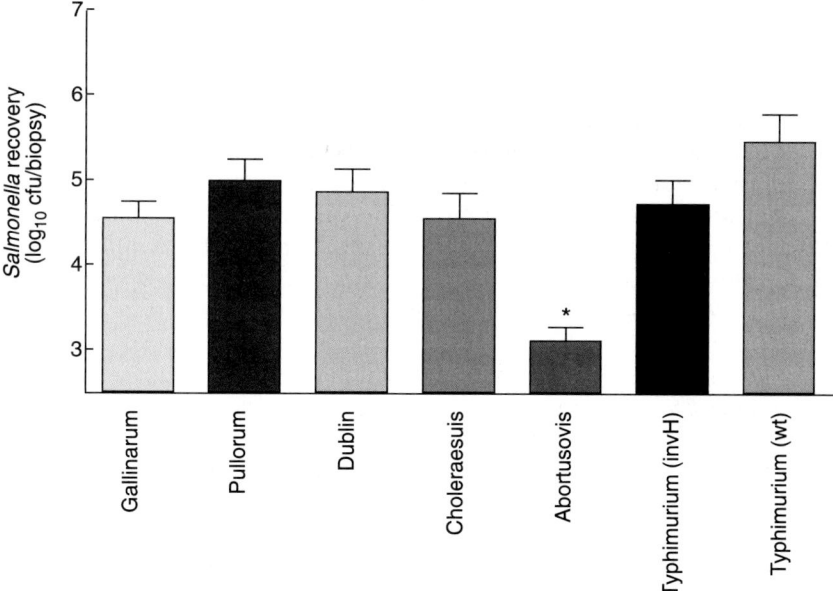

Fig. 16.2. Level of invasion of the jejunum of the chicken by different host-adapted *Salmonella* serovars (from Chadfield, *et al.*, 2003). The chicken-adapted *S.* Gallinarum and *S.* Pullorum do not invade the intestine at a higher level than broad host-range serovars or serovars associated with bovine and porcine hosts.

to ten-fold increase in production of the pro-inflammatory cytokine IL-6 which triggers the influx of heterophilic granulocytes to the *lamina propria* of the gut. However, invasion by S. Gallinarum of CKC causes no increase in IL-6 production. Bacterial factors suggested as being responsible for these recognition differences include flagella proteins (Iqbal *et al.*, 2005).

Following the initial invasion events in the pathogenetic process, macrophages have been considered to play a key role in the subsequent phase of the infection, but *in vitro* studies with chicken macrophages infected with S. Gallinarum have not demonstrated increased multiplication or survival intracellularly when compared with other *Salmonella* serovars (Chadfield *et al.*, 2003). However, earlier reports have shown that host adaptation is expressed primarily in different abilities to multiply within the tissues, particularly those of the reticuloendothelial system (RES). The exact site of multiplication is not known, but macrophages or heterophils of the liver and spleen have been suggested as a possible site (Barrow *et al.*, 1994; Wigley *et al.*, 2001).

Experimental depletion of heterophils has been convincingly shown to result in systemic septicaemia following S. Enteritidis infection, which is very similar to S. Gallinarum typhoid-like disease (Kotgut *et al.*, 1993). Thus, the data strongly suggest a role for heterophilic granulocytes in the gut in the protection against internal organ colonization and subsequent septicaemia of the broad host-range serovars, and perhaps as a site for multiplication in the organs of the host-adapted serovars.

Among the many broad host-range serovars only a few of the zoonotic serovars, notably S. Typhimurium and S. Enteritidis, are known to cause disease in chickens, where they are reported to invade and localize in organs (Gast, 2003). Why these serovars are more virulent than others has not been fully clarified. However, it has been demonstrated by the use of an *in vivo* chicken intestinal loop model that they are significantly better invaders of the intestine than are other, more exotic, serovars associated with, e.g. feedstuff (Aabo *et al.*, 2002) (see Fig. 16.3).

These results were supported by the use of Madin Derby Canine Kidney (MDCK) epithelial cell cultures. The molecular basis for this difference remains to be determined. The results indicated that the invasive properties of the non-host-adapted serovars may be correlated with general virulence.

ESCHERICHIA COLI

Escherichia coli infection in poultry (often refered to as colibacillosis) includes a variety of localized or systemic infections, where *E. coli* may be involved as the primary or secondary infectious agent (Barnes *et al.*, 2003). *E. coli* infections in mammals and humans often affect the intestine. From the abundant literature available on this topic, it appears as if this is either uncommon or unrecognized in avian infections, where the most notable infections are extra-intestinal and include omphalitis/yolksac infection, respiratory tract infections, colisepticaemia, salpingitis/peritonitis, swollen head syndrome, avian cellulitis, coligranuloma (Hjarre's disease), osteomyelitis/synovitis and panophthalmitis (Gross, 1994).

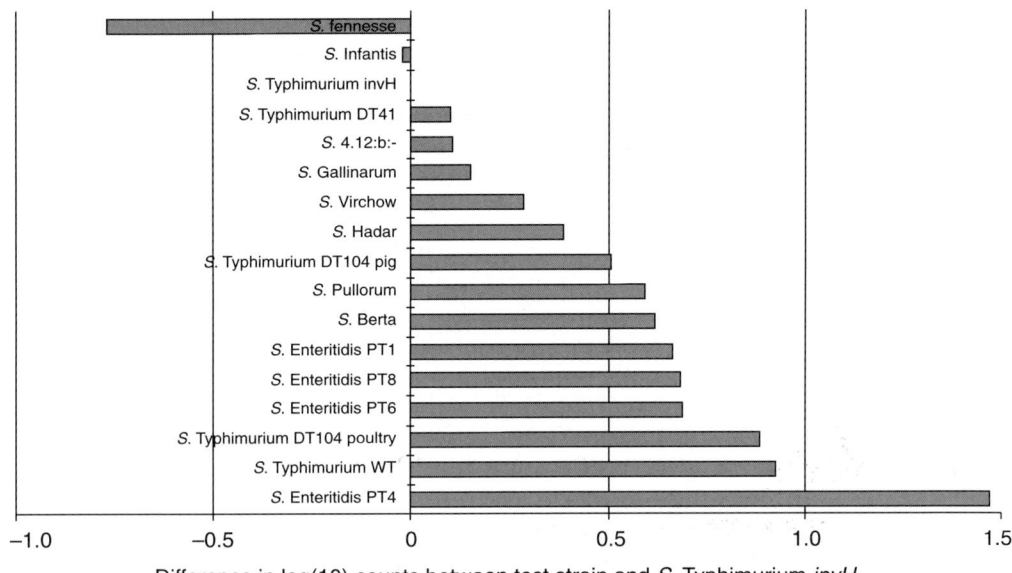

Fig. 16.3. Invasion of jejunum by different poultry-related *Salmonella* serovars using an intestinal loop model (from Aabo *et al.*, 2002). *Salmonella* Enteritidis and *S.* Typhimurium are significantly better invaders than are many other poultry-associated *Salmonella* serovars.

However, it has been indicated that *E. coli* may be implicated in the development of histological lesions of the intestine in pigeons and chickens (Wada *et al.*, 1995; Sueyoshi *et al.*, 1996, 1997). In the following, a more detailed description of some of the manifestations mentioned above will be given in relation to poultry and, in particular, to the intestine.

Intestinal disease

Escherichia coli infection in many mammals, including humans, manifests itself primarily as enteritis (Gross, 1994). As mentioned above, this is uncommon in poultry, where the presence of this bacterium in the intestines very seldom gives rise to clinical signs or gross lesions. However, diarrhoeal disease where *E. coli* was believed to be involved as the primary aetiological agent has been reported (Joya *et al.*, 1990). It has also been shown that *E. coli* may be involved in poult enteritis and mortality syndrome. This disease syndrome is characterized by enteritis with diarrhoea and high mortality in poults between 7 and 28 days of age (Guy *et al.*, 2000). There is evidence that the role of *E. coli* in this syndrome may be secondary to a coronavirus infection (Edens *et al.*, 1997; Guy *et al.*, 2000).

An interesting aspect of gut colonization of poultry by *E. coli* is the question of whether the intestine may serve as a site of entry into the bloodstream for certain clones. It has been shown experimentally that stress factors – such as food and water deprivation or exposure to heat – may lead to spread of pathogenic clones of *E. coli* from the gut of chickens and turkeys into the circulation, resulting in substantial mortality compared with that in non-stressed controls (Leitner and Heller, 1992). The extent to which this occurs under natural conditions, and the possibility that certain clones of *E. coli* require only minor reduction in immune response before they are capable of causing bacteraemia, are relevant topics for future investigations.

Recently, it has been demonstrated that some of the *E. coli* subtypes normally associated with mammals, e.g. enterotoxigenic (ETEC), verotoxigenic (VTEC) and attaching and effacing (AEEC) *E. coli* may be able at least to colonize the intestinal tract of birds and, in some cases, also to generate histological lesions (Stavric *et al.*, 1993; Schoeni and Doyle, 1994; Sueyoshi *et al.*, 1996, 1997; Zhao *et al.*, 1996; Best *et al.*, 2003). The most important aspect of these coliform infections, however, appears to be the ability of poultry to serve as a possibly reservoir for these human pathogens.

Acute septicaemia of chickens and turkeys

Only a few early reports describe an acute septicaemia in mature chickens and turkeys caused by *E. coli*, but such a disease syndrome is well recognized among those involved in poultry/turkey production (Sponenberg *et al.*, 1985; Chadfield *et al.*, 2006). Often, the birds originate from flocks with no previously recognized disease problems and are found dead, but in good bodily condition. Post-mortem findings include full crop and marked congestion of the carcass, accompanied by serohaemorrhagic exudations and swollen liver and spleen.

The liver may have a parboiled colour, while the spleen is often dark red; multiple petechiation throughout the viscera has also been observed. In more protracted cases, necrotic foci in the liver may also be seen. It should be noted that, in typical cases, no lesions are observed in the respiratory tract. Thus, the pathogenesis of this manifestation is unclear (Chadfield *et al.*, 2006).

However, many speculate that the respiratory tract is also an important port of entry for such septicaemic conditions (Edelman *et al.*, 2003). One major obstacle in the study of the septicaemic form of *E. coli* infection is the difficulty in reproducing the condition experimentally. Larsen *et al.* (1985) produced an acute fulminating septicaemia in 8-week-old poults by inoculation of *E. coli* (O1:K1) into the air sacs and by providing *E. coli* (O1:K1)-inoculated drinking water *ad libitum*. The mortality rates were 50 and 2.5%, respectively. However, the exact nature of the gross lesions of these birds was not clear, but the birds died within 1–2 days, underlining the acute nature of the disease.

This study also included birds inoculated with haemorrhagic enteritis virus (HEV), followed by challenge with *E. coli*. The overall conclusion of the study was that mortality resulting from experimental *E. coli* infection was significantly increased when *E. coli* (O1:K1) was presented to poults that had been orally infected with HEV 1 week earlier.

Field outbreaks of acute *E. coli* infection in 6–12-week-old poults have also been associated with HEV (Sponenberg, *et al.*, 1985), and current opinion appears to be that outbreaks of acute colibacillosis in birds of this age group are likely to be associated with previous infection by HEV or other predisposing agents (van den Hurk, 1994). However, in previously described outbreaks of colibacillosis in the form of septicaemia in 8–16-week-old turkeys there were no signs that HEV or other infections had previously occurred in the flocks (Chadfield *et al.*, 2006). The clinical course of the outbreaks was peracute/acute. The fact that most of these outbreaks have been demonstrated to be clonal indicates that some clones of *E. coli* may share as-yet unidentified factors which, under certain circumstances, enable them to act as primary infectious agents and subsequently to cause acute, fulminating disease.

Adhesive and invasive properties of several clinical isolates from different disease manifestations have been investigated by the use of different cell culture systems such as HeLa, Hep-2 and KPCC (Chadfield *et al.*, 2000; da Silveira *et al.*, 2002). Adhesion was common for most strains regardless of their origin. However, invasion of tissue cultures was observed in only a minority of the isolates.

A type-1 fimbria-expressing O78 *E. coli* isolate has been shown to adhere efficiently to several parts of the chicken intestine, by the use of immunohistochemical methods (Edelman *et al.*, 2003). In this study, it was speculated whether the type-1 fimbria are of importance for colonization of the gut as they may well be in the upper respiratory tract. By the use of an *in vivo* chicken intestinal model it was possible to quantify differences in intestinal wall association of *E. coli* isolates from various sources (Chadfield *et al.*, 2000).

In general, clinical isolates tested did not exhibit an enhanced ability to associate with the chicken intestine compared with isolates derived from healthy birds. *In vitro*, infection of epithelial cell (HEp-2) culture assays with the *E. coli* isolates did not directly correlate with the intestinal loop assay, but some isolates did invade the cell culture to some extent.

Further work is needed to establish the extent of intestinal adhesion/ invasion *in vivo* with regard to histopathology. In summary, it seems clear that many avian-associated *E. coli* are efficient in adhering to the epithelium of the gut and that they possess some invasive potential. Neither the damage to the epithelium required nor the level of immunosuppressive effect for invasion to happen, however, have been properly investigated.

PASTEURELLA MULTOCIDA

Although the respiratory tract is considered the major route of infection of *P. multocida* in fowl, colonization and invasion of the gastrointestinal tract with subsequent mortality has also been demonstrated (Lee *et al.*, 2000). Following oral challenge of 16-week-old chickens with 10^6 CFU, *P. multocida* could be isolated from the crop from 1–30 h post-infection (PI) and the microorganisms could also be detected in other parts of the GIT, including the jejunum and ileum, from 28 h PI.

The results showed that *P. multocida* is able to survive passage of the GIT and induce mortality, but the exact site or mechanism of invasion was not determined. Oral challenge of 4-week-old ducks also resulted in significant mortalities, ranging from 33–75% depending on the challenge strain used (Phlivanoglu *et al.*, 1999).

The fact that the Clemson University strain (CU) – used for vaccination in the USA – is delivered through the drinking water further supports the idea of some kind of intimate interaction of *P. multocida* and the intestinal mucosa. By the use of an *in vivo* chicken intestinal model it was demonstrated that *P. multocida* may invade the jejunum in the chicken gut and that quantification is possible (Christensen *et al.*, 2002a, b). In this investigation, confirmation of invasion was obtained by the use of *in situ* hybridization. In addition, significant differences between strains in the ability to invade were demonstrated. In general, counts obtained were high and comparable with those obtained with an invasive *Salmonella* serotype, *S.* Typhimurium (4/74). No obvious correlation between virulence and intestinal invasion was observed. *Pasteurella multocida* VP17 is almost avirulent for chickens, but was highly invasive.

In contrast, the highly virulent strain VP161 was less invasive. This would suggest that virulence of *P. multocida* in avian hosts has little correlation with invasion of the gut, but more likely involves mechanisms of macrophage survival and complements resistance, as observed with the host-adapted *Salmonella* serovars. However, invasion of epithelial cell monolayers has been reported to reflect virulence differences between strains. In a study by Lee *et al.* (1994), a virulent strain penetrated primary turkey kidney epithelial cells at ten times the level of the low-virulence vaccine strain. The invasive properties of *P. multocida* have also been clearly demonstrated by the use of chicken embryo fibroblast cells (Ali *et al.*, 2004).

One of the possible virulence factors of *P. multocida* which has repeatedly been correlated with the early stages in pathogenesis is the capsule (Christensen *et al.*, 1997). Some *in vitro* results would indicate that the capsule is of importance for adhesion and invasion (Lee *et al.*, 1994; Ali *et al.*, 2004), but this has been difficult to confirm with *in vivo* studies (Rhoades and Rimler, 1990). Recently, LPS has been demonstrated to be of importance for virulence but whether invasion is impaired in LPS mutant strains is not clear (Harper *et al.*, 2005).

There is little doubt that *P. multocida* strains under certain circumstances are able to invade the alimentary tract. Under natural conditions this may take place primarily in waterfowl, where the intestinal tract has been reported to be severely affected during outbreaks (Wobeser, 1981). In acute/peracute cases of fowl cholera, lesions are not always observed in the respiratory tract, but whether this means that *P. multocida* has entered through the GIT is not currently known.

CAMPYLOBACTER

Campylobacter exists as a commensal in avian hosts, characterized by colonization and persistence in the gut and the absence of gross pathology (Barrow and Page, 2000; Friis *et al.*, 2005). However, certain strains of *Campylobacter jejuni* can produce diarrhoea experimentally in 36–72 h-old chicks.

Under such circumstances distension of the intestinal tract, with accumulation of mucous and watery fluid and even blood, has been observed (Sanyal *et al.*, 1984; Welkos, 1984). In the case of watery diarrhoea, only a slight sub-mucosal oedema was observed but when mucoid diarrhoea was present the organisms were found to adhere to brush borders and penetrate into epithelial cells, causing a localised invasion. Formation of breaches in the continuity of the brush border was also observed. Occasionally, mononuclear cell infiltration of the *lamina propria* has been reported (Welkos, 1984).

Most of the lesions associated with diarrhoea have been correlated with the secretion of a cytotonic toxin that shares some immunological similarities with cholera toxin (Sanyal *et al.*, 1984). The capability of systemic invasion of some isolates of *Campylobacter jejuni* has been clearly documented by oral challenge experiments, where up to 50% of 36–72 h-old chicks harboured the organism in the spleen on day 5 PI. However, subsequent fast clearance from the visceral organs was observed (Sanyal *et al.*, 1984).

The invasive properties of different isolates of *Campylobacter jejuni* appear to differ significantly. Some isolates have been demonstrated not to invade eukaryotic cells at all, whereas others are highly invasive. Most of these invasion studies have included a variety of cell culture systems such as Caco-2, HeLa and Hep-2 cells. Results from such *in vivo* and *in vitro* investigations have indicated a correlation of the ability to colonize the GIT of the young chick with the invasive ability in Caco-2 cells (Hänel *et al.*, 2004).

In summary, it may be concluded that colonization properties of *Campylobacter* in the chicken gut are of major importance in the role of poultry as carriers for zoonotic infections, but not for the development of disease in poultry.

GRAM-POSITIVE ORGANISMS

In Denmark, increasing problems associated with streptococci and enterococci have been observed in broilers and broiler parent flocks resulting in increased mortality, uneven flocks and subsequent downgrading and increased condemnation (Chadfield *et al.*, 2004). Post-mortem lesions associated with recent outbreaks in Denmark have been accompanied or dominated by septicaemia and endocarditis. Many of the isolates were unclassified according to current diagnostic methods, and final classification required taxonomic investigations. For the same reasons, the pathogenesis and epidemiology of observed outbreaks have remained speculative.

However, *Enterococcus hirae* has been isolated frequently from such cases and, experimentally, valvular endocarditis has been induced (Chadfield *et al.*, 2005a). Streptococcal and enterococcal systemic disease in avian species has previously been associated with *E. faecalis*, *E. faecium*, *E. durans*, *S. equi* spp., *S. zooepidemicus* and *S. gallinaceus* (Chadfield *et al.*, 2004). In the literature, little can be found concerning the pathogenesis of these infections. Both avian streptococci and enterococci are reported as being transmitted through oral and aerosol routes, but the circumstances under which the gut may serve as the port of entry are unclear.

However, experimental studies to determine the potential route of infection demonstrated adhesion and invasion by *E. hirae* of intestinal epithelial cell culture and *in vivo* intestinal loop invasion of the chicken gut. *E. hirae* was capable of adhesion and invasion in both the *in vitro* and *in vivo* settings and, furthermore, was able to survive intracellularly in avian macrophages. This was in contrast to other *Streptococcus* spp. also previously associated with similar clinical disease in chickens, where no/minimal invasion or survival was apparent (Chadfield *et al.*, 2005b).

Knowledge concerning the port of entry of *Erysipelothrix rhusiopathiae*, the cause of erysipelas, is also scarce. *E. rhusiopathiae* is currently causing increasing problems in organic and free-range table egg production in many European countries (Hafez, 2002; Eriksson *et al.*, 2005; Købke *et al.*, 2005). It is suggested that the microorganism may enter through skin abrasions or via the alimentary tract (Brooke and Riley, 1999; Bricker and Saif, 2003) and, recently, the poultry red mite (*Dermanyssus gallinae*) has been demonstrated as being a possible vector of *E. rhusiopathiae* (Chirico *et al.*, 2003).

It is also commonly observed that turkeys dying from erysipelas also suffer from cannibalism in the head and lower back region (authors' observations), where this is not always the case with chickens. Experimentally induced mortality in turkeys has been obtained by oral challenge, demonstrating that the GIT under certain conditions may serve as a port of entry (Corstvet, 1967). However, which parts of the intestine are involved and the significance of gastointestinal invasion under natural conditions remain to be determined.

CONCLUSION

Invasion of the GIT appears to be a common or occasional feature of many pathogens (and opportunistic pathogens). However, it is noteworthy that, for important organisms such as *E. coli* and the Gram-positive cocci, very little is known concerning the circumstances under which the organisms actually do invade and cause disease and subsequent mortality. In this respect the immune status of the host seems to be of major importance.

In addition, knowledge concerning the role of influencing factors on invasion – including feed composition, gut flora, etc. – is even more limited. Another aspect which requires investigation in order to clarify specific mechanisms of pathogenesis is the role of the host. As shown in this chapter, invasion may not always reflect the nature of virulence. In order to understand the pathogenesis of a bacterial infection the host response clearly warrants elucidation with respect to the specific organism, as documented in the case of avian salmonella infections.

REFERENCES

Aabo, S., Christensen, J.P., Chadfield, M.S., Carstensen, B., Jensen, T.K., Bisgaard, M. and Olsen, J.E. (2000) Development of an *in vivo* model for the study of intestinal invasion by salmonella in chickens. *Infection and Immunity* 68, 7122–7125.

Aabo, S., Christensen, J.P., Chadfield, M.S., Carstensen, B., Olsen, J.E. and Bisgaard, M. (2002) Quantitative comparison of intestinal invasion of zoonotic serotypes of *Salmonella enterica* in poultry. *Avian Pathology* 31, 41–47.

Ali, H.A-h., Sawada, T., Hatakeyama, H., Katayama, Y., Ohtsuki, N. and Itoh, O. (2004) Invasion of chicken embryo fibroblast cells by avian *Pasteurella multocida*. *Veterinary Microbiology* 104, 55–62.

Amy, M., Velge, P., Senocq, D., Bottreau, E., Mompart, F. and Virlogeux-Payant, I. (2004) Identification of a new *Salmonella enterica* serovar Enteritidis locus involved in cell invasion and in the colonization of chicks. *Research in Microbiology* 155, 543–552.

Barnes, H.J., Vaillancourt, J.-P. and Gross, W.B. (2003) Colibacillosis. In: Saif, Y.M., Barnes, H.J., Glisson, J.R., Fadly, A.M., McDougald, L.R. and Swayne, D.E. (eds) *Diseases of Poultry* (11th edn). Iowa State University Press, Ames, Iowa, pp. 631–657.

Barrow, P.A. (2005) Salmonella infections and vaccines. Abstract book from the *14th Veterinary Poultry Congress*, 22–26 August 2005, Istanbul, Turkey, pp. 86–98 (keynote lecture).

Barrow, P.A. and Page, K. (2000) Inhibition of colonization of the alimentary tract in young chickens with *Campylobacter jejuni* by pre-colonization with strains of *C. jejuni. FEMS Microbiology Letters* 182, 87–91.

Barrow, P.A., Huggins, M.B. and Lovell, M.A. (1994) Host specificity of *Salmonella* infection in chickens and mice is expressed *in vivo* primarily at the level of the reticuloendothelial system. *Infection and Immunity* 62, 4602–4610.

Beal, R.K., Powers, C., Wigley, P., Barrow, P.A., Kaiser, P. and Smith, A.L. (2005) A strong antigen-specific T-cell response is associated with age and genetically dependent resistance to avian enteric salmonellosis. *Infection and Immunity* 73, 7509–7516.

Best, A., La Ragione, R.M., Cooley, W.A., O'Connor, C.D., Velge, P. and Woodward, M.J. (2003) Interaction with avian cells and colonization of specific pathogen-free chicks by Shiga-toxin negative *Escherichia coli* O157:H7 (NCTC 12900). *Veterinary Microbiology* 29, 207–222.

Bolton, A.J., Osborne, M.P., Wallis, T.S and Stephen, J. (1999a) Interaction of *Salmonella* Choleraesuis, *Salmonella* Dublin and *Salmonella* Typhimurium with porcine and bovine terminal ileum *in vivo. Microbiology* 145, 2431–2441.

Bolton, A.J., Martin, G.D., Osborne, M.P., Wallis, T.S. and Stephen, J. (1999b) Invasiveness of *Salmonella* Typhimurium, *Salmonella* Cholerasuis and *Salmonella* Dublin for rabbit terminal ileum *in vitro. Journal of Medical Microbiology* 48, 800–810.

Bricker, J.M. and Saif, Y.M (2003) Erysipelas. In: Saif, Y.M., Barnes, H.J., Glisson, J.R., Fadly, A.M., McDougald, L.R. and Swayne, D.E. (eds) *Diseases of Poultry* (11th edn). Iowa State University Press, Ames, Iowa, pp. 631–657.

Brooke, C.J. and Riley, T.V. (1999) *Erysipelothrix rhusiopathiae*: bacteriology, epidemiology and clinical manifestations of an occupational pathogen. *Journal of Clinical Microbiology* 48, 789–799.

Chadfield, M.S., Christensen, J.P. and Bisgaard, M. (2000) Comparative investigations of selected avian *E. coli* isolates for demonstration of cell association in the chicken intestine and *in vitro* epithelial cell cultures. Proceedings of the XXI World's Poultry Congress. Annual meeting of the Poultry Science Association and the 6th International Symposium on Marek's Disease. Montreal, Canada, pp. 91.

Chadfield, M.S., Brown, D.J., Aabo, S., Christensen, J.P. and Olsen, J.E. (2003) Comparison of intestinal invasion and macrophage response of *Salmonella* Gallinarum and other host-adapted *Salmonella enterica* serovars in the avian host. *Veterinary Microbiology* 92, 49–64.

Chadfield, M.S., Christensen, J.P., Juhl-Hansen, J., Christensen, H. and Bisgaard, M. (2004) Characterisation of Enterococcus *hirae* outbreaks in broiler flocks demonstrating increased mortality due to septicaemia and endocarditis and/or altered production parameters. *Avian Diseases* 49, 16–23.

Chadfield, M.S., Bojesen, A.M., Christensen, J.P., Juhl-Hansen, J., Nielsen, S.S. and Bisgaard, M. (2005a) Reproduction of sepsis and endocarditis by experimental infection of chickens with *Streptoccocus gallinaceus* and *Enterococcus hirae*. *Avian Pathology* 34, 238–247.

Chadfield, M.S., Christensen, J.P., Christensen, H., Bojesen, A.M., and Bisgaard, M. (2005b) *Enterococcus hirae* infections in chickens. *In vitro* characterisation and *in vivo* reproduction of infection. *Proceedings of the 2nd ASM Conference on Enterococci*, Helsingør, Denmark, 28–31 August 2005, pp. 89–90.

Chadfield, M.S., Christensen, J.P. and Bisgaard, M. (2006) Demonstration of clonality and virulence traits associated with *Escherichia coli* clinical outbreaks in turkey flocks. *Avian Pathology* (in press).

Chirico, J., Erikson, H., Fossum, O. and Jansson, D. (2003) The poultry red mite, *Dermanyssus gallinae*, a potential vector of *Erysipelothrix rhusiopathiae* causing erysipelas in hens. *Medical and Veterinary Entomology* 17, 232–234.

Corstvet, R.E. (1967) Pathogenesis of *Erysipelothrix insidiosa* in the turkey. *Poultry Science* 46, 1247.

Christensen, J.P. and Bisgaard, M. (1997) Avian Pasteurellosis: taxonomy of the organisms involved and aspects of pathogenesis. *Avian Pathology* 26, 461–483.

Christensen, J.P., Bojesen, A.M., Jensen, T.K. and Bisgaard, M. (2002a) Quantitative investigations of *in vivo* invasion of *Pasteurella multocida* using a chicken intestinal loop model. *Proceedings of the International Pasteurellaceae Society Conference*, Banff, Canada. 5–10 May 2002, p. 30.

Christensen, J.P., Jensen, T.K., Bojesen, A.M., Kabell, S. and Bisgaard, M. (2002b) Intestinal invasion of *Pasteurella multocida* investigated by an *in vivo* chicken intestinal loop model and *in situ* hybridization. *Proceedings from the XII International Congress of the World Veterinary Poultry Association*, Cairo, Egypt, 28 January–1 February, p. 214.

Cox, N.A., Bailly, J.S. and Berrang, M.E. (1996) Alternative routes for *Salmonella* intestinal tract colonization of chicks. *Journal of Applied Poultry Science* 5, 282–288.

Da Silveira, W.D., Ferreira, A., Brocchi, M., de Hollanda, L.M., de Castro, A.F.P., Yamada, A.T. and Lancellotti, M. (2002) Biological characteristics and pathogenicity of avian *Escherichia coli* strains. *Veterinary Microbiology* 85, 47–53.

Edelman, S., Leskela, S., Ron, E., Apajalahti, J. and Korhonen, T.K. (2003) *In vitro* adhesion of an avian pathogenic *Escherichia coli* O78 strain to surfaces of the chicken intestinal tract and to ileal mucus. *Veterinary Microbiology* 91, 41–56.

Edens, F.W., Parkhurst, C.R., Qureshi, M.A., Casas, I.A. and Havenstein, G.B. (1997) Atypical *Escherichia coli* strains and their association with poult enteritis and mortality syndrome. *Poultry Science* 76, 952–960.

Eriksson, H., Jansson, D., Båverud, V., Chirico, J. and Aspán, A. (2005) Methods for subtyping *Erysipelothrix rhusiopathiae*. Abstract book from the *14th Veterinary Poultry Congress*, 22–26 August 2005, Istanbul, Turkey, p. 418.

Friis, L.M., Pin, C., Pearson, B.M. and Wells, J.M. (2005) *In vitro* cell culture methods for investigating *Campylobacter* invasion mechanisms. *Journal of Microbiological Methods* 61, 145–160.

Gast, R.K. (2003) Paratyphoid infections. In: Saif, Y.M., Barnes, H.J., Glisson, J.R., Fadly, A.M., McDougald, L.R. and Swayne, D.E. (eds) *Diseases of Poultry* (11th edn). Iowa State University Press, Ames, Iowa, pp. 583–613.

Gross, W.G. (1994). Diseases due to *Escherichia coli* in poultry. In: Gyles, C.L. (ed.) Escherichia coli *in Domestic Animals and Humans*. CAB International, Wallingford, UK, pp. 237–261.

Guy, J.S., Smith, L.G., Breslin, J.J., Vaillancourt, J.P. and Barnes, H.J. (2000) High mortality and growth depression experimentally produced in young turkeys by dual infection with enteropathogenic *Escherichia coli* and turkey corona virus. *Avian Diseases* 44, 105–113.

Hafez, H.M. (2002) Emerging and re-emerging bacterial infections in poultry and their significance to the poultry industry. Oral presentation in *Proceedings, XII International Congress of the World Veterinary Poultry Association*, Cairo, Egypt, 28 January–1 February 2002, p. 67.

Hänel, I., Müller, J., Müller, W. and Schulze, F. (2004) Correlation between invasion of Caco-2 eukaryotic cells and colonization ability in the chick gut in *Campylobacter jejuni*. *Veterinary Microbiology* 101, 75–82.

Harper, M., Boyce, J.D., St. Michael, F., Wilkie, I.W., Cox, A.D. and Adler, B. (2005) Structural and functional analysis of *Pasteurella multocida* LPS mutants. *Proceedings. International Pasteurellaceae Society Conference*, Kohala Coast, Big Island, Hawaii, 23–26 October 2005, p. 72.

Henderson, C.S., Bounous, D.I. and Lee, M.D. (1999) Early events in the pathogenesis of avian salmonellosis. *Infection and Immunity* 67, 3580–3586.

Iqbal, M., Philbin, V.J., Withanage, G.S.K., Wigley, P., Beal, R.K., Goodchild, M.J., Barrow, P.A., McConnell, I., Maskell, D.J., Young, J., Bumstead, N., Boyd, Y. and Smith, A.L. (2005) Identification and functional characterization of chicken toll-like receptor 5 reveals a fundamental role in the biology of infection with *Salmonella enterica* serovar Typhimurium. *Infection and Immunity* 73, 2344–2350.

Joya, J.E., Tsuji, T., Jacaline, A.V., Arita, M., Tsukamoto, T., Honda, T. and Miwatani, T. (1990) Demonstration of enterotoxigenic *Escherichia coli* in diarrheic broiler chicks. *European Journal of Epidemiology* 6, 88–90.

Kaiser, P., Rothwell, L., Galyov, E.E., Barrow, P.A., Burnside, J. and Wigley, P. (2000) Differential cytokine expression in avian cells in response to invasion by *Salmonella* Typhimurium, *Salmonella* Enteritidis and *Salmonella* Gallinarum. *Microbiology* 146, 3217–3226.

Kogut, M.H., Tellez, G., Hargis, B.M., Corrier, D.E. and DeLoach, J.R. (1993) The effect of 5-fluorouracil treatment of chicks: a cell depletion model for the study of avian polymorphonuclear leukocytes and natural host defences. *Poultry Science* 72, 1873–80.

Købke, B., Eigaard, N.M., Christensen, J.P. and Bisgaard, M. (2005) Investigations on clonality and stability of clones of *Erysipelothrix rhusiopathiae* in turkeys and free range layer flocks affected by erysipelas. Abstract book from the *14th Veterinary Poultry Congress*, 22–26 August 2005, Istanbul, Turkey, p. 313.

Larsen, C.T., Domermuth, C.H., Sponenberg, D.P. and Gross, W.B. (1985) Colibacillosis of turkeys exacerbated by hemorrhagic enteritis virus; laboratory studies. *Avian Diseases*, 29, 729–732.

Lee, C.W., Wilkie, I.W., Townsend, K.M. and Frost, A.J. (2000) The demonstration of *Pasteurella multocida* in the alimentary tract of chickens after experimental oral infection. *Veterinary Microbiology* 72, 47–55.

Lee, M.D., Wooley, R.E. and Glisson, J.R. (1994) Invasion of epithelial cell monolayers by turkey strains of *Pasteurella multocida*. *Avian Diseases* 38, 72–77.

Leitner, G. and Heller, E.D. (1992) Colonization of *Escherichia coli* in young turkeys and chickens. *Avian Diseases* 36, 211–220.

Lillehoj, H.S. and Trout, J.M. (1996) Avian gut-associated lymphoid tissues and intestinal immune responses to *Eimeria* parasites. *Clinical Microbiology Reviews* 9, 349–360.

Pehlivanoglu, F., Morishita, T.Y., Aye, P.P., Porter Jr., R.E., Angrick, E.J., Harr, B.S. and Nersessian, B. (1999) The effect of route of inoculation on the virulence of raptorial *Pasteurella multocida* isolates in Pekin ducks (*Anas platyrhynchos*). *Avian Diseases* 43, 116–121.

Rhoades, K.R. and Rimler, R.B. (1990) *Pasteurella multocida* colonization and invasion in experimentally exposed turkey poults. *Avian Diseases* 34, 381–383.

Sanyal, S.C., Islam, K.M.N., Neogy, P.K.B., Islam, M., Speelman, P. and Huq, M.I. (1984) *Campylobacter jejuni* diarrhea model in infant chickens. *Infection and Immunity* 43, 931–936.

Schoeni, J.L. and Doyle, M.P. (1994) Variable colonization of chickens perorally inoculated with *Escherichia coli* O157:H7 and subsequent contamination of eggs. *Applied and Environmental Microbiology* 60, 2958–2962.

Sponenberg, D.P., Domermuth, C.H. and Larsen, C.T. (1985) Field outbreaks of colibacillosis of turkeys associated with hemorrhagic enteritis virus. *Avian Diseases* 29, 838–842.

Stavric, S., Buchanan, B. and Gleeson, T.M. (1993) Intestinal colonization of young chicks with *Escherichia coli* O157:H7 and other verotoxin-producing serotypes. *Journal of Applied Bacteriology* 74, 557–563.

Sueyoshi, M., Fukui, H. Tanaka, S., Nakazawa, M. and Ito, K. (1996) A new adherent form of an attaching and effacing *Escherichia coli* (eaeA+, bfp-) to the intestinal epithelial cells of chicks. *Journal of Veterinary Medical Science* 58, 1145–1147.

Sueyoshi, M., Nakazawa, M. and Tanaka, S. (1997) A chick model for the study of 'attaching and effacing *Escherichia coli*' infection. *Advances in Experimental Medicine and Biology* 412, 99–102.

Uzzau, S., Leori, G.S., Petruzzi, V., Watson, P.R., Schianchi, G., Bacciu, D., Mazzarello, V., Wallis, T.S. and Rubino, S. (2001) *Salmonella enterica* serovar-host specificity does not correlate with the magnitude of intestinal invasion in sheep. *Infection and Immunity* 69, 3092–3099.

Wada, Y., Kondo, H., Nakazawa, M. and Kubo, M. (1995) Natural infection with attaching and effacing *Escherichia coli* and adenovirus in the intestine of a pigeon with diarrhoea. *Journal of Veterinary Medical Science* 57, 531–533.

Wigley, P., Berchieri, A., Page, K.L., Smith, A.L. and Barrow, P.A. (2001) *Salmonella enterica* serovar Pullorum persists in splenic macrophages and in the reproductive tract during persistent disease-free carriage in chickens. *Infection and Immunity* 69, 7873–7879.

Wobeser, G.A. (1981) Avian cholera. In: Wobeser, G.A. (ed.) *Diseases of Wild Waterfowl*. Plenum Press, New York, pp. 47–59.

Welkos, S.L. (1984) Experimental gastroenteritis in newly-hatched chicks injected with *Campylobacter jejuni*. *Journal of Medical Microbiology* 18, 233–248.

van den Hurk, J., Allan, B.J., Riddell, C., Watts, T. and Potter, A.A. (1994) Effect of infection with hemorrhagic enteritis virus on susceptibility of turkeys to *Escherichia coli*. *Avian Diseases* 38, 708–716.

Zhao, S., Meng, J., Doyle, M.P., Meinerman, R., Wang, G. and Zhao, P. (1996) A low molecular weight outer-membrane protein of *Escherichia coli* O157:H7 associated with adherence to INT407 cells and chickens' caeca. *Journal of Medical Microbiology* 45, 90–96.

CHAPTER 17
Parasite genetics, protection and antigen identification

D.P. Blake,[1,2]** M.W. Shirley[2] and A.L. Smith[1]*

[1]*Enteric Immunology Group;* [2]*Eimerian Genomics Group, Institute for Animal Health, Newbury, UK; e-mails:* *adrian.smith@bbsrc.ac.uk;* **damer.blake@bbsrc.ac.uk*

ABSTRACT

The *Eimeria* species, causative agents of the disease coccidiosis, are largely controlled by prophylactic administration of anticoccidial drugs, but this approach is unsustainable. Live-parasite vaccines are available and used principally in the egg production sector. The development of subunit vaccines would revolutionize coccidiosis control, but this approach requires identification of the protective antigens expressed by the parasite. Traditionally, antigen selection is based upon recognition by the host immune system. However, many of the responses generated by infection are not protective and only a small subset of antigens is capable of stimulating protective immunity.

Recognition of this aspect of the host–parasite relationship has driven the development of a new approach based upon parasite genetics, DNA fingerprinting and selection by immunity. This approach was developed with *Eimeria maxima* and the results have significant implications for the development of vaccines against other parasitic pathogens. First, the relative ease in generating recombinant parasites that combine the phenotypes of drug resistance and strain-specific immunity indicates that few parasitic loci encode antigens that stimulate protective immunity. Secondly, the inheritance of relevant genomic regions can be followed using genetic fingerprinting of DNA from selected and non-selected parasites. Thirdly, these regions are highly likely to encode the protective antigens required for the development of an effective vaccine.

At present, the genetic markers group into four regions of the genome, supporting our hypothesis that few parasite antigens are capable of stimulating protective immunity. These genomic regions are under scrutiny for identification of candidate antigens for inclusion in future vaccines.

INTRODUCTION

Protozoan parasites are genetically complex organisms with genomes that encode several thousands of genes (Gardner *et al.*, 2002; Abrahamsen *et al.*, 2004; Xu *et al.*, 2004; Berriman *et al.*, 2005; El-Sayed *et al.*, 2005; Ivens *et al.*, 2005; Pain *et al.*, 2005). As such, the identification of genes that encode antigens capable of stimulating a protective immune response in the host is a difficult task (Richie and Saul, 2002). Many studies have suggested candidates for inclusion in subunit or recombinant vaccination strategies targeted against protozoan parasites (Profous-Juchelka *et al.*, 1988; Melby *et al.*, 2000; Smooker *et al.*, 2000; Shibui *et al.*, 2001). However, differentiation between an antigen capable of stimulating an immune response and one capable of stimulating genuine protective immunity has proved to be a major obstacle (Melby *et al.*, 2001), illustrated by the paucity of such vaccines currently in use (examples include those described by Wallach *et al.*, 1995).

Empirical approaches to the identification of suitable antigens involve a significant input of resources with no guarantee of success, suggesting a requirement for alternative, more selective, screens to identify vaccine candidates.

In response to this problem we recently proposed a genetics-led approach to reveal antigens encoded within the genome of *Eimeria maxima* negatively selected under strain-specific immunity (Blake *et al.*, 2004). Integral to our strategy are key traits exhibited by *E. maxima* related to strain-specific immunity, anticoccidial drug resistance and the brief diploid phase of the eimerian life cycle, which allows recombination to take place. These traits are discussed, as is their utilization in previous mapping studies and their application to the identification of immunoprotective antigens of *E. maxima*.

AVIAN COCCIDIOSIS

The *Eimeria* species, causative agents of the disease coccidiosis, are endemic protozoan pathogens of livestock. Seven *Eimeria* species infect the chicken and the most economically important are *E. acervulina*, *E. maxima* and *E. tenella*. Every chicken reared in the world is likely to be exposed to one or more of these protozoan parasites, necessitating prophylactic control to prevent disease. Even sub-clinical infection can compromise bird welfare and the economic viability of poultry production whilst clinical disease can prove to be devastating.

Current control is based largely on the use of chemotherapy, although resistance – commonly detected shortly after the introduction of a compound – has become widespread (Cuckler and Malanga, 1955; Joyner, 1957; Warren *et al.*, 1966; Hodgson *et al.*, 1969; Martin *et al.*, 1997), and legislation continues to reduce the range of products available (e.g. Regulation (EC) No. 1831/2003 of the European Parliament and of the Council of 22 September 2003 on additives for use in animal nutrition; available at http://europa.eu.int/eur-lex/pri/en/oj/dat/2003/l_268/l_26820031018en00290043.pdf).

Alternative control is effected through the use of vaccines. The majority of vaccines on the market are based upon either: (i) live wild-type; or (ii) attenuated parasite formulations. Such vaccines have been available since 1952 and 1992, respectively (Bedrnik *et al.*, 1995), but uptake is limited for a variety of reasons, including cost.

Immunity to the *Eimeria* species

The immunity generated through natural exposure to eimerian infection is specific to both the infecting *Eimeria* species and, in some examples, the particular strain (Joyner, 1969; Norton and Hein, 1976; Shirley and Bellatti, 1988; Fitz-Coy, S. 1992; Martin *et al.*, 1997; Smith *et al.*, 2002). Of those species that infect the chicken, *E. maxima* is the most immunogenic (Long *et al.*, 1976): a single sporocyst is capable of stimulating partially protective immunity against homologous challenge, whilst a dose of 20 or fewer oocysts can result in complete protection (> 99.99%) (Lee and Fernando, 1978; Blake *et al.*, 2006).

In contrast, immunity against infection by a heterologous strain can be incomplete or non-existent (depending on parasite strain and host genetics). For example, in a recent study immune protection arising from primary infection with the Houghton (H) or Weybridge (W) strains of *E. maxima* was only complete against challenge with the heterologous strain in one of seven chicken lines tested (see Fig. 17.1; Smith *et al.*, 2002). Complete absence of cross-protection was evident against heterologous infection in three of the seven lines of chicken (Fig. 17.1). The importance of parasite genotype on escape from immune selection in Line C chickens suggested a strong heritable trait associated with antigenic type, providing the opportunity to genetically map those loci integral to the trait 'escape from strain-specific immune killing'.

Drug resistance in the *Eimeria* species

As described above, anticoccidial resistance is common in the field. Experimentally, resistance to a wide variety of anticoccidial compounds can be selected in an incremental stepwise process, although the difficulty in generating resistance depends on the drug (Norton and Joyner, 1975). Variation in the rate of development of resistance to a specific compound can highlight the genetic complexity of the resistant genotype or the nature of the target. For example, resistance to the quinolone-based anticoccidials can occur within a single passage of the parasite, suggesting a requirement to modify few parasite loci (Chapman, 1997). In contrast, the development of resistance to ionophores requires a more prolonged exposure over multiple cycles of development, suggesting a requirement for a larger number of mutations.

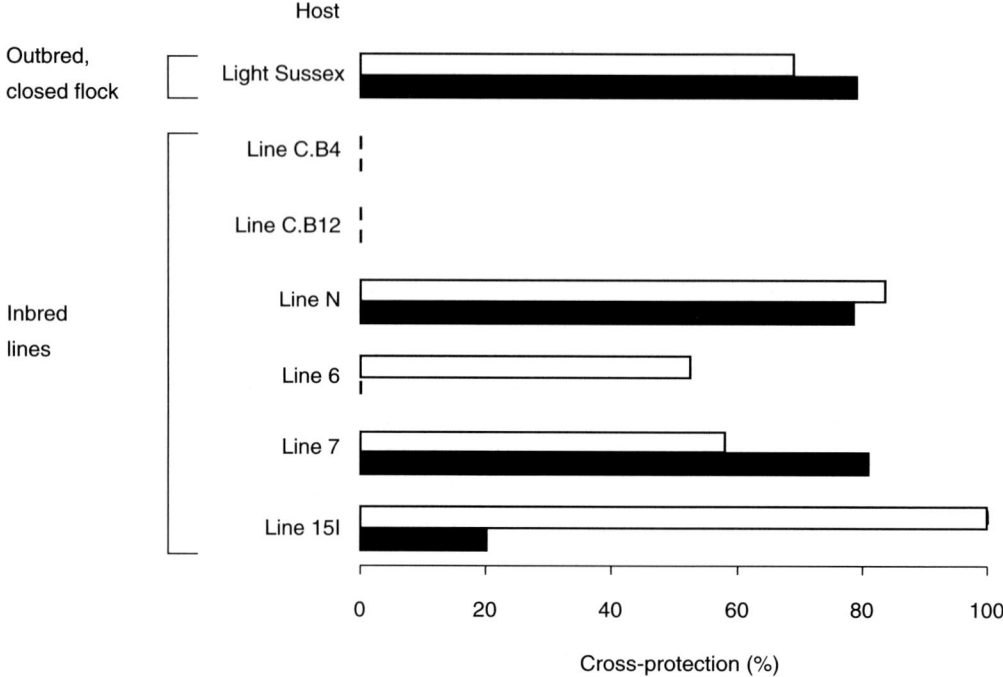

Fig. 17.1. The influence of host genotype on immune protection provided by primary infection with either the Houghton (white bars) or Weybridge (black bars) strains of *Eimeria maxima* against secondary infection with a heterologous strain. Protection is expressed in terms of the reduction in oocyst production as a percentage of a non-immunized control (adapted from Smith *et al.*, 2002).

Resistance to different levels of robenidine can be selected by a stepwise process of increasing drug concentrations, which suggests a role for multiple modifications of the genome (Norton and Joyner, 1975; Chapman, 1976). The Weybridge strain of *E. maxima*, used extensively in genetic studies at the Institute for Animal Health (see Fig. 17.2), has been rendered resistant to robenidine at VLA Weybridge (kindly donated by Janet Catchpole and Ralph Marshall; see also Blake *et al.*, 2004).

The genetic basis of resistance to many anticoccidial drugs remains unknown, but the strong selectivity of the resistance trait makes it a phenotype particularly amenable to genetic mapping. Indeed, Shirley and Harvey (2000) mapped resistance to arprinocid in *E. tenella* to a locus on chromosome 1. Genetic mapping studies of other apicomplexan parasites have located loci integral to chloroquine resistance in *Plasmodium falciparum* and *P. chabaudi* (Wellems *et al.*, 1991; Hunt *et al.*, 2004).

Genetic recombination and the *Eimeria*

Whilst the eimerian life cycle is largely haploid, the presence of sexual life stages gives rise to a diploid phase which allows genetic recombination to

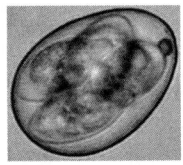

E. *maxima* H and W strains:
H = drug sensitive, antigenic type H
W = drug resistant, antigenic type W

Chickens infected with E. *maxima* H +
Woocysts, providing an opportunity
for cross-fertilization to occur.

Collect '***Progeny of the Cross***' (***PC***), a mix of:
  • Pure parental types (arising from self-fertilization)
  • Recombinants (arising from cross-fertilization)

Passage *PC* in the presence of drug-
and W strain-specific immunity.

Collect *Relevant Recombinants* (***RR***):
  • Recombinant parasites capable of escape from killing
    by drug AND strain-specific immunity

Clone-based mapping:

Passage single sporocysts under
selection to isolate clonal *RR*

Examine the inheritance of
polymorphic genetic markers
(e.g. AFLP markers) by all
clonal lines

Population-based mapping:

Examine the inheritance of
polymorphic genetic markers
(e.g. AFLP markers) by the *PC*,
RR and parental parasites

Investigate markers influenced
by selection (Fig. 17.3)

Fig. 17.2. A flow diagram illustrating the steps taken in the genetic mapping of strain-specific immunity and resistance to the anticoccidial drug robenindine in *Eimeria maxima*, utilizing either a clone-based or a population-based approach.

occur. Among the first experimental evidence supporting the occurrence of genetic recombination within the eimerian genome was the generation of multi-drug-resistant parasites by co-infecting chickens with two differentially resistant strains of *E. maxima* or *E. tenella* (Jeffers, 1974; Joyner and Norton, 1975, 1976). Further reports based upon the recombination of loci associated with precocious development and enzyme markers followed (Jeffers, 1976; Rollinson *et al.*, 1979; Sutton *et al.*, 1986).

Given evidence that a locus encoding a trait may be transferred genetically between strains, that locus may be mapped within the donor genome, as described for *E. tenella* (Shirley and Harvey, 2000). Observations of escape from strain-specific immunity by two strains of *E. maxima* in Line C chickens revealed a phenotype akin to that of resistance/susceptibility to anticoccidial treatment (Smith *et al.*, 2002).

We therefore hypothesized that loci integral to escape from strain-specific immune killing may be accessible through a genetic mapping strategy (Blake *et al.*, 2004). Previous observations of recombination between genetic markers of strain-specific immunity and resistance to pyrimethamine in the murine malarial parasite *Plasmodium berghei* supported the application of such an approach (Oxbrow, 1973). A similar approach for the murine malarial parasite *P. chabaudi chabaudi* has been independently reported elsewhere (Culleton *et al.*, 2005; Martinelli *et al.*, 2005).

Genetic mapping of loci encoding immunoprotective antigens of *E. maxima*

Generation and isolation of recombinant E. maxima

The phenotypically divergent H and W strains of *E. maxima*, characterized by differential escape from killing by strain-specific immunity and robenidine, were used in these studies (Fig. 17.1; Blake *et al.*, 2004). Oocysts excreted following concurrent infection with the *E. maxima* H and W strains were harvested in the usual manner (Long *et al.*, 1976).

These parasites, termed the *Progeny of the Cross* (PC), were expected to represent the progeny of self-fertilization by either strain, together with a pool of recombinant parasites arising from cross-fertilization (Fig. 17.2). Passage infection with the PC through a selective double barrier consisting of strain-specific immunity lethal to the W parent and dietary robenidine lethal to the H parent yielded a population of parasites recombinant at loci integral to escape from both barriers, proving in the process that these loci segregate independently. Those parasites capable of escape from the double barrier were termed *Relevant Recombinants* (RR).

Subsequent passage of a RR parasite population in the presence of either individual component of the double barrier revealed that escape from neither drug, nor immunity, was fixed in the population after a single selective passage (Table 17.1). An important feature of the eimerian life cycle is that meiosis occurs during sporulation, an environmental stage of the life cycle removed from selective pressures such as host immunity or chemotherapy. Hence, it

Table 17.1. Phenotypic characterisation of a *Relevant Recombinant* (RR) parasite population: escape from single and double barrier selection as a percentage of oocyst excretion following comparable infection in the absence of selection. Birds exposed to a mixed dose of the parental parasites (H + W) demonstrated the integrity of each barrier (from Blake *et al.*, 2004).

Parasite	Barrier	Escape (%)
RR	Drug	72.4[a]
RR	Immunity	39.4[b]
RR	Drug + Immunity	21.1[c]
H+W	Drug + Immunity	0.0[d]

Different superscript letters indicate significant differences ($p < 0.05$).

appears that escape from killing by drug or immune selection are both multilocus traits which may be disrupted by resegregation. However, RR were always isolated from the PC in the presence of double barrier selection, even when using an oocyst dose as low as 100/bird, implicating a limited number of genetic loci in escape from both barriers (Blake *et al.*, 2004). The finding that RR escape the drug barrier better than the immune barrier suggests that the larger proportion of those loci that encode molecules important for escape from double barrier selection encode antigens exposed to deleterious strain-specific selection.

Genetic mapping analyses of populations

Classically, genetic mapping strategies with protozoan parasites have been based upon the clonal progeny of a genetic cross (Wellems *et al.*, 1991; Shirley and Harvey, 2000; Su *et al.*, 2002; Hunt *et al.*, 2004). However, whilst achievable through passage of a single sporocyst, cloning the *Eimeria* species is not a trivial undertaking (Shirley and Harvey, 1996).

As an alternative we used an approach based upon populations of pure and recombinant parasites generated and isolated under a variety of selection regimes (Fig. 17.2). Central to this approach is the hypothesis that genetic markers not physically linked within the genome to loci under deleterious selection will be inherited by progeny parasites irrespective of the influence of that selection. In contrast, a genetic marker that is physically linked to a locus under deleterious selection close enough not to be separated by the majority of recombination events in a genetically diverse haploid population will be reduced, or lost altogether, from the subset of that population capable of escaping the selection (see Fig. 17.3).

Whole-genome genetic fingerprinting of parasite populations using the amplified fragment length polymorphism (AFLP) technique generated markers from total genomic DNA extracted from oocysts of either parent and multiple independent PC and RR lineages (Vos *et al.*, 1995; Blake *et al.*, 2003). Comparison between AFLP fingerprints revealed five patterns of marker inheritance (Fig. 17.4).

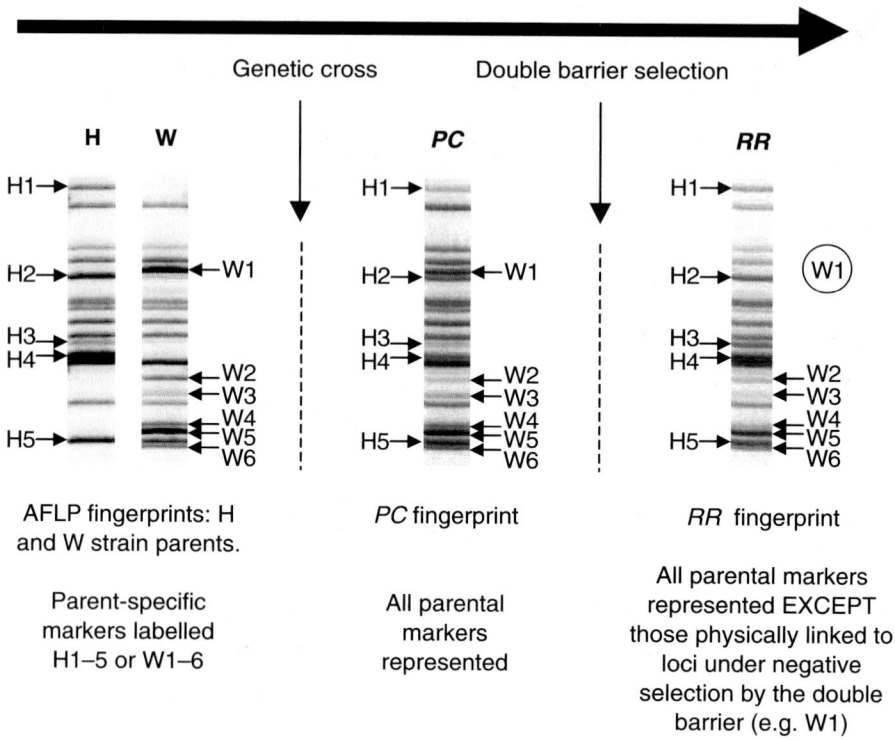

Genetic cross   **Double barrier selection**

H   W   **PC**   **RR**

AFLP fingerprints: H   *PC* fingerprint   *RR* fingerprint
and W strain parents.

                                                        All parental markers
Parent-specific   All parental   represented EXCEPT
markers labelled   markers   those physically linked to
H1–5 or W1–6   represented   loci under negative
                                                    selection by the double
                                                      barrier (e.g. W1)

Fig. 17.3. AFLP analyses of populations. The inheritance of AFLP markers amplified from the parental H and W strains of *Eimeria maxima* used in the production of a genetic cross was compared between the *Progeny of the Cross* (PC) and those recombinants within the PC capable of reproducing in the presence of barriers lethal to either individual parent (the *Relevant Recombinants*, RR). Genetic markers lost under selection by either strain-specific immunity or dietary robenidine are physically linked to those loci under negative selection.

By far the most common pattern represented a non-polymorphic marker found in both parental *E. maxima* strains which persisted in all progeny populations, irrespective of selection (e.g. pattern 1, Fig. 17.4). The second most common patterns were strain-specific markers inherited by all progeny populations (e.g. patterns 2 and 3, Fig. 17.4). The inheritance pattern of these markers indicates no physical linkage to loci under drug or immune selection.

In contrast, a subset of markers (currently a total of 38) was clearly correlated with the deleterious effects of double barrier selection (e.g. pattern 4, Fig. 17.4). The use of single barriers (drug or W strain-specific immunity) allowed clear attribution of the markers to either of the two phenotypes present in the double barrier. At present two markers display an inheritance pattern that relates to the drug resistant phenotype whereas 36 can be related to the immune-escape phenotype.

This bias in marker association supports our earlier observations with the capacity of RR to escape single barriers and suggests more independent loci encode strain-specific protective antigens than drug resistance. The markers associated with immunity have been subjected to further analysis. Using a combination of: (i) probing Southern blotted digested parental genomic DNA resolved by pulsed field gel electrophoresis; and (ii) a BAC library generated from the W strain parent, it is clear that the AFLP markers group in regions of the parasite genome. At present, markers can be linked to four regions of the genome, supporting the premise that strain-specific protective immunity is restricted to very few antigen-encoding loci.

CONCLUSION

For many years the *Eimeria* species have been subject to detailed study of variables that affect immune protection in hosts of differing genotype against

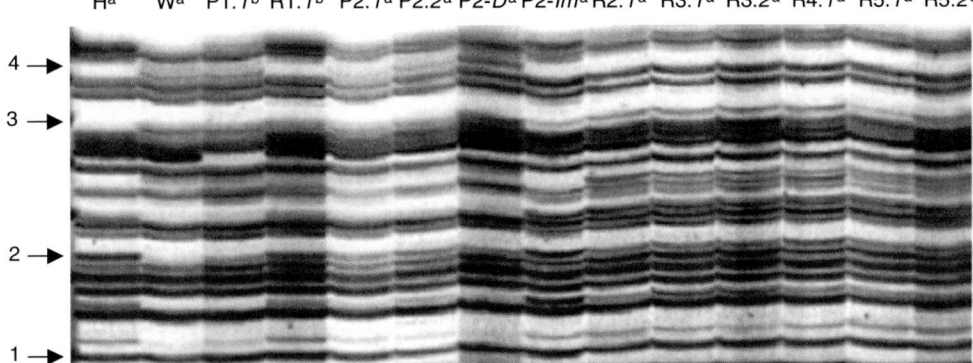

Fig. 17.4. AFLP analysis of a population-based DNA mapping panel. H, *Eimeria maxima* H strain parent; W, *E. maxima* W strain parent; P, *Progeny of the Cross* (PC) population; R, *Relevant Recombinant* (RR) population. Lane identifiers: first number is the independent population replicate number; *second number* is the generation number (under the selective conditions specified). -*D*, selected through the single barrier of dietary drug (66 ppm robenidine); -*Im*, selected through the single barrier of W strain-specific immunity. [a] Derived against a Line C host genetic background; [b] derived against a Line 151 host genetic background.

The characteristics of each of the five band patterns arising from AFLP analysis of the DNA mapping panel: 1, a non-polymorphic band consistently amplified from all parasite populations, regardless of selection regime. 2, a polymorphic band specific to the H strain; this band was retained by all of the RR populations and therefore did not mark a region of the genome negatively selected in the presence of dietary robenindine. 3, a polymorphic band specific to the W strain; this band was retained by all of the RR populations and therefore did not mark a region of the genome negatively selected in the presence of W strain-specific immunity. 4, a polymorphic band specific to the W strain; amplified from all PC progeny populations and retained following exposure to the single barrier of robenidine; this band was lost following exposure to the single barrier of W strain-specific immunity with or without robenidine; hence implicated as a marker associated with immune killing of the W strain. 5 (not shown), a polymorphic band specific to the H strain; amplified from all PC progeny populations and retained following exposure to the single barrier of W strain-specific immunity; this band was lost following exposure to robenidine; hence implicated as a marker associated with killing by robenidine.

homologous and heterologous parasite challenge (e.g. Rose and Long, 1962; Norton and Hein, 1976; Long and Millard, 1979; Shirley and Hoyle, 1981; Lillehoj, 1988; Bumstead and Millard, 1992; Fitz-Coy, 1992; Martin *et al.*, 1997; Smith *et al.*, 2002). More recently, techniques to characterize and manipulate nucleic acids have progressed rapidly (e.g. Jeffreys *et al.*, 1985; Vos *et al.*, 1995). Together, the combination of the discriminating strain-specific immune protection exhibited by certain host lines of chicken and the ability to generate genome-wide genetic markers in the absence of detailed sequence data provided an opportunity to utilize immunological characters as genetic markers in the search for immunoprotective antigens (Walker, 1968; Oxbrow, 1973; Vos *et al.*, 1995; Smith *et al.*, 2002). The development of a mapping strategy targeting populations, not clones, has expedited these studies (Blake *et al.*, 2004).

It is clear that many *Eimeria* antigens are recognized by the host immune response but it is equally apparent that few of these confer protection. The development of a genetic approach in identifying protective, antigen-encoding loci represents a major step forward in the development of new vaccines protective against the *Eimeria* species. Moreover, the proof of principle for the genetic approach described here offers the opportunity to apply this methodology more broadly in other parasitic diseases of veterinary and medical importance.

ACKNOWLEDGEMENTS

This work was primarily funded by the BBSRC (grant no. 201/S15343; M.W.S. and A.L.S.), with some support from DEFRA (grant code OD0534; A.L.S.).

The authors would like to thank members of the enteric immunology and parasitology groups for their help and input.

REFERENCES

Abrahamsen, M.S., Templeton, T.J., Enomoto, S., Abrahante, J.E., Zhu, G., Lancto, C.A., Deng, M., Liu, C., Widmer, G., Tzipori, S. *et al.* (2004) Complete genome sequence of the apicomplexan, *Cryptosporidium parvum. Science* 304, 441–445.

Bedrnik, P., Hiepe, T., Mielke, D. and Drossigk, U. (1995) Antigens and immunisation procedures in the development of vaccines against poultry coccidiosis. COST 89/820. In: Eckert, J., Braun, R., Shirley, M. and Coudert, P. (eds) *Biotechnology: Guidelines on Techniques in Coccidiosis Research.* European Commission, Luxembourg, pp. 176–189.

Berriman, M., Ghedin, E., Hertz-Fowler, C., Blandin, G., Renauld, H., Bartholomeu, D.C., Lennard, N.J., Caler, E., Hamlin, N.E., Haas, B. *et al.* (2005) The genome of the African trypanosome *Trypanosoma brucei. Science* 309, 416–422.

Blake, D.P., Smith, A.L. and Shirley, M.W. (2003) Amplified fragment length polymorphism analyses of *Eimeria* spp.: an improved process for genetic studies of recombinant parasites. *Parasitology Research* 90, 473–475.

Blake, D.P., Hesketh, P., Archer, A., Carroll, F., Smith, A.L. and Shirley, M.W. (2004) Parasite genetics and the immune host: recombination between antigenic types of *Eimeria maxima* as an entree to the identification of protective antigens. *Molecular and Biochemical Parasitology* 138, 143–152.

Blake, D.P., Hesketh, P., Archer, A., Carroll, F., Shirley, M.W. and Smith, A.L. (2006) The influence of immunizing dose size and schedule on immunity to subsequent challenge with antigenically distinct strains of *Eimeria maxima*. *Avian Pathology* 34, 489–494.

Bumstead, N. and Millard, B. (1992) Variation in susceptibility of inbred lines of chickens to seven species of *Eimeria*. *Parasitology* 104, 407–413.

Chapman, H. (1976) *Eimeria tenella*: experimental studies on the development of resistance to robenidine. *Parasitology* 73, 265–273.

Chapman, H. (1997) Biochemical, genetic and applied aspects of drug resistance in *Eimeria* parasites of the fowl. *Avian Pathology* 26, 221–244.

Cuckler, A.C. and Malanga, C.M. (1955) Studies on drug resistance in Coccidia. *Journal of Parasitology* 41, 302–311.

Culleton, R., Martinelli, A., Hunt, P. and Carter, R. (2005) Linkage group selection: rapid gene discovery in malaria parasites. *Genome Research* 15, 92–97.

El-Sayed, N.M., Myler, P.J., Bartholomeu, D.C., Nilsson, D., Aggarwal, G., Tran, A.N., Ghedin, E., Worthey, E.A., Delcher, A.L., Blandin, G. *et al.* (2005) The genome sequence of *Trypanosoma cruzi*, etiologic agent of Chagas disease. *Science* 309, 409–415.

Fitz-Coy, S. (1992) Antigenic variation among strains of *Eimeria maxima* and *E. tenella* of the chicken. *Avian Diseases* 36, 40–43.

Gardner, M.J., Hall, N., Fung, E., White, O., Berriman, M., Hyman, R.W., Carlton, J.M., Pain, A., Nelson, K.E., Bowman, S. *et al.* (2002) Genome sequence of the human malaria parasite *Plasmodium falciparum*. *Nature* 419, 498–511.

Hodgson, J.N., Ball, S.J., Ryan, K.C. and Warren, E.W. (1969) The incidence of drug-resistant strains of *Eimeria* in chickens in Great Britain, 1966. *British Veterinary Journal* 125, 31–35.

Hunt, P., Martinelli, A., Fawcett, R., Carlton, J., Carter, R. and Walliker, D. (2004) Gene synteny and chloroquine resistance in *Plasmodium chabaudi*. *Molecular and Biochemical Parasitology* 136, 157–164.

Ivens, A.C., Peacock, C.S., Worthey, E.A., Murphy, L., Aggarwal, G., Berriman, M., Sisk, E., Rajandream, M.A., Adlem, E., Aert, R. *et al.* (2005) The genome of the kinetoplastid parasite, *Leishmania major*. *Science* 309, 436–442.

Jeffers, T.K. (1974) Genetic transfer of anticoccidial drug resistance in *Eimeria tenella*. *Journal of Parasitology* 60, 900–904.

Jeffers, T. (1976) Genetic recombination of precociousness and anticoccidial drug resistance in *Eimeria tenella*. *Zeitschrift fur Parasitenkunde* 50, 251–255.

Jeffreys, A.J., Wilson, V. and Thein, S.L. (1985) Hypervariable 'minisatellite' regions in human DNA. *Nature* 314, 67–73.

Joyner, L. (1957) Induced drug-fastness to nitrofurazone in a laboratory strain of *Eimeria tenella*. *Journal of Parasitology* 69, 1415.

Joyner, L.P. (1969) Immunological variation between two strains of *Eimeria acervulina*. *Parasitology* 59, 725–732.

Joyner, L. and Norton, C. (1975) Transferred drug-resistance in *Eimeria maxima*. *Parasitology* 71, 385–392.

Joyner, L. and Norton, C. (1976) Further observations on transferred drug resistance in *Eimeria maxima*. *Parasitology, Abstracts* 73, III.

Lee, E. and Fernando, M. (1978) Immunogenicity of a single sporocyst of *Eimeria maxima*. *Journal of Parasitology* 64, 483–485.

Lillehoj, H. (1988) Influence of inoculation dose, inoculation schedule, chicken age, and host genetics on disease susceptibility and development of resistance to *Eimeria tenella* infection. *Avian Diseases* 32, 437–444.

Long, P. and Millard, B. (1979) Immunological differences in *Eimeria maxima*: effect of a mixed immunizing inoculum on heterologous challenge. *Parasitology* 79, 451–457.

Long, P., Joyner, L., Millard, B. and Norton, C. (1976) A guide to laboratory techniques used in the study and diagnosis of avian coccidiosis. *Folia Veterinaria Latina* 6, 201–217.

Martin, A., Danforth, H., Barta, J. and Fernando, M. (1997) Analysis of immunological cross-protection and sensitivities to anticoccidial drugs among five geographical and temporal strains of *Eimeria maxima*. *International Journal for Parasitology* 27, 527–533.

Martinelli, A., Cheesman, S., Hunt, P., Culleton, R., Raza, A., Mackinnon, M. and Carter, R. (2005) A genetic approach to the *de novo* identification of targets of strain-specific immunity in malaria parasites. *Proceedings of the National Academy of Sciences USA* 102, 814–819.

Melby, P., Ogden, G., Flores, H., Zhao, W., Geldmacher, C., Biediger, N., Ahuja, S., Uranga, J. and Melendez, M. (2000) Identification of vaccine candidates for experimental visceral leishmaniasis by immunization with sequential fractions of a cDNA expression library. *Infection and Immunity* 68, 5595–5602.

Melby, P., Yang, J., Zhao, W., Perez, L. and Cheng, J. (2001) *Leishmania donovani* p36(LACK) DNA vaccine is highly immunogenic but not protective against experimental visceral Leishmaniasis. *Infection and Immunity* 69, 4719–4725.

Norton, C. and Hein, H. (1976) *Eimeria maxima*: a comparison of two laboratory strains with a fresh isolate. *Parasitology* 72, 345–354.

Norton, C. and Joyner, L. (1975) The development of drug-resistant strains of *Eimeria maxima* in the laboratory. *Parasitology* 71, 153–165.

Oxbrow, A. (1973) Strain specific immunity to *Plasmodium berghei*: a new genetic marker. *Parasitology* 67, 17–27.

Pain, A., Renauld, H., Berriman, M., Murphy, L., Yeats, C.A., Weir, W., Kerhornou, A., Aslett, M., Bishop, R., Bouchier, C. *et al.* (2005) Genome of the host-cell transforming parasite *Theileria annulata* compared with *T. parva*. *Science* 309, 131–133.

Profous-Juchelka, H., Liberator, P. and Turner, M. (1988) Identification and characterization of cDNA clones encoding antigens of *Eimeria tenella*. *Molecular and Biochemical Parasitology* 30, 233–241.

Richie, T. and Saul, A. (2002) Progress and challenges for malaria vaccines. *Nature* 415, 694–701.

Rollinson, D., Joyner, L. and Norton, C. (1979) *Eimeria maxima*: the use of enzyme markers to detect the genetic transfer of drug resistance between lines. *Parasitology* 78, 361–367.

Rose, M. and Long, P. (1962) Immunity to four species of *Eimeria* in fowls. *Immunology* 5, 79–92.

Shibui, A., Ohmori, Y., Suzuki, Y., Sasaki, M., Nogami, S., Sugano, S. and Watanabe, J. (2001) Effects of DNA vaccine in murine malaria using a full-length cDNA library. *Research Communications in Molecular Pathology and Pharmacology* 109, 147–157.

Shirley, M. and Bellatti, M. (1988) Live attenuated coccidiosis vaccine: selection of a second precocious line of *Eimeria maxima*. *Research in Veterinary Science* 44, 25–28.

Shirley, M. and Harvey, D. (1996) *Eimeria tenella*: infection with a single sporocyst gives a clonal population. *Parasitology* 112, 523–528.

Shirley, M. and Harvey, D. (2000) A genetic linkage map of the apicomplexan protozoan parasite *Eimeria tenella*. *Genome Research* 10, 1587–1593.

Shirley, M. and Hoyle, S. (1981) The antigenicity of *Eimeria maxima* populations obtained from commercial farms. *Journal of Parasitology* 67, 587–588.

Smith, A., Hesketh, P., Archer, A. and Shirley, M. (2002) Antigenic diversity in *Eimeria maxima* and the influence of host genetics and immunisation schedule on cross-protective immunity. *Infection and Immunity* 70, 2472–2479.

Smooker, P., Setiady, Y., Rainczuk, A. and Spithill, T. (2000) Expression library immunisation protects mice against challenge with virulent rodent malaria. *Vaccine* 18, 2533–2540.

Su, C., Howe, D.K., Dubey, J.P., Ajioka, J.W. and Sibley, L.D. (2002) Identification of quantitative trait loci controlling acute virulence in *Toxoplasma gondii*. *Proceedings of the National Academy of Sciences USA* 99, 10753–10758.

Sutton, C.A., Shirley, M.W. and McDonald, V. (1986) Genetic recombination of markers for precocious development, arprinocid resistance, and isoenzymes of glucose phosphate isomerase in *Eimeria acervulina. Journal of Parasitology* 72, 965–967.

Vos, P., Hogers, R., Bleeker, M., Reijans, M., van de Lee, T., Hornes, M., Frijters, A., Pot, J., Peleman, J., Kuiper, M. and Zabeau, M. (1995) AFLP: a new technique for DNA fingerprinting. *Nucleic Acids Research* 23, 4407–4414.

Walker, P. (1968) Investigational problems and the mechanisms of inheritance in blood protozoa. In: Weinman, D. and Ristic, M. (eds) *Infectious Blood Diseases of Man and Animals: Diseases Caused by Protista*. Academic Press, New York, pp. 367–391.

Wallach, M., Smith, N.C., Petracca, M., Miller, C.M., Eckert, J. and Braun, R. (1995) *Eimeria maxima* gametocyte antigens: potential use in a subunit maternal vaccine against coccidiosis in chickens. *Vaccine* 13, 347–354.

Warren, E., Ball, S. and MacKenzie, D. (1966) The incidence of drug-resistant strains of *Eimeria* species in chickens in Great Britain, 1964/65. *British Veterinary Journal* 122, 534–543.

Wellems, T.E., Walker-Jonah, A. and Panton, L.J. (1991) Genetic mapping of the chloroquine-resistance locus on *Plasmodium falciparum* Chromosome 7. *Proceedings of the National Academy of Sciences USA* 88, 3382–3386.

Xu, P., Widmer, G., Wang, Y., Ozaki, L.S., Alves, J.M., Serrano, M.G., Puiu, D., Manque, P., Akiyoshi, D., Mackey, A.J. *et al.* (2004) The genome of *Cryptosporidium hominis. Nature* 431, 1107–1112.

Part VI
Immunological and pathogen control

Chapter 18
Developments and pitfalls of feed acidification in controlling gut pathogens in poultry, with emphasis on Salmonella

F. van Immerseel,[1]* I. Gantois,[1] L. Bohez,[1] L. Timbermont,[1] F. Boyen,[1] I. Hautefort,[2] J.C.D. Hinton,[2] F. Pasmans,[1] F. Haesebrouck[1] and R. Ducatelle[1]

[1]Department of Pathology, Bacteriology and Avian Diseases, Research Group, Veterinary Public Health and Zoonoses, Faculty of Veterinary Medicine, Ghent University, Ghent, Belgium; [2]Molecular Microbiology Group, Institute of Food Research, Norwich, UK; *e-mail: filip.vanimmerseel@UGent.be

ABSTRACT

Acidification of drinking water and feed has been used for years in poultry. Drinking water acidification increases the threshold of bacterial infection due to the bactericidal activity of the acids added. Feed supplementation with powder products also increases the threshold of bacterial infection due to its bacteriostatic/bactericidal activity in the crop. For drinking water and powder form feed supplementation, the choice of the acids used is dependent on their ability to kill bacteria. Currently, coated or impregnated acid products are also on the market, with the aim of allowing acids to penetrate further down the gastrointestinal tract.

These products thus aim to change the ratio of organic acids in the gut, and thus the concentration changes achieved are more subtle. It was shown that exposure of Salmonella to low concentrations of acetic acid increased invasion of the bacterium in intestinal epithelial cells, while propionic and butyric acid decreased invasion at non-growth inhibitory concentrations.

This is due to changes in virulence gene expression of the genes of the Salmonella Pathogenicity Island I. Gene expression profiling by microarray technology has shown that low concentrations of butyric acid result in a specific downregulation of Salmonella Pathogenicity Island I genes, directly responsible for invasion, while regular metabolic genes are not altered in expression, making the butyric acid effect highly specific.

It was also shown that supplementation of acetic acid-coated products in the feed increased colonization of the chicken gut by *Salmonella*, and that coated propionic and butyric acid products decreased colonization of the chicken intestinal tract. Moreover, it was shown that coated butyric acid decreased faecal shedding in a seeder model in broilers throughout the rearing period up to slaughter age. The possibility of differential effects of organic acids on virulence of other pathogens and on performance of the animals is under research.

It was concluded that, instead of empirically using organic acid compounds as additives in feed and water of poultry, well-formulated and well-balanced preparations of organic acids should be developed, based on scientific data.

USE OF ORGANIC ACIDS IN POULTRY FEED

Organic acids have been widely used as feed and drinking water additives for various reasons in different farm animal species. Many products are on the market, consisting of mixtures of organic acids in different concentrations and formulations. Organic and short-chain fatty acid (SCFA) mixtures have been produced aiming at enhancing digestibility and diet palatability, thus improving feed conversion and growth rates of animals, including pigs and poultry. Acid mixtures are sometimes included in feed to control mould, and claims even for egg production increases using acids have been made.

When searching the scientific, peer-reviewed literature however, information on these potential beneficial characteristics is lacking and it is impossible to define the optimal acid mixture for achieving certain aims. Whatever the claims, it is clear that SCFA, medium-chain fatty acids (MCFA) and other organic acids have a more or less pronounced antimicrobial activity, depending on both the concentration of the acid and the bacterial species that is exposed to the acid.

Indeed, for certain acids there is a large difference in minimal inhibitory concentrations (MIC) when different bacterial species are exposed, while for certain bacterial species there are large differences in MIC values when these bacteria are exposed to different organic acids (Nakai and Siebert, 2002).

Some organic acids that are used as feed additives, however, do not seem to have any significant antimicrobial activity in the concentration range that is suitable for use in feed. Short-chain fatty acids are bacteriostatic or bactericidal *in vitro*, provided that there are sufficient undissociated acid molecules present and that they are in contact with the bacteria for a sufficiently long time (Thompson and Hinton, 1997). Lowering the pH increases the concentration of undissociated molecules. Therefore, the products will be more efficient at low pH.

This was proved in *in vitro* experiments by a decreased growth rate of *Salmonella* Typhimurium at decreasing pH levels in the presence of volatile fatty acids (VFA) (Durant *et al.*, 1999). The SCFA diffuse into the bacterial cell in undissociated form. Inside the bacterial cell the acid dissociates, resulting in reduction of intracellular pH and anion accumulation (Van der Wielen *et al.*, 2000). Concentrations of VFA sufficient to cause growth inhibition of *Escherichia coli in vitro* immediately slowed the rates of RNA, DNA, protein, lipid and cell wall synthesis (Cherrington *et al.*, 1990). Since synthesis of macromolecules other than DNA is not completely inhibited, and since there was no loss in membrane integrity, bacterial mass increased without cell division in the presence of VFA (Cherrington *et al.*, 1990, 1991).

Another common field of application for SCFA or other fatty acids is in pathogen control, especially *Salmonella*. This is the main focus of commercial products, claiming a (partial) control of shedding or colonization of the chicken gut. Drinking water additives are widely applied, and it is clear that when the concentration of certain acids in the drinking water is high, the pH of the drinking water will be sufficiently low to kill the bacteria that are present. This will thus create a barrier at the drinking water level, and recontamination of the birds due to the presence of pathogens in the drinking water (e.g. from the litter) can be controlled in this way to some extent. Powder-form feed additives have also been used widely to reduce *Salmonella* contamination.

The action of these compounds will not take place in the feed itself due to the lack of water for carrying out their activity. Once taken up by the animal they can become active. It is, however, the case that their action will most likely be limited to the crop and the stomach, since it has been shown that one cannot detect the acids further down in the gastrointestinal tract due to resorption (Thompson and Hinton, 1997).

More recently, new technologies have allowed the development of formulations that release the acids further down the gastrointestinal tract. These formulations include acids coated on to or impregnated into microbeads. Whether this really is the case remains to be seen, but these kinds of preparations are surely capable of bringing probiotic strains into the gut of animals (Kailasapathy, 2002).

Considering the above-mentioned data, it is thus advisable that research is carried out on the use of acids for a specific purpose. Preferably, the acids should be tested one by one prior to the study of the more complicated effects of mixtures. In this way one can try to find the optimal composition and formulation designed for achieving a specific aim.

In this chapter, details are given on the evaluation of the SCFA formic, acetic, propionic and butyric acid for their ability to control *Salmonella* in poultry. A study was conducted to assess the antimicrobial activity of *Salmonella* and the effects of the single SCFA on *Salmonella* virulence gene expression and invasion into intestinal epithelial cells. Furthermore, the effects of the different single acids on early colonization of the caeca and internal organs in young chickens were evaluated. The effect of the formulation in *Salmonella* control was tested using (powder form and impregnated) butyric acid, and long-term shedding of *Salmonella* was followed in broilers using butyric acid as a feed additive.

These data should provide information on how to optimize SCFA preparations for *Salmonella* control in poultry. Before detailed studies on the use of SCFA in *Salmonella* control are presented, an introduction to this food-contaminating pathogen and its pathogenesis is discussed, in order to underline the importance of having well-developed solutions for *Salmonella* control in poultry and to inform the reader about some important characteristics of the *Salmonella* bacteria, which will be used below in the discussion on the use of acids in *Salmonella* virulence.

SALMONELLA: AN IMPORTANT PATHOGEN IN POULTRY

Public health consequences

Salmonella is a foodborne pathogen for humans. Poultry is the main source of *Salmonella*, causing human infections. The emergence of S. Enteritidis has raised concern in the last two decades, due to its frequent transmission to humans and its ability to cause severe disease in man. A major problem is the tropism of S. Enteritidis for the reproductive organs of the hen and the contamination of eggs. Nevertheless poultry meat can also be contaminated.

In Europe, contamination of poultry is low in the Scandinavian countries but higher in the Mediterranean region. While contamination of breeders is generally low, a considerably high level of positive layer flocks is detected. In addition, Southern Europe has the highest percentage of positive flocks, ranging from 5 to 10%. Concerning table eggs, high percentages of *Salmonella*-positive table eggs are reported in Southern Europe (between 3 and 8%), while in the rest of Europe less than 1% of the eggs were positive.

The serotype Enteritidis is predominant in layers (57.7%) and in table eggs (72.9%) (European Commission, 2002). Broilers are also frequently contaminated by *Salmonella*. Scandinavian countries reported very low percentages of positive flocks, while Western European countries reported higher levels of contamination, with large differences between different countries, ranging between 1 and 11% in 2002.

Contamination levels of poultry meat are dramatically high in Europe. In 2002, in all European countries except Scandinavia, about 10–15% of the poultry meat at the retail level was positive. Also, Enteritidis is one of the most frequently isolated serotypes in broilers (10.8%) and in broiler meat (11.1%), next to a wide range of other serotypes (European Commission, 2002).

In Europe most human *Salmonella* isolates are currently of the serotype Enteritidis. In England and Wales S. Enteritidis isolates became the most frequent, with a percentage of 70.7% of human isolates in 1997. In recent years, fortunately, there has been a general downward trend in the number of human cases in Europe. In 2002, more than 145,000 cases of human salmonellosis were reported in 15 member states of the EU, compared with a figure of more than 200,000 cases in 1997. In the EU and Norway, about 70% of all isolates in 2002 were of the serotype Enteritidis and 17% were of serotype Typhimurium (European Commission, 2002), making it very clear that these serotypes need be controlled.

The intestinal phase of infections with broad host-range *Salmonella* serotypes

Salmonella bacteria can colonize a wide range of host species, dependent on the serotype. *Salmonella* serotypes that are exclusively associated with one particular host species are referred to as being host-restricted, such as *S. Gallinarum* in poultry. Serotypes that usually induce a self-limiting gastroenteritis in a broad range of unrelated host species, whilst being capable of inducing systemic disease in a wide range of host animals, are termed unrestricted or broad host-range serotypes (Uzzau *et al.*, 2000). Examples are *S.* Enteritidis and *S.* Typhimurium, thus these are the serotypes leading to human infections.

Chickens are usually infected by oral uptake of bacteria from the environment. *Salmonella* bacteria are able to survive gastric acidity and can therefore pass through the stomach to reach the intestinal tract of the animal (Kwon and Ricke, 1998). In the chicken, the caeca are the predominant sites of *Salmonella* colonization (Desmidt *et al.*, 1997, 1998). The bacteria can adhere to intestinal epithelial cells by specific adhesion–receptor interactions in which bacterial fimbriae and mannosylated residues on the mucous and resorptive epithelial cells are involved (Dibb-Fuller *et al.*, 1999; Vimal *et al.*, 2000).

A crucial step in the pathogenesis of *Salmonella* is the process of invasion into intestinal epithelial cells. Upon contact with intestinal epithelial cells, *Salmonella* bacteria inject a set of bacterial proteins into host cells via the bacterial Pathogenicity Island I (SPI-1) Type-3 secretion system (TTSS) (Zhou and Galán, 2001). Type-3 secretion systems are specialized, needle-like structures located on the bacterial membrane whose core function is the injection of bacterial proteins into eukaryotic cells (Hueck, 1998; Galán and Collmer, 1999; Cornelis and Van Gijsegem, 2000).

The Pathogenicity Island I of *Salmonella* is a large genetic element on the chromosome encoding multiple proteins, necessary to assemble a complex TTSS apparatus (structural proteins), regulatory proteins as well as some effector proteins, injected by the needle complex into intestinal epithelial cells (Lostroh and Lee, 2001). The effector proteins of SPI-1 that are injected in the epithelial cells interact with host proteins of the intestinal epithelial cells.

The main effect of the injected proteins is the rearrangement of the cytoskeleton of the intestinal epithelial cells, in such a way that bacteria are engulfed by ruffles on the host cell membrane, resulting in uptake by the epithelial cell. This is an active process mediated by the *Salmonella* proteins that are injected in the intestinal epithelial cells. It is called invasion.

The proteins that are encoded by the *Salmonella* Pathogenicity Island I also play a role in the attraction of immune cells to the gut wall. Macrophages may take up bacteria penetrating through the caecal mucosa, which is the start of the systemic phase of the infection. This systemic phase of the infection and the mechanism of egg contamination are not discussed here, but more information on these topics can be read in review articles by Hensel (2000) and De Buck *et al.* (2004).

Prevention and control

In 2003, the European Parliament and the Council of the European Union introduced Regulation (EC) No 2160/2003 to ensure that proper and effective measures were taken to detect and control *Salmonella* and other zoonotic agents at all relevant stages of production, processing and distribution, particularly at the level of primary production – including in feed – in order to reduce their prevalence and the risk they pose to public health.

National control programmes should provide for the detection of zoonoses and zoonotic agents in accordance with certain minimal sampling requirements. These programmes should also define the respective responsibilities of competent authorities and specify the control measures to be taken following the detection of zoonoses and zoonotic agents, in particular to protect public health. For example, with broilers, national control programmes should include a policy that fresh poultry meat may not be placed on the market for human consumption unless *Salmonella* is absent in 25 g, starting from December 2010. Thus the infection needs to be controlled.

Different measures are available to control *Salmonella* in poultry. It is not within the scope of this chapter to develop this topic in detail but, for the interested reader, reviews on the use of vaccines and feed additives for control of *Salmonella* in poultry are available (Van Immerseel *et al.*, 2002, 2005). In the text below, the use of SCFA in *Salmonella* control is discussed, based on data generated at the Department of Pathology, Bacteriology and Avian Diseases of the Faculty of Veterinary Medicine, Ghent University, Belgium.

SHORT-CHAIN FATTY ACIDS AND THEIR EFFECTS ON *SALMONELLA* VIRULENCE, WITH EMPHASIS ON INVASION

In vitro analysis of the antimicrobial characteristics of SCFA was performed by making growth curves at different pH values by Durant *et al.* (1999) and by Van Immerseel *et al.* (2003a) for *S.* Typhimurium and *S.* Enteritidis, respectively. Propionic acid was slightly more antimicrobial than were the other acids. Antimicrobial activity of the SCFA was limited, if present at all, at pH 7. It is concluded, therefore, that the antimicrobial activity against *Salmonella* cannot be used as a criterion in choosing an appropriate SCFA for use as a feed additive to control *Salmonella* in poultry.

Effects of pre-incubation of *S.* Enteritidis in growth medium, supplemented with the different SCFA, on invasion of *Salmonella* into intestinal epithelial cells were evaluated by different authors. Van Immerseel *et al.* (2003b) used human intestinal epithelial cells (T84 cell line) for this purpose. The cells were grown in cell culture plates until a confluent layer of cells was reached.

Simultaneously, for each SCFA, three Luria-broth (LB) solutions were made with the following concentrations: 25, 50 and 100 mM. Luria-broth medium without supplements was used as control. The pH of all solutions was brought to 6 and 7 by addition of NaOH and HCl, respectively. Osmolarity of all solutions was adjusted to the same value (610 mmol/kg) by addition of NaCl in solutions with lower osmolarity. This was done because differences in osmolarity influence invasion of *Salmonella* (Galán and Curtiss, 1990).

The bacteria were grown for 20 h in LB, whereafter the suspension was diluted 1:50 in the SCFA solutions, the compositions of which were described above. After 4 h of incubation at 37°C suspensions were centrifuged and diluted in cell culture medium. The cfu/ml value was determined by plating 6×20 μl of a dilution series of the suspensions on BGA, whereafter the plates were incubated overnight at 37°C. The suspensions were held at 4°C until they were used in the assay. The bacterial suspensions were diluted to a density of 5.10^6 cfu/ml.

The growth medium of the epithelial cells was then replaced by 1 ml of the diluted bacterial suspensions. This was centrifuged for 10 min at 1500 rpm, whereafter the plates were incubated for 1 h at 37°C and 5% CO_2. Then cells were rinsed three times, cell culture medium with gentamicin (50 μg/ml) was added to kill extracellular bacteria and the plates were incubated for 1 h at 37°C and 5% CO_2. Hereafter, cells were rinsed three times with PBS and lysed with 1% Triton X-100 in distilled water. From this lysate, a tenfold dilution series was made. From each dilution, 6×20 μl were inoculated on BGA, to determine the cfu/ml intracellular *S. enteritidis*.

As can be seen in Fig. 18.1, a decreased percentage of invaded bacteria was detected when bacteria were pre-incubated in propionate- and butyrate-supplemented media at pH 6, relative to that of bacteria pre-incubated in acetate- and formate-supplemented medium. These differences were significant ($p < 0.05$) in all cases within the same concentration of the VFA. The propionate- and butyrate-exposed bacteria also invaded significantly less ($p < 0.05$) than did the control bacteria. When the test was performed at pH 7, similar results were detected (data not shown).

The use of primary chicken caecal epithelial cells resulted in the same conclusions (Van Immerseel *et al.*, 2004). Moreover, unpublished data from our laboratory showed that the effects of the SCFA on *Salmonella* invasion are valid for multiple serotypes, including Hadar and Infantis.

This phenomenon was explained by changes in expression patterns of genes of the *Salmonella* Pathogenicity Island I (SPI-1). As already discussed above, the SPI-1 genes are directly involved in the invasive phenotype. The gene island consists of regulatory genes, genes encoding structural components of the needle complex and genes encoding effector proteins that are inserted in the eukaryotic cells. The hilA gene is the crucial regulator of the SPI-1 and is a transcriptional activator that can be influenced by numerous other regulatory proteins.

The molecular mechanism of acetate-induced increases in invasion has been elucidated. Lawhon *et al.* (2002) showed that acetate, after conversion to acetyl-phosphate, activates the BarA/SirA two-component system by phosphorylation of BarA and subsequently SirA. Activation of SirA results in enhanced transcription of hilA, the key regulator of SPI-1.

Our laboratory (Gantois *et al.*, 2006) showed that hilA is downregulated when *Salmonella* is grown in butyrate-supplemented growth media. This was shown when plasmids were constructed with transcriptional fusions between

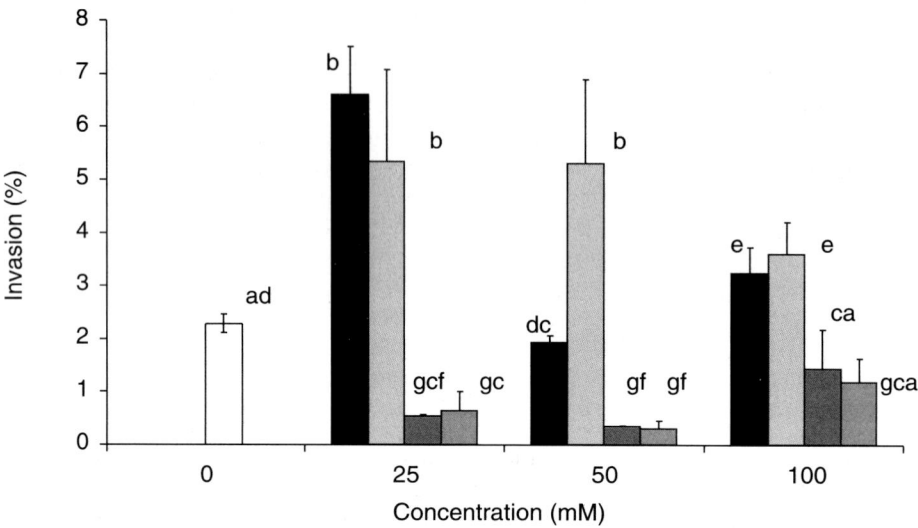

Fig. 18.1. *Salmonella* Enteritidis bacteria were pre-incubated for 4 h at 37°C in LB medium with short-chain fatty acids concentrations of 25, 50 and 100 mM. Column 1, formic acid; Column 2, acetic acid; Column 3, propionic acid; Column 4, butyric acid. Following this, an invasion assay was performed in 24-well cell culture plates. The pH of the growth medium was 6. Invasion was calculated as the percentage of invaded bacteria relative to the initial number of bacteria inoculated in one well of the cell culture plate (y-axis). Different concentrations of short-chain fatty acids are given in the x-axis. Values that do not share the same superscripts are statistically significant different from each other ($p < 0.05$).

the promoter of hilA and the reporter genes luxCDABE, and thus real-time luminescence measurement was used as a marker of hilA expression.

Furthermore, transcriptomic analysis of both *S.* Typhimurium and *S.* Enteritidis grown in media supplemented with low doses of butyric acid showed a specific downregulation of the whole SPI-1 without alterations in metabolic gene expression, compared with the control conditions. Also hilD, a positive regulator of hilA, and HU, a nucleoid-binding protein (and previously shown to be involved in virulence gene regulation), was downregulated (Gantois *et al.*, 2006).

These data suggest that butyric acid or one of its metabolic intermediates specifically interacted with one of the *Salmonella* proteins that regulate SPI-1. Most probably a simple molecular mechanism is involved, as for acetate, but the exact biochemical/molecular basis is still unclear.

To facilitate completion of the effects of SCFA on virulence, it has to be mentioned that pre-incubation of *Salmonella* with SCFA also influences acid resistance and intracellular survival in macrophages (Kwon and Ricke, 1998). For example, acid resistance is increased when the bacteria are in contact for a sufficiently long time with acetic acid. The exact relevance of these effects on *in vivo* colonization needs to be clarified.

Importance of invasion into intestinal epithelial cells *in vivo*

Since invasion of intestinal epithelial cells is influenced differently by the different SCFA, a study was carried out to assess the exact role of invasion in both the colonization of chicken gut and internal organs and shedding. The key regulatory protein of SPI-1, hilA, was therefore deleted in the *Salmonella* chromosome, and the mutant was shown to have lost its ability to invade epithelial cells *in vitro*. An *in vivo* study was thus carried out with the wild-type *S*. Enteritidis strain and its hilA mutant strain (Bohez *et al.*, 2006).

Therefore, 1-day-old chicks were randomly divided into two groups of 39 and orally inoculated with 10^8 colony-forming units (CFU) of *S*. Enteritidis 76Sa88 parent strain and *S*. Enteritidis ΔhilA mutant in a volume of 0.2 ml of LB broth. At days 2, 4, 6, 9, 15, 20 and 28 post-hatch, cloacal swabs were taken for the detection of *Salmonella*. Four chickens per group were euthanized at days 3, 4, 6, 9, 15, 20 and 28 post-hatch and samples of caeca, liver and spleen were taken for bacteriological analysis.

As shown in Fig. 18.2, cloacal swabs taken 1 day after infection revealed that all animals were positive after direct plating for both inoculated groups. All chicks infected with *S*. Enteritidis 76Sa88 parent strain shed the bacteria in faeces throughout the study. The number of animals shedding the ΔhilA mutant bacteria in the faeces was significantly lower compared with the number infected with the parent strain from day 5 post-inoculation onwards. The chickens did not shed the ΔhilA strain any more at 27 days post-inoculation.

Bacteriological analysis of the caeca showed peak values of about $\log_{10} 8$ CFU/g organ at 2 days post-infection for both *S*. Enteritidis 76Sa88 parent strain and *S*. Enteritidis ΔhilA mutant (Fig. 18.3). Subsequently, the numbers of the ΔhilA mutant bacteria in the caeca decreased. Statistically significant differences ($p < 0.005$) between both groups were observed from day 5 following infection, with the exception of day 8 post-infection. At 3 weeks post-infection, the parent strain still efficiently colonized the caeca (*c*. $\log_{10} 6$ CFU/g), whereas the birds infected with the ΔhilA mutant had largely cleared the infecting bacteria, as these were detected only after enrichment At the age of 28 days only two out of four chicks were positive at enrichment level (Bohez *et al.*, 2006).

It is thus concluded that intestinal epithelial cell invasion is important for long-term colonization of the chicken gut and the shedding pattern. Since SCFA have effects on invasion, this effect can have significance *in vivo*.

Use of short-chain fatty acids in *Salmonella* control *in vivo*

To study the use of SCFA to control *Salmonella in vivo* and to generate data on the differences between the acids with respect to *Salmonella* control, it is of utmost importance that these products are compared within the same experiment using standardized infection protocols.

In an experiment using SPF layer chickens, the different SCFA were compared using the micro-encapsulated form. For all products, 0.5% of micropearls was included in the feed, so that final in-feed concentrations of

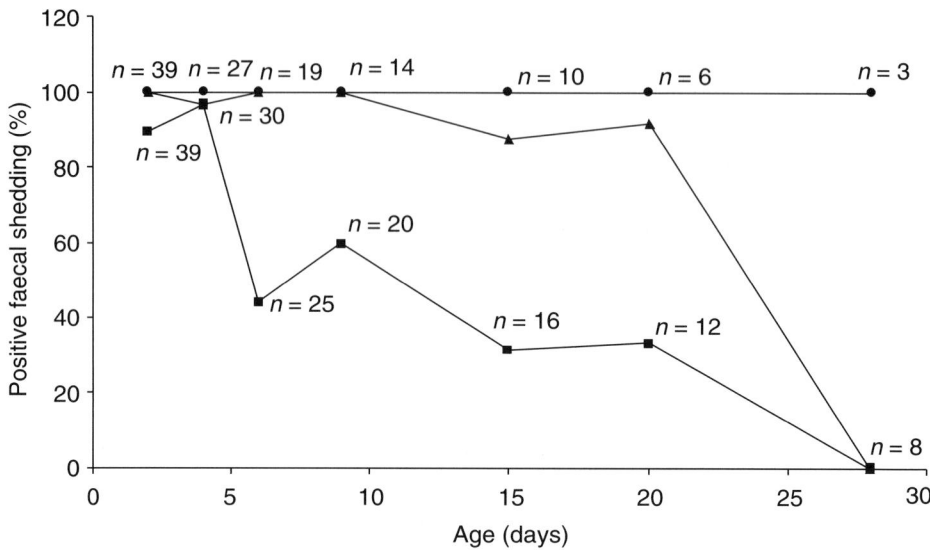

Fig. 18.2. Percentages of chickens having a *Salmonella*-positive cloacal swab. Animals were infected at day of hatch with 10^8 cfu *S.* Enteritidis 76Sa88 parent strain (● direct plating) and 10^8 cfu *S.* Enteritidis ΔhilA mutant (■ direct plating, ▲ after enrichment). The numbers of chickens are represented by *n*.

SCFA were 0.217% for formic acid, 0.242% for acetic acid, 0.275% for propionic acid and 0.156% for butyric acid. For this experiment, the chickens were randomly divided into five groups each of 20 chickens.

From the day of hatch, four groups received feed supplemented with the above-described feed additives, while one group received unsupplemented feed. The animals were orally inoculated with 2×10^3 cfu *S.* Enteritidis 76Sa88 on days 5 and 6. At day 7, cloacal swabs were taken from five animals per group to detect *Salmonella* bacteria. At day 8, chickens were euthanased by intravenous T61 injection. Samples of caecum, liver and spleen were taken for bacteriological analysis. One day after the second infection, 4/5 animals of the group that had received acetic acid as feed supplement had positive cloacal swabs, compared with none of the animals in any of the other groups.

Table 18.1 shows the number of animals classified according to caecal bacterial count of *Salmonella* (log-scaled intervals). In the control group, the caeca of 6/20 animals were positive only after enrichment, while 9/20 animals had $> 10^6$ cfu/g *Salmonella* in their caeca. In the group that had received formic acid as feed supplement, 10/20 animals had $> 10^6$ cfu/g *Salmonella* in the caeca, and only one animal was positive after enrichment. Thus, a higher number of animals in the formic acid-treated group had intermediate *Salmonella* titres, compared with the control animals.

In the group receiving acetic acid as feed supplement, 16/20 animals had a *Salmonella* count of $> 10^6$ cfu/g, from which 13 had $> 10^7$ cfu/g caecum. In

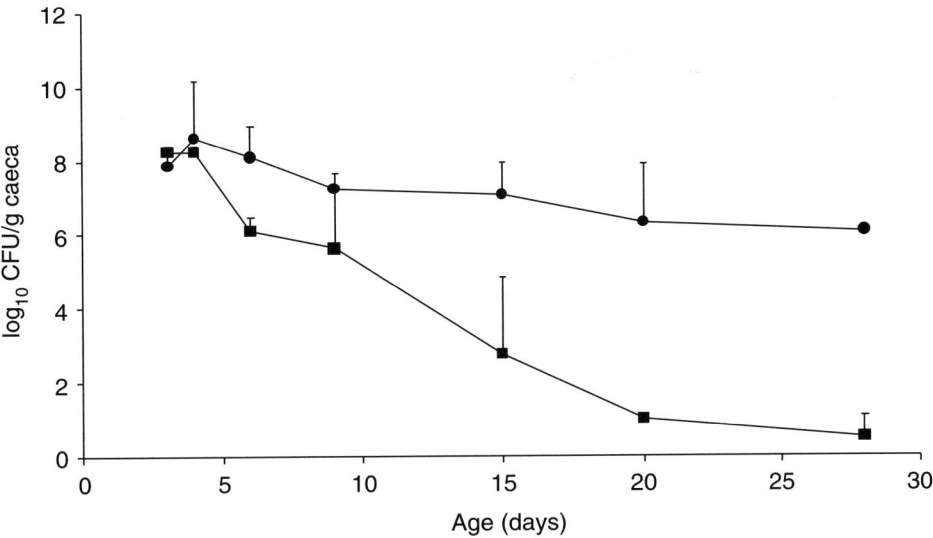

Fig. 18.3. Log_{10} cfu of *S.* Enteritidis 76Sa88 parent strain (●) and *S.* Enteritidis ΔhilA mutant (■) in the caeca of chicks dosed orally at 1 day of age and then euthanased at different time intervals. Each time point represents four chicks, with the exception of three chicks at days 20 and 28 in the *S.* Enteritidis 76Sa88 parent strain-infected group.

the propionic acid-supplemented group, seven animals had $> 10^6$ cfu of *Salmonella*/g caecum, while 8/20 animals were only positive after enrichment. Finally, in the group receiving butyric acid-impregnated microbeads, no animals had a caecal *Salmonella* count of $> 10^6$ cfu/g, while 11/20 animals were positive only after enrichment.

Table 18.2 shows the number of animals classified according to bacterial counts of *Salmonella* in liver and spleen. In the control and the butyric acid

Table 18.1. Colonization of the caeca at day 8 of life (inoculation with 10^3 cfu *S.* Enteritidis 76Sa88 on days 5 and 6) in chickens fed a diet supplemented with formic, acetic, propionic and butyric acids or no feed additives (control). Figures represent numbers of chickens in a group of 20 having a given number of *Salmonella* bacteria in the caeca.

	Control[a]	Formic acid[a]	Acetic acid[b]	Propionic acid[ac]	Butyric acid[c]
Negative	0	0	0	0	0
Positive after enrichment	6	1	1	8	11
10^2–10^3 cfu/g	0	1	1	1	2
10^3–10^4 cfu/g	0	4	0	1	1
10^4–10^5 cfu/g	3	2	0	2	3
10^5–10^6 cfu/g	2	2	2	1	3
10^6–10^7 cfu/g	8	7	3	3	0
$> 10^7$ cfu/g	1	3	13	4	0

[a,b,c] Groups not having equal superscripts are statistically significantly different.

Table 18.2. Colonization of the liver and spleen at day 8 of life (inoculation with 10^3 cfu *S.* Enteritidis 76Sa88 on days 5 and 6) in chickens fed a diet supplemented with formic, acetic, propionic and butyric acids or no feed additives. Figures represent numbers of chickens in a group of 20 having a given number of *Salmonella* bacteria in the liver (L) and spleen (S).

	Control		Formic acid		Acetic acid		Propionic acid		Butyric acid	
	L^a	S^a	L^{bd}	S^b	L^b	S^c	L^{ad}	S^a	L^a	S^a
Negative	0	8	0	0	0	0	1	5	0	7
Positive after enrichment	19	10	11	9	8	3	14	10	20	12
10^2–10^3 cfu/g	0	0	6	4	6	4	5	5	0	1
10^3–10^4 cfu/g	1	1	3	6	3	6	0	0	0	0
10^4–10^5 cfu/g	0	1	0	1	2	5	0	0	0	0
> 10^5 cfu/g	0	0	0	0	1	2	0	0	0	0

a,b,c,d Within each organ, groups not having equal superscripts are statistically significantly different.

groups only a few animals had a bacterial count of > 10^2 cfu/g of spleen, but five animals in the propionic acid group had *Salmonella* numbers between 10^2 and 10^3 cfu/g spleen. In the formic acid group, 11/20 animals had a *Salmonella* count of > 10^2 cfu/g, of which six had a count of > 10^3 cfu/g and one of more than 10^4 cfu/g. In the acetic acid group, 17/20 animals had > 10^2 cfu/g spleen, of which 13 had > 10^3 cfu/g, seven more than 10^4 cfu/g and two animals had even more than 10^5 cfu/g spleen. For the liver, comparable results were obtained.

Mean log cfu/g of caeca, liver and spleen for the different animal groups are shown in Fig. 18.4. The mean log cfu *S.* Enteritidis/g caecum was significantly higher in the acetic acid-treated group than in all other groups. The butyric acid-treated group had a significantly lower mean log cfu/g caecum than all other groups, except for the propionic acid-treated group. For liver and spleen, the acetic and formic acid-treated groups had significantly higher values of mean log cfu/g compared with all other groups.

Butyric acid thus decreases the invasion of *Salmonella in vitro* and decreases *Salmonella* colonization of the caeca in chickens early after infection of 1-day-old chicks. It was not yet known whether the formulation of the butyric acid feed additive affects the protective effect. Therefore, an infection experiment was performed with young chickens to study the differences between powdered and coated butyric acid as protection against *Salmonella* colonization of the caeca and internal organs of chickens.

In an experimental study, SPF chickens were randomly divided into four groups, each of 25 chickens. From the day of hatch, three groups received feed supplemented with different additives. The first group received the powdered form of butyric acid at a concentration of 0.63 g/kg. The second group received the coated product at a concentration of 2.5 g/kg. The third group received 0.315 g/kg of the powdered and 1.25 g/kg of the coated product.

All three groups thus received the same amount of the active product butyric acid (or sodium salt) i.e. 0.63g/kg. One group received unsupplemented

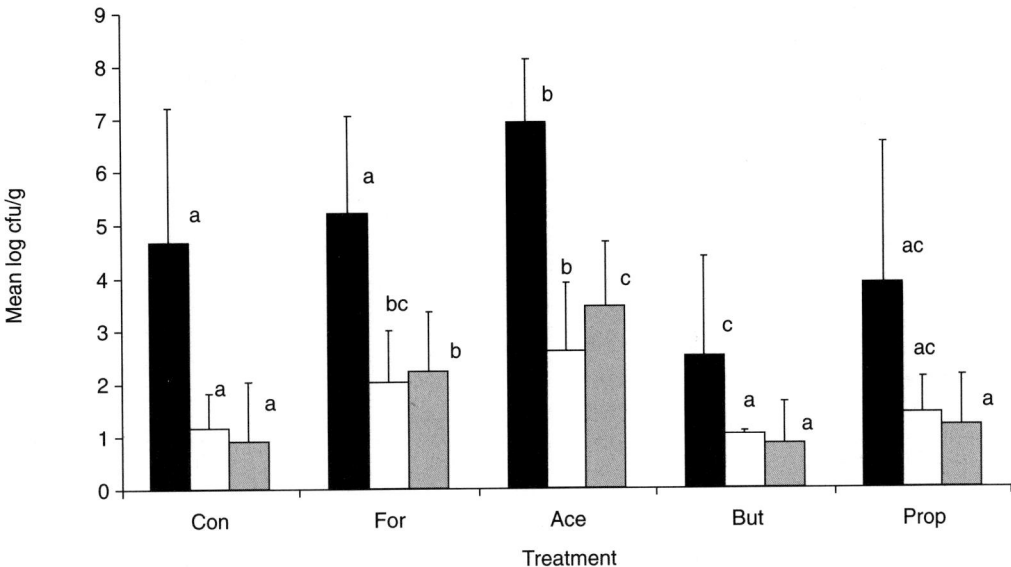

Fig. 18.4. The mean log cfu/g in caeca (black), liver (white) and spleen (grey) at day 8 of life of chickens that were orally inoculated with 10^3 cfu *S*. Enteritidis 76Sa88 on days 5 and 6. Chickens received diets supplemented with formic acid (For), acetic acid (Ace), propionic acid (Prop), butyric acid (But) and non-supplemented feed (Con). Between conditions, for each organ, values not having the same superscripts are statistically significantly different.

feed. The animals were orally inoculated with 10^6 cfu *S*. Enteritidis 76Sa88 on day 5. At day 6, cloacal swabs were taken to detect *Salmonella* bacteria. At day 8, chickens were euthanased and samples of caecum, liver and spleen were taken for bacteriological analysis.

Cloacal swabs taken 1 day post-infection showed that 10/25 animals had shed *Salmonella* in the control group, 8/25 animals in the group receiving the powdered form of butyric acid, 4/25 in the group receiving coated butyric acid and 3/25 in the group receiving the combined product as feed additive.

Caecal counts revealed more distinct differences. As can be seen from Table 18.5, in the group of animals receiving the coated form of butyric acid or the combination of the powdered and coated forms, a small number of animals had $> 10^6$ cfu/g caecum (4 and 3/25, respectively). The control group and the group receiving the powdered form of butyric acid had a rather higher number of animals with $> 10^6$ cfu/g caecum (10 and 12/25, respectively). Colonization of the caeca by groups that had received coated butyrate or the combined powdered/coated form was significantly lower than either the control group or the group that had received the powdered form of butyrate.

Concerning splenic colonization, there was a significantly lower colonization of the group that had received the combined butyrate product as feed additive compared with the control group or the group that had received

Table 18.3. Colonization of the caeca at day 8 of life (inoculation with 10^6 cfu S. Enteritidis 76Sa88 on day 5) in chickens fed a diet supplemented with butyric acid in powdered form, coated form, a combination of half-doses of powdered and coated forms (Combi) or no feed additives (Control). Concentration of the active product butyric acid was 0.63 g/kg feed in each group. Figures represent numbers of chickens in a group of 25 having a given number of Salmonella bacteria in the caeca.

	Control[a]	Powdered[a]	Coated[b]	Combi[b]
Negative	0	0	2	1
Positive after enrichment	6	8	12	8
$< 10^4$ cfu/g	2	1	3	2
10^4–10^5 cfu/g	3	1	0	6
10^5–10^6 cfu/g	4	3	4	5
$> 10^6$ cfu/g	10	12	4	3

Groups with different superscripts are significantly different.

the powdered form of butyrate (Table 18.4). Liver colonization was similar (data not shown).

From the data above, it can be concluded that butyric acid, impregnated in microbeads, decreases colonization of Salmonella in poultry and also reduces shedding. In these trials colonization was analysed only shortly after infection in layer-type chicks. In a further study the protective effect in broilers was examined and this was performed over a longer period.

For this, Ross broiler chickens were randomly divided into two groups of 50 chickens each. From the day of hatch, one of the groups received a wheat-based starter diet until day 10, after which the diet was changed to a wheat-based grower diet. In the other group the feed was supplemented with 2.5 g/kg

Table 18.4. Colonization of the spleen at day 8 of life (inoculation with 10^6 cfu S. Enteritidis 76Sa88 on day 5) in chickens fed a diet supplemented with butyric acid in powdered form, coated form, a combination of half-doses of powder and coated form (Combi) or no feed additives (Control). Concentration of the active product butyric acid was 0.63 g/kg feed in each group. Figures represent numbers of chickens in a group of 25 that were negative or had a given amount of Salmonella bacteria in the spleen.

	Control[a]	Powdered[a]	Coated[a]	Combi[b]
Negative	12*	10	12	15
Positive after enrichment	2	1	4	7
$< 10^3$ cfu/g	3	3	1	2
10^3–10^4 cfu/g	6	10	4	1
$> 10^4$ cfu/g	2	1	4	0

[a,b] Groups not having equal superscripts are statistically significantly different ($p < 0.05$).

of the coated butyric acid product. Ten out of 50 animals of both groups were orally inoculated with 10^5 cfu S. Enteritidis 76Sa88 at day 5 post-hatch. On days 6, 9, 13, 20, 27, 34 and 41 cloacal swabs were taken for bacteriological analysis. At day 42, the animals were killed by intravenous injection of an anaesthetic and samples of caeca were taken for bacteriological analysis.

In the control group, the percentage of chickens having *Salmonella*-positive cloacal swabs after direct plating increased to 35–40% at 1 and 3 weeks after seeder bird inoculation (days 12 and 27 of age). At slaughter, however, only 4.8% of birds were positive in direct plating. In the group of animals that had received butyric acid as feed additive, 15% of the animals had *Salmonella*-positive cloacal swabs after direct plating 1 and 3 weeks after seeder bird inoculation (days 12 and 27 of age), whereas the percentage of animals having *Salmonella*-positive cloacal swabs after direct plating decreased to 2.2% at slaughter age (Fig. 18.5). Differences between both groups were statistically significant at day 9 ($p < 0.1$) and days 12 and 27 ($p < 0.05$).

Analysis of total numbers of positive cloacal swabs (including those positive at enrichment level) showed an increase to 97% in the proportion of chickens having *Salmonella*-positive cloacal swabs at day 20 of age (day 15 post-inoculation of seeders) in the control group. Thereafter, a decrease to 36% was observed by slaughter age. In the group of animals that received butyric acid as feed additive the percentage of chickens having *Salmonella*-positive cloacal swabs increased to 48% at day 20 (day 15 post inoculation of seeders) and then decreased to 6% at slaughter age. Differences between both groups were statistically significant at all time points starting from day 9 of age (day 4 post-inoculation of seeder birds; $p < 0.05$).

Bacteriological analysis of caeca at slaughter age showed 68% of the animals as being positive for *Salmonella* at enrichment level (no direct positives) in both the control group and the group that had received butyric acid as a feed additive.

CONCLUSION

Organic acids, and in particular SCFA, are currently being used as feed additives in a wide range of applications. For most applications, combinations of acids are empirically chosen and placed on the market. Data presented in this chapter clearly illustrate that, for *Salmonella* control, both the nature of the acid and the formulation of the feed additive play a significant role. Indeed, while butyric acid decreased virulence gene expression and invasion of *Salmonella* in intestinal epithelial cells, acetic acid increased both virulence gene expression and invasion.

When coated organic acids were used as feed additives, the differences between acetic and butyric acid concerning effects on the potential of *Salmonella* to invade epithelial cells were reflected in differences in colonization of the caeca in chickens. Indeed, coated acetic acid in the feed resulted in a more than a 10,000-fold increase in the number of bacteria colonizing the caeca in comparison with coated butyric acid, and 100-fold increases

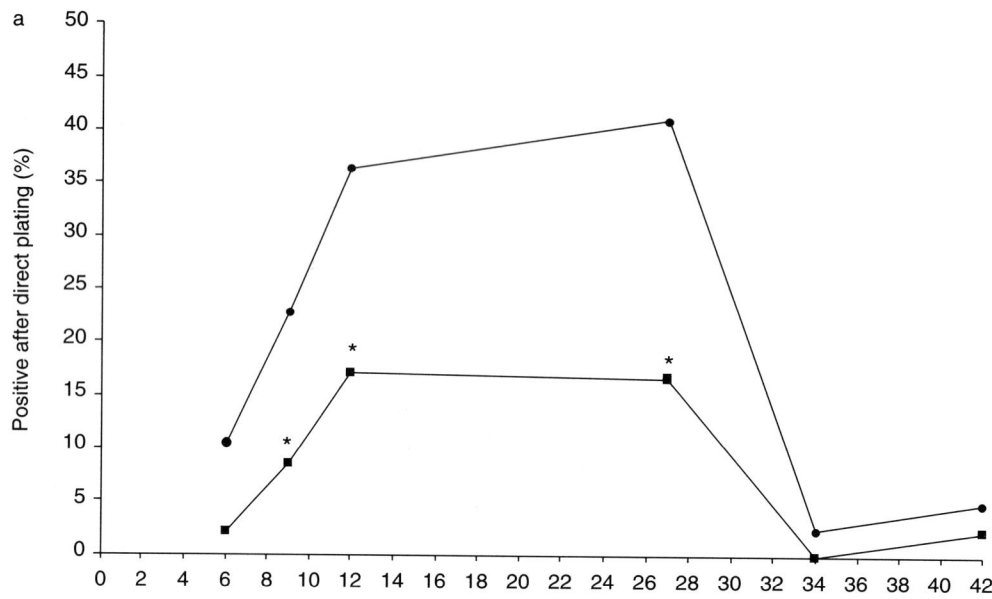

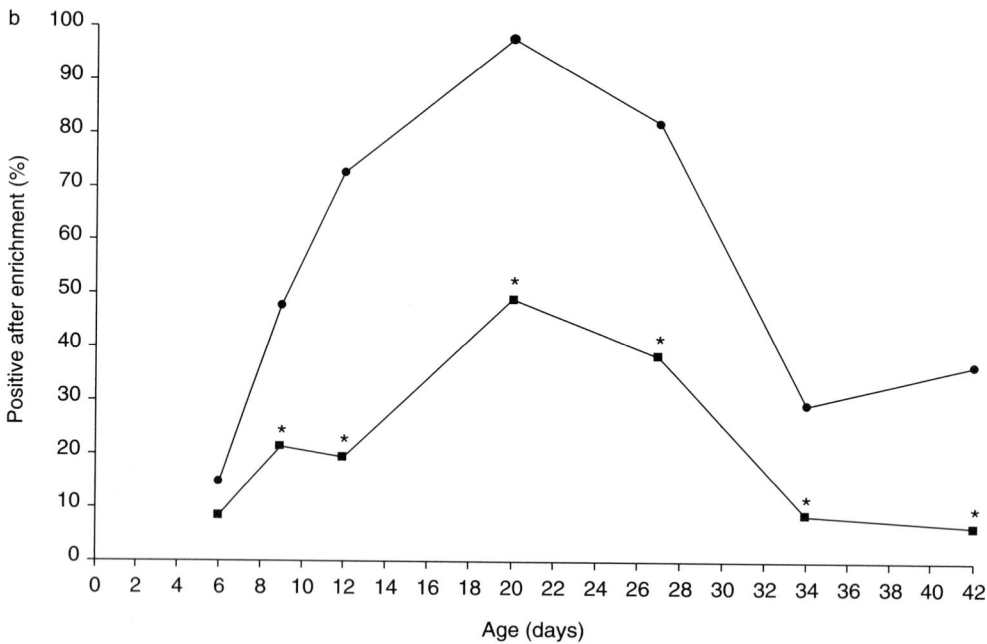

Fig. 18.5. The number of positive cloacal swabs during time (days) is shown after direct inoculation on brilliant green agar (a) and after enrichment (b). Infection was carried out using a seeder model, infecting ten animals with 10^5 cfu *S*. Enteritidis 76Sa88 at day 5 post-hatch, and housing these animals with 40 non-inoculated broilers. Animals were given feed supplemented with coated butyric acid at 0.63 g/kg butyric acid (■) or unsupplemented feed (●). * Denotes significant differences between groups.

compared with control animals. In the powdered form, butyric acid did not aid in controlling *Salmonella* colonization when animals were infected orally, in contrast to the coated form.

This clearly showed that the formulation is a very important feature that has to be optimized. Although the formulation of feed additives in control can certainly still be improved, it seems to be the case that coated propionic or butyric acid should be included in these preparations. Acetic acid needs to be avoided for this purpose, based on the above-mentioned results.

One aspect that certainly needs further attention in the future is the role of the microbiota in the endogenous production of SCFA and the way in which gut microbiota can be shifted to become butyric acid producing. It seems to be the case that *Salmonella* takes advantage of the large amount of acetic acid in the gut and responds with increased invasion of the gut cells. Butyric acid fermentation can be improved by manipulation of the feed, and by the addition of proper prebiotic substrates for certain microbiotic populations.

As an alternative to the use of SCFA preparations in the feed, simple feed composition changes could be used to improve *Salmonella* control. Also, the impact of SCFA on the virulence of other bacterial pathogens, such as *Clostridium perfringens*, needs to be clarified.

Finally, the effect of changes in acid composition in the gut on the gut epithelial cells can also have an influence on the outcome of an infection and on performance parameters. It has been shown in humans and mice that intestinal epithelial cell proliferation is increased and that gut damage is resolved faster when butyric acid is present, thus improving digestibility (Hodin, 2000).

REFERENCES

Bohez, L., Ducatelle, R., Pasmans, F., Botteldoorn, N., Haesebrouck, F. and Van Immerseel, F. (2006) *Salmonella* enterica serovar enteritidis colonization of the chicken caecum requires the hila regulatory protein. *Veterinary Microbiology* 116, 202–210.

Cherrington, C.A., Hinton, M. and Chopra, I. (1990) Effect of short-chain organic acids on macromolecular synthesis in *Escherichia coli*. *Journal of Applied Bacteriology* 68, 69–74.

Cherrington, C.A., Hinton, M., Pearson, G. and Chopra, I. (1991) Short-chain organic acids at pH 5.0 kill *Escherichia coli* and *Salmonella* spp. without causing membrane perturbation. *Journal of Applied Bacteriology* 70, 161–165.

Cornelis, G.R. and Van Gijsegem, F. (2000) Assembly and function of type III secretory systems. *Annual Reviews in Microbiology* 54, 735–774.

De Buck., J., Van Immerseel, F., Haeseboruck, F. and Ducatelle, R. (2004) Colonization of the chicken reproductive tract and egg contamination by *Salmonella*. *Journal of Applied Microbiology* 97, 233–245.

Desmidt, M., Ducatelle, R. and Haesebrouck, F. (1997) Pathogenesis of *Salmonella* Enteritidis phage type four after experimental infection of young chickens. *Veterinary Microbiology* 56, 99–109.

Desmidt, M., Ducatelle, R. and Haesebrouck, F. (1998) Immunohistochemical observations in the caeca of chickens infected with *Salmonella* Enteritidis phage type four. *Poultry Science* 77, 73–74.

Dibb-Fuller, M.P., Allen-Vercoe, E., Thorns, C.J. and Woodward, M.J. (1999) Fimbriae- and flagella-mediated association with and invasion of culture epithelial cells by *Salmonella* Enteritidis. *Microbiology* 145, 1023–1031.

Durant, J.A., Lowry, V.K., Nisbet, D.J., Stanker, L.H., Corrier D.E. and Ricke, S.C. (1999) Short-chain fatty acids affect cell-association and invasion of Hep-2 cells by *Salmonella* Typhimurium. *Journal of Environmental Science and Health B* 34, 1083–1099.

European Commission (2002) Trends and sources of zoonotic agents in animals, feedingstuffs, food and man in the European Union and Norway in 2002. *Health and Consumer Protection Directorate-General. Salmonella and Food-borne Diseases – Zoonoses Reports for 2002.*

Galán, J.E. and Collmer, A. (1999) Type III secretion machines: bacterial devices for protein delivery into host cells. *Science* 284, 1322–1328.

Galán, J.E., Curtiss, R. 3rd. (1990) Expression of *Salmonella* Typhimurium genes required for invasion is regulated by changes in DNA supercoiling. *Infection and Immunity* 58, 1879–1885.

Gantois, I., Ducatelle, R., Pasmans, F., Haesebrouck, F., Hautefort, I., Thompson, A., Hinton, J.C. and Van Immerseel, F. (2006) Butyrate specifically down-regulates *Salmonella* pathogenicity island 1 gene expression. *Applied Environmental Microbiology* 72, 946–949.

Hensel, M. (2000) *Salmonella* Pathogenicity Island 2. *Molecular Microbiology* 36, 1015–1023.

Hodin, R. (2000) Maintaining gut homeostasis: the butyrate NF-κB connection. *Gastroenterology* 118, 798–801.

Hueck, C.J. (1998) Type III secretion systems in bacterial pathogens of animals and plants. *Microbiology and Molecular Biology Reviews* 62, 379–433.

Kailasapathy, K. (2002) Microencapsualtion of probiotic bacteria: technology and potential applications. *Current Issues in Intestinal Microbiology* 32, 39–48.

Kwon, Y.M. and Ricke, S.C. (1998) Induction of acid resistance of *Salmonella* Typhimurium by exposure to short-chain fatty acids. *Applied and Environmental Microbiology* 64, 3458–3463.

Lawhon, S.D., Maurer, R., Suyemoto, M. and Altier, C. (2002) Intestinal short-chain fatty acids alter *Salmonella* Typhimurium invasion gene expression and virulence through BarA/SirA. *Molecular Microbiology* 46, 1451–1461.

Lostroh, C.P. and Lee, C.A. (2001) The *Salmonella* Pathogenicity Island-1 type III secretion system. *Microbes and Infection* 3, 1281–1291.

Nakai, S.A. and Siebert, K.J. (2002) Validation of bacterial growth inhibition models based on molecular properties of organic acids. *International Journal of Food Microbiology* 2678, 1–7.

Thompson, J.L. and Hinton, M. (1997) Antibacterial activity of formic acid and propionic acid in the diet of hens on *Salmonellae* in the crop. *British Poultry Science* 38, 59–65.

Uzzau, S., Brown, D.J., Wallis, T., Rubino, S., Leori, G., Bernard, S., Casadesús, J., Platt, D.J. and Olsen, J.E. (2000) Host-adapted serotypes of *Salmonella enterica*. *Epidemiology and Infection* 125, 229–255.

Van der Wielen, P.W., Biesterveld, S., Notermans, S., Hofstra, H., Urlings, B.A. and van Knapen, F. (2000) Role of volatile fatty acids in development of the cecal microflora in broiler chickens during growth. *Applied and Environmental Microbiology* 66, 2536–2540.

Van Immerseel, F., Cauwerts, K., Devriese, L.A., Haesebrouck, F. and Ducatelle, R. (2002) Feed additives to control *Salmonella* in poultry. *World's Poultry Science Journal* 58, 501–513.

Van Immerseel, F., De Buck, J., Pasmans, F., Velge, P., Bottreau, E., Fievez, V., Haesebrouck, F. and Ducatelle, R. (2003a) Invasion of *Salmonella* Enteritidis in avian intestinal epithelial cells *in vitro* is influenced by short-chain fatty acids. *International Journal of Food Microbiology* 85, 237–248.

Van Immerseel, F., De Buck, J., Pasmans, F., Rychlik, I., Volf, J., Sevcik, M., Haesebrouck, F. and Ducatelle, R. (2003b) The choice of volatile fatty acids for supplementation of poultry feed to reduce *Salmonella*. *14th European Symposium on Poultry Nutrition,* Lillehammer, Norway, pp. 160–161.

Van Immerseel, F., De Buck, J., De Smet, I., Pasmans, F., Haesebrouck, F. and Ducatelle, R. (2004) Interaction of butyric and acetic acid-treated *Salmonella* with chicken primary caecal epithelial cells *in vitro*. *Avian Diseases* 48, 384–391.

Van Immerseel, F., Methner, U., Rychlik, I., Nagy, B., Velge, P., Martin, G., Foster, N., Ducatelle, R. and Barrow, P.A. (2005) Vaccination and early protection against non-host-specific *Salmonella* serotypes in poultry; exploitation of innate immunity and microbial metabolic activity. *Epidemiology and Infection* 133, 959–978.

Vimal, D.B., Khullar, M., Gupta, S. and Ganguly, N.K. (2000) Intestinal mucus: the binding sites for *Salmonella* Typhimurium. *Molecular and Cellular Biochemistry* 204, 107–117.

Zhou, D. and Galán, J. (2001) *Salmonella* entry into host cells: the work in concert of type III secreted effector proteins. *Microbes and Infection* 3, 1293–1298.

CHAPTER 19
Competitive exclusion in poultry production

C. Schneitz

Orion Corporation, Orion Pharma Animal Health, Turku, Finland;
e-mail: carita.schneitz@orionpharma.com

ABSTRACT

The concept of Competitive Exclusion or CE, sometimes also described as the 'Nurmi concept', involves treatment of newly hatched chickens with mixed bacterial preparations originating from the gut of healthy adult birds to control intestinal disturbances. The treatment has also been given successfully to older birds after therapeutic doses of antibiotics to regenerate the intestinal microflora. CE treatment is usually applied by spraying, but can also be given via the first drinking water.

The concept of CE was originally developed to control *Salmonella* in growing chickens but has, with time, been expanded to involve several other enteropathogens like chicken and human pathogenic *Escherichia coli*, *Clostridium perfringens*, *Listeria* and *Campylobacter*. The chicken intestinal flora has also been shown to be efficacious in other avian species like turkey, quail and pheasant. The mechanisms by which the intestinal bacteria inhibit the invasion and proliferation of enteropathogens include competition for receptor sites, elaboration of inhibitory compounds and competition for nutrients or stimulation of the immune system.

In addition to pathogen control, it has been shown in field studies that CE treatment enhances the growth and decreases the mortality of birds and improves the feed conversion. The nutritional effects of the treatment have been confirmed in well-controlled laboratory studies. Increased bacterial resistance to antibiotics in humans has caused an increase in public and official interest in abandonment of the use of growth-promoting antibiotics. Competitive exclusion is a reliable alternative because the treatment is fully biological, it does not leave any residues and only one treatment on the day of hatch is normally enough.

© CAB International 2006. *Avian Gut Function in Health and Disease* (ed. G.C. Perry)

INTRODUCTION

In 1971 the Finnish broiler industry suffered from a severe outbreak of *Salmonella infantis*. More than 80% of the broiler flocks were *Salmonella* positive. At the same time the incidence of human cases caused by the same serotype increased dramatically. The problem with *Salmonella*-infected poultry flocks could not be solved using antibiotics, so research was started at the National Veterinary and Food Research Institute under the leadership of Professor Esko Nurmi to find a different approach.

In 1952 Milner and Schaffer had infected chicks with *S.* Typhimurium and observed that the resistance of young chickens to *Salmonella* increased with age, but the researchers had no explanation for this. The theory of Nurmi was that newly hatched chickens are sensitive to infection by *Salmonella* and other enteropathogens because of delayed establishment of normal intestinal microflora due to lack of contact with the brooding hen. Nurmi and his research group were the first to demonstrate experimentally that administration of intestinal contents from a healthy adult hen to newly hatched chickens protected the latter from colonization by *Salmonella*. This finding was published in *Nature* in 1973 (Nurmi and Rantala), and triggered worldwide research around this subject.

The competitive exclusion concept can be defined as 'the early establishment of an adult-type intestinal microflora to prevent subsequent colonization by enteropathogens'. The concept was originally designed to prevent the colonization of foodborne *Salmonellae* in the gut of young chickens but has, with time, been extended also to cover other human and poultry pathogens like chicken and human pathogenic *E. coli*, *Campylobacter*, *Cl. perfringens* and *Listeria*. Improvement in bird performance has also been shown in well-controlled laboratory-scale studies and also in the field (reviewed by Pivnick and Nurmi, 1982; Schleifer, 1985; Mead and Impey, 1987; Stavric and D'Aoust, 1993; Mead, 2000; Schneitz and Mead, 2000; Schneitz, 2005).

In this chapter, the development and applicability of the CE concept in poultry production – as well as factors affecting the efficacy of the treatment in field conditions – are reviewed and discussed.

THE DEVELOPMENT OF THE CE CONCEPT

The phenomenon by which the mature intestinal microflora protects the host against invading pathogens is called competitive exclusion (CE), or the 'Nurmi concept' (Pivnick and Nurmi, 1982). Greenberg was the first to use the term 'competitive exclusion', in 1969 (Greenberg, 1969). He studied the intestinal flora of maggots of blowflies and claimed that competitive exclusion of *S.* Typhimurium from the maggots was so effective that the organism survived in the gut only if the normal microbiota was simplified or eliminated.

Luckey (1963) had demonstrated a similar phenomenon in higher animals. Colonization resistance, a term synonymous with CE, was introduced by van der Waaij *et al.* in 1971 when examining the intestinal flora of mice. The term competitive exclusion was used in relation to poultry for the first time by Lloyd *et al.* in 1974.

Mixed cultures

In their first study, Nurmi and Rantala used diluted material from the crop, small intestine and caeca of adult, healthy cocks (Nurmi and Rantala, 1973). The results of this study showed that 1-day-old chicks became very well protected against relatively high doses of *S. infantis*. Later, the same workers successfully used broth cultures of the alimentary tract of adult chickens (Rantala and Nurmi, 1973).

Although little is known about the essential properties of the protective organisms, the one certainty is that protection depends upon the oral administration of viable bacteria (Mead, 2000). According to Rantala (1974), the protective material has to be cultured under anaerobic conditions to be effective, and the results of our unpublished studies support this proposition. However, according to the results of recent studies, aerobically incubated caecal contents gave equal protection to that of anaerobically incubated contents against *S. Typhimurium*, *S. infantis*, *S. agona* and *S.* Enteritidis in newly hatched chickens (de Oliveira *et al.*, 2000; Filho *et al.*, 2003). However, this does not nullify the previous conclusions, because static broth cultures develop very low redox potential due to the many facultative anaerobic strains dwelling in the inoculum.

Constantly effective CE preparations and products have always been more or less unknown mixed cultures. Pure culture preparations that are equal in efficacy to mixed cultures have not yet been developed. In addition, they apparently lose their efficacy upon storage (Gleeson *et al.*, 1989; Stavric, 1992).

Since 1987, five commercial CE products have been developed: AviFree®, Aviguard®, Broilact®, MSC® and Preempt® or CF-3® (DeLoach 29). These are all mixed cultures derived from the caecal contents and/or wall of domestic fowl. According to the knowledge of the author, MSC® and Preempt® have now been withdrawn from the market. These commercial CE products have been reviewed recently (Schneitz, 2005) and will not be discussed here.

Pure cultures

The use of isolates from native intestinal microflora differs in some respects from the application of conventional probiotics in *Salmonella* control in poultry. Probiotics are used to enhance the growth performance of food animals or to control conditions such as scouring, and are given via feed or water over a longer period of time (Mead, 2000). The anti-*Salmonella* activity of several probiotics containing *Bacillus*, *Enterococcus* and *Lactobacillus* spp. was tested in laboratory-scale trials by Hinton and Mead (1991). No protection could be shown. Similar results were presented by Stavric *et al.* (1992) while testing the efficacy of a probiotic containing *L. acidophilus* and *Bifidobacterium bifidum*, and a commercial yoghurt fermented with *L. acidophilus* and *Bifidobacterium* spp.

The competitive exclusion properties involving one single dose on the day of hatch have recently been tested on several pure cultures. Shin *et al.* (2002) tested the ability of a *L. fermentum* culture to prevent colonization by *S.* Typhimurium. The results showed that chicks inoculated with *L. fermentum* on day 1 and challenged with *S.* Typhimurium on day 3 had decreased caecal colonization of *Salmonella*, but the reduction was only 1.02 or 0.15 $\log_{10}$ units. La Ragione *et al.* (2004) tested the efficacy of *L. johnsonii* F19785 in reducing the colonization and shedding of *S.* Enteritidis, *E. coli* O78:K80 and *C. perfringens* in poultry when given on the day of hatch as a single dose.

According to the results there were no significant effects against *S.* Enteritidis, whereas colonization of the small intestine by *E. coli* O78K80 was reduced significantly. No differences could be observed in caecal and colonic counts between treated and control birds. However, the test organism was able to reduce the counts of *Cl. perfringens* when given to 20-day-old chicks as a single dose. When *L. salivarius* CTC2197 was given to newly hatched chickens together with *S.* Enteritidis as a single dose, the pathogen was completely removed from the birds after 21 days (Pascual *et al.*, 1999).

A *L. reuteri* pure culture produced by Biogaia Biologics AB is marketed as a CE product. The protective effect of *L. reuteri* against enteropathogens is based on its ability to produce reuterin, a broad-spectrum antibiotic (Talarico *et al.*, 1988). Very little information is available concerning the use of this organism in poultry and its ability to prevent intestinal colonization with enteropathogens. However, it is claimed that *in ovo* treatment with *L. reuteri* reduces chick mortality caused by *Salmonella* (Dunham *et al.*, 1994). In another study, however, *L. reuteri* given *in ovo* to turkey poults had only a minimal effect against *S.* Typhimurium (Edens *et al.*, 1991, 1994, 1997).

Hofacre *et al.* (2000) tested the efficacy of a commercially produced, freeze-dried *L. acidophilus* culture against *S. kedougou* in 1-day-old turkey poults. The water mixture of the culture was both sprayed on the birds and given in the drinking water during the first 4 days of the rearing period – as suggested by the manufacturer – but no protection against *Salmonella* could be shown.

THE MECHANISM OF CE TREATMENT

The mechanism of the protective effect of CE is unknown. One of the most important factors, evidently, is competition for receptor sites on the intestinal mucosa. Seuna (1979), Soerjadi *et al.* (1981) and Stavric (1987) showed that protection of chicks against oral *Salmonella* challenge starts to become apparent within only 1–2 h of CE treatment. A similar observation was made by Mead *et al.* (1989b) when studying the effect of CE treatment on the transmission of *S.* Enteritidis in chicken transport boxes.

Besides competition for the adherence sites on the mucosa, there are many other factors which could be involved in the process, e.g.: (i) local immunity;

(ii) pH and Eh (redox potential); (iii) peristalsis, diet and body temperature of the host; and (iv) inhibitory substances such as bacteriocins, H2S, deconjugated bile acids and fatty acids (Meynell, 1963; Savage, 1977; Barnes *et al.*, 1979; Mead and Barrow, 1990; Corrier *et al.*, 1995a, b; van der Wielen *et al.*, 2000).

van der Wielen *et al.* (2000) showed that viable counts of *Bacteroides* spp. and *Eubacterium* spp. were established after 14 days at stable levels in the caecal contents of chicks, and propionic and butyric acids were detected in 12–15 day-old chickens. Simultaneously, a decrease in the numbers of *Enterobacteriaceae* was noticed. Additionally, pure cultures of *Enterobacteriaceae* isolated from the caecal contents were grown in the presence of volatile fatty acids. Growth rates and maximal optical density decreased when the strains grew in the presence of increasing volatile fatty acid concentrations. Hume *et al.* (1998) have shown that chicks challenged 4 h after CE treatment are relatively well protected against *Salmonella* colonization. The increase of the caecal propionic acid concentrations was also significant within 1 day post-treatment.

THE APPLICATION OF CE TREATMENT

For prophylactic use, chicks should be *Salmonella*-free prior to treatment. Seuna (1979) showed that treatment given only 1 h before the *Salmonella* challenge also afforded protection against higher doses of the challenge organism. The *Salmonella*-reducing effect of CE treatment has also been shown in those flocks that were *Salmonella* positive already in the hatchery.

The results of a field study conducted by Palmu and Camelin (1997) showed that, of 34 CE-treated flocks, 13% tested *Salmonella* positive at 1 day-old but only 6% of them were found positive at 45 days-of-age. Of the 34 control flocks 25% were positive at day-old and 42% at 45. The CE treatment also significantly reduced contamination of neck skin samples taken at the processing plant.

Bolder *et al.* (1995) also found that treatment of *Salmonella*-positive chicks with a CE preparation reduced the level of infection. Reynolds *et al.* (1997) found that combined treatment of 11 infected flocks with enrofloxacin and CE resulted in a long-term reduction of *Salmonella* in two, and short-term reduction in birds from another five, trials.

Both drinking-water administration and spraying in the hatchery are suitable means for dosing birds under field conditions. Initially, the only way of administering CE preparations in the field was via the first drinking water. A slaughterhouse in Finland monitored the effect of CE treatment by comparing the incidence of *Salmonella* infection in treated and untreated flocks. During a 3-year period from 1986 to 1988 they gave CE treatment to 400 broiler flocks and left 192 untreated. Of the treated flocks, 6.5% were *Salmonella* positive at the time of slaughter, while the corresponding figure for the untreated flocks was 21% (Nurmi *et al.*, 1990).

The drinking-water method was used even more successfully in Sweden by Wierup *et al.* (1988, 1992), when of 179 CE-treated flocks only one was

Salmonella positive at slaughter. The same method was used by Martin *et al.* in 2000.

However, there are several disadvantages to this method. One is that sometimes some chicks fail to drink and protection may spread unevenly among the flock (Schneitz *et al.*, 1991). Then, the most oxygen-sensitive anaerobic organisms in the treatment preparation are likely to die, and the product becomes ineffective before all the chicks have received an adequate dose (Seuna *et al.*, 1978). Additionally, chicks may be exposed to *Salmonella* during transportation to the farm, and even earlier if there is vertical transmission from infected breeders. In the former case, treatment on the farm via drinking water would be too late.

The idea of aerosol use as a method of administering CE prepartions was first created by Pivnick and Nurmi (1982). Later, Goren *et al.* (1984b, 1988) developed a method of spray application to treat newly hatched chicks in the hatchery, either in the hatchers themselves or in delivery boxes. Spraying in the hatcher followed by drinking water administration on the farm was used by Blankenship *et al.* (1993) to assure maximum efficacy of the treatment.

Manual spray application employing a hand-held garden spray device (Schneitz *et al.*, 1990) preceded automated spray cabinets; this latter method offers a much more rapid and even treatment to each box of chickens (Schneitz, 1992). Spray application does not have any adverse effects on the health or performance of the birds during grow-out (Corrier *et al.*, 1995a).

The possibility of chicks becoming infected in the hatchers has encouraged researchers to look for a method of administration that would enable the birds to be treated prior to hatch (Cox *et al.*, 1990, 1991). Cox and Bailey (1993) developed an *in ovo* method in which the CE preparation was introduced into either the air cell or amnion of the egg a few days before hatching. However, the use of a caecal culture resulted in depressed hatchability when the material was introduced into the air cell. Introducing the preparation into the amnion killed all the embryos (Cox *et al.*, 1992; Cox and Bailey, 1993). Similar results were obtained with the commercial CE product Broilact® by Meijerhof and Hulet (1997).

However, the results of our recent unpublished studies have shown that it is possible to develop effective mixed culture preparations that do not affect the hatchability.

PATHOGEN AND HOST SPECIFICITY OF CE TREATMENT

Studies carried out by several research groups around the world have shown that the CE concept applies to all serotypes of *Salmonella* that are capable of intestinal colonization in the chick.

In addition to the ability to control *Salmonella* infections in poultry, it has been shown experimentally that CE treatment also protects chicks against pathogenic *E. coli* (Soerjadi *et al.*, 1981; Weinack *et al.*, 1981, 1982, 1984; Stavric *et al.*, 1992; Hakkinen and Schneitz, 1996; Hofacre *et al.*, 2002), *Yersinia enterocolitica* (Soerjadi-Liem *et al.*, 1984b) and *C. jejuni* (Soerjadi *et al.*, 1982; Soerjadi-Liem *et al.*, 1984a; Hakkinen and Schneitz, 1999; Stern *et al.*, 2001).

In addition, CE treatment decreases mortality due to necrotic enteritis and hepatitis and reduces levels of caecal *Cl. perfringens*, which is considered to be one of the causative factors in necrotic enteritis (Barnes *et al.*, 1980; Snoeyenbos *et al.*, 1983; Elwinger *et al.*, 1992; Craven *et al.*, 1999; Kaldhusdal *et al.*, 2001). In a 7-day study, CE treatment also significantly reduced caecal colonization by *L. monocytogenes* (Hume *et al.*, 1998), though according to another study, most chicks were able to eliminate the organism from the body within 9 days without any treatment (Husu *et al.*, 1990).

Protection of newly hatched chicks against *Salmonella* colonization, using material from adult birds of the same species, seems to be independent of breed, strain or sex of bird, even though individual differences exist with respect to protective capability. Chickens can be protected to some extent by the microflora of a few other species of birds (Snoeyenbos *et al.*, 1979; Weinack *et al.*, 1982; Impey *et al.*, 1984), but material from other animals, e.g. horse and cow, has been shown to be ineffective (Rantala and Nurmi, 1973).

Weinack *et al.* (1982), Impey *et al.* (1984) and Schneitz and Nuotio (1992) showed that native chicken flora (Broilact®) and turkey microflora provided reciprocal protection in chicks and turkey poults. Cox *et al.* (2001) demonstrated that CE cultures generated from mucosal scrapings of turkeys effectively controlled *Salmonella* in turkeys during brooding. Hofacre *et al.* (2000) showed that fresh turkey caecal material was significantly more protective in turkeys than the commercial CE product, Aviguard®. Bamba *et al.* (1997) showed that the *S.* Typhimurium count decreased significantly in artificially challenged Japanese quails that had been treated on day one with chicken caecal contents. In a recent study, pheasant chicks were successfully protected against *S. infantis* by Broilact® (Schneitz and Renney, 2003).

EFFECTS OF CE TREATMENT ON BIRD PERFORMANCE

CE treatment has also been shown to enhance bird performance parameters. According to Goren *et al.* (1984b), an improvement in growth rate was observed in commercial broiler flocks sprayed with an undefined CE preparation. Corrier *et al.* (1995a) reported an improvement in the efficiency of feed utilization in broiler flocks that were given CE treatment on the day of hatch. An improvement in bird performance in terms of higher body weight, better feed conversion and lower mortality was reported by Abu-Ruwaida *et al.* (1995). Higher bodyweight and lower mortality were also noticed by Bolder *et al.* (1995).

To explain the nutritional effects of the treatment, a laboratory-scale study was conducted with Broilact®. Broiler chicks were treated orally on the day of hatch and ileal and caecal samples were taken at 12 and 31 days-of-age. The results of the study showed that CE treatment decreased the viscosity of the ileal contents and increased the faecal dry matter content significantly. It also improved the ME value of the feed by 1.6%, increased the concentration of propionic acid in the caeca and decreased that of butyric acid in the ileal contents significantly (Schneitz *et al.*, 1998).

These results were supported by another study performed by Bilal *et al.* (2000); the results of this study showed that Broilact® improved the growth rate and increased the faecal dry matter content of broilers numerically, but the difference was not significant. At the end of the experiment, total digestibility at 35 days-of-age in the group treated with Broilact® was significantly higher than in the other groups.

FACTORS THAT MAY AFFECT THE EFFICACY OF CE TREATMENT

Because of the fact that the use of growth-promoting antibiotics was banned within the European Union from 1 January 2006, their possible effect on CE treatment is not discussed here. On the other hand, anticoccidials which are still commonly used have not been found to decrease the efficacy of CE treatment.

With regard to therapeutic use of antibiotics, the results of a pilot-scale study indicated that 5 days' (days 1–5) treatment of newly hatched chicks with either furazolidone or trimethoprim-methoxasole sulphate did not eliminate the CE effect of Broilact® (Bolder and Palmu, 1995). CE treatment has also been given successfully to older birds after therapeutic doses of antibiotics to regenerate the intestinal microflora (Johnson, 1992; Uyttebroek *et al.*, 1992; Humbert *et al.*, 1997; Reynolds *et al.*, 1997).

The effect of *in ovo* administration of gentamicin or ceftiofur on the efficacy of CE application has been speculated upon. McReynolds *et al.* (2000) reported that *in ovo* administration of either ceftiofur or gentamicin resulted in marked depressions in caecal propionate levels, an indicator of Preempt® establishment, in chicks treated orally on day 1 with the CE product compared with non-injected control chicks. On the other hand, Bailey and Line (2001) demonstrated that gentamicin, at a rate of 0.4 mg/egg administered *in ovo* on day 18, had no adverse effect on the CE product MSC®.

Other factors that can reduce the efficacy of CE treatment include stress and disease. Starving chicks for the first 24 h of life also has a negative effect (Goren *et al.*, 1984a) whereas, in older birds, the protective flora is more difficult to disrupt (Snoeyenbos *et al.*, 1985). With the 1 day-old chick, physiological stress induced by high or low environmental temperatures or removal of feed and water either interfered with the colonization of protective organisms or reduced the protection provided by these organisms; however, there was no obvious effect at 2 weeks of age (Weinack *et al.*, 1985).

Lafont *et al.* (1983) studied CE-treated chicks that were carrying low numbers of *Salmonellae* in their intestines and administered oocytes of *Eimeria tenella* at a level known to produce caecal coccidiosis. The birds then shed large numbers of *Salmonellae* for more than 2 weeks. Exposure of CE-treated chicks to aerosols of *Mycoplasma gallisepticum* and/or infectious bronchitis virus increased the number of birds shedding pathogenic *E. coli* or *S. Typhimurium*, following a challenge 2 days after protective treatment (Weinack *et al.*, 1984).

Induced moulting of White Leghorn layers and subjecting market-age broilers to feed withdrawal have also been shown to increase both the numbers of *Salmonellae* in the gastrointestinal tract and the proportion of infected individuals (Holt and Porter, 1993; Holt *et al.*, 1995; Macri *et al.*, 1997; Raminez *et al.*, 1997; Corrier *et al.*, 1999).

To evaluate the effect of pre-slaughter feed withdrawal on *Salmonella* caecal contamination on broiler chickens and the ability of CE treatment to prevent it, the following pilot-scale trial was conducted:

At 1 day-of-age 180 broiler chicks were challenged with S. Enteritidis, approximately 10^4 per bird, divided between six small, environmentally controlled houses, 30 per house, two houses per treatment. Commercial rearing conditions were simulated. At 42 days-of-age, the birds in four houses were placed on feed withdrawal. For the first 6 h two of the four houses had access to regular tap water and two houses received Broilact®-suspended water (approx. 1 mg of lyophilized Broilact® per bird). Two houses had access to feed and water. Six h post-feed withdrawal, all birds were loaded into transport boxes and placed in a pick-up truck, where they were held until the total feed withdrawal time was 12 h, after which time the birds were killed.

Caeca were collected from 20 birds per house and *Salmonella* was determined from the caecal contents using conventional methods. *Salmonella*-positive birds were found neither in the Broilact®-treated houses nor in the houses that were not on feed withdrawal, whereas 30% (12/40) of the untreated chicks that were on feed withdrawal were *Salmonella* positive. In accordance with the findings of Corrier *et al.* (1999), the results of this trial indicate that birds can become infected with *Salmonella* by consuming contaminated rearing house litter during feed withdrawal. Administration of CE treatment material in the drinking water during feed withdrawal can prevent the birds from becoming infected. However, more work is needed to prove the results of this trial.

Seuna (1979) showed that treatment given only 1 h before the *Salmonella* challenge afforded protection also to higher doses of the challenge organism. However, commercial hatchery environments, as well as the breeder birds, may be contaminated with *Salmonella* (Cox *et al.*, 1990, 1991), which can also reduce the efficacy of CE treatment (Bailey *et al.*, 1998).

DISCUSSION AND CONCLUSIONS

The understanding of the significance of the intestinal flora in protection of the host animal against enteropathogens arose from the work of Elie Metchnikoff, a Russian biologist working at the Pasteur Institute at the turn of the 19th century (Metchnikoff, 1908). He suggested that fermented milk containing so-called 'lactic acid bacteria' was responsible for the health of Bulgarian farmers. It was assumed that the introduction of benign, non-toxigenic gut microflora created conditions that favoured good health. Milner and Schaffer had already observed, in 1952, that the bird's natural resistance to *Salmonella* infection increased with age (Milner and Shaffer, 1952), but nearly 20 years passed before attention was drawn to the conditions under which chicks were hatched and reared, having no contact with the brooding hen.

The observation that oral administration of adult intestinal microflora to newly hatched chicks increased their resistance to colonization by food poisoning *Salmonellae* was first presented in Finland and published in *Nature* in 1973 by Nurmi and Rantala. Several research groups around the world have confirmed the validity of the CE concept in poultry production and a number of profound reviews on the topic have been written over the years (Pivnick and Nurmi, 1982; Schleifer, 1985; Mead and Impey, 1987; Stavric and D'Aoust, 1993; Mead, 2000; Schneitz and Mead, 2000; Schneitz, 2005).

The widest experience concerning the effect of this treatment in poultry production is probably in Finland, which continues to be the only country where the entire chain, from grandparents to meat birds, are given CE treatment on the day of hatch. The use of CE began in 1976 when the National Veterinary and Food Research Institute started the production and delivery of CE broth cultures.

Broilact® was developed under leadership of Professor Nurmi and launched in 1987 as a broth culture, but from 1994 onwards it has been sold freeze-dried. CE treatment has been found to contribute significantly to the decrease in incidence of *Salmonella*-contaminated flocks and carcasses in Finland (Hirn *et al.*, 1992). The incidence of *Salmonella*-contaminated broiler flocks has stayed, on average, below 1.0% (National Food Agency).

The incidence of *Campylobacter*-infected broiler flocks was tested during, June, July and August in 2004 and, on average, 7.3% of the flocks were found to be positive. The number of infected flocks was highest during July and August, being 8.8 and 10.1%, respectively. In June the percentage of infected flocks was 3.1%. Outside of the summer period *Campylobacter*-positive poultry flocks are seldom found in Finland (National Veterinary and Food Research Institute). It has been suggested that the long-term use of CE cultures in Finnish broiler flocks has contributed to the low incidence of *Campylobacter* in broilers in Finland (Aho and Hirn, 1988).

Five different commercial CE products have been developed since 1987: AviFree®, Aviguard®, Broilact®, MSC® and Preempt®. They are all mixed cultures derived from the caecal contents and/or wall of healthy, adult chickens. Of these commercial products, MSC® and Preempt® have been withdrawn from the market.

To standardize the methods used to evaluate different CE preparations, Mead *et al.* (1989a) described a recommended assay. Newly hatched chicks are treated orally on day 1, challenged orally with *Salmonella* 24 h later and examined 5 days post-challenge to determine both the proportion of positive birds in treated and control groups and the levels of *Salmonella* carriage in infected individuals. The efficacy of the treatment is determined by the calculation of an Infection Factor (IF) value, which is the geometric mean of the number of *Salmonellae*/g of caecal contents for all chicks in a particular group.

There would be obvious advantages in being able to use CE preparations under commercial conditions with completely defined strain composition.

Inclusion of any potential poultry pathogen could be avoided with certainty and the quality control of the treatment product during manufacture would be alleviated (Mead, 2000). However, consistent protection of newly hatched chicks against *Salmonella* has been obtained only with undefined mixed cultures. Fully effective defined preparations have not yet been developed. The work is difficult because the mechanism of CE is poorly understood, as are the bacteria involved in the process. Only defined cultures containing large numbers of strains from different genera have been comparable to undefined cultures in their efficacy (Stavric, 1992).

Competitive exclusion is a very effective measure for protection of newly hatched chicks, turkey poults, quails and pheasants – and possibly other game birds, too, against *Salmonella* and other enteropathogens. The results of studies performed by several research groups around the world and the 10-year field experience in Finland show that both manual and automated spray application are effective means of dosing newly hatched chicks, though spray application is preferred (Mead, 2000).

Obviously, CE treatment is effective against all host non-specific *Salmonellae* able to colonize the alimentary tract of the bird. There are also indications that current treatment products may be of value in controlling infections with other enteropathogens, including *Cl. perfringens* and chicken and human pathogenic *E. coli*. Protection against *Campylobacter* has also been shown with two of the existing CE products (Hakkinen and Schneitz, 1999; Stern *et al.*, 2001).

However, to improve results a different composition of protective bacteria may be needed, because of the specific location of *Campylobacter* in the mucous layer of the caecal crypts (Beery *et al.*, 1988). Furthermore, the treatment efficacy is unaffected by breed, strain or sex of the bird. Both improvement in bird growth and feed utilization and a reduction in the mortality of chicks are also of great importance.

The efficacy of the CE treatment may be affected by antimicrobials, stress, disease, moulting, feed withdrawal and infected breeders and contaminated hatchery area. To overcome the problem with hatchery contamination, the *in ovo* application method was developed by Cox and Bailey (1993). However, the use of a caecal culture containing highly proteolytic organisms and abundant gas-formers resulted in depressed hatchability when the material was introduced into the air cell. Introducing the preparation into the amnion killed all the embryos. Adverse effects on hatchability can be avoided by excluding strongly proteolytic and gas-forming organisms (unpublished results).

WHO have suggested that there should be a special category for the licensing of CE products termed normal gut flora, and the European Union is currently reviewing its position on CE, according to Wray and Davies (2000).

Removal of antibiotic growth promoters is likely to increase the variability of broiler performance and lead to the use of therapeutic doses of antibiotics under prescription (Bedford, 2000). According to present knowledge on the efficacy of the competitive exclusion concept, commercial, safety-approved CE products could be integrated into management programmes for commercial poultry, as has already been done in Finland.

REFERENCES

Abu-ruwaida, A.S., Husseini, M. and Banat, I.M. (1995) *Salmonella* exclusion in broiler chicks by the competitive action of adult gut microflora. *Microbios* 83, 59–69.

Aho, M. and Hirn, J. (1988) Prevelance of campylobacteria in the Finnish broiler chicken chain from the producer to the consumer. *Acta Veterinaria Scandinavica* 29, 451–462.

Bailey, J.S. and Line, E. (2001) *In ovo* gentamicin and Mucosal Starter Culture to control *Salmonella* in broiler production. *Journal of Applied Poultry Research* 10, 376–379.

Bailey, J.S., Cason, J.A. and Cox, N.A. (1998) Effect of *Salmonella* in young chicks on competitive exclusion. *Poultry Science* 77, 394–399.

Bamba, H., Toyoshima, K. and Kamiya, M. (1977) Protective properties of the treatments of cecal contents on *Salmonella* Typhimurium colonization in the bowel of growing quail chicks. *Research Bulletin of the Aichi-ken Agricultural Research Centre* 29, 355–358.

Barnes, E.M., Impey, C.S. and Srevens, J.H. (1979) Factors affecting the incidence and anti-*Salmonella* activity of the anaerobic caecal flora of the young chick. *Journal of Hygiene, Cambridge* 82, 263–283.

Barnes, E.M., Impey, C.S. and Cooper, D.M. (1980) Manipulation of the crop and intestinal flora of the newly hatched chick. *American Journal of Clinical Nutrition* 33, 2426–2433.

Bedford, M. (2000) Removal of antibiotic growth promoters from poultry diets: implications and strategies to minimise subsequent problems. *World's Poultry Science Journal* 56, 347–365.

Beery, J.T., Hugdahl, M.B. and Doyle, M.P. (1988) Colonization of gastrointestinal tracts of chicks by *Campylobacter jejuni*. *Applied and Environmental Microbiology* 54, 2365–2370.

Bilal, T., Özpinar, H., Kutay, C., Eseceli, H. and Abas, I. (2000) The effects of Broilact® on performance and feed digestibility of broilers. *Archiv für Geflügelkunde* 64, 134–138.

Blankenship, L.C., Bailey, J.S., Cox, N.A., Stern, N.J., Brewer, R. and Williams, O. (1993) Two-step mucosal competitive exclusion flora to diminish *Salmonellae* in commercial broiler chickens. *Poultry Science* 72, 1667–1672.

Bolder, N.M. and Palmu, L. (1995) Effect of antibiotic treatment on competitive exclusion against *Salmonella* Enteritidis PT4 in broilers. *The Veterinary Record* 137, 350–351.

Bolder, N.M., Vereijken, P.F.G., Putirulan, F.F. and Mulder, R.W.A.W. (1995) The effect of competitive exclusion on the *Salmonella* contamination of broilers (a field study). In: Briz, R.C. (ed.) *Proceedings of the 2nd annual meeting of EC COST Working Group*, No. 2. Graficas Imprinter, Zaragoza, Spain, pp. 89–97.

Corrier, D.E., Nisbet, D.J., Scanlan, C.M., Hollister, A.G., Caldwell, D.J., Thomas, L.A., Hargis, B.M., Tomkins, T. and Deloach, J.R. (1995a) Treatment of commercial broiler chickens with a characterized culture of cecal bacteria to reduce *Salmonellae* colonization. *Poultry Science* 74, 1093–1101.

Corrier, D.E., Nisbet, D.J., Scanlan, C.M., Hollister, A.G. and Deloach, J.R. (1995b) Control of *Salmonella* Typhimurium colonization in broiler chicks with continuous-flow characterized mixed culture of cecal bacteria. *Poultry Science* 74, 916–924.

Corrier, D.E., Byrd, J.A., Hargis, B.M., Hume, M.E., Bailey, R.H. and Stanker, L.H. (1999) Presence of *Salmonella* in the crop and caeca of broiler chickens before and after pre-slaughter feed withdrawal. *Poultry Science* 78, 45–49.

Cox, N.A. and Bailey, J.S. (1993) Introduction of bacteria *in ovo*. USA Patent No. 5,206,015.

Cox, N.A., Bailey, J.S., Mauldin, J.M. and Blankenship, L.C. (1990) Presence and impact of *Salmonellae* contamination in commercial broiler hatcheries. *Poultry Science* 69, 1606–1609.

Cox, N.A., Bailey, J.S., Mauldin, J.M. and Blankenship, L.C. (1991) Extent of *Salmonellae* contamination in breeder hatcheries. *Poultry Science* 70, 416–418.

Cox, N.A., Bailey, J.S., Mauldin, J.M. and Blankenship, L.C. (1992) *In ovo* administration of a competitive exclusion treatment to broiler embryos. *Poultry Science* 71, 1781–1784.

Cox, N.A., Bailey, J.S. and Stern, N.J. (2001) Effectiveness on an undefined mucosal competitive exclusion treatment to control *Salmonella* in turkeys during brooding. *Journal of Applied Poultry Research* 10, 319–322.

Craven, S.E., Stern, N.J., Coc, N.A., Bailey, J.S. and Berrang, M. (1999) Cecal carriage of *Clostridium perfringens* in broiler chickens given Mucosal Starter Culture. *Avian Diseases* 43, 484–490.

De Oliveira, G.H., Berchieri, A.J.R. and Barrow, P.A. (2000) Prevention of *Salmonella* infection by contact using intestinal flora of adult birds and/or a mixture of organic acids. *Brazilian Journal of Microbiology* 31, 116–120.

Dunham, H.J., Edens, F.W., Casas, I.A. and Dobrogosz, W.J. (1994) Efficacy of *Lactobacillus reuteri* as a probiotic for chickens and turkeys. *Microbial Ecology in Health and Disease* 7, 52–53.

Edens, F.W., Parkhurst, C.R. and Joyce, K. (1991) *Lactobacillus reuteri* and whey reduce *Salmonella* colonization in the caeca of turkey poults. *Poultry Science* 70, 158.

Edens, F.W., Casas, I.A., Parkhurst, C.R. and Joyce, K. (1994) Reduction of egg-borne *E. coli*-associated chick mortality by in-hatcher exposure to *Lactobacillus reuteri*. *Poultry Science* 73, 79.

Edens, F.W., Parkhurst, C.R. and Casas, I.A. (1997) Principles of *ex ovo* competitive exclusion and *in ovo* administration of *Lactobacillus reuteri*. *Poultry Science* 76, 179–196.

Elwinger, K., Schneitz, C., Berndtson, E., Fossum, O., Teglöf, B. and Engström, B. (1992) Factors affecting the incidence of necrotic enteritis, caecal carriage of *Clostridium perfringens* and bird performance in broiler chicks. *Acta Veterinaria Scandinavica* 33, 369–378.

Filho, R.L.A., Sampaio, H.M., Barros, M.R., Gratâo, P.R. and Cataneo, A. (2003) Use of cecal microflora cultured under aerobic or anaerobic conditions in the control of experimental infection of chicks with *Salmonella* Enteritidis. *Veterinary Microbiology* 92, 237–244.

Gleeson, T.M., Stavric, S. and Blanchfield, B. (1989) Protection of chicks against *Salmonella* infection with a mixture of pure cultures of intestinal bacteria. *Avian Diseases* 33, 636–642.

Goren, E., De Jong, W.A., Doornenbal, P., Koopman, J.P. and Kennis, H.M. (1984a) Protection of chicks against *Salmonella infantis* infection induced by strict anaerobically cultured intestinal microflora. *The Veterinary Quarterly* 6, 22–26.

Goren, E., De Jong, W.A., Doornenbal, P., Koopman, J.P. and Kennis, H.M. (1984b) Protection of chicks against *Salmonella* infection induced by spray application of intestinal microflora in the hatchery. *The Veterinary Quarterly* 6, 73–79.

Goren, E., De Jong, W.A., Doornenbal, P., Bolder, N.M., Mulder, R.W.A.W. and Jansen, A. (1988) Reduction of *Salmonella* infection of broilers by spray application of intestinal microflora: a longitudinal study. *The Veterinary Quarterly* 10, 249–255.

Greenberg, B. (1969) *Salmonella* suppression by known populations of bacteria in flies. *Journal of Bacteriology* 99, 629–635.

Hakkinen, M. and Schneitz, C. (1996) Efficacy of a commercial competitive exclusion product against chicken pathogenic *Escherichia coli* and *E. coli* O157:H7. *The Veterinary Record* 139, 139–141.

Hakkinen, M. and Schneitz, C. (1999) Efficacy of a commercial competitive exclusion product against *Campylobacter jejuni*. *British Poultry Science* 40, 619–621.

Hinton, M. and Mead, G.C. (1991) *Salmonella* control in poultry: the need for the satisfactory evaluation of probiotics for this purpose. *Letters in Applied Microbiology* 13, 49–50.

Hirn, J., Nurmi, E., Johansson, T. and Nuotio, L. (1992) Long-term experience with competitive exclusion and salmonellas in Finland. *International Journal of Food Microbiology* 15, 281–285.

Hofacre, C.L., Primm, N.D., Vance, K., Goodwin, M.A. and Brown, J. (2000) Comparison of a lyophilized chicken-origin competitive exclusion culture, a lyophilized probiotic, and fresh turkey cecal material against *Salmonella* colonization. *Journal of Applied Poultry Research* 9, 195–203.

Hofacre, C.L., Johnson, A.C., Kelly, B.J. and Froyman, R. (2002) Effect of a commercial competitive exclusion culture on reduction of colonization of an antibiotic-resistant pathogenic *Escherichia coli* in day-old broiler chickens. *Avian Diseases* 46, 198–202.

Holt, P.S. and Porter, J.R. (1993) Effect of induced molting on the recurrence of a previous *Salmonella* Enteritidis infection. *Poultry Science* 72, 2069–2078.

Holt, P.S., Macri, P. and Porter, J.R. (1995) Microbial analysis of the early *Salmonella* Enteritidis infection in molted and unmolted hens. *Avian Diseases* 39, 55–63.

Humbert, F., Carraminana, F., Lalande, F. and Salvat, G. (1997) Bacteriological monitoring of *Salmonella* Enteritidis carrier birds after decontamination using enrofloxacin, competitive exclusion and movement of birds. *The Veterinary Record* 141, 297–299.

Hume, M.E., Byrd, J.A., Stanker, L.H. and Ziprin, R.L. (1998) Reduction of caecal *Listeria monocytogenes* in Leghorn chicks following treatment with a competitive exclusion culture (Preempt™). *Letters in Applied Microbiology* 26, 432–436.

Husu, J.R., Beery, J.T., Nurmi, E. and Doyle, M.P. (1990) Fate of *Listeria monocytogenes* in orally dosed chicks. *International Journal of Food Microbiology* 11, 259–269.

Impey, C.S., Mead, G.C. and George, S.M (1984) Evaluation of treatment with defined and undefined mixtures of gut microorganisms for preventing *Salmonella* colonization in chicks and turkey poults. *Food Microbiology* 1, 143–147.

Johnson, C.T. (1992) The use of an antimicrobial and competitive exclusion combination in *Salmonella*-infected pullet flocks. *International Journal of Food Microbiology* 15, 293–298.

Kaldhusdal, M., Schneitz, C., Hofshagen, M. and Skjerve, E. (2001) Reduced incidence of *Clostridium perfringens*-associated lesions and improved performance in broiler chickens treated with normal intestinal bacteria from adult fowl. *Avian Diseases* 45, 149–156.

Lafont, J.P., Brée, A., Naciri, M., Yvoré, P., Guillot, J.F. and Chaslus-Dancla, E. (1983) Experimental study of some factors limiting 'competitive exclusion' of *Salmonella* in chickens. *Research in Veterinary Science* 34, 16–20.

La Ragione, R.M., Narbad, A., Gasson, M.J. and Woodward, M.J. (2004) *In vivo* characterization of *Lactobacillus johnsonii* F19785 for use as a defined competitive exclusion agent against bacterial pathogens in poultry. *Letters in Applied Microbiology* 38, 197–205.

Lloyd, A.B., Cumming, R.B. and Kent, R.D. (1974) Competitive exclusion as exemplified by *Salmonella* Typhimurium. *Proceedings of the Australian Poultry Science Convention*, Hobart, Tasmania, pp. 185–186.

Luckey, T.D. (1963) *Germ-free Life and Gnotobiology*. Academic Press, New York.

Macri, N.P., Porter, J.R. and Holt, P.S. (1997) The effects of induced molting on the severity of acute intestinal inflammation caused by *Salmonella* Enteritidis. *Avian Diseases* 41, 117–124.

Martin, C., Dunlap, E., Caldwell, S., Barnhart, E., Keith, N. and Deloach, J.R. (2000) Drinking water delivery of a defined competitive exclusion culture (Preempt®) in 1-day-old broiler chicks. *Journal of Applied Poultry Research* 9, 88–91.

McReynolds, J.L., Caldwell, D.Y., Barnhart, E.T., Deloach, J.R., McElroy, A.P., Moore, R.W., Hargis, B.M. and Caldwell, D.J. (2000) The effect of *in ovo* or day-of-hatch subcutaneous antibiotic administration on competitive exclusion culture (Preempt®) establishment in neonatal chickens. *Poultry Science* 79, 1524–1530.

Mead, G.C. (2000) Prospects for 'competitive exclusion' treatment to control salmonellas and other foodborne pathogens in poultry. *The Veterinary Journal* 159, 111–123.

Mead, G.C. and Barrow, P.A. (1990) *Salmonella* control in poultry by 'competitive exclusion' or immunization. *Letters in Applied Microbiology* 10, 221–227.

Mead, G.C. and Impey, C.S. (1987) The present status of the Nurmi Concept for reducing carriage of food-poisoning *Salmonellae* and other pathogens in live poultry. In: Smulders, F.J.M. (ed.) *Elimination of Pathogenic Organisms from Meat and Poultry*. Elsevier Science Publishers BV, Amsterdam, pp. 57–77.

Mead, G.C., Barrow, P.A., Hinton, M.H., Humbert, F., Impey, C.S., Lahellec, C., Mulder, R.W.A.W., Stavric, S. and Stern, N.J. (1989a) Recommended assay for treatment of chicks to prevent *Salmonella* colonization by 'competitive exclusion'. *Journal of Food Protection* 52, 500–502.

Mead, G.C., Schneitz, C., Nuotio, L.O. and Nurmi, E.V. (1989b) Treatment of chicks using competitive exclusion to prevent transmission of *Salmonella* Enteritidis in delivery boxes. *19th International Congress of the World Veterinary Poultry Association*, Brighton, UK, p. 115 (poster abstract).

Meijerhof, R. and Hulet, R.M. (1997) *In ovo* injection of competitive exclusion culture in broiler hatching eggs. *Journal of Applied Poultry Research* 6, 260–266.

Metchnikoff, E. (1908) *Prolongation of Life*. G.P. Putnam and Sons, New York.

Meynell, G.G. (1963) Antibacterial mechanisms of the mouse gut. II. The role of Eh and volatile fatty acids in the normal gut. *British Journal of Experimental Pathology* 44, 209–219.

Milner, K.C. and Shaffer, M.F. (1952) Bacteriologic studies of experimental *Salmonella* infections in chick. *Journal of Infectious Diseases* 90, 81.

Nurmi, E.V. and Rantala, M. (1973) New aspects of *Salmonella* infection in broiler production. *Nature* 241, 210.

Nurmi, E.V., Schneitz, C. and Mäkelä, P.H. (1987) Process for the production of a bacterial preparation for the prophylaxis of intestinal disturbances in poultry. *USA Patent No. 4,689,226.*

Nurmi, E.V., Nuotio, L., Schneitz, C., Hakkinen, C. and Hirn, J. (1990) Prevention of *Salmonella* and *Campylobacter* infections in poultry by competitive exclusion. *Public Health of Poultry Meat and Egg Consumption, International Symposium, 6–7 June, 1990, Wiesbaden, Germany.*

Palmu, L. and Camelin, I. (1997) The use of competitive exclusion in broilers to reduce the level of *Salmonella* contamination on the farm and at the processing plant. *Poultry Science* 76, ·1501–1505.

Pascual, M., Hugas, M., Badiola, J.I., Monfort, J.M. and Garriga, M. (1999) *Lactobacillus salivarius* CTC2197 prevents *Salmonella* Enteritidis colonization in chickens. *Applied and Environmental Microbiology* November, 4981–4986.

Pivnick, H. and Nurmi, E. (1982) The Nurmi Concept and its role in the control of *Salmonella* in poultry. In: Davies, R. (ed.) *Developments in Food Microbiology* 1. Applied Science Publishers Ltd., Barking, UK, pp. 41–70.

Raminez, G.A., Sarlin, L.L., Caldwell, D.J., Yezak, C.R., Hume, D.E., Corrier, D.E., Deloach, J.R. and Hargis, B.M. (1997) Effect of feed withdrawal on the incidence of *Salmonella* in the crops and caeca of market age broiler chickens. *Poultry Science* 76, 654–656.

Rantala, M. (1974) Cultivation of a bacterial flora able to prevent the colonization of *Salmonella infantis* in the intestines of broiler chickens, and its use. *Acta Pathologica Microbiologica Scandinavica.* Section B 82, 75–80.

Rantala, M. and Nurmi, E. (1973) Prevention of the growth of *Salmonella infantis* in chicks by the flora of the alimentary tract of chickens. *British Poultry Science* 14, 627–630.

Reynolds, D.J., Davies, R.H., Richards, M. and Wray, C. (1997) Evaluation of combined antibiotic and competitive exclusion treatment in broiler breeder flocks infected with *Salmonella enterica* serovar Enteritidis. *Avian Pathology* 26, 83–95.

Savage, D.C. (1977) Microbial ecology of the gastrointestinal tract. *Annual Review of Microbiology* 31, 107–133.

Schleifer, J.H. (1985) A review of the efficacy and mechanism of competitive exclusion for the control of *Salmonella* in poultry. *World's Poultry Science Journal* 41, 72–83.

Schneitz, C. (1992) Automated droplet application of a competitive exclusion preparation. *Poultry Science* 71, 2125–2128.

Schneitz, C. (2005) Competitive exclusion in poultry: 30 years of research. *Food Control* 16, 657–667.

Schneitz, C. and Mead, G.C. (2000) Competitive exclusion. In: Wray, C. and Wray, A. (eds) *Salmonella in Domestic Animals*. CAB International, Wallingford, UK, pp. 301–322.

Schneitz, C. and Nuotio, L. (1992) Efficacy of different microbial preparations for controlling *Salmonella* colonization in chicks and turkey poults by competitive exclusion. *British Poultry Science* 33, 207–211.

Schneitz, C. and Renney, D.J. (2003) Effect of a commercial competitive exclusion product on the colonization of *Salmonella infantis* in day-old pheasant chicks. *Avian Diseases* 47, 1448–1451.

Schneitz, C., Hakkinen, M., Nuotio, L., Nurmi, E. and Mead, G. (1990) Droplet application for protecting chicks against *Salmonella* colonization by competitive exclusion. *The Veterinary Record* 126, 510.

Schneitz, C., Nuotio, L., Kiiskinen, T. and Nurmi, E. (1991) Pilot-scale testing of the competitive exclusion method in chickens. *British Poultry Science* 32, 877–880.

Schneitz, C., Kiiskinen, T., Toivonen, V. and Näsi, M. (1998) Effect of Broilact® on the physico-chemical conditions and nutrient digestibility in the gastrointestinal tract of broilers. *Poultry Science* 77, 426–432.

Seuna, E. (1979) Sensitivity of young chickens to *Salmonella* Typhimurium var. Copenhagen and *S. infantis* infection and the preventive effect of cultured intestinal microflora. *Avian Diseases* 23, 392–400.

Seuna, E., Raevuori, M. and Nurmi, E. (1978) An epizootic of *Salmonella* Typhimurium var. Copenhagen in broilers and the use of cultured chicken intestinal flora for its control. *British Poultry Science* 19, 309–314.

Shin, J.W., Kang, J.K.J., Jang, K.-I. and Kim, K.Y. (2002) Intestinal colonization characteristics of *Lactobacillus* spp. isolated from chicken caecum and competitive inhibition against *Salmonella* Typhimurium. *Journal of Microbiology and Biotechnology* 12, 576–582.

Snoeyenbos, G.H., Weinack, O.M. and Smyser, C.F. (1979) Further studies on competitive exclusion for controlling salmonellas in chickens. *Avian Diseases* 24, 904–914.

Snoeyenbos, G.H., Weinack, O.M. and Soerjadi, A.S. (1983) Our current understanding of the role of native microflora in limiting some bacterial pathogens of chickens and turkeys. *Australian Veterinary Poultry Association and International Union of Immunological Societies, Proceedings No. 66, Disease Prevention and Control in Poultry Production*. Sydney, Australia, pp. 45–51.

Snoeyenbos, G.H., Weinack, O.M., Soerjadi-Liem, A.S., Miller, B.M., Woodward, D.E. and Weston, C.R. (1985) Large-scale trials to study competitive exclusion of *Salmonella* in chickens. *Avian Diseases* 29, 1004–1011.

Soerjadi, A.S., Stehman, S.M., Snoeyenbos, G.H., Weinack, O.M. and Smyser, C.F. (1981) Some measurements of protection against paratyphoid *Salmonella* and *Escherichia coli* by competitive exclusion in chickens. *Avian Diseases* 24, 706–712.

Soerjadi, A.S., Snoeyenbos, G.H. and Weinack, O.M. (1982) Intestinal colonization and competitive exclusion of *Campylobacter fetus* subsp. *jejuni* in young chicks. *Avian Diseases* 26, 520–524.

Soerjadi-Liem, A.S., Snoeyenbos, G.H. and Weinack, O.M. (1984a) Comparative studies on competitive exclusion of three isolates of *Campylobacter fetus* subsp. *jejuni* in chickens by native gut microflora. *Avian Diseases* 28, 139–146.

Soerjadi-Liem, A.S., Snoeyenbos, G.H. and Weinack, O.M. (1984b) Establishment and competitive exclusion of *Yersinia enterocolitica* in the gut of monoxenic and holoxenic chicks. *Avian Diseases* 28, 256–260.

Stavric, S. (1987) Microbial colonization control of chicken intestine using defined cultures. *Food Technology* 41, 93–98.

Stavric, S. (1992) Defined cultures and prospects. *International Journal of Food Microbiology* 15, 245–263.

Stavric, S. and D'Aoust, J.-Y. (1993) Undefined and defined bacterial preparations for the competitive exclusion of *Salmonella* in poultry – a review. *Journal of Food Protection* 56, 173–180.

Stavric, S., Buchanan, B. and Gleeson, T.M. (1992a) Competitive exclusion of *Escherichia coli* O157:H7 from chicks with anaerobic cultures of faecal microflora. *Letters in Applied Microbiology* 14, 191–193.

Stavric, S., Gleeson, T.M., Buchanan, B. and Blanchfield, B. (1992b) Experience of the use of probiotics for *Salmonella* control in poultry. *Letters in Applied Microbiology* 14, 69–71.

Stern, N.J., Cox, N.A., Bailey, J.S., Berrang, M.E. and Musgrove, M.T. (2001) Comparison of mucosal competitive exclusion and competitive exclusion treatment to reduce *Salmonella* and *Campylobacter* spp. colonization in broiler chickens. *Poultry Science* 80, 156–160.

Talarico, T.L., Casas, I.A., Chung, T.C. and Dobrogosz, W.J. (1988) Production and isolation of reuterin, a growth inhibitor produced by *Lactobacillus reuteri*. *Antimicrobial Agents and Chemotherapy* 32, 1854–1858.

Uyttebroek, E., Devriese, L.A., Desmidt, M., Ducatelle, R. and Haesebrouck, F. (1992) Efficacy of early *versus* delayed treatment of *Salmonella* Enteritidis infection in replacement pullets. *Proceedings, Posters, of International Symposium on Salmonella and Salmonellosis* (poster abstract). Imprimerie Guivarch, Saint-Brieuc, France, p. 176.

van der Waaij, D., Berghuis-de Vries, J.M. and Lekkerkerk-van der Wees, J.E.C. (1971) Colonization resistance of the digestive tract in conventional and antibiotic-treated mice. *Journal of Hygiene* 69, 405–511.

van der Wielen, P.W.J.J., Biesterveld, S., Notermans, S., Hofstra, H., Urlings, B.A.P. and van Knapen, F. (2000) Role of volatile fatty acids in development of the cecal microflora in broiler chickens during growth. *Applied and Environmental Microbiology* 66, 2536–2540.

Weinack, O.M., Snoeyenbos, G.H., Smyser, C.F. and Soerjadi, A.S. (1981). Competitive exclusion of intestinal colonization of *Escherichia coli* in chicks. *Avian Disease* 25, 696–705.

Weinack, O.M., Snoeyenbos, G.H., Smyser, C.F. and Soerjadi, A.S. (1982) Reciprocal competitive exclusion of *Salmonella* and *Escherichia coli* by native intestinal microflora of the chicken and turkey. *Avian Diseases* 26, 585–595.

Weinack, O.M., Snoeyenbos, G.H., Smyser, C.F. and Soerjadi-Liem, A.S (1984) Influence of *Mycoplasma gallisepticum*, infectious bronchitis, and cyclophosphamide on chickens protected by native intestinal microflora against *Salmonella* Typhimurium or *Escherichia coli*. *Avian Diseases* 28, 416–425.

Weinack, O.M., Snoeyenbos, G.H., Soerjadi-Liem, A.S. and Smyser, C.F. (1985) Influence of temperature, social and dietary stress on development and stability of protective microflora in chickens against *S.* Typhimurium. *Avian Diseases* 29, 1177–1183.

Wierup, M., Wold-Troell, M., Nurmi, E. and Hakinen, M. (1988) Epidemiological evaluation of the *Salmonella*-controlling effect of a nationwide use of a competitive exclusion culture in poultry. *Poultry Science* 67, 1026–1033.

Wierup, M., Wahlström, H. and Engström, B. (1992) Experience of a 10-year use of competitive exclusion treatment as part of the *Salmonella* control programme in Sweden. *International Journal of Food Microbiology* 5, 287–291.

Wray, C. and Davies, R.H. (2000) Competitive exclusion: an alternative to antibiotics. *The Veterinary Journal* 159, 107–108.

CHAPTER 20
Campylobacters and their bacteriophage in poultry

P.L. Connerton and I.F. Connerton*

*Division of Food Sciences, School of Biosciences, University of Nottingham, Loughborough, UK; *e-mail: ian.connerton@nottingham.ac.uk*

ABSTRACT

Mounting levels of resistance to antimicrobial agents by clinically important bacterial pathogens have led scientists to review alternative methods for their control, including bacteriophage therapy. Bacteriophage are viral parasites that are ubiquitous in the environment. Although they are specific to their host bacteria, as a group they infect almost all bacterial genera, including *Campylobacter*. Those specific to *Campylobacter* are frequently found in the intestinal contents of poultry, along with their hosts. They are particularly prevalent in free-range and organic birds exposed to the environment. Bacteriophage may be transferred from the intestinal contents of birds to the surface of poultry meat, at slaughter. Here they are able to survive for more than 10 days and are therefore naturally present on food for human consumption.

Experimental trials have indicated the possibility of their use as therapeutic agents. Reductions in *Campylobacter* numbers in chicken caecal contents of between $\log_{10} 2$ and $\log_{10} 5$ colony-forming units (CFU)/g have been achieved. Phage intervention, carried out just prior to slaughter, could therefore reduce the numbers of campylobacters entering the human food chain. A possible drawback of phage therapy is the acquisition of resistance by the target campylobacters, although recent research has indicated that this may not be as problematic as initially suggested. Phage may also be applied to the surface of poultry meat to reduce *Campylobacter* numbers. This review seeks to explore what is known about the *Campylobacter* phage present in poultry and their potential as therapeutic agents.

INTRODUCTION

Bacteriophage were first reported in 1896 by Ernest Hankin, who noted that something in the waters of the Ganges had an antibacterial effect against *Vibrio cholerae* (Hankin, 1896). The word phage comes from the Greek 'phagein' – to eat. The discovery that phage were, in fact, viruses was made independently by Frederick Twort in 1915 (Twort, 1915) and Felix D'Herelle in 1917 (D'Herelle, 1922). It was Felix D'Herelle who first suggested bacteriophage therapy as a means of treating or preventing bacterial infection. Bacteriophage offer several intrinsic advantages in that they are specific to their target host bacteria and will not directly cause wholesale changes in non-target microflora that may be present.

Unlike antimicrobial treatments, these bystanders will not be placed under strong selective pressure to develop resistance that has the potential to spread from one environment to another. The acquisition of antibiotic resistance is not a barrier to the application of bacteriophage, as their natural diversity and rapid replication give them the potential to combat many types of resistance. Bacteriophage are already present in the environment to such an extent that their use will not constitute the addition of any new biologically active product, with the consequent benefit of the associated low risk of unwanted side effects, such as an allergic response.

Bacteriophage, like other viruses, are composed of nucleic acid surrounded by a protein coat (the capsid). The first stage of the bacteriophage life cycle usually involves absorption to a specific cell surface component of their host bacteria, followed by penetration of the nucleic acid through the cell wall. The nucleic acid then directs synthesis of further phage nucleic acid and protein capsids using the host cell biosynthetic apparatus. Once the phage particles have been assembled the cell envelope is ruptured (lysed) to release the phage progeny. Phage that undergo this type of life cycle without any other intermediate form are called lytic phage.

However, a group of phage called the temperate phage undergo an alternative life cycle where the nucleic acid can become incorporated into the host cell genome, where it remains and is replicated along with its host bacterium. This process is called lysogeny. Lysogenized phage can sometimes be induced to re-enter the lytic life cycle. When this happens they may carry additional excised host bacterial DNA, which can be introduced into a new host following infection by the phage. For this reason lysogenic phage are not suitable for phage therapy as they have the potential to transfer genetic material from one bacterium to another. These genetic traits may include virulence factors and antibiotic resistance. It is for this reason that lytic phage should be carefully selected in order to take advantage of the benefits of bacteriophage therapy.

CAMPYLOBACTER PHAGE

The first reports of the isolation of *Campylobacter* phage were in the 1960s, where phage against *Vibrio coli* (now known as *C. coli*) or *V. fetus* (now known as *C. fetus*) were isolated from cattle and pigs (Fletcher and Bertschinger, 1964; Firehammer and Border, 1968; Fletcher, 1968). In the early 1980s *C. jejuni* that were reported as having been isolated from aborted sheep fetuses were found to induce lysogenic phage when treated with mitomycin C (Bryner *et al.*, 1982). *Campylobacter* phage were also reported to play a role in the auto-agglutination of cells, which interfered with attempts to serotype *Campylobacter* isolates (Ritchie *et al.*, 1982). Poultry manure was one of the main sources of bacteriophage isolated in order to develop phage typing systems for *C. jejuni* and *C. coli* (Grawjewski *et al.*, 1985; Salama *et al.*, 1989; Khakhria and Lior, 1992; Frost *et al.*, 1999).

More recently, studies of the incidence of *Campylobacter* phage in UK poultry and in retail poultry products have been carried out (Atterbury *et al.*, 2003 a, b, 2005; El-Shibiny *et al.*, 2005) as a prelude to investigating their use in phage therapy aimed at reducing campylobacters in poultry, and their subsequent entering of the human food chain from this source (Connerton and Connerton, 2005).

Characteristics of *Campylobacter* phage

Some of the phage used in the UK phage typing system (Frost *et al.*, 1999) were characterized and subdivided into three groups based on their genome size, determined by pulsed field gel electrophoresis (PFGE; Sails *et al.*, 1998). Group I phage were those with head diameters of 140.6 and 143.8 nm and genome sizes in excess of 320 kb (Sails *et al.*, 1998). However, in our hands, using slightly different PFGE running conditions, the genome sizes of these same phage (NCTC 12676 and NCTC 1277) were found to be approximately 530 and 560 kb, respectively (our unpublished observations).

Group II phage had head diameters of 99 nm and average genome size of 184 kb (Sails *et al.*, 1998). The Group III phage had average head diameters of 100 nm and genomes sizes averaging 138 kb. *Campylobacter* phage are quite large compared to those that infect other genera, particularly those in Group I. This is especially notable considering the small physical size (0.2–0.5 μm diameter and approximately 5 μm length) of campylobacters.

In common with many other types of phage, genomic DNA from *Campylobacter* phage is refractory to digestion by many restriction enzymes. All of the phage isolated from poultry and characterized in our laboratory have been members of either Group II or Group III (Atterbury *et al.*, 2003a; Connerton *et al.*, 2004; El-Shibiny *et al.*, 2005; Loc Carrillo *et al.*, 2005). The implication of this is that the Group I phage are uncommon in UK poultry. The Group II and III phage were all lytic phage typical of the *Myoviridae* family (International Committee on the Taxonomy of Viruses, 2004). Lysogenic phage have, however, been isolated from *C. fetus* (Ritchie *et al.*, 1982) and from *C. jejuni* isolated from aborted sheep fetuses (Bryner *et al.*, 1982).

Recent genome sequence data (Fouts *et al.*, 2005) showed the presence of Mu-like phage sequences in *C. jejuni* RM1221 and that these sequences were present in other *C. jejuni* strains but not in the prototype genomic sequence of *C. jejuni* 11168 (Parkhill *et al.*, 2000). The majority of the open reading frames within the prophage regions are hypothetical, so it is unknown what possible function the genes encoded by these sequences have, if any.

Survival of *Campylobacter* phage

The ability of phage to survive processing is an important aspect of their potential use in the biocontrol of *Campylobacter* in poultry production. Procedures have been developed to recover *Campylobacter* bacteriophage from chilled and frozen retail poultry and have been validated using a characterized *Campylobacter* phage NCTC 12674 (Atterbury *et al.*, 2003a). Survival experiments demonstrated that bacteriophage can be recovered up to 10 days following inoculation, which is well beyond the stated shelf life of the product.

Phage have the advantage that they are fairly robust in nature and therefore can be simply added to drinking water and in feed provided the intended targets are intestinal bacteria. However, some phage may be sensitive to the low pH encountered in the stomach or proventriculus (Leverentz *et al.*, 2003; Loc Carrillo, 2004). These problems can be overcome through the use of antacid or by selection of appropriate low pH-tolerant phage. Antacids such as Maalox (aluminium and magnesium hydroxide) or calcium carbonate have been used to improve the ability of phage to survive low acidity in digestive systems (Smith *et al.*, 1987b; Koo *et al.*, 2001).

Incidence of *Campylobacter* phage in UK poultry

The incidence of *Campylobacter* phage, in intensively reared chickens, was determined between August and September 2002, from 22 farms belonging to three national poultry producers based in the UK (Atterbury *et al.*, 2005). *Campylobacter jejuni* was isolated from 63% (129/205) of the caeca of these birds and lytic bacteriophage of *C. jejuni* were isolated from 20% (41/205). Enumeration of campylobacters determined that the mean number in the presence of bacteriophage was $\log_{10}$ 5.1 cfu/g of caecal contents, while it was $\log_{10}$ 6.9 cfu/g when phage were absent (Atterbury *et al.*, 2005). This $\log_{10}$ 1.8 cfu/g difference was a significant ($p < 0.001$) reduction in numbers comparing caecal contents with or without phage. Clearly, the presence of phage already influence the numbers of campylobacters in poultry without any therapeutic intervention.

Organic flocks are generally close to 100% colonized by *Campylobacter* (Heuer *et al.*, 2001). The incidence of bacteriophage in *Campylobacter*-positive organic birds was found to be 51% (El-Shibiny *et al.*, 2005). Presumably this figure is higher than that in intensively reared birds (20%; Atterbury *et al.*, 2005) due to organic birds having greater exposure to the environment and therefore to a greater range of *Campylobacter* types and their phage.

The study of a broiler chicken barn, in which flocks were naturally infected with *Campylobacter* and bacteriophage, indicated that the phage and its host *Campylobacter* could be carried over from one flock to the next (Connerton *et al.*, 2004). The role of developing bacteriophage resistance was investigated and the phage-sensitive hosts were found to be succeeded by unrelated campylobacters that were insensitive to the resident phage. This succession occurred due to incursion of new genotypes rather than *de novo* development of resistance. The implication from this work was that while campylobacters like other bacteria can mutate to become resistant to phage, this may have negative consequences in terms of fitness. The role of developing resistance will be discussed further below.

BACTERIOPHAGE THERAPY

Using bacteriophage to control bacterial pathogens is not a new idea, but Western countries have been somewhat slow to embrace bacteriophage therapy. This is primarily because antibiotics were, in the past, deemed to be superior and were developed and refined on a large scale. There was also a lack of consistent evidence of phage efficacy. However, former Soviet Union countries have embraced phage therapy and developed procedures over many years (Sulakvelidze *et al.*, 2001).

The failure of Western scientists to achieve consistent results initially may have resulted from a general lack of understanding of phage biology and replication dynamics. Phage replication is critically dependent on bacterial density. There is predicted to be a distinct threshold above which phage numbers increase and below which they decrease, termed the phage proliferation threshold (Payne and Jansen, 2003). The outcome of phage therapy also depends on the various life history parameters, including the inoculum size and the inoculum timing (Payne and Jansen, 2001; Weld *et al.*, 2004).

From the previous sections it is clear that the use of bacteriophage to reduce the numbers of campylobacters entering the food chain at farm level is a feasible strategy. Reductions in the numbers of campylobacters in chickens could lead to a measurable reduction in carcass contamination (Connerton and Connerton, 2005). With more than 80% of birds in the UK harbouring these organisms as a part of their normal intestinal flora (Newell and Wagenaar, 2000; Corry and Atabay, 2001), a reduction of *Campylobacter* from poultry sources is certainly a desirable objective. The fact that the bacteria are present in the intestines of poultry at very high densities – ranging from $\log_{10} 4$ to $\log_{10} 8$ (Rudi *et al.*, 2004) – is a factor that makes phage treatment feasible.

In addition, bacteriophage can readily be isolated from poultry excreta, which means that their potential therapeutic application in animals would not introduce any new biological entity into the food chain. A growing number of studies have successfully used phage to treat animal diseases (reviewed by Barrow, 2001; Joerger, 2003; Payne and Jansen, 2003). These include applications to treat *Salmonella* (Berchieri *et al.*, 1991; Sklar and Joerger, 2001) and *E. coli* (Barrow *et al.*, 1998; Huff *et al.*, 2003, 2004) infections of young chickens.

The use of phage to control *Campylobacter* in chickens differs from these previous studies as campylobacters may be considered a natural part of the commensal microflora of poultry species, and not specifically as pathogens.

However, as noted earlier, concerns have been raised that campylobacters will simply become resistant to bacteriophage, rendering this strategy ineffective in the long term (Barrow, 2001). This will be discussed further in the next section.

To test the efficacy of phage therapy for *Campylobacter* it was first necessary to design and evaluate experimental models of *Campylobacter* infection in chickens (Newell and Wagenaar, 2000). This is particularly important as strains of *C. jejuni* are variable in their ability to colonize chickens (Ringoir and Korolik, 2003). Once the colonization model has been established it was important to test candidate bacteriophage for their efficacy *in vitro* prior to use in experimental birds. This has been done in our laboratory with phage isolates from broiler chickens (Loc Carrillo *et al.*, 2005).

Following the *in vitro* experiments, bacteriophage were administered to the *Campylobacter*-colonized birds at three different doses ($\log_{10} 5$, $\log_{10} 7$ and $\log_{10} 9$ PFU) in antacid suspension. The numbers of campylobacters in the caecal contents of phage-treated and untreated birds were determined at 24 h intervals over 5 days. The bacteriophage-treated birds showed a marked reduction in numbers of campylobacters, particularly for the first 48 h after treatment. The reduction in caecal *C. jejuni* numbers varied with the phage from $\log_{10} 2$ to $\log_{10} 5$ (per g of caecal contents) compared with controls. The success of treatment was dependent on the bacteriophage and the colonization strain used, with some bacteriophage being considerably more virulent than others.

A *Campylobacter* bacteriophage isolated from poultry meat rather than poultry excreta was found to be ineffective in a similar trial, despite the fact that the phage was virulent *in vitro* on laboratory media (Atterbury, 2003). Optimization of dose and selection of appropriate phage were found to be the key elements in the use of phage therapy in reducing campylobacters in broiler chickens.

Wagenaar *et al.* (2005) reported the use of the *Campylobacter*-typing phage 69 (NCTC 12669) to prevent, as well as to reduce, *Campylobacter* colonization of broiler chickens. The administration of phage resulted in a $\log_{10} 3$ decline in *C. jejuni* caecal counts. Preventative phage treatment delayed the onset of *C. jejuni* colonization, and the peak titres remained $\log_{10} 2$ lower than the controls. There were no adverse health effects from the phage treatments. A second phage (NCTC 12671) was added to the original, in an attempt to counter any tendency towards development of phage resistance in the *Campylobacter* population. However, this strategy did not result in any significant increase in efficacy. The use of these model systems provides valuable data and indicates the potential of phage therapy to reduce campylobacters.

ACQUISITION OF PHAGE RESISTANCE

Bacteria rapidly mutate to become resistant to bacteriophage *in vitro* on laboratory media (Adams, 1959). However, bacteriophage constantly evolve to circumvent host barriers to infection. This leads to an evolutionary balance that allows both host and prey to proliferate, otherwise one or the other would not survive.

The numbers of phage-resistant campylobacters present in the experimental birds following bacteriophage therapy was determined and found to be < 4% (Loc Carrillo *et al.*, 2005). Interestingly, these resistant isolates appeared to be compromised in their ability to colonize a further set of experimental birds, rapidly reverting back to the phage-sensitive phenotype in the absence of phage. In contrast, when phage resistant isolates were obtained by mixing host and phage *in vitro* and culturing on laboratory media, the frequency of resistance was 11% and the resistant phenotype was maintained on subculture (Loc Carrillo *et al.*, 2005). This level of mutation means that phage-resistant types can soon dominate phage-challenged laboratory cultures, but this does not happen in chickens.

Other studies have noted decreased fitness when bacteria acquire phage resistance through mutation (Smith *et al.*, 1987a; Park *et al.*, 2000). This is often because phage have evolved to target essential structures present on their outer surface as their receptors to gain entry to the cell. If bacteria alter these structures through mutation it may prevent phage entry but can also be detrimental to their fitness, such that they do not survive in competitive or environmental challenge. This may not always be apparent when grown on laboratory media that provide an ideal growth medium.

In the wider environment, field studies indicate that phage may enable the succession of new phage-insensitive bacterial types rather than the selection and proliferation of phage-resistant mutants. The use of a cocktail of phage with diverse specificity will prevent this type of succession occurring following phage therapy.

PHAGE TREATMENT OF POULTRY MEAT

Bacteriophage have been successfully applied as a decontamination technique to reduce *C. jejuni* and *Salmonella* Enteriditis on poultry meat under experimental conditions (Atterbury *et al.*, 2003b; Goode *et al.*, 2003). The most effective treatments involved using high doses of phage. Campylobacters are generally believed to be unable to multiply under refrigeration conditions, so the effects are likely to be limited to 'lysis from without' or by prevention of regrowth once suitable conditions for growth are provided, i.e. when contaminated food is consumed. Phage inoculated on to the surface of skin contaminated with *Campylobacter* exhibit a control effect even in the absence of host growth.

CONCLUSIONS AND FURTHER WORK

Research in any area is driven by the need to solve a particular problem. Much of the initial studies of *Campylobacter* phage were carried out with the aim of producing reliable and practical phage-typing methods for epidemiological tracking of strains. At the time this work was undertaken, this aspect was a particular problem. In recent years phage typing has largely been superseded by genetic methods (reviewed by Wassenaar and Newell, 2000) such as multilocus sequence typing (MLST), *fla* typing, pulsed-field gel electrophoresis (PFGE) and amplified fragment length polymorphism (AFLP), all of which are highly reproducible and widely available.

However, the rising level of resistance to various antimicrobials has fuelled a new wave of research into both *Campylobacter* bacteriophage and the bacteriophage of other important pathogens. There are many reviews on the subject (Barrow, 2001; Sulakvelidze *et al.*, 2001; Summers, 2001; Huff *et al.*, 2005), but relatively few actual experimental trials to prove the efficacy of phage in real situations.

As *Campylobacter* is a particularly good target for phage therapy, we anticipate that over the next few years more experimental ground work will be performed with this organism. This will advance the possibility of using phage therapy to reduce the numbers of campylobacters entering the human food chain and causing enteritis. It has been calculated that a 2 $\log_{10}$ decline in *Campylobacter* contamination of food would result in a 30-fold decrease in cases of *Campylobacter* foodborne disease (Rosenquist *et al.*, 2003). Such a decrease has been shown to be feasible by our experiments (Loc Carrillo *et al.*, 2005) and by those of Wagenaar *et al.* (2005).

We anticipate that different classes of phage may also be used in a variety of ways to study *Campylobacter* through the development of vectors for gene transfer and transposition. Moreover, the determination of phage genome sequences may provide new opportunities for devising measures to specifically target the colonization of poultry by campylobacters, and possibly for the treatment of human infection. Genes encoding new enzymes and proteins may be discovered and an understanding of phage-mediated biological processes may be more forthcoming.

REFERENCES

Adams, M.H. (1959) *Bacteriophages.* Interscience Publishers Inc., New York.

Atterbury, R.J. (2003) Bacteriophage control of campylobacters in retail poultry. PhD Thesis, University of Nottingham, UK.

Atterbury, R.J., Connerton, P.L., Dodd, C.E.R., Rees, C.E.D. and Connerton, I.F. (2003a) Isolation and characterization of *Campylobacter* bacteriophages from retail poultry. *Applied and Environmental Microbiology* 69, 4511–4518.

Atterbury, R.J., Connerton, P.L., Dodd, C.E.R., Rees, C.E.D. and Connerton, I.F. (2003b) Application of host-specific bacteriophages to the surface of chicken skin leads to a reduction in recovery of *Campylobacter jejuni*. *Applied and Environmental Microbiology* 69, 6302–6306.

Atterbury, R.J., Dillon, E., Swift, C., Connerton, P.L., Frost, J.A., Dodd, C.E.R., Rees, C.E.D. and Connerton, I.F. (2005) Correlation of *Campylobacter* bacteriophage with the reduced presence of their hosts in broiler chicken ceca. *Applied and Environmental Microbiology* 71, 4885–4887.

Barrow, P.A. (2001) The use of bacteriophages for treatment and prevention of bacterial disease in animals and animal models of human infection. *Journal of Chemical Technology and Biotechnology* 76, 677–682.

Barrow, P., Lovell, M. and Berchieri, A. (1998) Use of lytic bacteriophage for control of experimental *Escherichia coli* septicemia and meningitis in chickens and calves. *Clinical and Diagnostic Laboratory Immunology* 5, 294–298.

Berchieri, A., Lovell, M.A. and Barrow, P.A. (1991) The activity in the chicken alimentary tract of bacteriophages lytic for *Salmonella* Typhimurium. *Research in Microbiology* 142, 541–549.

Bryner, J.H., Ritchie, A.E. and Foley, J.W. (1982) Techniques for phage typing *Campylobacter jejuni*. In: Newell, D.G. (ed.) *Campylobacter: Epidemiology, Pathogenisis and Biochemistry,* MTP Press Ltd., Lancaster, UK, pp. 52–56.

Connerton, P.L. and Connerton, I.F. (2005) Natural *Campylobacter* control with bacteriophage. *World Poultry (Salmonella & Campylobacter Special)*, 25–26.

Connerton, P.L., Loc Carrillo, C.M., Swift, C., Dillon, E., Scott, A., Rees, C.E.D., Dodd, C.E.R., Frost, J. and Connerton, I.F. (2004) A longitudinal study of *Campylobacter jejuni* bacteriophage and their hosts from broiler chickens. *Applied and Environmental Microbiology* 70, 3877–3883.

Corry, J.E.L. and Atabay, H.I. (2001) Poultry as a source of *Campylobacter* and related organisms. *Journal of Appied Microbioogy* 90, 96S–114S.

D'Herelle, F. (1922) *The Bacteriophage: its Role in Immunity*. Williams and Wickens Co./Waverley Press, Baltimore, Ohio.

El-Shibiny, A., Connerton, P.L. and Connerton, I.F. (2005) Enumeration and diversity of campylobacters and bacteriophages isolated during the rearing cycles of free-range and organic chickens. *Applied and Environmental Microbiology* 71, 1259–1266.

Firehammer, B.D. and Border, M. (1968) Isolation of temperate bacteriophages from *Vibrio fetus*. *American Journal of Veterinary Research* 29, 2229–2235.

Fletcher, R.D. (1968) Activity and morphology of *Vibrio coli* phage. *American Journal of Veterinary Research* 26, 361–364.

Fletcher, R.D. and Bertschinger, H. (1964) A method of isolation of *Vibrio coli* from swine fecal material by selective filtration. *Zentralblatt für Veterinaeromed Reiche B* 11, 469–474.

Fouts, D.E., Mongodin, E.F., Mandrell, R.E., Miller, W.G., Rasko, D.A., Ravel, J., Brinkac, L.M., DeBoy, R.T., Parker, C.T., Daugherty, S.C., Dodson, R.J., Durkin, A.S., Madupu, R., Sullivan, S.A., Shetty, J.U., Ayodeji, M.A., Shvartsbeyn, A., Schatz, M.C., Badger, J.H., Fraser, C.M. and Nelson, K.E. (2005) Major structural differences and novel potential virulence mechanisms from the genomes of multiple *Campylobacter* species. *Public Library of Science: Biology* 3, e15.

Frost, J.A., Kramer, J.M. and Gillanders, S.A. (1999) Phage typing of *Campylobacter jejuni* and *Campylobacter coli* and its use as an adjunct to serotyping. *Epidemiology and Infection* 123, 47–55.

Goode, D., Allen, V.M. and Barrow, P.A. (2003) Reduction of experimental *Salmonella* and *Campylobacter* contamination of chicken skin by application of lytic bacteriophages. *Applied and Environmental Microbiology* 69, 5032–5036.

Grawjewski, B.A., Kusek, J.W. and Gelfand, H.M. (1985) Development of a bacteriophage typing scheme for *Campylobacter jejuni* and *Campylobacter coli*. *Epidemiology and Infection* 104, 403–414.

Hankin, E.H. (1896) L'action bactericide des eaux de la Jumna et du Gange sur le vibrion du cholera. *Annales de l'Institut Pasteur* 10, 511.

Heuer, O.E., Pedersen, K., Andersen, J.S. and Madsen, M. (2001) Prevalence and antimicrobial susceptibility of thermophilic *Campylobacter* in organic and conventional broiler flocks. *Letters in Applied Microbiology* 33, 269–274.

Huff, W.E., Huff, G.R., Rath, N.C., Balog, J.M. and Donoghue, A.M. (2003) Bacteriophage treatment of a severe *Escherichia coli* respiratory infection in broiler chickens. *Avian Diseases* 47, 1399–1405.

Huff, W.E., Huff, G.R., Rath, N.C., Balog, J.M. and Donoghue, A.M. (2004) Therapeutic efficacy of bacteriophage and Baytril (enrofloxacin) individually and in combination to treat coli-bacillosis in broilers. *Poultry Science* 47, 1944–1947.

Huff, W.E., Huff, G.R., Rath, N.C., Balog, J.M. and Donoghue, A.M. (2005) Alternatives to antibiotics: utilization of bacteriophage to treat colibacillosis and prevent foodborne pathogens. *Poultry Science* 84, 655–659.

International Committee on the Taxonomy of Viruses (2004) *Myoviridae* (http:// www.ncbi.nlm.nih.gov/ICTVdb/Ictv/fs_myovi.htm, accessed July 2005).

Joerger, R.D. (2003) Alternatives to antibiotics: bacteriocins, antimicrobial peptides and bacterio-phages. *Poultry Science* 82, 640–647.

Khakhria, R. and Lior, H. (1992) Extended phage-typing scheme for *Campylobacter jejuni* and *Campylobacter coli*. *Epidemiology and Infection* 108, 403–414.

Koo, J., Marshall, D.L. and DePaola, A. (2001) Antacid increases survival of *Vibrio vulnificus* and *Vibrio vulnificus* phage in a gastrointestinal model. *Applied and Environmental Microbiology* 67, 2895–2902.

Leverentz, B., Conway, W.S., Camp, M.J., Janisiewicz, W.J., Abuladze, T., Yang, M., Saftner, R. and Sulakvelidze, A. (2003) Biocontrol of *Listeria monocytogenes* on fresh-cut produce by treatment with lytic bacteriophages and a bacteriocin. *Applied and Environmental Microbiology* 69, 4519–4526.

Loc Carrillo, C.M. (2004) Bacteriophage control of campylobacters in poultry production. PhD Thesis, University of Nottingham, UK.

Loc Carrillo, C.M., Atterbury, R.J., El-Shibiny, A., Connerton, P.L., Dillon, E., Scott, A. and Connerton, I.F. (2005) Bacteriophage therapy to reduce *Campylobacter jejuni* colonization of broiler chickens. *Applied and Environmental Microbiology* 71 (11), 6554–6563.

Newell, D.G. and Wagenaar, J.A. (2000) Poultry infections and their control at farm level. In: Nachamkin, I. and Blaser, M.J. (eds) *Campylobacter*, 2nd edn. ASM press, Washington DC, pp. 497–510.

Park, S.C., Shimamura, I., Fukunaga, M., Mori, K.I. and Nakai, T. (2000) Isolation of bacterio-phages specific to a fish pathogen, *Pseudomonas plecoglossicida*, as a candidate for disease control. *Applied and Environmental Microbiology* 66, 1416–1422.

Parkhill, J., Wren, B.W., Mungall, K., Ketley, J.M., Churcher, C., Basham, D., Chillingworth, T., Davies, R.M., Feltwell, T., Holroyd, S., Jagels, K., Karlyshev, A.V., Moule, S., Pallen, M.J., Penn, C.W., Quail, M.A., Rajandream, M.A., Rutherford, K.M., van Vliet, A.H., Whitehead, S. and Barrell, B.G. (2000) The genome sequence of the foodborne pathogen *Campylobacter jejuni* reveals hypervariable sequences. *Nature* 10, 665–668.

Payne, R.J. and Jansen, V.A. (2001) Understanding bacteriophage therapy as a density-dependent kinetic process. *Journal of Theoretical Biology* 208, 37–48.

Payne, R.J. and Jansen, V.A. (2003) Pharmacokinetic principles of bacteriophage therapy. *Clinical Pharmacokinetics* 42, 315–325.

Ringoir, D.D. and Korolik, V. (2003) Colonisation phenotype and colonization potential differences in *Campylobacter jejuni* strains in chickens before and after passage *in vivo*. *Veterinary Microbiology* 92, 225–235.

Ritchie, A.E., Bryner, J.H. and Foley, J.W. (1983) Role of DNA and bacteriophage in *Campylobacter* auto-agglutination. *Journal of Medical Microbiology* 16, 333–340.

Rosenquist, H., Nielsen, N.L. Sommer, H.M., Norrung, B. and Christensen, B.B. (2003) Quantitative risk assessment of human campylobacteriosis associated with thermophilic *Campylobacter* species in chickens. *International Journal of Food Microbiology* 83, 87–103.

Rudi, K., Hoidal, H.K., Katla, T., Johansen, B.K., Nordal, J. and Jakobsen, K.S. (2004) Direct real-time PCR quantification of *Campylobacter jejuni* in chicken fecal and cecal samples by integrated cell concentration and DNA purification. *Applied and Environmental Microbiology* 70, 790–797.

Sails, A.D., Wareing, D.R.A., Bolton, F.J., Fox, A.J. and. Curry, A. (1998) Characterisation of 16 *Campylobacter jejuni* and *C. coli* typing bacteriophages. *Journal of Medical Microbiology* 47, 123–128.

Salama, S.M., Bolton, F.J. and Hutchinson, D.N. (1989) Improved method for the isolation of *Campylobacter jejuni* and *Campylobacter coli* bacteriophages. *Letters in Applied Microbiology* 8, 5–7.

Sklar, I.B. and Joerger, R.D. (2001) Attempts to utilize bacteriophage to combat *Salmonella enterica* serovar Enteritidis infection in chickens. *Journal of Food Safety* 21, 15–30.

Smith, H.W., Huggins, M.B. and Shaw, K.M. (1987a) The control of experimental *Escherichia coli* diarrhoea in calves by means of bacteriophages. *Journal of General Microbiology* 133, 1111–1126.

Smith, H.W., Huggins, M.B. and Shaw, K.M. (1987b) Factors influencing the survival and multiplication of bacteriophages in calves and in their environment. *Journal of General Microbiology* 133, 1127–1135.

Sulakvelidze, A., Alavidze, Z. and Morris, J.G. (2001) Bacteriophage therapy. *Antimicrobial Agents and Chemotherapy* 45, 649–659.

Summers, W.C. (2001) Bacteriophage therapy. *Annual Reviews in Microbiology* 55, 437–451.

Twort, F.W. (1915) An investigation on the nature of ultramicroscopic viruses. *Lancet* 2, 1241.

Wagenaar, J.A., Van Bergen, M.A.P., Mueller, M.A., Wassenaar, T.M. and Carlton, R.M. (2005) Phage therapy reduces *Campylobacter jejuni* colonization in broilers. *Veterinary Microbiology* 109 (3/4), 275–283.

Wassenaar, T.M. and Newell, D.G. (2000) Genotyping of *Campylobacter* spp. *Applied and Environmental Microbiology* 66, 1–9.

Weld, R.J., Butts, C. and Heinemann, J.A. (2004) Models of phage growth and their applicability to phage therapy. *Journal of Theoretical Biology* 227, 1–11.

CHAPTER 21
Breeding for disease resistance

S.C. Bishop

Roslin Institute, Midlothian, UK; e-mail: stephen.bishop@bbsrc.ac.uk

ABSTRACT

Breeding for increased resistance has a major role to play in disease control in poultry and in the maintenance of flock health. This is because of: (i) the urgent need for new and sustainable strategies to assist in disease control; (ii) the existence of genetic variation in resistance to many economically important diseases in poultry; and (iii) the advent of molecular tools that enable breeders to exploit these genetic effects.

Avian diseases for which genetic variation in either resistance to infection or the ability of the host to tolerate infection include Marek's disease, infectious laryngotracheitis, avian leukosis, infectious bursal disease, avian infectious bronchitis, Rous sarcoma, Newcastle disease, *E. coli*, pullorum, fowl typhoid, salmonellosis, coccidiosis and *Ascaridia galli*. Marek's disease is the classic example where genetics has assisted in disease control.

Because poultry units usually attempt to minimize exposure to infection, breeding strategies will generally be aimed at minimizing epidemic risks (or epidemic severity should an epidemic arise) rather than towards helping birds cope with endemic diseases. Therefore, selection will usually need to be undertaken on birds that are not exposed to infection, and genetic markers will generally be required. Commonly, genetic markers will only be associated with a proportion of the between-animal variation, therefore several markers may be required, and to minimize risks the breeding strategy should be combined with other disease control approaches.

Utilizable genetic markers have been identified for several diseases, including Marek's disease and salmonellosis. Much of current research aims to elucidate the genetic control of resistance to infectious diseases, especially zoonotic infections, and to refine known markers. In particular, it is predicted that the integration of gene mapping and microarray studies will assist in this task.

© CAB International 2006. *Avian Gut Function in Health and Disease* (ed. G.C. Perry)

INTRODUCTION

This chapter considers the role of genetics and breeding for disease resistance as part of a wider strategy towards disease control and the maintenance of flock health. General principles in breeding for disease resistance will be outlined, and applications to poultry diseases will be given where appropriate, although lessons from other species will be taken on board. Lastly, future research requirements to enable breeding for disease resistance in practice will be considered.

First, it is useful to define what is meant by disease and disease resistance, as two separate concepts are often confused: infection and disease itself. Infection is defined in this chapter as the colonization of a host animal by a parasite, where 'parasite' is a general term to describe an organism with a dependence upon a host. Parasites will include pathogens or microparasites with a direct dependence upon the host, such as viruses, bacteria and protozoa, as well as macroparasites that complete some part of their life cycle external to the host, such as helminths, flies or ticks. Disease describes the side effects of infection by a parasite or pathogen. Disease may take several forms – acute, sub-acute, chronic or subclinical – which may or may not be debilitating.

In terms of disease resistance, it is also necessary to distinguish between resistance *per se* and tolerance. An individual host may be infected, but suffer little or no harm. This is known as tolerance. In contrast, resistance is the ability of the individual host to resist infection or control the parasite life cycle, e.g. reduce proliferation of the virus or bacterium within the host. The distinction between resistance and tolerance is important when considering the impact of selecting for disease resistance, as described below. In general terms, if genetic improvement is made in host resistance to infection there will be a reduction in the transmission of infection to other animals. Conversely, genetic improvement of tolerance may reduce clinical signs of disease, but may not reduce transmission of infection to other animals.

Breeding for disease resistance should be considered as part of a wider disease control strategy. Such strategies may include interventions or decisions affecting the animal, the pathogen or the environment. For example, the animal may be influenced by means of vaccination, the selection of resistant animals or the culling of infected animals. Pathogen-based interventions include the use of chemotherapies, e.g. antibiotics or anthelmintics. Environmental interventions include biosecurity, sanitation, improved husbandry, etc.

However, despite the range of disease control options, there are many continuing challenges relating to animal health and disease. Many previously used control strategies are now less available due to either legislation (e.g. antibiotic usage) or evolution of the pathogen in avoiding the control strategy (e.g. antibiotic or anthelmintic resistance); new issues arise continually and some disease control problems simply remain unsolved.

DISEASE GENETICS: GENERAL PRINCIPLES

Host genetic variation in disease resistance, i.e. in the ability of the host animal either to control the extent of infection or to tolerate infection, appears to be ubiquitous. For example, Bishop (2004) and Gibson (2002) give partial summaries of more than 50 diseases for which there is documented or strong anecdotal evidence of genetic variation in host resistance or tolerance among the major domestic livestock species. These examples include all major livestock species, as well as all categories of parasite or pathogen.

However, it is often not clear whether the observed genetic variation is for resistance to infection, tolerance of infection or a combination of both. Nevertheless, it is probably safe to conclude that for almost every disease that has been intensively and carefully investigated, evidence has been found for host genetic variation in either resistance or tolerance. Further examples are likely to be found as experimental studies become more sophisticated and exploit new genomic tools, and also as data collection from commercial populations increases.

Many processes may control tolerance or resistance. For example:

- The host may have an appropriately targeted immune response against the pathogen. This may enable the animal to successfully combat the infection or avoid pathogenic effects of disease.
- The animal may have non-immune response genes that preclude infection, or limit infection in target organs. Examples include the binding protein genes that allow specific strains of *E. coli* – e.g. those with K88 or F18 fimbriae – to attach to the gut of the pig, resulting in pre-weaning or post-weaning diarrhoea. Absence of appropriate binding proteins results in resistance to infection.
- The animal may have physical attributes which make infection by the pathogen difficult. An example is the role of skin thickness in conferring resistance to ticks, or attributes of the mucosal surface.
- The animal may have behavioural attributes which enables it to avoid infection. An example is the hygienic behaviour of honeybees, their dominant natural defence against various brood diseases.

The genetic architecture of resistance or tolerance varies dramatically according to the disease. In some cases, genetic control may be due predominantly to a single gene. Examples include resistance to *E. coli* diarrhoea in pigs and the PrP gene which is associated with resistance to scrapie in sheep. However, in other cases, resistance or tolerance is due to the combined effects of several or many genes. This genetic architecture will also determine whether resistance and tolerance are qualitative phenomena, as in the case of *E. coli* diarrhoea in pigs, or quantitative phenomena (i.e. they show continuous variation from one extreme to the other). Continuous variation is expected when resistance or tolerance is due to the combined effects of several genes.

GENETICS OF DISEASE RESISTANCE IN POULTRY

There are a large number of diseases for which host genetic variation has been demonstrated in poultry, including: Marek's disease, infectious laryngotracheitis, avian leucosis, infectious bursal disease, avian infectious bronchitis, Rous sarcoma, Newcastle disease, *E. coli*, pullorum, fowl typhoid, salmonellosis, *Campylobacter*, coccidiosis and *Ascaridia galli*. Knowledge of genes underlying genetic variation in resistance to these diseases is likely to increase considerably in the next few years, due to utilization of the recently published chicken genome sequence and the accompanying dense SNP map. A selection of these diseases is now considered.

Marek's disease

Marek's disease is a lymphoproliferative disease of chickens caused by the MD virus, an oncogenic α-herpesvirus. It is characterized by immunosuppression and the development of tumours in various organs. The economic impact of MD on the world poultry industry is great and it is the most serious chronic disease in the poultry industry.

Losses caused by mortality, reduced egg production and meat contamination were estimated to approach $1 billion (Purchase, 1985). Infections with MDV can kill 40% or more of unvaccinated flocks. The growing frequency of highly virulent MDV strains in the field and the increasing inability of MDV vaccines to provide high levels of protection raise the need for genetic strategies to control MD.

Genetic differences in resistance to Marek's disease have been known for more than 60 years, with conclusive evidence documented by Cole (1968), who showed strong responses to selection after only two generations. Genetic management of Mareks's disease is well established in intensive poultry production systems. Selection for resistance based on both response to infection (Friars *et al.*, 1972) and specific B alleles within the MHC (major histopatibility complex) (Bacon, 1987) has been used for many years to assist in disease management. Concurrent vaccination strategies, whilst critical in disease control, have possibly had deleterious consequences as evolution of ever more pathogenic strains of virus has followed each new vaccine (Witter, 1998). Arguably, intensive poultry production would no longer have been possible in many circumstances without genetic selection.

However, genetic control of resistance of MD is considerably more complex than the MHC effects, as one research group (led by Cheng) has identified up to 14 Quantitative Trait Loci (QTL) associated with resistance to MD (Vallejo *et al.*, 1998; Yonash *et al.*, 1999). Bumstead (1998a) identified an additional and separate resistance QTL. The QTL described by Yonash *et al.* (1999) are for nine different traits describing the proliferation of tumours, survival and viraemia, with an average of four QTL per trait (range 2–7). This team of researchers has now also detected resistance QTL in commercial populations (Cheng, 2005), with an indication that some QTL are common across populations.

It is difficult, by linkage mapping, to refine the intervals containing QTL sufficiently tightly to enable straightforward identification of candidate genes. Hence, other techniques must be used to achieve this. For example, Liu *et al.* (2001) described an early approach using gene expression microarray techniques in an attempt to resolve some of the QTL identified by Cheng's group. Subsequently, using a combination of microarray and virus–host protein–protein interaction screens, Cheng (2005) reported several novel host–pathogen interactions, with the identification of the growth hormone gene as a candidate MD resistance gene.

Avian leukosis

Avian leukosis is an economically important virus infection that can cause cancer-like disease and other production problems in meat-type chickens. In particular, subgroup J avian leukosis (ALV-J) is an emerging economically important viral infection. Poultry companies that produce primary breeding stock are trying to remove this virus from their chickens because it is commonly transmitted through the egg to progeny. The main approach is to test hens and remove all those that are positive for virus. However, there is no agreement on which tests are best and when the tests should be performed.

In common with Marek's disease and Rous sarcoma, MHC B alleles play a major role in resistance to ALV viraemia and subsequent tumour development (Bacon, 1987). There is also a relationship between MHC genotype and the risk of becoming a virus shedder (Yoo and Sheldon, 1992). Additionally, resistance to infection per se for at least some strains of ALV (subgroups A and B) is inherited as a single dominant gene (described by Bumstead, 1998b), this gene coding for the viral receptor (Young *et al.*, 1993). This gene potentially allows for direct DNA-based selection for resistance to infection. However, the chromosomal location of this gene is yet to be refined. More recent studies have confirmed an effect of various B haplotypes on immune response to infection; however, line differences – explained by other genetic factors – appear to have a stronger influence (Williams *et al.*, 2004; Mays *et al.*, 2005).

Infectious bursal disease

Infectious bursal disease (IBD or Gumboro disease) is an immunosuppressive viral disease, with clinical effects in birds over 2 weeks of age. In the clinical disease, morbidity can be seen in nearly 100% of the flock and mortality can range from 0 to over 50% with some very virulent IBDV strains. The World Organisation for Animal Health (OIE) estimates that IBD is present in more than 95% of the Member Countries. The acute clinical form of IBD caused by vvIBDV isolates has been observed in over 80% of these countries. It has been reported in Europe, Asia, Africa, South America and Central America.

Line differences in mortality have been observed with crossing experiments, indicating the likely involvement of a single gene (Bumstead *et al.*, 1993). Subsequently, it was shown that MHC alleles were associated with both response to vaccination and with disease severity in deliberately infected birds (previously vaccinated). Additionally, large breed differences have recently been reported in mortality rate following challenge with vvIBDV (Hassan *et al.*, 2004).

Rous sarcoma

The Rous sarcoma virus is a retrovirus causing cancer in chickens. Whilst Rous sarcoma is not an economically important disease in chickens, the study of this virus has made major contributions to cancer genetics. The possibility of selecting for RSV resistance was noted by Gyles and Brown (1971), when they developed lines in which the sarcomas would progress or regress. In other words, this genetic control is not for susceptibility to infection, but the ability to control the effects of infection.

Many studies have linked genetic differences between chickens in the progression of Rous sarcomas to specific MHC B alleles (e.g. Collins *et al.*, 1977; Bacon *et al.*, 1981; Hala *et al.*, 1998; Kaufman and Venugopal, 1998), a result often found with virus-induced conditions. However, genes outside the MHC complex tumours in chickens are likely to be involved. The results of Young *et al.* (1993) indicate a gene for resistance to infection to Rous sarcoma as well as ALV, this gene coding for the virus receptor. More recently, successful divergent selection for tumour progression has been reported in a population that was homozygous for the B-19 MHC (Pinard-van der Laan *et al.*, 2004).

Newcastle disease

Exotic Newcastle disease is a contagious and fatal viral disease affecting all species of birds and is probably one of the most infectious diseases of poultry in the world. Newcastle disease is endemic in many countries of Asia, the Middle East, Africa and Central and South America. Some European countries are considered free of Newcastle disease.

It has long been realized that genetic variation exists between birds in their ability to cope with infection, specifically in their mortality post-infection (Gordon *et al.*, 1971). Additionally: (i) genetic variation has been shown to exist for antibody response to NDV vaccines in turkeys (Sacco *et al.*, 1994), with heritabilities being relatively high, i.e. > 0.3; (ii) lines of turkeys previously selected for various performance traits differed in their Newcastle disease mortality (Tsai *et al.*, 1992), with increased meat production apparently being adversely correlated with survival; (iii) large breed differences have been reported in mortality rate following challenge with NDV (Hassan *et al.*, 2004); and (iv) QTL for antibody response to NDV have been located (Yonash *et al.*, 2001).

Therefore, the possibility of selecting for resistance to Newcastle disease exists, although considerable refinements need to be made before it becomes a practical reality. This selection may be of limited utility in the USA or in Western Europe, where the disease is successfully addressed by eradication strategies; however, it may be of immense importance in countries where the disease is endemic.

Escherichia coli infections in chickens

Escherichia coli infections are of primary importance, due to effects on the mortality and morbidity of chickens and, possibly more important, to the zoonotic potential of various strains of *E. coli*. Additionally, concerns about antibiotic resistance and the need to reduce antibiotic usage potentially impinge upon the control of various *E. coli* infections.

There has been substantial work performed on the antibody response to *E. coli* vaccination. Divergent selection (high and low) for antibody response following vaccination was successful, resulting in substantial differences between the two lines, with antibody response itself being highly heritable (heritability > 0.4) (Heller *et al.*, 1992; Yonash *et al.*, 1996). Critically, these lines were simultaneously exposed to *E. coli*, infectious bursal disease and Newcastle disease, and they showed similar divergence in their antibody response to each of these antigens (Yunis *et al.*, 2002b), indicating that overall immunocompetence to a variety of challenges can be increased.

Currently, the implications of these results in terms of mortality and morbidity following infectious challenge are unclear (Yunis *et al.*, 2002a). Initial studies seeking genetic markers for *E. coli* antibody response have been successful, with significant associations found both within the MHC complex (Lavi *et al.*, 2005) and throughout the genome (Yunis *et al.*, 2002c).

Salmonellosis in chickens

As with *E. coli* infections, salmonellosis is of primary concern to the poultry industry because of pathogenic effects to poultry and, through contaminated products, the introduction of food poisoning to humans. Additionally, concerns regarding antibiotic resistance and antibiotic or vaccine residues in meat are important. Increasing pathogen virulence, intensive husbandry practices and the failure of pathogen eradication have led to the recognition of the need to use genetics to help control the disease.

Considerable research has shown genetic differences in resistance, particularly resistance to *Salmonella enteritidis*, but also to other *Salmonella* species. Line differences in resistance are well documented (Bumstead and Barrow, 1993), and recently QTL have been found for resistance or antibody response (Mariani *et al.*, 2001; Yunis *et al.*, 2002c; Tilquin *et al.*, 2005). Additionally, associations have been shown with a wide variety of candidate genes (Kramer *et al.*, 2003; Malek *et al.*, 2004), including NRAMP (e.g. Hu *et al.*, 1997; Girard-Santosuosso *et al.*, 2002; Lamont *et al.*, 2002; Liu *et al.*, 2003) and MHC polymorphisms (Liu *et al.*, 2002).

In summary, there is compelling evidence of genetic variation in resistance, with many clues as to the underlying genes. However, based on published results, there is not a uniformly useful genetic test as yet, and selection may need to be based on several genes rather than on a single gene.

Coccidiosis in chickens

Coccidiosis is caused by *Eimeria* infections and is one of the most economically important diseases in modern poultry production. Replication of the asexual and/or sexual stages of the parasite in the intestine causes this disease. It has been estimated that the total cost of coccidial infections in the UK (*c.*780 million broilers) is at least £42 m per annum, of which 74% is due to subclinical effects on weight gain and feed conversion and 24% to the cost of prophylaxis and therapy of commercial birds. In intensively reared poultry, infections have always been controlled primarily by prophylactic in-feed medication, but vaccination with live, virulent or attenuated parasites has a role to play. In Europe, control is threatened by the withdrawal of anti-coccidial drugs.

Chickens are infected by seven species of *Eimeria*. Genetic differences between inbred lines of chickens in resistance to each of these species have been demonstrated (Bumstead and Millard, 1992) and between the Institute for Animal Health Light Sussex chickens and these inbred lines (Smith *et al.*, 2002). Pinard-van der Laan *et al.* (1998) also observed large differences between various outbred lines, both in their resistance and also in the impact that infection had on performance, and this author is currently refining QTL for resistance. A QTL associated with oocyst shedding has been reported by Zhu *et al.* (2003).

Ascaridia galli

One of the most common parasitic roundworms of poultry, *Ascaridia galli*, occurs in chickens and turkeys. Adult worms are about 3.5–8 cm long and about the size of an ordinary pencil lead. Thus, they can be seen easily with the naked eye. Heavily infected birds may show droopiness, emaciation and diarrhoea. The primary damage is reduced efficiency of feed utilization, but death has been observed in severe infections. Specimens of this parasite are found occasionally in eggs. The worm apparently wanders from the intestine up the oviduct, and is included in the egg contents as the egg is formed. *Ascaridia* infections are observed with a high prevalence in free-range (and hence organic) chickens. Available drugs remove only the adult parasite, but it is probably the immature form that produces the most severe damage.

Whilst breed differences in resistance have been known for many years, only recently has genetic variation been more formally quantified. Faecal egg counts differ between modern breeds (Permin and Ranvig, 2001; Gauly *et al.*, 2002; Schou *et al.*, 2003), and they are repeatable and heritable (heritability between 0.1 and 0.2) (Gauly *et al.*, 2002).

GENERALIZED IMMUNITY

So far, consideration has been given to specific diseases. However, it sometimes becomes more difficult to identify overriding diseases of importance in production systems where most endemic diseases are under control. In this case, a more generalized approach may be desirable, e.g. looking for general resistance to a variety of diseases, or generalized immunity.

Therefore, generalized resistance to a wide range of diseases, covering different types of parasites or pathogens, is probably too great an expectation. Mechanisms of disease resistance tend to be specific to individual classes of disease, and therefore it is unlikely to expect animals or breeds to be resistant to a wide range of diseases unless they have had a long evolutionary relationship with this disease. However, resistance to diseases caused by related parasites does tend to be correlated.

For example, the genetic correlation between the resistance of ruminants to different species or genera of nematode parasites does tend to be strongly genetically correlated (e.g. Bishop *et al.*, 2004), indicating that many of the mechanisms controlling resistance may be common to different types of nematode parasites, and there is an opportunity to improve generalized resistance to different classes of diseases. This is most often addressed through the concept of generalized immunity.

Generalized immunity may be considered as a combination of immune responses to a variety of immune system challenges, the aim being to identify animals that are better able to respond to a variety of challenges. The goals of generalized immunity are to: (i) genetically improve the overall health status and performance of animals, in the face of unknown infectious challenges; and (ii) attempt to provide animals with protection against accidental infection from existing or new diseases. The generalized immunity approach may also be desirable if the breeder does not wish to base a genetic management strategy upon a single mode of resistance, if selection for this mode has been shown to adversely affect resistance to other diseases (although examples of this phenomenon are rare in outbred populations). An achievable goal through generalized immunity may be to breed animals better able to cope with subclinical infections and hence having higher productivity.

The immune measurements used in generalized immunity may be specific challenges, but more often they are non-specific challenges such as sheep red blood cells, or even simply measures of the innate immune response, e.g. natural killer (NK) cells measured in animals known to be facing subclinical challenges. However, it must be realized that, at the immunological level, the distinction between selecting for resistance to a specific disease (if resistance is immunologically controlled) and generalized immunity may not be clear. For example, allelic variants that affect immunity to many pathogens may be part of the innate immune system and could be selected for both when selecting for specific resistance and also when selecting for generalized immunity.

Defining success when improving generalized immunity is difficult; the goal may best be defined as enhanced performance in an environment with subclinical challenges. To this end, progress has been made in pig studies. For example, phenotypic relationships have been demonstrated between performance and proportions of peripheral mononuclear cells (Clapperton *et al.*, 2005a) or of acute-phase proteins (Clapperton *et al.*, 2005b). Additionally, the same authors have demonstrated that these immunological measurements are heritable and genetically related to performance attributes (Clapperton *et al.*, 2005b, 2006). Such measurements may well allow the selection of chickens with enhanced generalized immunity.

BREEDING FOR DISEASE RESISTANCE

General principles

Genetic markers, immunological tests or indicator traits may provide the means by which a breeder can select animals for enhanced resistance. However, the breeding goals have to be put into the context of the overall husbandry system and the overall disease management strategy. This will determine the precise goals, the constraints within which the breeder is working and the actual tools available. For example, poultry units usually attempt to minimize exposure to infection, therefore breeding strategies will generally be aimed at minimizing epidemic risks (or epidemic severity should an epidemic arise) rather than helping birds to cope with endemic diseases.

As a consequence, selection will usually need to be undertaken on birds that are not exposed to infection, and genetic markers will generally be required. Commonly, genetic markers will be associated with only a proportion of the between-animal variation, therefore several markers may be required, and to minimize risks the breeding strategy should be combined with other disease control approaches.

Because breeding for resistance in poultry is often associated with risk management, the goal of the breeding programme may be quantified in terms of decreased epidemic risks. Thus, evaluating the success of the breeding programme becomes an epidemiological question, and needs to be undertaken using epidemiological tools.

A first step in this process is the definition of the transmission pathways of infection (Bishop and MacKenzie, 2003), as this will encapsulate much of the disease dynamics and define possible outcomes from the disease control strategy. Typical pathways are shown in simplified form in Fig. 21.1. In most cases, only some of these pathways will be relevant to any particular disease, and the relevant pathways may well be more complex than shown here. For example, intermediate hosts are not shown in the diagram, and they would be a part of pathway c.

If the disease comes from a reservoir of infection outside the host population of interest, transmission of infection from this reservoir may be 'continuous' or sporadic. Once infection is in the host population it may follow several pathways. Typically, viral or bacterial infections will be transmitted by direct animal-to-animal contact along pathway b, whereas macroparasitic (for example, nematode or arthropod) infections will be transmitted along pathways

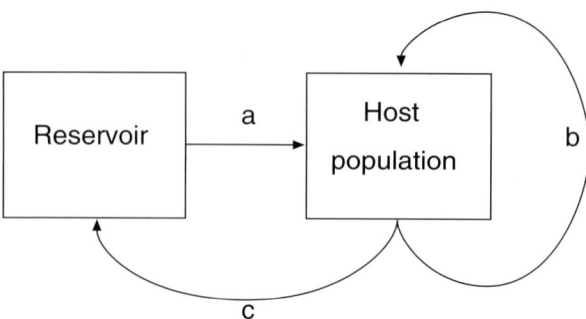

Fig. 21.1. Summary of pathways for diseases of domestic livestock.

a and c via some external host, vector or reservoir. Diseases where pathway a is important and continuous, and pathways b and c are non-existent or trivial in terms of the impact of the disease, include many endemic tropical diseases. Diseases where pathway b is critical, with sporadic infection from the reservoir (pathway a), include most viral diseases affecting livestock.

The consequences of these infection pathways have been explored by Bishop and Stear (1997, 1999) in the case of ruminant nematode infections, and by MacKenzie and Bishop (1999, 2001a, b) for viral infections. These impacts have been put into a broader context by Bishop and Stear (2003). Reducing transmission of infection along pathways a and c will lead to so-called Type III epidemiological effects, in which a virtuous cycle of reduction in infection and disease consequence can be achieved. Reducing transmission along pathway b will lead to Type II epidemiological effects, in which the outcome of the reduction of transmission is a reduction in the frequency and/or severity of epidemics.

A conclusion to be drawn from this approach is that outcomes of selection should be measured at the population – rather than at the individual animal – level. Moreover, the outcomes are nonlinear and depend upon the starting point. For example, a moderate improvement in animal resistance to viral disease might either solve the disease problem or make no impact at all, depending on the nature of the disease and the initial resistance of the host population.

One parameter which helps to summarize this concept is the reproductive ratio, R_0, which is defined as the average number of secondary cases of infection resulting from one primary case introduced into a population of susceptible individuals (Anderson and May, 1992). For example, if the primary animal infects five other animals, then R_0 is five. R_0 has direct application in terms of defining genetically resistant populations of animals. In the case of genes that determine complete resistance, the number of resistant animals that the population as a whole must contain is a simple function of R_0, as the requirement is simply to reduce R_0 below 1.0.

The concept of describing the outcomes of selection for disease resistance in terms of R_0 has been explored by MacKenzie and Bishop (1999) and developed by Bishop and MacKenzie (2003). As described above, one aim of a selection programme can be to reduce R_0 towards or even below 1.0. As R_0 reduces, the likely rate transmission of infection in a population decreases, as does the risk of an epidemic arising in the first instance. When R_0 is below 1.0, the risk of a major epidemic is zero, but short-term minor epidemics can still arise.

Translating this theoretical framework into practice is challenging. However, Nath *et al*. (2004) detailed a process in which parameters describing the transmission of infection through a population could be equated to host–pathogen interactions, and hence to QTL or genes describing resistance. Further, they demonstrated which QTL were likely to have a greater impact in terms of ameliorating diseases. This then translates into a decision tree for determining which types of QTL should be given priority in investigations and in breeding programmes.

Genetic improvement which results in a reduction in the clinical signs of disease – that is, improved tolerance of infection – will be effective in reducing the incidence or the impact of disease in the target population. However, it may not decrease the prevalence of the pathogen. Hence, the disease incidence in other populations in the same environment will not be affected. In worst-case scenarios, the presence of infection in the environment may be masked by the absence of symptoms in the carriers of the pathogen.

The sustainability of breeding for enhanced disease resistance is often challenged, with the assertion that the pathogen will evolve in the face of the changing host population. Pathogen evolution is of course a risk, but the risks are not easily quantified. More relevant is the question of whether pathogen evolution is more likely when genetic strategies are used than when other strategies are used. This question is addressed in detail by Gibson and Bishop (2005).

In summary, pathogen evolution risks are probably less than, and certainly not greater than, the risks imposed by other disease control strategies such as chemical control or vaccination. Resistance will pose greater selection pressure than tolerance, where the pathogen can complete its life cycle unhindered, but this advantage of tolerance has to be balanced against the possible disadvantages described above. A general conclusion is that to minimize pathogen evolution risks, the breeding strategy should be combined with other disease control approaches, thus lessening the overall risks and presenting a greater challenge to the pathogen. Such an approach will increase the effective lifespan of all the control strategies.

Into practice

Despite the large number of well-documented examples of genetic variation in resistance, there are surprisingly few cases of structured commercial breeding programmes for disease resistance in any species. The best-known examples are for nematode resistance in sheep, mastitis in dairy cattle and sheep and for Marek's disease.

For nematode resistance there are breeding programmes underway in New Zealand, Australia and the UK; mastitis resistance, either in terms of clinical cases of mastitis or reduced somatic counts, is incorporated into many dairy cow and sheep breeding programmes; and aspects of Marek's disease resistance are a feature of modern chicken breeding. Breeding for resistance to post-weaning *E. coli* diarrhoea in pigs, where the causal mutation or closely linked markers are known, has begun in some selection programmes. Additionally, scrapie resistance selection is now underway in several countries in Western Europe, although arguably this is imposed upon breeders rather than applied by choice.

The reasons for the relative lack of convincing examples of breeding schemes for disease resistance are many and varied. They probably include:

1. A belief by many practitioners – veterinarians, scientists and animal producers – in the adequacy of existing disease control strategies, even when these are failing.
2. The lack of awareness of the opportunities. This is partly a function of the interdisciplinary nature of the skills required to exploit these opportunities, requiring animal health experts to be familiar with genetics, and geneticists to be familiar with animal health issues.
3. The lack of infrastructure and tools to exploit the genetic opportunities. Exploitation often requires considerable information, infrastructure (technology), cooperation and will. Breeding for resistance to diseases that are absent from production units will require genetic markers; it is a daunting task to find unambiguous genetic markers that are in sufficiently close linkage disequilibrium with the causative mutation for use across populations. Likewise, breeding for resistance to an endemic disease requires considerable effort in terms of data collection protocols and database structures.

However, the opportunities are now becoming widely recognized and the tools to tackle many of the diseases of importance are becoming available. Success still requires effort and cooperation between geneticists and animal health specialists. Clearly, the simplest situation occurs in structured breeding industries such as those seen in the poultry industry, where specific markers or alleles can be fixed within breeding lines, and then propagated down through the breeding pyramid. As described above, resolving these genetic markers to the extent where they can be easily used is a large undertaking and, as described by Cheng (2005), this will often require a combination of genomics approaches, including both gene mapping and gene expression studies. Additionally, several genetic markers will most probably be required, as each polymorphic gene is likely to explain only a proportion of the genetic variation between host animals in their resistance. The approach of Nath *et al.* (2004) will help choose appropriate genetic markers from amongst a panel of candidates.

Finally, having implemented a breeding programme, it is necessary to regularly monitor and appraise the success of that programme. It is critical that a breeding programme for disease resistance is regularly assessed in terms of whether or not it is meeting its objectives of helping to improve animal health or control the targeted disease problem, and therefore whether or not it is contributing to sustainable disease control and livestock production.

ACKNOWLEDGEMENTS

I wish to thank the BBSRC for funding.

REFERENCES

Anderson, R.M. and May, R.M. (1992) *Infectious Diseases of Humans. Dynamics and Control.* Oxford University Press, Oxford, UK.

Bacon, L.D. (1987) Influence of the major histocompatability complex on disease resistance and productivity. *Poultry Science* 66, 802–811.

Bacon, L.D., Crittenden, L.B., Witter, R.L., Fadly, A. and Motta, J. (1981) B-haplotype influence on Marek's disease, Rous sarcoma, and lymphoid leukosis virus-induced tumors in chickens. *Poultry Science* 60, 1132–1139.

Bishop, S.C. (2004) Disease resistance: genetics. In: Pond, W.G. and Bell, A.W. (eds) *Encyclopedia of Animal Science.* Marcel Dekker, Inc., Ithaca, New York, pp. 288–290.

Bishop, S.C. and MacKenzie, K.M. (2003) Genetic management strategies for controlling infectious diseases in livestock populations. *Genetics, Selection, Evolution* 35 (1), S3–S17.

Bishop, S.C. and Stear, M.J. (1997) Modelling responses to selection for resistance to gasto-intestinal parasites in sheep. *Animal Science* 64, 469–478.

Bishop, S.C. and Stear, M.J. (1999) Genetic and epidemiological relationships between productivity and disease resistance: gastointestinal parasite infection in growing lambs. *Animal Science* 69, 515–524.

Bishop, S.C. and Stear, M.J. (2003) Modeling of host genetics and resistance to infectious diseases: understanding and controlling nematode infections. *Veterinary Parasitology* 115 (2), 147–166.

Bishop, S.C., Jackson, F., Coop, R.L. and Stear, M.J. (2004) Genetic parameters for resistance to nematode infections in Texel lambs. *Animal Science* 78, 185–194.

Bumstead, N. (1998a) Genomic mapping of resistance to Marek's disease. *Avian Pathology* 27, S78–S81.

Bumstead, N. (1998b) Genetic resistance to avian viruses. *Office International des Épizooties, Scientific and Technical Review* 17, 249–255.

Bumstead, N. and Barrow, P. (1993) Resistance to *Salmonella gallinarum, S. pullorum* and *S. Enteritidis* in inbred lines of chickens. *Avian Diseases* 37, 189–193.

Bumstead, N. and Millard, B.J. (1992) Variation in susceptibility of inbred lines of chickens to 7 species of *Eimeria. Parasitology* 104, 407–413.

Bumstead, N., Reece, R.L. and Cook, J.K.A. (1993) Genetic differences in susceptibility of chicken lines to infection with infectious bursal disease virus. *Poultry Science* 72, 403–410.

Cheng, H.H. (2005) Integrated genomic approaches to understanding resistance to Marek's disease. In: Lamont, S.J., Rothschild, M.F. and Harris, D.L. (eds) *Proceedings of the Third International Symposium on Genetics of Animal Health,* Iowa State University, Ames, Iowa, 13–15 July 2005.

Clapperton, M., Bishop, S.C., Cameron, N.D. and Glass, E.J. (2005a) Leucocyte sub-sets in Large White pigs: associations with weight gain and food intake. *Livestock Production Science* 96, 249–260.

Clapperton, M., Bishop, S.C., Cameron, N.D. and Glass, E.J. (2005b) Associations of acute-phase protein levels with growth performance and with selection for growth performance in Large White pigs. *Animal Science* 81, 213–220.

Clapperton, M., Bishop, S.C., Cameron, N.D. and Glass, E.J. (2006) Selection for lean growth and food intake levels leads to correlated changes in innate immune traits in Large White pigs. *Animal Science* 82, 1–11.

Cole, R.K. (1968) Studies on the genetic resistance to Marek's disease. *Avian Diseases* 12, 9–28.

Collins, W.H., Briles, W.E., Zsigray, R.M., Dunlop, W.R., Corbett, A.C., Clark, K.K., Marks, J.L. and McGrail, T.P. (1977) The B locus (MHC) in the chicken: association with the fate of RSV-induced tumors. *Immunogenetics* 5, 333–343.

Friars, G.W., Chambers, J.R., Kennedy, A. and Smith, A.D. (1972) Selection for resistance to Marek's disease in conjunction with other economic traits in chickens. *Avian Diseases* 16, 2–10.

Gauly, M., Bauer, C., Preisinger, R. and Erhardt, G. (2002) Genetic differences of *Ascaridia galli* egg output in laying hens following a single dose infection. *Veterinary Parasitology* 103, 99–107.

Gibson, J.P. (2002) Role of genetically determined resistance of livestock to disease in the developing world: potential impacts and researchable issues (Appendix). In: Perry, B.D., Randolph, T.F., McDermott, J.J., Sones, K.R. and Thornton, P.K. (eds) *Investing in Animal Health Research to Alleviate Poverty*. International Livestock Research Institute, Nairobi, Kenya, p.14.

Gibson, J.P. and Bishop, S.C. (2005) Use of molecular markers to enhance resistance of livestock to disease: a global approach. *Office International des Épizooties, Scientific and Technical Review* 24 (1), 343–353.

Girard-Santosuosso, O., Lantier, F., Lantier, I., Bumstead, N., Elsen, J.M. and Beaumont, C. (2002) Heritability of susceptibility to *Salmonella* Enteritidis infection in fowls and test of the role of the chromosome carrying the NRAMP1 gene. *Genetics, Selection, Evolution* 34, 211–219.

Gordon, C.D., Beard, C.W., Hopkins, S.R. and Siegel, H.S. (1971) Chick mortality as a criterion of selection towards resistance or susceptibility to Newcastle disease. *Poultry Science* 50, 783–789.

Gyles, N.R. and Brown, C.J. (1971) Selection in chickens for retrogression of tumours caused by Rous sarcomas. *Poultry Sciences* 50, 901–905.

Hala, K., Moore, C., Plachy, J., Kaspers, B., Bock, G. and Hofmann, A. (1998) Genes of chicken MHC regulate the adherence activity of blood monocytes in Rous sarcomas progressing and regressing lines. *Veterinary Immunology and Immunopathology* 66, 143–157.

Hassan, M.K., Afify, M.A. and Aly, M.M. (2004) Genetic resistance of Egyptian chickens to infectious bursal disease and Newcastle disease. *Tropical Animal Health and Production* 36, 1–9.

Heller, E.D., Leitner, G., Friedman, A., Uni, Z., Gutman, M. and Cahaner, A. (1992) Immunological parameters in meat-type chicken lines divergently selected by antibody-response to *Escherichia coli* vaccination. *Veterinary Immunology and Immunopathology* 34, 159–172.

Hu, J.X., Bumstead, N., Barrow, P., Sebastiani, G., Olien, L., Morgan, K. and Malo, D. (1997) Resistance to salmonellosis in the chicken is linked to NRAMP1 and TNC. *Genome Research* 7, 693–704.

Kaufman, J. and Venugopal, K. (1998) The importance of MHC for Rous sarcoma virus and Marek's disease virus – some Payne-ful considerations. *Avian Pathology* 27, S82–S87.

Kramer, J., Malek, M. and Lamont, S.J. (2003) Association of twelve candidate gene polymorphisms and response to challenge with *Salmonella* Enteritidis in poultry. *Animal Genetics* 34, 339–348.

Lamont, S.J., Kaiser, M.G. and Liu, W. (2002) Candidate genes for resistance to *Salmonella* Enteritidis colonization in chickens as detected in a novel genetic cross. *Veterinary Immunology and Immunopathology* 87, S423–S428.

Lavi, Y., Cahaner, A., Pleban, T. and Pitcovski, J. (2005) Genetic variation in major histocompatibility complex class I alpha 2 gene among broilers divergently selected for high or low early antibody response to *Escherichia coli*. *Poultry Science* 84, 1199–1208.

Liu, H.-C, Cheng, H.H., Tirunagaru, V., Sofer, L. and Burnside, J. (2001) A strategy to identify positional candidate genes conferring Marek's disease resistance by integrating DNA microarrays and genetic mapping. *Animal Genetics* 32, 351–359.

Liu, W., Miller, M.M. and Lamont, S.J. (2002) Association of MHC class I and class II gene polymorphisms with vaccine or challenge response to *Salmonella* Enteritidis in young chicks. *Immunogenetics* 54, 582–590.

Liu, W., Kaiser, M.G. and Lamont, S.J. (2003) Natural resistance-associated macrophage protein 1 gene polymorphisms and response to vaccine against or challenge with *Salmonella* Enteritidis in young chicks. *Poultry Science* 82, 259–266.

MacKenzie, K. and Bishop, S.C. (1999) A discrete-time epidemiological model to quantify selection for disease resistance. *Animal Science* 69, 543–551.

MacKenzie, K. and Bishop, S.C. (2001a) Developing stochastic epidemiological models to quantify the dynamics of infectious diseases in domestic livestock. *Journal of Animal Science* 79, 2047–2056.

MacKenzie, K. and Bishop, S.C. (2001b) Utilising stochastic genetic epidemiological models to quantify the impact of selection for resistance to infectious diseases in domestic livestock. *Journal of Animal Science* 79, 2057–2065.

Malek, M., Hasenstein, J.R. and Lamont, S.J. (2004) Analysis of chicken TLR4, CD28, MIF, MD-2, and LITAF genes in a *Salmonella* Enteritidis resource population. *Poultry Science* 83, 544–549.

Mariani, P., Barrow, P.A., Cheng, H.H., Groenen, M.A.M., Negrini, R. and Bumstead, N. (2001) Localization to chicken Chromosome 5 of a novel locus determining salmonellosis resistance. *Immunogenetics* 53, 786–791.

Mays, J.K., Bacon, L.D., Pandiri, A.R. and Fadly, A.M. (2005) Response of white leghorn chickens of various B haplotypes to infection at hatch with subgroup J avian leukosis virus. *Avian Diseases* 49, 214–219.

Nath, M., Woolliams, J.A. and Bishop, S.C. (2004) Identifying critical parameters in the dynamics and control of microparasite infection using a stochastic epidemiological model. *Journal of Animal Science* 82, 384–396.

Permin, A. and Ranvig, H. (2001) Genetic resistance to *Ascaridia galli* infections in chickens. *Veterinary Parasitology* 102, 101–111.

Pinard-van der Laan, M.H., Monvoisin, J.L., Pery, P., Hamet, N. and Thomas, M. (1998) Comparison of outbred lines of chickens for resistance to experimental infection with coccidiosis (*Eimeria tenella*). *Poultry Science* 77, 185–191.

Pinard-van der Laan, M.H., Soubieux, D., Merat, L., Bouret, D., Luneau, G., Dambrine, G. and Thoraval, P. (2004) Genetic analysis of a divergent selection for resistance to Rous sarcomas in chickens. *Genetics, Selection, Evolution* 36, 65–81.

Purchase, H.G. (1985) Clinical disease and its economic impact. In: Payne, L.N. (ed.) *Marek's Disease, Scientific Basis and Methods of Control*. Martinus Nkjhoff Publishing, Boston, Massachusetts, pp. 17–42.

Sacco, R.E., Nestor, K.E., Saif, Y.M., Tsai, H.J., Anthony, N.B. and Patterson, R.A. (1994) Genetic analysis of antibody responses of turkeys to Newcastle disease virus and *Pasteurella multocida* vaccines. *Poultry Science* 73, 1169–1174.

Schou, T., Permin, A., Roepstorff, A., Sorensen, P. and Kjaer, J. (2003) Comparative genetic resistance to *Ascaridia galli* infections of 4 different commercial layer-lines. *British Poultry Science* 44, 182–185.

Smith, A.L., Hesketh, P., Archer, A. and Shirley, M.W. (2002) Antigenic diversity in *Eimeria maxima* and the influence of host genetics and immunization schedule on cross-protective immunity. *Infection and Immunity* 70, 2472–2479.

Tilquin, P., Barrow, P.A., Marly, J., Pitel, F., Plisson-Petit, F., Velge, P., Vignal, A., Baret, P.V., Bumstead, N. and Beaumont, C. (2005) A genome scan for quantitative trait loci affecting the *Salmonella* carrier-state in the chicken. *Genetics, Selection, Evolution* 37, 539–561.

Tsai, H.J., Saif, Y.M., Nestor, K.E., Emmerson, D.A. and Patterson, R.A. (1992) Genetic variation in resistance of turkeys to experimental infection with Newcastle disease virus. *Avian Diseases* 36, 561–565.

Vallejo, R.L., Bacon, L.D., Liu, H.-C., Witter, R.L., Groenen, M.A.M., Hillel, J. and Cheng, H.H. (1998) Genetic mapping of quantitative trait loci affecting susceptibility to Marek's disease-induced tumours in F2 intercross chickens. *Genetics* 148, 349–360.

Williams, S.M., Reed, W.M., Bacon, L.D. and Fadly, A.M. (2004) Response of white leghorn chickens of various genetic lines to infection with avian leukosis virus subgroup J. *Avian Diseases* 48, 61–67.

Witter, R.L. (1998) The changing landscape of Marek's disease. *Avian Pathology* 27, S46–S53.

Yonash, N., Leitner, G., Waiman, R., Heller, E.D. and Cahaner, A. (1996) Genetic differences and heritability of antibody response to *Escherichia coli* vaccination in young broiler chicks. *Poultry Science* 75, 683–690.

Yonash, N., Bacon, L.D., Witter, R.L. and Cheng, H.H. (1999) High-resolution mapping and identification of new quantitative trait loci (QTL) affecting susceptibility to Marek's disease. *Animal Genetics* 30, 126–135.

Yonash, N., Cheng, H.H., Hillel, J., Heller, D.E. and Cahaner, A. (2001) DNA microsatellites linked to quantitative trait loci affecting antibody response and survival rate in meat-type chickens. *Poultry Science* 80, 22–28.

Yoo, B.H. and Sheldon, B.L. (1992) Association of the major histocompatibility complex with avian leukosis virus infection in chickens. *British Poultry Science* 33, 613–620.

Young, J.A.T., Bates, P. and Varmus, H.E. (1993) Isolation of a gene that confers susceptibility to infection by subgroup A avian leucosis and sarcoma viruses. *Journal of Virology* 67, 1811–1816.

Yunis, R., Ben-David, A., Heller, E.D. and Cahaner, A. (2002a) Antibody responses and morbidity following infection with infectious bronchitis virus and challenge with *Escherichia coli*, in lines divergently selected on antibody response. *Poultry Science* 81, 149–159.

Yunis, R., Ben-David, A., Heller, E.D. and Cahaner, A. (2002b) Genetic and phenotypic correlations between antibody responses to *Escherichia coli*, infectious bursa disease virus (IBDV), and Newcastle disease virus (NDV), in broiler lines selected on antibody response to *Escherichia coli*. *Poultry Science* 81, 302–308.

Yunis, R., Heller, E.D., Hillel, J. and Cahaner, A. (2002c) Microsatellite markers associated with quantitative trait loci controlling antibody response to *Escherichia coli* and *Salmonella* Enteritidis in young broilers. *Animal Genetics* 33, 407–414.

Zhu, J.J., Lillehoj, H.S., Allen, P.C., Van Tassell, C.P., Sonstegard, T.S., Cheng, H.H., Pollock, D., Sadjadi, M., Min, W. and Emara, M.G. (2003) Mapping quantitative trait loci associated with resistance to coccidiosis and growth. *Poultry Science* 82, 9–16.

PART VII

Monitoring and practical experience

CHAPTER 22
The EU perspective on the monitoring of zoonoses and zoonotic agents

S. Idei

European Commission Health and Consumer Protection Directorate, Brussels; e-mail: Sarolta.Idei@cec.eu.int

ABSTRACT

The mission statement of the Directorate General of Health and Consumer Protection (DG SANCO) is to ensure a high level of protection of consumer health and safety.

The monitoring of zoonoses along the entire food chain and in humans is necessary to assess related risks. Zoonoses present, particularly at the level of primary production, must be adequately controlled.

The Commission's objective is to recast the existing Community legislation in order to better harmonize the monitoring of zoonoses and to broaden the scope of control measures of zoonoses, starting with the reduction of *Salmonella* occurence in different animal populations.

INTRODUCTION

Objective of the monitoring of zoonoses and zoonotic agents

The protection of human health against diseases and infections transmissible between animals and humans is of paramount importance. Zoonoses can be transmitted through food, contact with domestic and wild animals and pet animal populations. Zoonoses may cause not only severe human suffering but also economic losses to food business operators.

A brief history

Council Directive 92/117/EEC (1992) laid down compulsory monitoring for four pathogens.

Monitoring was mostly based on national systems. Monitoring according to this Directive included the collection of data on animals, food, feed and on human cases of zoonoses, while control meant rules on *Salmonella* control in breeding flocks of *Gallus gallus*. Member States were required to submit to the Commission the national measures that they had taken to achieve the objectives of the Directive, and to draw up plans for the control of *Salmonella* in poultry flocks. Detailed Community minimum measures to eradicate *S.* Enteritidis and *S.* Typhimurium in breeding flocks of poultry were established. These two serotypes of *Salmonella* account for over 70% of the total *Salmonella* serotypes isolated from humans.

According to Council Directive 92/117/EEC (1992), the Commission approved a number of programmes in Member States to control *Salmonella* in poultry. Within those approved programmes, the eradication of some *Salmonella* serotypes (*S.* Enteritidis and *S.* Typhimurium) from fowl breeding flocks (*Gallus gallus*) is co-financed by the Community.

A number of other control and eradication programmes, covering diseases/infections which may be transmitted directly or indirectly to humans, are also co-financed by the Community, in particular brucellosis in large and small ruminants, as well as tuberculosis in cattle.

THE ZOONOSES REPORT

The Annual Report on Zoonoses provides the basis for evaluation of the current situation on zoonoses in the Community as a whole. However, since the data collection is not harmonized, the comparison between Member States is not always possible. In general, many cases of zoonoses in humans remain unreported and, therefore, the data collected do not give the entire picture of the current situation. On the other hand, trends can be assessed and, probably because of control measures, a decreasing trend of human cases of salmonellosis is observed.

The Directive required the Member States to collect, evaluate and report to the Commission each year data on specific zoonoses and zoonotic agents. *The Annual Community Report on Trends and Sources of Zoonoses and Zoonotic Agents in Animals, Feedingstuffs, Food and Man in the EU and Norway* was prepared on the basis of national reports and provides the basis for evaluation of the current situation on zoonoses in the Community as a whole.

This data collection covered tuberculosis due to *Mycobacterium bovis*, brucellosis, salmonellosis and trichinellosis, but the Directive also foresaw that the collection could be extended to certain additional zoonoses. The Council Directive 92/117/EC (1992) appointed the Institut für Veterinärmedizin in Berlin (currently Bundesinstitut für Risikobewertung) as a Community Reference Laboratory for the Epidemiology of Zoonoses (CRL-E). This legislation provided the legal basis for data collection, both on zoonoses already targeted by the Community surveillance and control programmes and on other zoonoses, in order to assess the priorities for Community actions.

The European Commission could follow the development of the epidemiological situation in order to be able to propose appropriate measures when necessary.

During recent years the Commission, in collaboration with CRL-E and the Member States, has further improved the zoonoses monitoring and reporting system based on Council Directive 92/117/EEC (1992) and this system, together with the *Annual Community Summary Report*, have gradually improved. By 2004, the reporting system had progressed to covering 11 zoonoses, as well as covering antimicrobial resistance in certain zoonotic agents, foodborne outbreaks and some demographic data. However, not all Member States delivered data in all these areas.

As a consequence, the Scientific Committee on Veterinary Measures relating to Public Health (SCVPH) recommended, in its opinion on zoonoses of 12 April 2000, to improve in particular the measures in place to monitor and control foodborne zoonoses, with special emphasis being placed on zoonotic agents of public health priority (Directive 2003/99 EC, 2003).

THE STATE OF PLAY

Council Directive 92/117/EEC (1992) was repealed and replaced – with effect from 12 June 2004 – by Directive 2003/99/EC (2003) on the monitoring of zoonoses and zoonotic agents and by Regulation (EC) No. 2160/2003/EC (2003) on the control of *Salmonella* and other specified foodborne zoonotic agents.

The Directive on the monitoring on zoonoses and zoonotic agents

Directive 2003/99/EC (2003) of the European Parliament and of the Council on the monitoring of zoonoses and zoonotic agents lays down obligations for Member States as well as duties for food business operators, and requires the collection of data on the occurrence of zoonoses and zoonotic agents in animals, food, feed and humans in order to obtain comparable data to evaluate related risks and determine the trends and sources of zoonoses.

Food business operators should: (i) keep the results and arrange for the preservation of any relevant isolate for a period to be specified by the competent authority; and (ii) communicate the results or provide isolate to the competent authority on request.

The monitoring is mandatory for eight zoonoses (brucellosis, campylobacteriosis, echinococcosis, listeriosis, salmonellosis, trichinellosis, tuberculosis due to *Mycobacterium bovis* and verotoxigenic *Escherichia coli* – later referred to as A-list zoonoses), whilst other zoonoses should be included in the monitoring and reporting according to the epidemiological situation in each Member State (later referred to as B-list zoonoses). Foodborne outbreaks and antimicrobial resistance in *Salmonella* and *Campylobacter* are also to be reported on a mandatory basis. The epidemiological investigation of foodborne outbreaks is also required.

The system used is based on that of the Member State, and in a few cases it is harmonized to the extent that the results from the monitoring are comparable between the Member States. The objective is to collect relevant and comparable data to support risk assessment activities. It provides the legal basis for: (i) harmonization of monitoring schemes; (ii) measurement of the burden of the disease and its attribution to different sources; (iii) assessment of exposure and trends; and (iv) identification of the need for measures or evaluation of the effects of those measures.

Risk-based management requires many data. The Directive on monitoring is intended to be a tool for collection of relevant and comparable data.

If data collected through routine monitoring are not sufficient, coordinated monitoring programmes concerning one or more zoonoses and/or zoonotic agents may be established, especially when specific needs are identified, to assess risks or to establish baseline values related to zoonoses or zoonotic agents at the level of Member States or/at Community level.

The European Food Safety Authority

From the start of 2005, the national reports on monitoring of zoonoses and zoonotic agents during 2004, prepared by Member States, have been transmitted to the European Food Safety Authority (hereinafter the EFSA), which assesses them and provides a Community summary report on trends and sources of zoonoses. The EFSA was legally created from the European Parliament and Council Regulation No 178/2002 of 28 January 2002 (2002). The core activity of the EFSA is to provide independent scientific advice on food safety issues throughout the food chain and to issue scientific opinion upon request from, e.g. the European Commission.

From 2005, the EFSA has been entrusted to prepare the yearly *European Union (EU) Summary Report*. The first report prepared by the EFSA, covering the data from 2004, was due by November 2005. The Commission has established a mandate to the EFSA concerning the monitoring of zoonoses. The prioritization on monitoring schemes will be decided by the Commission and Member States.

Reporting of data on zoonotic cases in humans

The new Directive (Directive 2003/99/EC, 2003) on monitoring highlights the importance of cooperation at national level between authorities in animal/feed/food/human health sectors and the continuous need for the interpretation of data on zoonoses and zoonotic agents at national level for yearly reporting.

Results from the investigations of foodborne outbreaks will be reported under the zoonoses Directive, while data on zoonotic cases in humans will be collected through the Community Legislation on Communicable Diseases (Decision 2119/98/EC of the European Parliament, 1998), and the Council has established a network for epidemiological surveillance and control of communicable disease in the Community managed by the Commission.

Decision No 2119/98/EC (1998) gives the legal basis for setting up the European Network for the Epidemiological Surveillance and Control of Communicable Diseases in humans in the European Community. The purpose of establishing such a network is to promote the collaboration between different Member States to strengthen the coordination on infectious diseases.

In future, this network will be coordinated by the European Centre for Disease Prevention and Control (ECDC) (Regulation (EC) No. 851/2004, 2004). The ECDC is a new EU agency that has been created to help strengthen Europe's defences against infectious diseases and, furthermore, provides the framework for coordination at EU as well as international level. It will work in partnership with national health protection bodies across Europe to strengthen and develop continent-wide disease surveillance and early warning systems.

The mission of the ECDC is defined in Regulation (EC) No 851/2004 (2004) of the European Parliament and of the Council of 21 April 2004. The Centre shall identify, assess and communicate current and emerging threats to human health from communicable diseases. In the case of other outbreaks of illness of unknown origin, which may spread within or to the Community, the Centre shall act on its own initiative until the source of the outbreak is known. In the case of an outbreak, which clearly is not caused by a communicable disease, the Centre shall act only in cooperation with the competent authority upon request from that authority.

Reference laboratories

Guidance and assistance for analyses and testing is given by reference laboratories.

To assist the Commission and Member States in the efficient detection and monitoring of biological hazards present in food, especially of animal origin, six Community Reference Laboratories (hereinafter CRLs) have been designated in the area of biological risks to coordinate the work of the National Reference Laboratories (hereinafter NRLs).

At present there are five operational CRLs in the field of biological risks, because CRL-Epidemiology in Berlin was operational only up to the end of 2004. These are divided into distinct operational areas: (i) milk and milk products (Paris); (ii) *Salmonella* (Bilthoven, Netherlands); (iii) marine biotoxins (Vigo, Spain); (iv) bacteriological and viral contaminants of molluscs (Weymouth, UK); and (v) TSE (Weybridge, UK).

The CRLs dealing with biological risks were appointed at European Community (EC) level, and their tasks and organization are defined in EC legislation. They receive financial assistance from the EC. They represent EC Centres of excellence on analytical methods.

The tasks of the CRLs are to provide NRLs with details of analytical methods and to coordinate application of methods by the NRLs, by organizing comparative testing (ring trials/proficiency testing). Particular tasks are: (i) to coordinate research on new methods; (ii) to conduct training for NRLs; (iii) to assist the Commission; (iv) to cooperate with laboratories in third countries; and (v) to help NRLs implement Quality Assurance.

Proficiency testing (PT) is a core mission for CRLs. Most CRLs organize PTs regularly. The CRLs usually produce the reference materials (RMs) for the PTs. They hold regular workshops and meetings to discuss PT results and carry out assessments for the implementation of reference methods.

Member States (MSs) shall organize the designation of one or more NRLs for each CRL area of responsibility. The MSs that have more than one NRL for one area of responsibility should ensure efficient coordination. The NRLs have the duty, in their area of competence, to collaborate with the relevant CRL and to coordinate, on a national level, the activities of laboratories responsible for the analysis of samples taken by official controls. NRLs are responsible for organizing comparative tests between the official national laboratories and to ensure the appropriate follow-up of the test results where necessary.

The NRLs should ensure dissemination of information obtained by the relevant CRL to the competent authorities (CA) and to official national laboratories. NRLs should provide scientific and technical assistance to the CA. NRLs and CRLs are important components for official feed and food control by coordinating laboratory activities with field laboratories.

CRLs need information on zoonoses and zoonotic agents for development, validation and standardization of diagnostic and analytical methods for detection, identification and typing of zoonotic agents. In particular, information on the matrices to be considered and need for rapid methods would be of interest. The information is also essential for keeping the CRLs abreast of developments in epidemiology. The data on zoonoses might sometimes indicate a need to establish additional CRLs.

Microbiological criteria

The Commission has introduced other tools to control foodborne pathogens along the food chain, particularly the revision of the legislation related to microbiological criteria (hereinafter MC) for foodstuffs in Community legislation.

A scientific opinion from the Standing Committee on Veterinary Public Health (SCVPH) on the evaluation of MC for food products of animal origin concluded that MC should be: (i) relevant and effective in relation to human health protection; (ii) based on formal risk assessment where possible; and (iii) harmonized and uniform. The revision of MC started in 2001.

It is necessary to establish MC and temperature control requirements based on scientific risk assessment. Food business operators should comply with MC for foodstuffs, sampling and analysis. MC are only one part of the overall strategy on food safety. According to the Codex Standard principles, MC shall include the foodstuffs, micro-organisms, analytical method, sampling plan, frequency, limits, stage of application and actions to be taken in case of unsatisfactory results. MC have to be applied by food business operators in the framework of their Hazard Analysis and Critical Control Points (HACCP) procedures. Two kinds of criteria are applied: (i) the criteria defining the safety of foodstuffs; and (ii) the criteria indicating the acceptability of the process.

Test results are dependent on the analytical method used. Therefore, a reference method (hereinafter RM) should be associated with each criterion. Other methods can be used if they provide equivalent results, and this includes alternative methods. RM are generally EN/ISO methods.

The legal basis for the draft Commission Regulation on microbiological criteria for foodstuffs (SANCO/4198/2001 Revision 21, 2001) is the Regulation (EC) No. 852/2004 (2004) of the European Parliament and of the Council of 29 April 2004 on the hygiene of foodstuffs. This draft Regulation received a favourable final vote by Member States in the Standing Committee of Food Chain and Animal Health of 23 September 2005.

Regulation (EC) No. 2160/2003 (2003) on the control of *Salmonella*

Regulation (EC) No. 2160/2003 (2003) on the control of *Salmonella* and other specified foodborne zoonotic agents is a framework legislation that prescribes that controls on zoonoses and zoonotic agents cover the whole food chain, particularly at the level of primary production.

The primary aim of this Regulation is to ensure that effective measures are taken to decrease the occurrence of *Salmonella* serotypes of significance for public health (and other zoonotic agents in the future where appropriate), particularly at the level of primary production by establishing reduction targets for *Salmonella*. Progressively, different categories of poultry and pigs will be covered by Community *Salmonella* reduction targets.

Implementation of Regulation (EC) No. 2160/2003 (2003)

Community targets and control programmes

The first *Salmonella* reduction target has been established by Commission Regulation (EC) No. 1003/2005 (2005). The Community target for the reduction of *S.* Enteritidis, *S.* Hadar, *S. infantis*, *S.* Typhimurium and *S.* Virchow in breeding flocks of *Gallus gallus* shall be a reduction of the maximum percentage of adult breeding flocks, comprising at least 250 birds remaining positive to 1% or less by 31 December 2009.

This target was set on the basis of the collection of EU data on the prevalence of the five top serotypes in poultry breeding flocks, whereby the samples were investigated by bacteriology, according to the provisions on monitoring of Council Directive 92/117/EEC (1992). The Commission made publicly available the EU report on *Salmonella* in breeding flocks of *Gallus gallus* in 2004 via its website.

For certain zoonoses and zoonotic agents, specific requirements for controls should be laid down, which are based on Community targets for the reduction of the occurrence of zoonoses and zoonotic agents (Regulation (EC) No. 2160/2003, 2003).

National programmes should be operational 18 months after the setting of the Community targets: this means 1 January 2007 in the case of poultry breeding flocks. In order to ensure achievement of the targets, Member States should set up specific control programmes, which are to be approved by the Commission (Regulation (EC) No. 2160/2003, 2003).

Commission regulation of the use of specific control methods for the control of Salmonella

Commission Regulation (EC) No. 1091/2005 (2005) on the use of control methods for the control of *Salmonella* specifies control methods in the framework of the national programmes for the control of *Salmonella* and introduces measures on the use of antimicrobials. The EFSA recommended that the use of antimicrobials should be discouraged due to public health risks associated with development, selection and spread of resistance. The use of antimicrobials should be subject to formally defined conditions that would ensure protection of public health, and must be fully justified in advance and recorded by the competent authority.

Baseline surveys on Salmonella in animal populations

IN LAYING HENS. A baseline study on *Salmonella* in laying hens was launched from 1 October 2004 for a period of 1 year, with the financial assistance of the Community (Commission Decision 2004/665/EEC, 2004).

The objective of this study was to estimate, across the EU, the prevalence of *Salmonella* spp. in flocks of laying hens (*Gallus gallus*) for table egg production, at the end of their production period. The results shall be used to set Community targets.

IN BROILERS. The Commission Decision of 1 September 2005 (2005/636/EC, 2005) has been adopted and outlines a baseline survey on *Salmonella* in broilers. A 1-year study with the financial assistance of the Community should be implemented by Member States from 1 October 2005. The final results of this survey shall also be used to set Community targets.

In 2006 similar baseline surveys will be drafted for turkeys and pigs.

REFERENCES

Commission Decision 2004/665/EEC (2004) This Decision of 22 September concerns a baseline study on the prevalence of *Salmonella* in laying flocks of *Gallus gallus* (notified under document number C 3512).
Commission Decision 2005/636/EC (2005) This Decision of 1 September concerns a financial contribution by the Community towards a baseline survey on the prevalence of *Salmonella* spp. in broiler flocks of *Gallus gallus* to be carried out in the Member States.
Commission Regulation (EC) No. 1003/2005 (2005) This Regulation of 30 June implements Regulation (EC) No. 2160/2003 of the European Parliament and of the Council as regards a

Community target for the reduction of the prevalence of certain *Salmonella* serotypes in breeding flocks of *Gallus gallus*, and amends Regulation (EC) No. 2160/2003.

Commission Regulation (EC) No. 1091/2005 (2005) This Regulation of 12 July implements Regulation (EC) No. 2160/2003 of the European Parliament and of the Council as regards requirements for the use of specific control methods in the framework of the national programmes for the control of *Salmonella*.

Council Directive 92/117/EEC (1992) This Directive concerns measures for protection against specified zoonoses and specified zoonotic agents in animals and products of animal origin in order to prevent outbreaks of foodborne infections and intoxications (17 December).

Decision No. 2119/98/EC (1998) This Decision of the European Parliament and the Council established a network for epidemiological surveillance and control of communicable diseases in the Community.

Directive 2003/99/EC (1999) This Directive of the European Parliament and of the Council of 17 November on the monitoring of zoonoses and zoonotic agents amended Council Decision 90/424/EEC and repealed Council Directive 92/117/EEC.

Draft Commission Regulation SANCO/4198/2001 Revision 21 (2001) Concerns microbiological criteria for foodstuffs.

Regulation (EC) No. 178/2002 (2002) This Regulation of the European Parliament and Council laid down the general principles and requirements of food law, established the European Food Authority and set down procedures in matters of food.

Regulation (EC) No. 2160/2003 (2003) This Regulation of 17 November of the European Parliament and of the Council concerns the control of *Salmonella* and other specified foodborne zoonotic agents.

Regulation (EC) No. 851/2004 (2004) This Regulation of the European Parliament and of the Council of 21 April established a European centre for disease prevention and control.

Regulation (EC) No. 852/2004 (2004) This Regulation of the European Parliament and of the Council of 29 April concerns the hygiene of foodstuffs.

CHAPTER 23
Gastrointestinal problems: the field experience and what it means to the poultry farmer

S.A. LISTER

Crowshall Veterinary Services, Attleborough, UK;
e-mail: salister@crowshall.co.uk

ABSTRACT

Farmers are interested in profit, and solutions to problems that affect that profit.

Gut problems and the loss of intestinal integrity are potentially very important in maximizing that profit, as the intestinal tract is the interface between a whole range of feed ingredients taken in by the bird and their conversion into saleable meat or eggs.

As a result, the intestinal tract is arguably the most important organ of the body. Although making up only 5% of total body weight the intestines use up to 30% of whole body oxygen consumption and protein turnover.

Added to the direct commercial penalty caused by loss of intestinal integrity are the equally important aspects of reduced health through secondary disease and adverse welfare (e.g. poor environment, pododermatitis), which may be sequelae to gut dysfunction.

Furthermore, the current draft EU Broiler Directive, which will determine future broiler stocking density and define standards of poultry production, will concentrate on welfare outcomes as an audit of performance. One of these audits will be an assessment of footpad dermatitis, which can be related to litter conditions and hence intestinal function/dysfunction. This further strengthens the need to understand and respond to gut problems in modern poultry production.

As with so many syndromes affecting poultry production, the causes of the loss of intestinal integrity are many, varied and interrelated.

The art of solving the farmer's problems is centred around diagnosing and interpreting the relative contribution of these various factors in specific farming situations. Once this has been achieved, control measures concentrating on environmental manipulation, the role of specific disease agents, nutritional effects and the use of novel products for maintaining gut flora, together with the structured use of therapeutic antibiotics and other medicines, can be formulated.

© CAB International 2006. *Avian Gut Function in Health and Disease* (ed. G.C. Perry)

INTRODUCTION

The bird's intestine is the industry's method of transforming a whole range of feed ingredients into money. Anything that damages intestinal integrity damages profit. There are a wide range of factors that can disturb intestinal integrity, including nutritional, infectious, environmental and management changes. The loss of antimicrobial digestive enhancers (ADEs) has only heightened the requirement for a holistic approach to maintaining intestinal integrity and, hence, enabling birds to show their full potential.

THE INTESTINES: A BALANCE BETWEEN STRUCTURE AND FUNCTION

The intestinal tract is arguably the most important organ of the bird and, although making up only around 5% of total bodyweight, it uses up to 30% of whole body oxygen consumption and protein turnover.

Physiological damage to the digestive system can result from insults from a variety of infectious agents, toxins and other noxious chemicals and effects. This damage will result in inflammation, loss of or damage to absorptive epithelial cells and shortening of the all-important villi. Due to the dynamic nature of the organ there can be rapid repair to the villi, but frequently repopulation is with immature and poorly absorptive epithelial cells. The end result is reduced absorption of nutrients leading to poor growth and/or stunting, as well as difficulties in maintaining fluid balance, leading to diarrhoea and wet litter.

DIARRHOEA, WET LITTER AND THE CONSEQUENCES

Common clinical signs associated with enteric disease include some or all of the following:

● Diarrhoea.
● Dehydration.
● Depression.
● Weakness.
● Reduced appetite.
● Huddling.
● Vocalization.
● Emaciation.
● Feed refusal.

The physiological damage and loss of ability to regulate fluid intake and loss can result in wet litter and difficulties in maintaining a satisfactory environment, air quality and conditions underfoot as dry friable litter.

Wet litter can manifest in a number of ways:

● A general increase in moisture in the litter, with dry matter falling below 50%; good litter conditions usually have a dry matter content approaching 75%.

- Wet litter restricted to service areas and under drinker lines.
- Increased passing of wet droppings, scouring or diarrhoea.
- Increased faecal output and caecal droppings.
- Increased passing of poorly digested feed.

The consequences of diarrhoea, malabsorption and ensuing wet litter can be grouped into three main areas:

1. Commercial effect
- Loss of weight gain and performance.
- Unevenness.
- Increased respiratory and leg problems associated with deterioration in litter conditions and air quality.
- Increased downgrading at processing.
- Increased littering-down plus associated labour costs. Possible increased use of therapeutic antibiotics.

2. Health issues
- Increased risk of coccidiosis and other protozoal disease.
- Increased bacterial challenge and imbalance.
- Respiratory challenges and air sac damage.
- Leg problems.
- Public health concerns from dirty birds presented for processing.

3. Welfare issues
- Pododermatitis (foot pad lesions) and hock burn.
- Other leg problems
- Breast blisters.
- Soiled feathering.
- Secondary disease, especially respiratory.

The recent draft EU Council Directive laying down minimum rules for the protection of chickens kept for meat production (COM (2005) 221 final) sets out potentially punitive consequences of a high incidence of pododermatitis in broiler flocks. It is proposed that each consignment of broilers will be assessed and scored for pododermatitis lesions. Flocks showing an unacceptable incidence of lesions may be required to reduce stocking density for future broiler crops on that site.

Although the cause of pododermatitis in broiler flocks is a complex issue, wet litter and its causation by environmental/management effects, together with intrinsic factors affecting intestinal integrity, present challenges for producers.

CAUSES OF LOSS OF INTESTINAL INTEGRITY

The major reasons revolve around the following aspects:

1. What you are doing
Producers may be influencing the situation, e.g. lack of control of disease, poor biosecurity or poor environmental control through inadequate or inappropriate ventilation or poor drinker management.

2. What others are doing

This may relate to unwanted challenges from poor regional biosecurity or nutritional effects due to inappropriate diets, errors in specification, unwanted interaction, etc.

3. What we are forced to do

The loss of antimicrobial digestive enhancers due to retail pressure and/or legislation is bound to have an effect on intestinal integrity and the ability of the intestine to recover from disease and other toxic challenges.

4. Factors outside our control

Ambient temperature and climatic conditions can affect the producer's ability to manage a wet litter problem, especially in very cold weather where there may be difficulties in maintaining suitable minimum ventilation or where relative humidity is excessive.

These reasons are important in the ability to cope with wet litter problems, but the actual precipitating causes can be even more numerous and varied.

There are a number of ways of categorizing these causes, but they can be broadly divided into external or internal effects.

External causes include:

1. Feed
- Interruptions, physical quality, ingredients.
- Minerals/electrolytes (e.g. potassium, sodium, chloride).
- Protein – quality and quantity or digestibility.
- Fats – carbohydrate source (e.g. wheat versus maize).
- Mycotoxins.
- Other anti-nutritive factors.

2. Drugs
- Anticoccidial choice.
- Feed interactions.
- Feed errors.

3. Environmental
- Ventilation (especially minimum ventilation).
- Structural insulation.
- Water and water systems.
- Drinker management.
- Litter type.
- Ambient temperature.

Internal causes include:

1. Nutrition
- Physical quality.
- Effects of digestibility and viscosity of intestinal contents.
- Presence/absence of antimicrobial digestive enhancers.

2. Disease
- Coccidiosis.
- Other parasites.

3. Viruses
- Coronaviruses.
- Rotaviruses.
- Enteroviruses.
- Adenoviruses.
- Reoviruses.
- Astroviruses.
- Parvoviruses.
- Caliciviruses.
- Picornaviruses.

4. Intrinsic effects
- Malasorption syndrome.
- Immunosuppression.
- Bacteria and dysbacteriosis.

Feed and various feed ingredients are commonly blamed as triggers for wet litter/scouring episodes. This is frequently prompted by a temporal relationship with observed clinical problems. However, on most commercial sites, feed deliveries are such a frequent occurrence that it would be surprising if an episode of scouring did not occur close to the time of delivery. In the case of broilers, feed interruption, sudden changes in physical quality and changes in raw material matrix are known to precipitate conditions such as necrotic enteritis.

Changes in other factors such as mineral/electrolyte composition, protein quality, quantity and digestibility and carbohydrate source (wheat versus maize) have also been linked to problems with intestinal stability. Bacteria in the intestines gain their energy for growth from components of the diet. Different bacteria have different preferences for substrates, such that the chemical composition of the digesta can have a potent effect on the composition of the microbial community of the intestines.

The physical presentation of feed, pellet versus mash, grist size, whole wheat feeding and other digestibility factors which can have potent effects on viscosity and time of feed passage are all areas requiring further research, especially in a 'post-ADE' era.

The role of disease agents in their ability to cause insult to the intestinal tract have long been studied and full details can be found in other publications.

NOTHING IS SIMPLE!

It can be seen from the foregoing discussion that there are a vast number of possible factors which can have an adverse effect on intestinal integrity. Whereas, in the past, disease control was more related to 'simple' monofactor diseases which require diagnosis and then prevention or treatment, current syndromes are, by their nature, multifactorial in their pathology and impact. As a result, disease management is more focused on evaluating the extent and role of different components and subsequent management of the syndrome, rather than on finding simple treatment options.

Mediation of the effects of these numerous factors is often through their interruption of bacterial gut homeostasis.

The chick hatches with a sterile gut which is receptive to rapid colonization, leading to the gradual development of a balanced gut microflora.

The intestinal tract and its bacterial community are as metabolically powerful as any other organ in the body. The gut wall is the interface between a massive bacterial soup and healthy sterile tissue such as the bloodstream. Bacteria, including *Clostridia*, *Bacillus* and *Staphylococcus*, are known to produce toxins that will damage intestinal cells. Bacteria also compete with the bird for nutrients, notably energy and amino acids. In addition, they can produce microbial metabolites that increase gut mucosal turnover and reduce growth efficiency.

Changes in the chemical and physical composition of digesta can affect the composition of the microbial community in the intestinal tract.

This 'soup' of intestinal bacteria is very complex, such that probably less than 10% can actually be cultured and identified using current techniques. Those that have been identified and investigated are probably the less fastidious and more easily cultured species. This may also suggest they are the species least sensitive and hence influenced by changes in composition of the intestinal environment. Recent work highlighting the DNA profiles of the bacterial flora of the intestines offers the possibility of a better understanding of changes relating to certain diseases or dietary inputs and their effects on digestion and performance.

The role of antibiotic digestive enhancers and the impact of their removal is of concern to the poultry industry. It is known that feeding antibiotics to germ-free birds does not induce a growth response. On the other hand, infecting germ-free birds with intestinal bacteria from normal birds results in growth depression. Hence, as is well understood by the industry – but perhaps not by consumers – in-feed antibiotic digestive enhancers are growth permitting rather than growth promoting.

The absence of these products from poultry diets can lead to an imbalance of bacteria, when the intestinal tract is exposed to some or many of the factors discussed above. This condition, loosely described as 'dysbacteriosis', can be a sequel to a number of conditions. There is usually no specific illness or mortality, but feed intake is reduced and water intake may change. There is significant diarrhoea and frequently abnormal, foamy caecal droppings. This can lead to severe wet litter and poor performance.

THE FUTURE?

This chapter has highlighted the importance of intestinal integrity to profitable and healthy production and a complexity of factors that can damage that integrity. However, the industry is more interested in solutions rather than in causes.

It is clear that the complexity of causation is mirrored by the difficulties in finding answers. It is imperative to maintain competitiveness and, to alleviate possible health and welfare concerns, that progress is made. The art is establishing that progress.

Getting a good diagnosis

In evaluating and managing such complex interactions of factors, it is necessary to establish an assessment of which factors are involved and their relative importance and contribution. This can be achieved by a partnership approach between producers and their veterinarians to utilize detailed post-mortem data and the results of further laboratory tests. This should also involve maximizing the benefit of new, accurate, targeted and cost-effective techniques for diagnosis such as Polymerase Chain Reaction (PCR) technology. This also requires the maintenance of centres of excellence, adequately funded, to undertake this type of work.

Therapy

Targeted therapy can be undertaken when specific bacterial or parasitic involvement has been identified. Antibiotics such as amoxicillin or tylosin can be used to treat necrotic enteritis due to *Clostridium perfringens* infection and may also help to re-establish gut microflora. Electrolytes can help to ensure effective fluid balance during scouring episodes.

Vaccination

There are currently no commercially available vaccines to control enteric viral infections in broilers. Experience in other animal species suggests this approach may not be likely to achieve great success, but work on identifying specific viral involvement and candidate vaccines is in progress and should be encouraged. Bacterial vaccines such as that for *Clostridium perfringens*, the cause of necrotic enteritis, are similarly useful developments for investigation.

Stockmanship

Undoubtedly, many of the factors discussed in this chapter are outwith the control of the individual producer. However, it is equally true that operating the highest standards of management and stockmanship in all areas of production will reduce the impact of these factors on health, welfare and performance.

Biosecurity

In the absence of effective tools to treat or prevent infectious agents, the role of biosecurity in cleansing and disinfection and ongoing actions during the life of the flock must remain a priority.

The aim is to prevent the introduction of primary or contributory infectious agents. Important areas to consider include:

- Use of protective clothing.
- Use of footdips, utilizing effective disinfectants at the manufacturer's recommended dilution rates, regularly replenished.
- Use of hand sanitizers.
- Terminal cleansing and disinfection utilizing effective products with known efficacy against targeted agents, properly applied.
- Ensuring a down time sufficient to allow procedures to be undertaken effectively.
- Following a full 'all in, all out' approach – do not re-use old litter.

Immunomodulation

The role of products designed to boost the avian immune system should help to reduce the effects of infectious agents on the intestinal mucosa, and this is an area requiring further work.

Feed husbandry and ration quality

As a non-nutritionist commenting on nutritional aspects, my requirement is for stable, smooth rations with as few changes as possible to allow a consistent microbial gut flora and hence offer the best opportunity for maintaining intestinal integrity. This may be achieved by attention to:

- High-quality feed ingredients.
- Limited raw material matrix.
- Pellet quality.
- Grist size.
- Feed form (e.g. whole wheat feeding reduces gizzard pH and reduces gut transit time, although the exact mechanisms of this benefit are not fully understood).
- Anti-nutritive factors.
- Implications of enzymes releasing elements then made 'available' to the bird; possible unwanted interactions or releases.
- Factors affecting gut viscosity and feed passage time.
- Longer term, fundamental nutritional manipulation to reduce pressure on the avian intestinal tract.

Oral treatments and feed additives

In the absence of anitmicrobial digestive enhancers, it is hoped that other 'non-antibiotic' alternatives may be identified. Some examples include:

Undefined microbial cultures

These treatments tend to relate to competitive exclusion (CE) products. The WHO definition for CE products is an 'undefined preparation of live obligate and facultatively anaerobic bacterial originating from normal healthy individuals of an avian species which is free from specific pathogenic micro-organisms and its quality control'. As a result, they represent a 'safe' bacterial soup without defining exactly which bacteria are, or are not, present. These products have a proven track record for exclusion of *Salmonella* but their role in deterring some of the other bacteria associated with dysbacteriosis and general bacterial imbalance is less clear cut.

Defined microbial cultures or products

This covers the area of probiotics, which can be defined as 'mono- or mixed cultures of living microorganisms which beneficially affect the host by improving the properties of indigenous microflora'. This does not exclude CE products, but tends to include more specific cultures such as *Lactobacillus* or *Bifidobacter*.

Prebiotics

These are food ingredients that stimulate selectively the growth and activity of certain bacteria, including *Bifidobacter* and *Lactobacillus* in the gut, and therefore benefit health as probiotics. There is also some suggestion that they may reduce the growth of pathogens such as *Clostridium*.

Enzymes

Supplementary enzymes are aimed at improving digestibility of nutrients and reducing gut fermentation. However, are the 'released' elements always good substrates for the microflora of the gut? Their activity may not be selective in helping specific bacteria.

Organic acids

These are known to reduce bacterial contamination of animal feeds, but it is not clear whether their direct *in vivo* activity has any significant effect on gut pH and bacterial flora other than in the crop.

Plant products

A number of products available on the market have hinted at antimicrobial benefits, although practical experience again suggests equivocal results.

CONCLUSIONS

The maintenance of intestinal integrity must be a cornerstone of effective poultry production. There are a multitude of different challenges, be they infectious, nutritional, managemental or environmental, which can disturb that integrity.

It is clear that there are no easy or single answers, but it is important that producers try to do everything they are doing just a little bit better, and for all 'stakeholders' to take a holistic view by looking at the whole range of factors involved and to attempt to influence the things they can influence to aid progress.

POSTER 1
Structure and function of the ostrich (Struthio camelus *var.* domesticus) *gut: nutritional impact*

R.G. Cooper*

Department of Physiology, University of Zimbabwe, Mount Pleasant Drive, Harare, Zimbabwe

Data collected from Zimbabwean farms indicated that protein (69 ± 2 g/day for maintenance; 140 ± 1 g/day for 1.35 kg egg mass; in 103 kg body mass; *n*, 8) and energy requirements (12 ± 1 MJ/day for maintenance; 7 ± 1 MJ/day for egg production; *n*, 8) form an important part of layer intake. Breeder birds require increased levels of calcium and phosphorus during lay. Correct nutritional proportions are necessary for high fertility after the attainment of sexual maturity at 24 months. Impaction due to ingestion of sticks, etc. leads to impaired peristaltic contractility and digestibility and damage to mucosae, with consequences for reproductive performance.

The structure of the ostrich digestive system differs from that of poultry. Post-mortem examinations demonstrated the lack of a crop, with the oesophagus emptying directly into the proventriculus without any distinct demarcation.

Function was associated with secretion of protein from glands combined with entrapped sloughed cells and other cellular debris, resulting in a dark, tough lining of the proventriculus and ventriculus. The green-brown colour was attributed to the refluxed bile pigments from the duodenum. Measurements of the small intestine showed it to be significantly shorter than the rectum (660 ± 3 cm and 853 ± 2 cm, respectively; *n*, 10), presumably to enable digestion of bulky food and to facilitate fluid absorption. The round gizzard was muscular and contained pebbles for facilitating the grinding of food. The spiral fold in the caecum was noted.

In conclusion, accurate formulation of diets is important for maximum sustainability of reproductive health and performance.

---

*Current address: Department of Applied Physiology, School of Health and Policy Studies, University of Central England, Birmingham, UK

POSTER *2*
Effect of metabolic efficiency on kinetic differences in Salmonella *Enteritidis shedding rates*

Ellen Van Eerden

Adaptation Physiology Group, Animal Sciences Group, Wageningen University, PO Box 338, 6700 AH Wageningen, Netherlands

It has been presumed that genetic selection for production efficiency might have negative consequences on other physiological processes. If immunity is compromised because of selection, efficient animals might respond less adequately to infectious challenges, e.g. shedding bacteria in higher numbers or for a longer period of time, or both.

Therefore, we compared shedding rates in both efficient and non-efficient pullets after a *Salmonella* Enteritidis (SE) challenge. Efficiency status was calculated as residual feed intake, the difference between actual and expected feed intake based on metabolic body weight and growth. The terms R– and R+ indicate efficient and non-efficient animals, respectively. After 1 week of cage adaptation 29 R– and 32 R+ pullets were challenged orally with 10^8 cfu SE. Shedding was determined in fresh faeces obtained daily between days 0 and 27.

The results show that in the first week after inoculation, shedding rates of *Salmonella* were higher in R– pullets compared with R+ pullets, and also the number of bacteria excreted were higher in the R– group. However, the rate and level of shedding decreased gradually in the R– group, whereas they remained relatively high in the R+ group. Preliminary conclusion: kinetics of shedding differs between R– and R+ pullets. Apart from being efficient growers, R– pullets seem to be at least as capable as , or even better than, R+ pullets of mounting an adequate response to a *Salmonella* challenge.

Poster 3
Live vectored delivery of therapeutic and prophylactic proteins to the chicken gut

S.A. Sheedy,[1,2] J.I. Rood[2] and R.J. Moore[1,2]

[1]*CSIRO Livestock Industries, Australian Animal Health Laboratory (AAHL), PO Bag 24, Geelong, Victoria 3220, Australia;* [2]*ARC Centre for Structural and Functional Microbial Genomics, Department of Microbiology, Monash University, Victoria 3800, Australia*

Successful implementation of biological vectoring strategies has the potential to reduce the use of conventional antibiotics and chemicals, and hence significantly to reduce animal losses and the cost of treatment of diseases such as necrotic enteritis to the poultry industry. In this study we have identified live *Escherichia coli* vectors from commercial broiler (meat) chickens and have used them to deliver agents that may have effects on the health status of the bird. After oral inoculation into chickens, genetically marked vector strains were able to persist in the gastrointestinal tract of the chicken for a considerable time and caused no ill-health effects.

Genes encoding biological agents – such as cytokine IL6 and the bacteriocin Pisicolin 126 – were introduced into the vector strains and the expression of functional protein detected *in vitro*. The constructs were delivered to the chicken gut via the *E. coli* vector and effective delivery of the agent assessed by its effect on the immune system and the bacterial load. A potential necrotic enteritis vaccine antigen was also delivered to the chicken gut via the selected live bacterial vector. This project has shown that biological agents could be successfully delivered to the gastrointestinal tract of the chicken via the selected live bacterial vectors where they had a detectable effect on the immune system; however, the specific agents selected may not necessarily control disease in chickens under these conditions.

POSTER 4
The development and use of molecular tools to monitor the gut microflora of poultry

Valeria A. Torok,[1] Kathy Ophel-Keller[1] and R.J. Hughes[2]

The Australian Poultry Cooperative Research Centre; [1]SARDI, Field Crops Pathology Unit, Plant Research Centre, GPO Box 397, Adelaide SA 5064, Australia; [2]SARDI, Pig and Poultry Production Institute, University of Adelaide, Roseworthy SA 5371, Australia

Gut microbiology and its role in animal health has become increasingly important, particularly now that the use of antibiotics in animal feeds to promote growth and/or to prevent enteric disease is limited due to legislation in some countries and to consumer pressure generally. The microorganisms that colonize the gastrointestinal tract during the early post-hatch period form a synergistic relationship with their poultry host. Gastrointestinal microorganisms have a highly significant impact on uptake and utilization of energy and other nutrients, and on the response of poultry to anti-nutritional factors (such as non-starch polysaccharides), pre- and probiotic feed additives and feed enzymes.

Microorganisms can also directly interact with the lining of the gastrointestinal tract, which may alter the physiology of the tract and immunological status of the bird. The gastrointestinal tract contains a complex population of bacteria, which can have both negative and positive effects on their host. However, the complexity of these interactions is not yet fully understood. This report describes the development and application of terminal restriction fragment length polymorphism (T-RFLP), a microbial profiling technique for examining the chicken intestinal microflora based on high-throughput, high-resolution fingerprinting of bacterial gene regions. This tool is being used to examine diet-induced changes in the microbial community of the chicken gut and will contribute to an increased knowledge of the chicken gut microbiota and, hence, a better understanding of its role in chicken nutrition.

POSTER 5
Effects of a Clostridium perfringens infection on the normal intestinal microflora of broiler chickens

L. Bjerrum,[1] K. Finster[2] and K. Pedersen[1]

[1] *Danish Institute for Food and Veterinary Research, Hangøvej 2, 8200 Aarhus N, Denmark;* [2] *Department of Microbial Ecology, Institute of Biological Sciences, University of Aarhus, Bldg. 540, 8000 Aarhus C, Denmark*

The effect of a *Clostridium perfringens* infection on the composition of the bacterial community of the caecum and ileum of broiler chickens was investigated. Broiler chickens were kept in isolator facilities. At day 17–20 one group was given feed mixed 1:1 with fresh BHI broth cultures containing approximately 10^8 cfu/ml of a rifampicin-resistant *Cl. perfringens* type A strain. In addition, the birds were given ten times the prescribed dose of a coccidial vaccine (Paracox-8) as a predisposing factor for necrotic enteritis. Control birds were given pure BHI broth.

At days 17, 19, 20 and 24, five birds were randomly selected and killed by cervical dislocation. The numbers of *Cl. perfringens* were quantified in the contents of ileum and caecum by plate counts. In addition, material was collected for molecular analysis of the microbial community, by PCR-denaturing gradient gel electrophoresis (DGGE), using universal 16S rDNA primers. Birds in the group given *Cl. perfringens* and coccidial vaccine developed subclinical necrotic enteritis, whereas no clinical or pathological signs were found in control birds. The administered *Cl. perfringens* strain was found in both ileum and caecum at levels of up to 10^7 cfu/g content. Control chickens also contained *Cl. perfringens*, but these were naturally occurring strains.

DGGE profiles from the total bacterial community indicated that the community composition was influenced by the infection, as different banding patterns were found in the control group compared those in infected birds. Specific DGGE bands were excised from the gel and identified by sequence analysis to obtain knowledge about the organisms involved.

POSTER 6
Development of the adaptive immune system in broiler gut-associated lymphoid tissue during the first two weeks post-hatch

A. Friedman and E. Bar-Shira

Section of Immunology, Department of Animal Sciences, Faculty of Agricultural, Food and Environmental Sciences, The Hebrew University of Jerusalem, PO Box 12, Rehovot, Israel

The chick's gastrointestinal tract undergoes dramatic changes within the first few days of life. These include a rapid increase in digestive structures and their function. Concomitant with the development of intestinal tissue, a rapid development of the gut-associated lymphoid tissue (GALT) occurs. Although this lymphoid system works within and in concert with the digestive tract parenchyma, there is scant information describing the normal development and immunological function of the avian GALT in the immediate post-hatch period. Using FACS analysis, RT-PCR and ELISA, we show a unique biphasic pattern of lymphocyte colonization of the intestine and gradual maturation of T cell functions by 2 weeks post hatch.

The first maturation phase of the adaptive GALT occurs during the first week of life and is characterized by establishment of lymphocytic populations in the intestine. The second phase of maturation, which takes place during the second week of life, is characterized by functional maturation which is demonstrated by increases in chIL2 and chIFN and the ability to mount a humoral response to enterically administered proteins.

Evidence is provided to show that maturation of GALT in the hindgut precedes that of the small intestine, is dependent upon development of intestinal tissue and is driven by exposure to microflora. Our data strongly suggest that, in the intestinal milieu, cellular immune responses mature earlier and are a prerequisite for humoral responses. Hence, the lack of antibody response in young chicks is primarily due to immaturity of T lymphocytes.

POSTER 7
Expression of innate immune functions in developing broiler gut-associated lymphoid tissue in the immediate post-hatch period

E. Bar-Shira and A. Friedman

Section of Immunology, Department of Animal Sciences, Faculty of Agricultural, Food and Environmental Sciences, The Hebrew University of Jerusalem, PO Box 12, Rehovot, Israel

In mucosal surfaces the innate immune system, including the epithelial lining, is the first line of defence against invasive microorganisms prior to induction of adaptive immunity. Previous studies from our laboratory have showed that the development of this gut-associated lymphoid tissue (GALT) is concomitant with the structural and functional development of intestinal tissue, and that its full adaptive immune maturation is obtained by the end of the second week post-hatch. As the maturation of the innate limb in the chick's GALT has not been described we hypothesized that, in addition to protection by maternal antibodies, it functions in providing immediate protection to post-hatch enteric challenges.

We tested the development and protective potential of the innate GALT in newly hatched chicks. RT-PCR was used to study expression of functional genes representing different activities of innate immune cells, and their distribution was studied by light microscopy. Polymorphonuclear cells increased in number during the first 2 weeks post-hatch. Surprisingly, we found that these cells completed their maturation in the intestine, as shown by both microscopic observation and by expression of presenilin 1 and gallinacin.

Expression of the pro-inflammatory genes IL1β, IL8 and K203 was basal at hatch but increased rapidly therafter, so that by day 2 levels were similar to those observed on day 7. The rapid increase in expression of these genes is related to the exposure to environmental microflora and demonstrates the capability of the enteric immune system to respond rapidly to inflammatory stimuli of external origin. In conclusion, innate cells in GALT complete their maturation in the intestine and are capable of mounting an immediate response to inflammatory stimuli encountered post-hatch.

POSTER 8
Temporal expression of immunoglobulin transporter genes in broiler gut epithelial barriers during the immediate pre- and post-hatch period

I. Bromberger and A. Friedman

Section of Immunology, Department of Animal Sciences, Faculty of Agricultural, Food and Environmental Sciences, The Hebrew University of Jerusalem, PO Box 12, Rehovot, Israel

Recent studies have described two types of immunoglobulin transporters in chicks: (i) the *Gallus gallus* polymeric Ig receptor (GG-pIgR) that binds polymeric IgA (pIgA); and (ii) the chicken yolk sac IgY receptor (FcRY) that binds IgY. However, receptor-mediated transport of antibodies through gut epithelial tissue has yet to be described in poultry. As the contents of the yolk sac are absorbed from the gut as well as from the yolk sac membrane, we hypothesized that the FcRY gene is expressed in epithelial cells of the intestinal lining. In contrast to IgY, most IgA found in chicks is not of maternal origin. Thus, while the FcRY gene is expected to be expressed in intestinal epithelial cells, pIgR gene expression is expected to be minimal in these cells at hatch and to increase with age.

To investigate these hypotheses we determined FcRY and pIgR gene expression in intestine, liver and cloacal bursal tissue slices sampled from pre- and post-hatched chicks. As expected, we found constant levels of FcRY gene expression in the tested tissues during the immediate pre- and post-hatch period. The expression of pIgR gene, however, was low in these tissues during late embryonic development, but increased significantly during the first few days post-hatch. The increase was found to correlate with increasing IgA levels in bile, intestinal washings and cloacal bursal secretions.

These findings are the first to demonstrate expression of immunoglobulin transporters in the chicken gut and gut-associated organs. Furthermore, the differential temporal expression of both transporters indicates that epithelial barriers in the gut are coordinated with the development of the immune system in terms of antibody availability. Interestingly, they indicate the function of the cloacal bursa as an active site of antibody production.

POSTER *9*
Strategy for the development of a competitive exclusion product for poultry which meets the regulatory requirements for registration in the EU

V. Klose,[1] M. Mohnl,[1] R. Plail[1] and G. Schatzmayr[2]

[1]*BOKU, University of Natural Resources and Applied Life Sciences, Vienna; c/o Dep. IFA-Tulln, Division of Environmental Biotechnology, Konrad Lorenz Straße 20, 3430 Tulln, Austria;* [2]*Biomin GmbH, Industriestraße 21, 3130 Herzogenburg, Austria*

In the course of a project which was aimed towards the development of a defined competitive exclusion (CE) product for young chickens with EU marketing potential, a large pool of diverse gut bacteria was isolated from different intestinal compartments from various healthy broilers. Around 500 chicken isolates were collected and classified by combining multiple methods along the route of a polyphasic approach. A reduced number of 121 representative strains could be selected for probiotic screening.

On the basis of the collected results, a multispecies combination comprising five probiotic strains of different species affiliation originating from the crop, jejunum, ileum and caeca of broilers was chosen for the final product. A complex and stepwise risk assessment was performed to guarantee the safety of each single strain and to meet the current regulatory demands in the European Community. New insights obtained in this study are presented and discussed in the context of the main quality and food safety aspects.

POSTER *10*
Development of in vitro *assays for the characterization of putative probiotic mechanisms for* Lactobacillus salivarius *and* Enterococcus faecium

A.J. Carter,[1,2,3] R.M. La Ragione[1] and M.J. Woodward[1]

[1]*Department of Food and Environmental Safety, Veterinary Laboratories Agency (Weybridge), New Haw, Addlestone, Surrey KT15 3NB, UK;* [2]*Food Safety Research Group, School of Biomedical and Molecular Science, University of Surrey, Guildford, Surrey GU2 7XH, UK;* [3]*Probiotics International Ltd., Stoke-sub-Hamdon, Somerset TA14 6QE, UK*

We report initial studies on the competition for pathogen receptor sites on mucosal epithelia and immunomodulation of host cells by two probiotic strains.

We used *in vitro* tissue culture to demonstrate the ability of *Lactobacillus salivarius* and *Enterococcus faecium* to adhere to human and avian epithelia. Both strains adhered to HEp-2 cells. *E. faecium* associated at approximately 1×10^1 cfu/ml lower than control strain EPEC O127. whereas *L. salivarius* associated 50-fold lower than EPEC O127. We have also conducted competition assays describing the effect of *L. salivarius* and *E. faecium* association on the association of *Salmonella* Entertidis to HEp-2 cells, the results of which will be presented.

In additional experiments using tissue derived from 1 day-old SPF chicks, *L. salivarius* associated with crop tissue at 5×10^6 cfu/ml after an original inoculum of 5×10^8 cfu/ml. This was significantly higher than the level of association observed in other tissue types. *E. faecium* and *L. salivarius* also elicited tissue-specific tropism to caecal tissue. *L. salivarius* formed micro-colonies when adhered to HEp-2 cells, as shown by scanning electron microscopy and Giemsa staining. whereas *E. faecium* did not appear to form microcolonies. No invasion of HEp-2 cell lines, by either *L. salivarius* or *E. faecium*, was observed by transmission electron microscopy. Modulation of cytokine TNF-α, IL8, IL6 and IL10 responses induced by both strains in Caco-2 and murine macrophage RAW is currently under investigation.

POSTER 11
Molecular characterization of two novel potential probiotics: Lactobacillus salivarius *and* Enterococcus faecium

A.J. Carter,[1,2,3] R.M. La Ragione[1] and M.J. Woodward[1]

[1]Department of Food and Environmental Safety, Veterinary Laboratories Agency (Weybridge), New Haw, Addlestone, Surrey KT15 3NB, UK; [2]Food Safety Research Group, School of Biomedical and Molecular Science, University of Surrey, Guildford, Surrey GU2 7XH, UK; [3]Probiotics International Ltd., Stoke-sub-Hamdon, Somerset TA14 6QE, UK

Acquisition and carriage of mobile antibiotic resistance elements by probiotic bacteria is a concern regarding the safety of these commercial products. *Enterococcus* species have been shown to possess elements conferring resistance to glycopeptide antibiotics, resistance to which is thought to be significantly under-reported in *E. faecium*. Many *Lactobacillus* species are intrinsically resistant to glycopeptide antibiotics but may also acquire mobile resistance elements which, as yet, is unproven. Transfer of resistance from commensal and probiotic bacteria to pathogens is a possibility.

In our analysis of two novel probiotic strains, we utilized 16S rRNA sequencing to confirm the identity of *L. salivarius* to the species level and PCR amplification of species-specific markers in the *ddl* gene locus to speciate. We have also employed a PCR-based method to demonstrate that both *L. salivarius* and *E. faecium* do not possess *van*A or *van*B, the genetic markers for acquired glycopeptide resistance. Also, to investigate intrinsic resistance, we are currently sequencing the coding region of the proposed D-Ala D-Lac ligase of *L. salivarius* and intend to compare this to published database sequences.

Molecular strain typing will become an increasingly important tool for the characterization of probiotics, especially in the light of EU requirements. We have developed a rapid, PCR-based, strain typing method using the repetitive Gram-positive BOX motif. This BOX PCR method was compared to PFGE profiles generated for our two probiotic strains.

POSTER 12
Fermented liquid feed for organic layers

R.M. Engberg,[1] S. Steenfeldt,[1] N.F. Johansen[2] and B.B. Jensen[1]

[1]*Danish Institute of Agricultural Sciences, Research Centre Foulum, PO Box 50, 8830 Tjele, Denmark;* [2]*Danish Agricultural Advisory Service, Udkærsvej 15, 8200 Aarhus N, Denmark*

In order to investigate the effect of fermented liquid feed on the activity and composition of the gastrointestinal microflora, an experiment was carried out with 24 layers (Hyline Brown, 19 weeks old) housed individually in cages. Over a period of 25 days, the birds received either an organic mash feed that was fed dry (group 1), after natural fermentation (group 2) or after fermentation using *Lactobacillus pentosus* (DSM 14025, Chr. Hansen A/S) as a starter culture (group 3). Fermented feeds were characterized by high lactic acid concentrations ($\approx$ 260 mmol/kg feed) and moderate amounts of acetic acid (20–30 mmol/kg feed), high numbers of lactic acid bacteria (10^9–10^{10}/g feed) and a pH of about 4.5.

After 25 days, the birds were killed and the contents of crop, gizzard, ileum and caeca were collected. In crop contents the pH was about 0.5 units lower in birds receiving fermented feed as compared to hens receiving dry feed (4.36 versus 4.89). The pH in caecal contents was about 0.5 units higher following intake of fermented feed (7.0 versus 6.5), indicating reduced microbial fermentation at this location. Fermented feed increased the numbers of anaerobic bacteria, lactic acid bacteria and the concentrations of lactic acid in the crop, gizzard and ileum. In caecal contents the number of *Clostridium perfringens* was about ten times lower in birds fed with fermented feed. In conclusion, fermented feed seems to improve gastrointestinal health due to both acidification of the upper digestive tract and reduced bacterial fermentation in the lower digestive tract.

Poster 13
Bacteriocins reduce duodenal crypt depth and reduce Campylobacter *colonization in turkey poults*

K. Cole,[1] M.B. Farnell,[2] A.M. Donoghue,[2] N.J. Stern,[3] E.A. Svetoch,[4] B.N. Eruslanov,[4] Y.N. Kovalev,[4] V.V. Perelygin,[4] V.P. Levchuck,[4] I. Reyes-Herrera,[1] P.J. Blore[1] and D.J. Donoghue[1]

[1]*Department of Poultry Science, University of Arkansas, Fayetteville, Arkansas, USA;* [2]*Poultry Production and Product Safety Research Unit,, ARS, USDA, Fayetteville, Arkansas, USA;* [3]*Poultry Microbiological Safety Research Unit, ARS, USDA, Athens, Georgia, USA;* [4]*State Research Center for Applied Microbiology, Obolensk, Russian Federation*

Campylobacter is a leading cause of human foodborne illness. Recent evidence has demonstrated that bacteriocins produced by *Lactobacillus* and *Paenibacillus* bacteria inhibit *C. jejuni* in broiler chicks. *Campylobacter coli* is the most prevalent *Campylobacter* isolate found in turkeys. The objective of the present study was to evaluate the efficacy of bacteriocins against *C. coli* colonization and to evaluate the influence of bacteriocins on the gastrointestinal architecture in poults.

Turkey poults were allocated to three treatment groups (total 90 poults; 30/treatment, three trials). Three days post-hatch, all poults were orally challenged with 1.0×10^6 cfu of *C. coli*. On days 10–12 post-hatch, the treatment groups were fed a commercial diet supplemented with purified, micro-encapsulated bacteriocins (500 mg/kg feed). On day 13 post-hatch, caeca were enumerated for *Campylobacter* and the duodenum collected for morphometric analysis.

Campylobacter was undetectable in poults supplemented with bacteriocins in all trials, whereas untreated controls were colonized by 5–6 $\log_{10}$ cfu/g of caecal contents. Duodenum crypt depth was decreased in poults treated with bacteriocin compared to controls ($p < 0.05$). Intestinal crypts are an important niche for *Campylobacter* colonization in poultry. The dynamic reduction in crypt depth in poults supplemented with bacteriocin may provide clues as to how bacteriocins inhibit *Campylobacter in vivo*.

POSTER *14*
Effect of acetic or citric acid administration in drinking water on performance, growth characteristics and ileal microflora of broiler chickens

H. Kermanshahi, M.R. Akbari and J.M. Nadaf

Ferdowsi University of Mashhad, PO Box 91775-1163, Mashhad, Iran

Two experiments were conducted. In each experiment, 300 day-old male Ross broiler chicks were divided into 25 groups and a completely randomized design, including five treatments and five replicates, was used. The treatments in the first experiment consisted of five different levels (0.0, 0.1, 0.2, 0.3 and 0.4%) of acetic acid in drinking water, while in the second experiment citric acid was used at the same levels. Treatments were administered from 1–21 days of age. All groups were fed a medicine-free, practical maize/soybean-based diet.

On days 14 and 28, one chicken of each replicate was weighed and killed by cervical dislocation. Ileal contents were used for microbial evaluation. On day 49, one bird from each pen was killed to weigh gastrointestinal tract, liver, pancreas and abdominal fat. The number of total aerobes and coliforms per gram of ileal contents were enumerated on the appropriate bacteriological media. Neither acetic nor citric acid had any significant influence on feed intake, weight gain, feed:gain ratio or on the weights of body, gastrointestinal tract, liver and pancreas ($p > 0.05$).

Also, the differences between treatments for total aerobe and coliform counts were not significant ($p > 0.05$). Although acetic acid had no significant effect on the weight of abdominal fat, citric acid caused a significant reduction in it ($p \leq 0.05$). Under the conditions of this study, addition of acetic or citric acid to drinking water at the used levels did not affect the performance and ileal microbial counts of chickens. However, citric acid may reduce abdominal fat.

POSTER 15
Gut microflora and its relation to broiler performance

P.J.A. Wijtten,[1] L.L.M. de Lange,[2] H. Panneman,[3] K.C.J. de Goffau[4] and P. van Vugt[1]

[1]*Provimi B.V., Veerlaan 17–23, 3072 AN Rotterdam, Netherlands;* [2]*De Heus, Rubensstraat 175, 6717 VE Ede, Netherlands;* [3]*Dr. Van Haeringen Laboratorium B.V., Agro Business Park 100, 6708 PW Wageningen, Netherlands;* [4]*Storteboom Group, Provincialeweg 70, 9864 PG Kornhorn, Netherlands*

The gut microflora is of major importance for animal performance and health. A novel molecular technique, Microbial Community Profiling and Characterization (MCPC), creates the possibility of obtaining detailed knowledge of the composition and dynamics of the intestinal microflora. We have performed an extensive field survey and a controlled experiment to assess the potential and usability of this method. The effects of feed ingredients on the intestinal microflora and the relation of the microflora to production parameters were studied.

The field survey consisted of 23 flocks on different farms. At approximately 8, 28 and 35 days of age, ten broilers per flock were sacrificed and swab samples of the crop, ileal and caecal contents were taken. In the controlled experiment the effects of dietary ingredients (avilamycin, whey protein, sucrose and lactose and combinations thereof) were tested. Each treatment was assigned to six pens. Crop and ileal swab samples of six birds per pen were taken at 8 and 28 days of age. The bacterial composition of the samples was determined using the MCPC method, a modification of the T-RFLP (Terminal Restriction Fragment Length Polymorphism) method.

Both experiments showed that lower levels of specific peaks in the bacterial profile are correlated with higher performance levels of the birds. Moreover, it was found that a combination of lactose, sucrose and whey protein significantly decreased the level of those bacterial peaks and improved weight gain and feed conversion ratio. These results indicate that the bacteria responsible for these peaks in the bacterial profile negatively affect broiler performance.

POSTER 16
Organic acids in poultry diets: improving the decision-making process

S. Keller,[1] D.S. Parker,[2] P. Buttin[2] and C. Schasteen[3]

[1]*Röthel GmbH, Schwänheit 10, 34281 Gudensberg, Germany;* [2]*Novus Europe, 200 Ave Marcel Thiry, 1200 Brussels, Belgium;* [3]*Novus International Inc., 20 Research Park Drive, St Charles, Missouri 63304, USA*

The EU-wide ban on the use of in-feed, sub-therapeutic levels of antibiotics has focused interest on the role of organic acids in animal feeds. Apart from optimizing feed hygiene there is also the potential for encouraging a favourable microflora in the digestive tract.

There are a number of factors to be considered when developing a strategy for acid use:

1. The level of inclusion that will ensure a consistent response – in some cases the actual concentration and acid composition in commercial blends is not stated, providing no information upon which to make decisions on the optimum inclusion rate.

2. Single acid or blend – blends of methionine hydroxy analogue (Alimet®)* and formic acid, for example, have been shown at acidic pH *in vitro* to be significantly more effective against *Salmonella* Enteriditis than the individual acids under the same conditions.

3. The use of free acid or salt – whereas the use of acid salts has advantages for handling, data from *in vitro* experiments show that acid salts may be less effective at modifying the microbial population than inclusion of the free acid in feed.

4. Encapsulated or protected acids – this approach will only be cost-effective if the protective matrix is truly acid resistant and also if the blend of acids released is active at the pH of the target intestinal site. This will be primarily determined by the pK_a value of the organic acids present in the matrix and the target microflora.

This poster reports on developments in *in vitro* testing systems which provide a means of evaluating individual organic acids and blends for use in poultry rations.

*Alimet® is a registered trademark of Novus International Inc., and is registered in the USA and in other countries.

POSTER 17
Mintrex™ Zn and Mintrex™ Cu organic trace minerals improve intestinal strength and immune response to coccidiosis infection and/or vaccination in broilers

J.D. Richards, T.R. Hampton, C.W. Wuelling, M.E. Wehmeyer and J.J. Dibner

Novus International Inc., 20 Research Park Drive, St Charles, Missouri 63304, USA

Mintrex™* Zn and Mintrex™ Cu are organic zinc and copper sources, respectively, with 2-hydroxy-4(methylthio) butanoic acid (HMTBa) as the organic ligand. Both were tested for the ability to improve intestinal breaking strength (IBS) and immune response to a coccidiosis challenge and/or vaccination.

In experiment one, Cobb 500 broilers were fed diets that were zinc deficient (35 ppm), supplemented with 70 ppm zinc sulphate or 70 ppm zinc sulphate + 35 ppm organic zinc (zinc-methionine or Mintrex Zn).

Half of the birds were vaccinated with a three-species (*Eimeria tenella*, *maxima* and *acervulina*) coccidiosis vaccine (Advent™,** Coccidiosis Control) at hatch, and challenged with all three species on day 24. Mintrex Zn and zinc-methionine both significantly increased IBS in both vaccinated and challenged birds. Only Mintrex Zn increased IBS in both unvaccinated and unchallenged birds. Only Mintrex Zn birds exhibited significantly improved post-vaccination antibody response to an *E. tenella* antigen.

In experiment two, broilers were fed a low copper (9 ppm) diet or a diet supplemented with 25 ppm copper from either copper sulphate, a copper proteinate, copper lysine or Mintrex Cu. All birds were vaccinated with Advent, and challenged as above. Mintrex Cu birds exhibited greater IBS than all other treatments. *Eimeria tenella* lesion scores were significantly reduced in Mintrex Cu-treated birds and were numerically lower than with the other copper sources. Only Mintrex Cu gave a significant improvement in anticoccidial (MIC2) antibody response. Therefore, Mintrex Zn and Mintrex Cu provide significant immune and intestinal health benefits to broilers.

* Mintrex™ is a trademark of Novus International, Inc.
** Advent® is a trademark of Viridus Animal Health, LLC, and is registered in the USA and in other countries.

POSTER *18*
The antimicrobial and acidification effects of Activate WD, an organic acid blend containing Alimet® with other organic acids, in poultry drinking water

C. Schasteen,[1] D. Parker,[2] S. Keller,[3] Jennifer Wu,[1] M. Davis[1] and S.P. Fiene[1]

[1]*Novus International Inc., 20 Research Park Drive, St Charles, Missouri 63304, USA;* [2]*Novus Europe, 200 Ave Marcel Thiry, 1200 Brussels, Belgium;* [3]*Röthel GmbH, Schwänheit, 34281 Gudensberg, Germany*

The antimicrobial properties of DL-2-hydroxy-4-(methylthio)butanoic acid (HMTBA or Alimet®) have previously been documented in culture broth. We have also evaluated the effect of formic, fumaric, propionic, butyric, lactic, HMTBA and HMTBA-containing organic acid blends (Activate™) on *Salmonella* growth in feed using a simulation of conditions (moisture and pH) encountered in the stomach and proximal to the small intestine.

The addition of HMTBA to individual or a blend of organic acids resulted in a more-than-additive increase in antibacterial effect against *Salmonella* in this assay. We have also done field testing of our HMTBA-containing organic acid blend for the drinking water, Activate WD. Three trials (one 22K broiler house treated and one untreated per trial) at different locations were conducted with Activate WD (0.04–0.08%, pH ~3) included in the drinking water for the last 10–14 days prior to processing.

Cloacal swabs of birds were positive for *Salmonella* prior to starting Activate WD, while swabs taken after treatment and prior to transporting the birds to the processing facility were found to be *Salmonella* negative. This does not guarantee the chickens have no *Salmonella* present in their intestine, but indicates that they were not shedding. Liveweight increases were seen in all three trials, with an average increase of 140.7 g/bird compared to untreated houses. The water intake of the Activate WD-treated birds was no different to that of the untreated controls. The presence of HMTBA-containing organic acid blends in the feed and/or water provides antimicrobial and acidification benefits.

*Alimet® is a registered trademark of Novus International Inc., and is registered in the USA and in other countries.

POSTER 19
Microbial diversity of the dominant bacterial communities in Ross 308 broiler chickens reared in different European countries

A. Knarreborg, K. Hoyebye, N. Milora and T.D. Leser

Department of Health and Nutrition, RDA, Chr. Hansen A/S, 10–12 Boege Allé, 2970 Hoersholm, Denmark

Baseline studies of the structure and variation of the dominant bacterial community are necessary to be able to detect changes in the intestinal microbiota of broiler chickens caused by probiotics, Trials, with identical experimental designs, were conducted in four European countries. Each trial involved Ross 308 male chicks, in eight replicate pens with an approximated final stocking density of 32 kg BW/m^2 and a trial duration of 0–35 days. The chickens were offered country-specific standard diets devoid of antimicrobials.

At slaughter, ileal contents from three birds/pen were collected for genetic profiling of the dominant microbial composition. PCR-Denaturing Gradient Gel Electrophoresis analyses (DGGE) using universal primers were performed. The method was refined by adjusting the DNA concentration before and after PCR amplification. Further, a molecular standard ladder was applied. The banding profiles were compared using the Dice similarity coefficient (D_{sc}) with the BioNumerics software package. When all samples were compared, the profiles clustered according to trial location with average D_{sc} of 0.60. Within each trial location, distinct profile separations were obtained between pens, indicating a smaller difference within pens (average $D_{sc} \approx 0.80$) than between pens (average $D_{sc} \approx 0.50$).

In conclusion, the dominant bacterial community in the ileum of broiler chickens was clearly governed by external factors, e.g. feeding, management and climatic conditions. However, with the relatively high homogeneity found within trial locations, effects on the dominant part of the microbiota by probiotics that improve animal health and growth performance are likely to be detected using the present experimental approach.

POSTER *20*
The effect of exogenous enzyme supplementation of barley- and oat-based diets on fermentation in the broiler chicken gastrointestinal tract

D. Józefiak,[1] A. Rutkowski,[1] O. Højberg,[2] B.B. Jensen[2] and R.M. Engberg[2]

[1]*Department of Animal Nutrition and Feed Management, August Cieszkowski Agricultural University, ul.Wolynska 33, 60-637 Poznan, Poland;* [2]*Department of Animal Nutrition and Physiology, Danish Institute of Agricultural Sciences, PO Box 50, 8830 Tjele, Denmark*

The objective of this experiment was to estimate the effect of different dietary fibre fractions from barley and oat and microbial enzyme supplementation (ES) on the fermentation in the broiler chicken gastrointestinal tract. The experiment, lasting 5 weeks, was performed on 384 1 day-old Cobb 500 cockerels, divided randomly into four experimental groups of 12 cages (replicates) each. Barley and oats constituted the majority of carbohydrates in the diets. Experimental diets were either non-supplemented or supplemented with a commercial β-glucanase (1g/kg feed Avizyme® 1100, Danisco, UK, which contained 100 U/g endo 1,3(4)-β-glucanase; 300 U/g endo 1,4-β-xylanase and 800 U/g protease).

The fermentation in the different gastrointestinal segments (crop, gizzard, ileum, caeca) was estimated in terms of the concentrations of short-chain fatty acids (SCFA) and lactic acid (LA) and pH value. Irrespective of the type of cereal, ES increased the lactic acid concentration and lowered the pH value in the crop content. No such changes in fermentation were observed in the contents of gizzard and ileum. The total concentration of SCFA was found to be the highest in the caeca, followed by the crop, ileum and gizzard.

In all gastrointestinal segments, the pH value depended on the cereal type ($p < 0.05$). The location of the crop anterior to the rest of the gastrointestinal tract allows it to influence the composition of the intestinal flora of the host. The observed higher LA and acetic acid concentrations in the crop of the ES birds suggest a stimulation of the growth and activity of lactic acid bacteria at this location. In conclusion, it can be stated that barley- and oat-based diets for broiler chickens can be modified by β-glucanase supplementation leading to acidification of the crop contents.

POSTER *21*
The effect of exogenous enzyme supplementation of barley- and oat-based diets on microbiotic populations in broiler chicken caeca

D. Józefiak,[1] A. Rutkowski,[1] O. Højberg,[2] B.B. Jensen[2] and R.M. Engberg[2]

[1]Department of Animal Nutrition and Feed Management, August Cieszkowski Agricultural University, ul.Wolynska 33, 60–637 Poznan, Poland; [2]Department of Animal Nutrition and Physiology, Danish Institute of Agricultural Sciences, PO Box 50, 8830 Tjele, Denmark

The objective of this experiment was to estimate the effect of different dietary fibre fractions from barley and rye and microbial enzyme supplementation (ES) on the microbiotic populations in broiler chicken caeca. The experiment, lasting 5 weeks, was performed on 384 1 day-old Cobb 500 cockerels, divided randomly into four experimental groups of 12 cages (replicates) of eight birds each. Barley and rye constituted the majority of carbohydrates in the diets. The experimental diets were either non-supplemented or supplemented with a commercial β-glucanase or xylanase (1g/kg feed Avizyme® 1100, Danisco UK, which contained 100 U/g endo 1,3(4)-β-glucanase; 300 U/g endo 1,4-β-xylanase and 800 U/g protease) or xylanase (1g/kg feed Avizyme® 1300, Danisco UK, which contained 2500 U/g endo 1,4-β-xylanase and 800 U/g protease).

Bacterial populations were estimated by fluorescence *in situ* single cell hybridization (FISH). Microbial diversity was determined by the terminal restriction fragment length polymorphism (T-RFLP). The *Clostridium coccoides–Eubacterium rectale* group constituted the majority of the total microbiota in all treatments (9.5–10.6%). This was followed by *Bacteroides* sp. (6.2–8.4%); *Lactobacillus* sp./*Enterococcus* sp. (1.8–3.9%); *Enterobacteriaceae* sp. (0.2–0.4%) and *Bifidobacterium* sp. (0.2–3.7%). The ES reduced the number of *Enterobacteriaceae* sp. in chickens receiving both diets (0.4 versus 0.2%). In chickens fed barley, the ES increased the numbers of *Bifidobacterium* sp. (0.61 versus 3.77%) and reduced the numbers of *Lactobacillus* sp./*Enterococcus* sp. (3.1 versus 1.8%).

POSTER 22
Effects of specific oligosaccharides (Bio-Mos®) on gut microbial populations and immunity

A. Kocher,[1] L. Nollet[1] and L.A. Tucker[2]

[1]*Alltech Biotechnology Centre, Dunboyne, Co. Meath, Republic of Ireland;*
[2]*Cundy Technical Services, Henderson, Auckland, New Zealand*

Immunity is recognized as being intimately associated with feed-influenced gut bacterial populations. Recent studies indicate how commercial oligosaccharide products (Bio-Mos™, Alltech Inc., USA) influence bacterial populations and immune function, demonstrated via vaccine efficacy.

A small-scale trial using 300 mixed-sex Cobb broilers examined three diets: negative control, positive AGP control (Avilamycin 10 ppm) and MOS (Bio-Mos, 2 kg/t 0–21 days, 1 kg/t 22–42 days). Each diet was fed to 100 birds housed in floor pens. Performance was measured at 0–42 days. At 42 days caeca were excised and contents used to determine bacterial populations.

MOS significantly improved body weight and FCR of broiler chickens at 21 and 42 days compared to both the untreated control and the Avilamycin-fed groups ($p < 0.001$), and decreased mortality by up to 4%. MOS significantly increased *Lactobacilli* and *Bifidobacteria* ($p < 0.05$) and decreased *Clostridium perfringens* and coliforms compared to both controls.

The impact of MOS supplementation on the efficacy of Newcastle Disease vaccination was examined. Forty-eight mixed-sex Ross broiler chickens, housed in floor pens, were fed either a negative control or Bio-Mos (1 kg/t) diet from 8–48 days. Birds were vaccinated with NewVac by the ocular–conjunctival route at 15 and 33 days. Antibody titres were analysed via ELISA (ELI-NDV, Pasteur Institute) or IHA-ND (Romvac, Romania) kits (ten birds/treatment). Data showed that birds receiving Bio-Mos had significantly higher antibody titres as determined by IHA at 21 days ($p = 0.044$), and as determined by both methods at 48 days ($p = 0.045$ and 0.009 for ELISA and IHA, respectively).

This research demonstrated that specific oligosaccharides (Bio-Mos) can influence bacterial populations and impact immunity via vaccination.

POSTER *23*

Plant extracts from Australian native plants as alternatives to antibiotic growth promoters in feed for broiler chickens

J.K. Vidanarachchi,[1] L.L. Mikkelsen,[1] I. Sims,[2] P.A. Iji[1] and M. Choct[3]

[1]School of Rural Science and Agriculture, University of New England, Armidale, NSW 2351, Australia; [2]Carbohydrate Chemistry, Industrial Research Limited, PO Box 31-310, Lower Hutt, New Zealand; [3]Australian Poultry CRC, University of New England, PO Box U242, Armidale, NSW 2351, Australia

The aim of the present study was to investigate the effect of water-soluble carbohydrate extracts from the Australian cabbage tree (*Cordyline australis*), Acacia tree (*Acacia pycnantha*) and the seaweed Wakame (*Undaria pinnatifida*) on growth performance and intestinal microflora of broiler chickens.

One-day-old male broiler chickens (Cobb 500) were fed one of eight experimental diets (*n*, 6, eight birds per replicate) from days 1–42.

The diets were: (i) a non-supplemented control diet; (ii) the control diet supplemented with 45 ppm Zn-bacitracin; or (iii) each of the plant extracts at 5 g/kg or 10 g/kg feed. Weight and feed intake per pen were recorded every week. On day 35, two birds from each replicate were killed and digesta from the ileum and caeca collected for bacterial enumeration. The plant extracts did not improve body weight gain, but both levels of Acacia extract significantly improved ($p < 0.05$) FCR compared with the negative control group. The antibiotic treatment group (positive control) significantly improved ($p < 0.05$) body weight gain and feed conversion.

The plant extracts had no effect on the total number of anaerobic bacteria, but significantly increased ($p < 0.05$) the number of *Lactobacilli*, especially in the caeca. Cabbage tree extract at 10 g/kg and seaweed extract at 5 g/kg significantly reduced the number of coliform bacteria in the ileum. The number of *Clostridium perfringens* was significantly reduced in the ileum by Acacia extract, but this effect did not persist into the caeca. The antibiotic-treated group reduced the number of *Cl. perfringens* in both ileum and caeca compared with the negative control group. The number of *Bifidobacteria* was unaffected and below detection levels (10^3 cfu/g digesta) in most samples.

POSTER 24
Effects of oxidation and vitamin E inclusion on resistance to haemolysis, intestinal microbiota, faecal coccidial counts and epithelial structure of broiler chickens

J.A. Choque-López, E.G. Manzanilla, A. Gomez de Segura, M.D. Baucells and A.C. Barroeta

Department of Animal Sciences and Food, Faculty of Veterinary Medicine, Universitat Autònoma de Barcelona, 08193 Bellaterra, Barcelona, Spain

A trial was carried out to study the effect of vitamin E supplementation or oxidized fat on chicken health status. A total of 45 commercial Ross broiler chickens were allocated to one of three dietary treatments (15 animals/treatment) from 4–18 days of age. Experimental diets were: (i) SO, 6% sunflower oil; (ii) OE, oil plus 400 ppm of α-tocopherol acetate; and (iii) OO, 6% sunflower oil oxidized by heating at 185°C for 18 h. Resistance to haemolysis, ileal microbiota, faecal coccidial concentrations and epithelial structure was studied, from five chickens/treatment, at days 11 and 18 of age.

Resistance to haemolysis was increased by vitamin E inclusion at days 11 ($p = 0.008$) and 18 ($p = 0.150$) but was not affected by oxidation of the fat. Concerning microbiota and coccidial counts, no significant differences were found at day 11, but at day 18 each diet promoted a totally different microbial population. In particular, the OO diet promoted lower counts for all studied bacteria: *Enterococcus* ($p = 0.017$), *Enterobacterium* ($p = 0.012$), *Lactobacillus* ($p = 0.007$) and Gram-negative anaerobes ($p = 0.14$).

On the other hand, the OE diet promoted only a decrease in *Enterococcus* ($p = 0.017$). Concerning coccidiosis, both oxidation and vitamin E inclusion promoted an increase in coccidia, counts being more pronounced in the case of the OE diet ($p = 0.0005$). Epithelial structure, villous height, crypt depth and intra-epithelial lymphocytes were not affected by the treatments.

POSTER 25
Molecular epidemiology, gastrointestinal ecology and interventions for commensal human foodborne bacterial pathogens in the chicken at the Poultry Microbiological Safety Research Unit, ARS, USA

B.S. Seal

Poultry Microbiological Safety Research Unit, Russell Research Center, Agricultural Research Service, USDA, 950 College Station Road, Athens, Georgia 30605, USA

Campylobacter spp., *Salmonella* spp. and *Clostridium perfringens*, the three leading causes of human bacterial foodborne illness, are commonly associated with normal poultry gastrointestinal flora. Additionally, *Listeria monocytogenes* is considered to be the deadliest bacterial foodborne pathogen. Our research unit determined, using DNA sequence correlated to rep-PCR analysis, speciation of *Campylobacter* spp., serotyping of *Salmonella* spp. and source-tracking for *Cl. perfringens*.

Using suppressive–subtractive hybridization and proteomics, genes associated with membrane protein glycosylation were identified that might promote colonization of *Campylobacter jejuni* in the chicken. Microbial ecology of the chicken gastrointestinal tract, reproductive tract, internal organs and exogenous biofilms was assayed utilizing real-time PCR and biophotonics methods.

Furthermore, investigators in the unit have developed bacteriocins (antibacterial peptides) and identified bacteriophages as potentially effective intervention strategies for alternatives to antibiotics in animal feeds. This was accomplished by assaying bacteria, such as *Paenibacillus polymyxa*, for anti-*Campylobacter* spp. replication and by isolating lytic bacteriophage specific for *Campylobacter* spp. and *Cl. perfringens* from poultry processing plants and sewage. The bacteriophages have DNA genomes that are approximately 50 kb in size. Investigators have also improved cultural methods for *Campylobacter* spp. obtained from poultry by developing a chromogenic selective plating medium, Campy-Line agar.

POSTER 26
Rep-PCR subtyping of type A Clostridium perfringens *isolates from healthy and necrotic broilers and their environs reared under drug-free conditions*

G.R. Siragusa,[1] A. Aryaie[2] and M.G. Wise[1]

[1]*Agricultural Research Service, USA Department of Agriculture, Russell Research Center, 950 College Station Road, Athens, Georgia 30605, USA;* [2]*Department of Microbiology, University of Georgia, Athens, Georgia 30605, USA*

A library of *Clostridium perfringens* (Cp) isolates from healthy and necrotic broiler chickens from 13 different drug-free farms (no antibiotic growth promotants), darkling beetles and larvae were examined for genetic relationships using rep-PCR (repetitive element, DiversiLab system) molecular subtyping. Two major groupings were observed: Group I consisted of 73% (eight of 11) Cp of necrotic origin; Group II was composed of 54% (20/37) Cp isolates of necrotic origin.

Within Group I, isolates were distributed between intestinal, caecal and larval origin from five different growout houses and ranged in similarity from 80 to 99% (Pearson's Correlation index). From Group II, isolates were within 70% similarity and of caecal, intestinal and beetle origin; clustering of isolates from necrotic birds, healthy birds and both was observed.

It appears that, outside of the two major subgroupings I and II and the three Group II minor clades, four main groupings were observed within this collection. Although from different growout houses representing 13 different farms, at least one of the clades (labelled IIB) consisted only of Cp isolates from healthy broiler intestinal/caecal origin and a second clade (labelled IIA1a) was composed of 70% Cp isolates of necrotic origin. These were not affiliated with one single house or farm. While the significance of these cladistic observations remains to be studied, it is our hypothesis that focusing study on Cp subtypes of necrotic enteritis linkage will lead to a better understanding of those Cp strains most involved in necrotic enteritis in drug-free rearing operations.

POSTER 27
Quantitative assessment of the gut microbial flora of broilers with and without antibiotic growth promotants, determined by quantitative, real-time PCR

M.G. Wise and G.R. Siragusa

Agricultural Research Service, USA Department of Agriculture, Russell Research Center, 950 College Station Road, Athens, Georgia, 30605, USA

The contribution of major bacterial groups to the broiler intestinal ecosystem was determined from samples of chickens reared with and without antibiotic growth promotants (AGPs). Using 16S rRNA-DNA primers and quantitative, real time PCR, the following groups were estimated in terms of relative abundance of signature sequences: the *Clostridium leptum* subgroup, the *Cl. coccoides* subgroup, the *Bacteroides* group, *Bifidobacterium* ssp.*, Enterobacteriaceae*, the *Lactobacilli*, the *Cl. perfringens* subgroup, *Enterococcus* spp., *Veillonella* spp., *Atopobium* spp., *Campylobacter* spp. and the domain bacteria.

Broiler chickens were obtained at 7, 14 and 21 days of age from each of two broiler farms (conventional and drug free). Ileal and caecal samples were removed, iced and frozen (–80°C) until DNA was extracted. Ileal communities were dominated by the *Lactobacilli* and *Enterobacteriaceae*, with enterics much more numerous in drug-free birds than in those reared with AGPs at the 7 and 14 day time periods.

With caecal samples, successional patterns were similar in conventional and drug-free birds: the *Enterobacteriaceae* dominated at day 7 and were replaced by obligate anaerobes of the *Cl. leptum* subgroup, the *Cl. coccoides* subgroup and the *Bacteroides* group by day 14. Numerically lesser groups were more abundant in drug-free caeca. It was observed that days 14 and 21 drug-free caecal samples were positive for *Campylobacter* spp. sequence, while no *Campylobacter*-positive samples were found in the conventional house after day 7. We plan to utilize this method in conjunction with a necrotic enteritis model to determine the gut profile during the time period preceding necrotic enteritis.

POSTER *28*
Dietary probiotic changes the luminal, mucosally associated and caecal microflora

A. Smirnov, R. Perez and Z. Uni

Department of Animal Sciences, Faculty of Agricultural, Food and Environmental Sciences, The Hebrew University of Jerusalem, PO Box 12, Rehovot, Israel

The commensal microbial populations have an important metabolic and protective function in the chicken gastrointestinal tract. In this study, the effect of dietary probiotic on the relative amounts of several jejunal luminal bacterial (JLB), mucosally associated bacterial (MAB) and caecal bacterial (CB) species were examined. For the experiment, 1 day-old Cobb chicks were fed diets containing either probiotic (Pro, Primalac 2 g/kg) or antibiotic growth promoter (AGP, Virgniamycin 20 mg/kg). A control group was fed a diet with no additives. Samples were taken at 7 and 28 days of age.

Relative amounts of the bacterial species were analysed by PCR amplification of the bacterial 16s ribosomal DNA (rDNA). *Lactobacillus* spp. were present in JLB and MAB fractions, with no differences between the treatments. However, *Lactobacillus* spp. in the CB fraction were increased at 7 days by 31% and at 28 days by 44% in the Pro group compared with the controls ($p < 0.05$). *Bifidobacterium* spp. at age 7 days were present only in the JLB fraction of all Pro and AGP chickens, and only in a few control chicks. *Bifidobacterium* spp. at day 28 were present in all fractions of Pro and control chicks and were undetectable in the AGP group. *Clostridium* spp. at 7 days were detected in only the JLB and CB fractions, while at 28 days all fractions contained *Clostridium* spp., with decreased amounts in the MAB and JLB of the AGP group compared with the controls (73% and 70%, respectively; $p < 0.05$).

Clostridium perfringens were detected only at 7 days in the JLB fraction, with no differences between the treatments. The results showed that in the early stages of chicken growth all bacterial species examined were present in the jejunal luminal content, while mucosa and caeca were colonized gradually. This might be related to the developmental changes occurring in the intestine. Probiotic supplementation led to early colonization and increased amount of *Lactobacillus* spp. in the caeca, which might have a beneficial effect on the chicken gut function and health.

POSTER *29*
Mucin dynamics in the chicken jejunum following dietary mannoligosaccharides (Bio Mos) and antibiotic growth promoter (Virginiamycin) supplementation

A. Smirnov, R. Perez, M. Thein and Z. Uni

Department of Animal Sciences, Faculty of Agricultural, Food and Environmental Sciences, The Hebrew University of Jerusalem, PO Box 12, Rehovot, Israel

The mucous layer plays an important protective role in the chicken small intestine. In this study, the effect of mannoligosaccharides and antibiotic growth promoter on mucin dynamics in the chicken jejunum was examined. Cobb chickens at day 1 were fed diets containing either mannoligosaccharides (Bio Mos 2 g/kg) or antibiotic growth promoter (AGP, Virginiamycin 20 mg/kg). A control group was fed a diet with no additives. The jejunum was sampled at age 7, 14 and 28 days. Mucin gene expression was determined by semiquantitative RT-PCR. The mucous-adherent layer thickness was measured according to Corne's method. Feeding BioMos led to an increase in mucin mRNA expression at day 7 (by 146%) and day 28 (by 154%) compared with the controls ($p < 0.05$).

Following enhanced mucin gene expression an increased thickness of the mucous-adherent layer was found at days 7, 14 and 28 (81, 22 and 60% greater than controls, respectively). Dietary AGP caused a 30% decrease at 14 days and a 30% increase in mucin gene expression at 28 days compared to the control, while mucous-adherent layer thickness was greater by 45% at 7 days and by 32% at 28 days compared with the controls ($p < 0.05$). Results show that both Bio Mos and AGP had a stimulating effect on the mucin dynamics in the chicken small intestine, either directly or by the changes in the intestinal microflora. A change in mucin dynamics may modify the mucous defensive barrier and have important physiological implications.

POSTER *30*
Characterization of bacterial community in the ileum of broiler chickens

W. McBurney,[1] G.W. Tannock[1] and V. Ravindran[2]

[1]*Department of Microbiology, University of Otago, Dunedin, New Zealand;*
[2]*Institute of Food, Nutrition and Human Health, Massey University,
Palmerston North, New Zealand*

In common with the gut microflora of other animal species, a large proportion of the bacteria that reside in the distal gut of broilers have not yet been cultivated in the laboratory. This has necessitated the application of nucleic acid-based, culture-independent methods of analysis of the gut microflora. In this study, denaturing gradient gel electrophoresis (DGGE) of DNA fragments obtained by polymerase chain reaction (PCR) amplification was used to define the microfloral profile in the ileal contents of healthy broilers.

The birds were fed maize–soy diets without or with an in-feed antibiotic (zinc bacitracin; 100 mg/kg diet) in a 6-week trial, and contents from the terminal ileum were obtained on days 1 and 2, and thereafter at weekly intervals. Bacterial DNA was extracted from each ileal sample and the V2–V3 regions of the 16S ribosomal RNA gene were amplified by PCR using bacterial primers HDA1-GC and HDA2. The 16S rDNA fragments in the PCR products were separated by DGGE to generate a profile of the bacterial community in ileal samples. DNA fragments of interest were cut from the DGGE gel and sequenced to permit bacterial identification.

The microfloral profile progressed from a simple collection of bacterial species containing *Enterococci* and *Escherichia coli* at day 1 to a profile in which *Lactobacilli* were predominant from week 3. Particularly noticeable in birds aged 1 week of age were DNA fragments representing Gram-positive cocci (*Enterococci/Streptococci*), the appearance of *L. aviarius* in the ileum from week 4 and *L. salivarius* from week 3. The segmented, filamentous ileal organism that attaches to the ileal mucosa was particularly apparent in birds aged 1 week. The impact of zinc bacitracin on the composition of the ileal microflora was negligible except at day 2, when *Clostridium perfringens* was detected in the untreated birds but not in the treated birds.

POSTER *31*
Performance and development of digestive organs and gut morphology of broilers fed diets based on maize or wheat

V. Ravindran and Y.B. Wu

Institute of Food, Nutrition and Human Health, Massey University, Palmerston North, New Zealand

The performance of broiler chickens fed diets containing viscous grains (wheat) is generally poorer than those fed diets containing non-viscous grains (maize), and the hypothesis that these differences in performance are related to differences in relative digestive tract size and intestinal morphology was tested in this study. Two experimental diets containing wheat and maize as the cereal base were formulated. All three diets were formulated to contain similar levels of metabolizable energy and amino acids. The wheat-based diet was supplemented with a commercial xylanase. Each diet was fed to four replicate groups (46 birds/replicate) from days 1–35 post-hatching. Body weights and feed intake were recorded at weekly intervals, and the feed per gain was calculated on a pen basis.

The weights of digestive organs and tracts were obtained from two birds per replicate pen at weekly intervals. On day 21, jejunal segments were removed from two birds per replicate pen for gut morphological measurements. Weight gain and feed intake of birds were unaffected ($p > 0.05$) by the cereal type, but feed efficiency of birds fed the maize-based diet was better ($p < 0.05$) than for those fed the wheat-based diet (1.60 versus 1.66 g/g).

In general, the type of cereal base had no effects ($p > 0.05$) on the development or relative weights of the proventriculus, gizzard, liver, pancreas and small intestine. The villus height tended ($p = 0.07$) to be greater and the goblet cell numbers lower ($p < 0.05$) in birds fed the maize-based diet compared with those fed the wheat-based diet. No significant differences ($p > 0.05$) were observed in epithelial thickness and crypt depth between the two dietary types. These results show that the differences in feed efficiency of broilers fed maize- and wheat-based diets may be explained, at least in part, by differences in gut morphology.

POSTER *32*
Effect of Virginiamycin on microbiota in five intestinal locations in broiler chicken using chaperonin-60 sequencing and quantitative PCR

A.G. Van Kessel,[1] T.J. Dumonceaux,[1] J.E. Hill[2] and S.M. Hemmingsen[2]

[1]*Department of Animal and Poultry Science, University of Saskatchewan, 51 Campus Drive, Saskatoon S7N 5AB, Canada;* [2]*National Research Council Canada, Plant Biotechnology Institute, Saskatoon, Canada*

The inclusion of antibiotic growth promoters (AGP) at subtherapeutic levels in poultry feeds can positively affect animal health and growth. These effects are presumed to be indirect and to result from changes in the host gastrointesinal microbiota. To improve our understanding of the intestinal microbiota and the effect of AGP, bacterial flora were characterized in five different gastrointestinal tract locations (duodenal loop, mid-jejunum, proximal ileum, ileocaecal junction and caecum) in twenty 47 day-old broiler chickens selected from a larger study in which maize/soybean-based diets, with and without virginiamycin, were fed.

Ten chaperonin-60 universal target (*cpn*60 UT) libraries were prepared, corresponding to each intestinal location within diet, and a total of 10,932 nucleotide sequences determined. A total of 370 distinct *cpn*60 UT sequences were identified, which ranged in frequency of recovery from 1 to 4753. Small intestinal libraries were dominated by Lactobacillales-like species, accounting for 90% of sequences. Sequence diversity was much higher in the caecum and included members of the Clostridiales (68% of sequences), Lactobacillales (25% of sequences) and Bacteroidetes (6% of sequences) families.

The effects of virginiamycin were investigated by quantitative PCR (qPCR) enumeration of 15 individual bacterial targets selected based on frequency of sequence recovery. Virginiamycin increased the abundance of many of the targets in the proximal gastrointestinal tract (duodenal loop to proximal ileum), with fewer targets affected in the distal regions (ileocaecal junction and caecum). These findings provide improved profiling of the composition of the chicken intestinal microbiota and indicate that microbial responses to AGP are most prominent in the proximal small intestine.

POSTER *33*
Effect of low-protein diets of differing digestibility on necrotic enteritis in broiler chickens

**T.E. Warren, D.C. Wilkie, J. Dahiya,
A.G. Van Kessel and M.D. Drew**
*Department of Animal and Poultry Science, University of Saskatchewan,
51 Campus Drive, Saskatoon S7N 5AB, Canada*

An experiment was performed to determine the effect of protein digestibility on intestinal *Clostridium perfringens* populations in broiler chickens. Canola protein concentrate (CPC) was autoclaved for 40 min and this treatment reduced the ileal digestibility of crude protein from 770 to 700 g/kg. Male broiler chickens (n, 144) were fed a commercial medicated starter diet on days 1–14. From day 14 onwards the birds were switched to one of three iso-energetic experimental diets. A control diet was formulated to contain 186 g/kg crude protein, 11 g/kg total lysine and 8.4 g/kg ileal digestible lysine, with all other amino acids balanced for ideal protein using CPC as the primary protein source.

A second diet was formulated to contain total amino acids equal to the control diet using the heat-damaged CPC as the primary protein source. A third diet based on heat-damaged CPC was formulated to contain ileal digestible amino acids equal to the control diet. Birds were orally gavaged with 1.0 ml of a culture of *Cl. perfringens* on days 14–21 and euthanased on day 28.

The birds were dull and depressed for the first 4–6 days after gavaging, with 9.7% mortality. No significant differences ($p > 0.05$) were observed in the ileal or caecal *Cl. perfringens* or *Lactobacilli* counts or growth performance during the experiment. The results suggest that the difference in protein digestibility due to heat treatment of CPC did not affect the intestinal *Cl. perfringens* populations and may not be a contributing factor in the development of necrotic enteritis in broiler chickens.

POSTER *34*
The effect of hen egg antibodies on Clostridium perfringens *colonization in the gastrointestinal tract of broiler chickens*

D.C. Wilkie, A.G. Van Kessel, T. Dumonceaux and M.D. Drew

Department of Animal and Poultry Science, University of Saskatchewan, 51 Campus Drive, Saskatoon S7N 5AB, Canada

Two feeding trials were performed to assess the ability of hen egg antibodies (HEA) to reduce intestinal colonization by *Clostridium perfringens* in broiler chickens. Antibodies against *Cl. perfringens* or cholera toxin (negative control) were obtained from the eggs of laying hens hyperimmunized using a *Cl. perfringens* bacterin or cholera toxin. A preliminary experiment was conducted to determine the *in vivo* activity of the administered antibody along the length of the intestine and, although antibody activity declined from proximal to distal regions of the intestine, it remained detectable in the caecum.

In the first experiment there was no significant reduction in the number of *Cl. perfringens* in the birds fed the diet amended with the anti-*Cl. perfringens* HEA compared with the birds that received the anti-cholera toxin HEA ($p > 0.05$). In the second experiment there was a significant decrease in *Cl. perfringens* intestinal populations 72 h after treatment, as assessed by cultured-based enumeration ($p < 0.05$), but not by molecular enumeration using real-time, quantitative PCR.

Antibody-mediated agglutination of *Cl. perfringens* cells may have contributed to the lower counts obtained using culture-based enumeration. In trial 2, intestinal lesion scores were higher ($p < 0.05$) in the birds that had received the anti-*Cl. perfringens* HEA. This work suggests that HEA raised against a *Cl. perfringens* bacterin did not reduce the level of *Cl. perfringens* intestinal colonization and, conversely, may exacerbate the severity of necrotic enteritis.

POSTER *35*
Effect of dietary glycine on intestinal populations of Clostridium perfringens *and* Lactobacillus *in broiler chickens*

J.P. Dahiya,[1] D. Hoehler,[2] D.C. Wilkie,[1] A.G. Van Kessel[1] and M.D. Drew[1]

[1]*Department of Animal and Poultry Science, University of Saskatchewan, 51 Campus Drive, Saskatoon S7N 5AB, Canada;* [2]*Degussa Corporation, 1701 Barrett Lakes Boulevard, Kennesaw, Georgia 30144, USA*

Previously we reported a significant correlation between gut populations of *Clostridium perfringens*, the causative agent of necrotic enteritis (NE), and dietary glycine level. To confirm these results, an experiment was conducted to examine the effect of dietary glycine on gut *Cl. perfringens* populations, α toxin production and necrotic enteritis lesion scores in broiler chickens. From 14 days of age, 24 groups of four birds were given six different ideal, protein-balanced diets (19 or 23% CP) formulated to contain 5.0, 7.5, 10.0, 15.0, 20.0 or 40.0 g/kg glycine. Diets were isocaloric (13.4 MJ/kg ME) and met or exceeded NRC nutrient requirements.

All birds were orally challenged with *Cl. perfringens* Type A broth culture daily on days 14–21 and killed on day 28. The majority of birds showed clinical signs of necrotic enteritis for the first 5–7 days after gavaging and there was 4.16% mortality. The lesion score was significantly higher ($p < 0.05$) in birds fed 40.0 g/kg glycine diets compared with those fed the other diets.

Glycine supplementation resulted in a significant ($p < 0.05$) quadratic response for *Cl. perfringens* growth in the caecum with maximum response at 35.1 g/kg dietary glycine concentration. *Lactobacillus* counts in the ileum and caecum declined significantly ($p < 0.05$) in birds fed 40 g/kg glycine diets compared with those in the other treatment groups. We conclude that glycine may be an important determinant of *Cl. perfringens* numbers in the intestinal tract and of the levels of morbidity and mortality due to necrotic enteritis in broiler chickens.

POSTER *36*
Effect on dietary AME and nitrogen retention in broilers previously fed diets containing phytase

V. Pirgozliev,[1] O. Oduguwa,[2] T. Acamovic[1] and M.R. Bedford[3]

[1]*Avian Science Research Centre, Scottish Agricultural College, West Mains Road, Edinburgh EH9 3JG, UK;* [2]*University of Agriculture, Abeokuta, Nigeria;* [3]*Zymetrics Inc., Chestnut House, Beckhampton, Marlborough, Wiltshire SN8 1QJ, UK*

Dietary constituents interact with the epithelial tissue of the gastrointestinal tract (GIT) and with the microflora, therein causing variation in the excretion of endogenous material which may influence the health and wellbeing of poultry. Enzyme supplementation may improve performance and microbial profiles and may reduce endogenous losses, improving the health of the avian GIT.

A precision feeding experiment was conducted with broilers, which had previously been fed diets containing phytase, to study the effects on dietary apparent metabolizable energy (AME), nitrogen retention and endogenous losses. Sixty-four female Ross broilers, weighing approximately 2.5 kg and previously fed with four experimental diets: (i) control diet based on a maize/soy diet; (ii) control + 250 IU phytase; (iii) control + 500 IU phytase; and (iv) control + 2500 IU phytase were used. Thirty-two birds (eight from each of the enzyme treatments) were fed a standard withdrawal diet, while another 32 received glucose. Birds from each of the previous treatments were replicated eight times in a randomized block design.

The AME and N retention was significantly higher ($p < 0.001$) for broilers which had been previously fed diets supplemented with 250 and 500 IU phytase compared with birds fed diets containing 0 and 2500 IU phytase. Interestingly, and in line with our hypothesis, there was a negative, linear relationship between sialic acid excreted ($p < 0.001$; $R^2 = 0.60$) and AME. The result of this study suggests that moderate phytase supplementation prior to consumption of diets without supplemental enzymes may improve gut health, reduce endogenous losses of the birds and improve nutrient utilization.

POSTER 37
Potential for natural products to control growth of Clostridium perfringens in chickens

J. Zrustova,[1] K. Svoboda,[2] M. Ritter[1] and J.D. Brooker[1]

[1]*Animal Health Group,* [2]*Life Sciences Group, Scottish Agricultural College, West Mains Rd, Edinburgh EH9 3JG, UK*

Clostridium perfringens causes necrotic enteritis, a disease that affects the poultry industry both subclinically and as overt disease. Enteric infection with *Cl. perfringens* is currently controlled by antimicrobial ionophores, but the incidence may increase with their proposed ban. We have investigated the use of certain plant secondary products as potential agents to reduce the risk of *Cl. perfringens* colonization in chickens.

Studies *in vitro* were carried out with *Cl. perfringens* Type A, and commensal gut microflora from chickens, using a fermentation model that we developed previously. Four essential oils, three condensed tannins and 14 different natural product feed additives were tested.

The results show that the fermentative activity of *Cl. perfringens* was inhibited by all essential oils tested, with the most effective being lemon myrtle (*Backhousia citriadora*), with an MIC of 0.05% v/v of culture medium. This was at least twice as effective as tea tree (*Leptospermum*) oil. In contrast, the fermentative activity of commensal chicken gut microflora was not affected until at least 0.1% v/v, and the MIC was > 0.2% v/v. Condensed tannins were also effective as anticlostridial agents, with MICs of 0.6–1% w/v of the culture medium; feed intake impairment *in vivo* did not occur until at least 3% w/v condensed tannin. Of 14 natural product feed additives tested, three were effective at preventing growth of *Cl. perfringens*.

The data show that several essential oils, tannins and other natural products have the potential to reduce the risk of enteric disease, and application *in vivo* now needs to be tested.

POSTER *38*
The supplementation of chicken diets with Bacillus subtilis

J. McLean,[1] T. Acamovic,[1] S. Hansen,[2] M. Moerk-Jensen[2] and I. Nevison[3]

[1]*Avian Science Research Centre, Scottish Agricultural College, West Mains Road, Edinburgh EH9 3JG, UK;* [2]*Chr. Hansen A/S, Boege Alle 10–12, 2970 Hoersholm, Denmark;* [3]*BioSS, Hannah Research Institute, Ayr KA6 5HL, UK*

The effects of dietary supplementation of *Bacillus subtilis* on performance of chickens fed wheat/soybean-based diets over a 35-day grower period were assessed. Day-old male chickens (Ross 308, *n*, 1600; four dietary treatments × eight replicates) in a randomised block design in a controlled-environment house were fed *ad libitum* with a starter feed (days 0–14) and grower feed (days 14–35). Diets were designed to be identical within phase (calculated ME; 12.2 and 12.8 MJ/kg and determined CP; 234 and 228 g/kg DM; starter and grower, respectively). Diets were supplemented with the *B. subtilis* spores at 0, 500, 1000 and 4000 g/tonne (0, 0.8×10^6, 1.6×10^6 and 6.4×10^6 CFU/g of diet).

Mortality (5% overall) was low and was not influenced by dietary treatment. The determined spore content of the diets approximated that expected. The feed and caecal spore counts were positively correlated (correlation coefficient 0.998) and regression analysis indicated a highly significant linear relationship ($R^2 = 0.99$, $p < 0.01$).

Supplementation of the wheat/soybean meal-based diets with *B. subtilis* spores improved overall (0–35 days) feed intake (6.8%, $p < 0.001$) and growth (5.1%, $p = 0.002$), although FCR was slightly poorer (1.6%, $p = 0.003$) compared to the birds fed the unsupplemented control diet. The modified European Efficiency Factor tended to increase with *Bacillus* supplementation compared to the unsupplemented control group (395 versus 388, $p = 0.384$). The beneficial effects on intake and body weight were consistent throughout the study and were apparent in the first week of life.

POSTER *39*
Diet affects gut structure and absorptive capacity: gaining strength from a universal method

E. Clarke and C. Drakley

Department of Animal Physiology, School of Biosciences, University of Nottingham, Sutton Bonington Campus, Loughborough, Leicestershire LE12 5RD, UK

The growth rate to market of broilers is now so rapid that any detrimental change in the growth and absorptive capacity of the gut can, potentially, have a huge effect on performance. Many mainstream dietary components contain harmful levels of anti-nutritive factors (ANF), from the β-glucanase in barley to the trypsin inhibitors of soy. Enzymes (and other supplements) are increasingly being used to counteract the effects of these ANFs and enable broilers to grow at their full potential. But how are these detrimental and positive effects of diet measured? Various methods exist, including: (i) direct measurements of the size of the absorptive surface of the villi (length, width height measurements, etc.); (ii) epithelial cell migration; (iii) measurement of crypt cell proliferation rate by radio-labelled thymidine; or (iv) metaphase arrest.

The importance of gut health is well accepted, but difficulty in defining the effects of diet on gut health means that this important area is often neglected in mainstream poultry nutrition. However, it could be a really powerful tool in the assessment of dietary impact on performance. At present there is no one universally accepted method, so students and researchers new to the area have no clear guidance on the best approach, often resulting in yet another hybrid method being born. This poster aims to kick-start debate around the main published approaches, leading to a universally accepted method of allowing the results of different studies to be compared directly.

POSTER *40*
Development of a model for inducing subclinical necrotic enteritis in chickens

R.M. McDevitt, J. Brooker, T. Acamovic and N.H.C. Sparks

Avian Science Research Centre, Scottish Agricultural College, West Mains Road, Edinburgh EH9 3JG, UK

Subclinical necrotic enteritis (NE), caused by *Clostridium perfringens*, affects the health and productivity of chickens. The correlation between *Cl. perfringens* colonization and small intestinal damage was evaluated in 168 chickens. Birds were fed a diet of wheat, soy meal, plus fish meal (100 g/kg) and carboxymethyl cellulose (7 g/kg). Birds were divided into four treatments: (i) basal diet (C–); (ii) basal diet plus the in-feed antibiotic monensin (C+); and (iii) two treatments of C– plus an inflammation agent (IA) – either dextran sulphate (DS) or DS plus trinitrobenzene sulphonic acid (TNBS).

All chickens received a background dose of *Cl. perfringens* (10^4 cfu) at 4 days of age. The DS and DS + TNBS birds were dosed with IA from 11–13 days of age. Lesion score and *Cl. perfringens* numbers were evaluated in all groups at 2, 5, 7, 15 and 22 days post-dosing. Overall, 27% of C+ birds had lesions, compared with 50% of C–, 56% of DS and 66% of DS + TNBS birds. Mean lesion score was three times higher in DS (0.80) and DS + TNBS (0.85) compared with C+ (0.24) birds, and was intermediate in C– (0.57). *Clostridium perfringens* numbers were greater in DS and DS + TNBS birds compared with either C+ or C–. Lesion score and *Cl. perfringens* numbers were positively correlated in this model of subclinical NE.

POSTER 41
Effect of dietary dried whey and probiotic on performance and humoral immune response in broilers

M. Torki and B. Molanapour

Animal and Poultry Science Department, Agricultural Faculty, Razi University, Imam Avenue, Kermanshah, Iran

Lactose is fermented to lactic acid and volatile fatty acids (VFA), which may stimulate the colonization of *Lactobacilli* in the intestinal tract. Dried whey contains approximately 65% lactose. This study was conducted to evaluate the effect of dietary supplementation of dried whey and probiotic on broiler performance and antibody response to Newcastle disease virus. A basal diet was supplemented with dried whey and probiotic in a 3×2 factorial arrangement (0, 1.5 and 3% dried whey with or without probiotic). Probiotic Bioplus 2B® contains *Bacillus subtilis* and *B. licheniformis* (1.6×10^9 CFU/g for each) at 0.1 and 0.04%, respectively.

During days 7–21 and 22–49 of age, 252 7-day-old Ross broiler chicks, a mixture of both sexes, were utilized for the 42-day experimental period. The chicks were randomly allocated to 18 pens containing 14 chicks, each with three replicates and assigned to receive one of the six dietary treatments. Chicks were immunized (eye dropper) against NDV (B1) at 12 d, followed by a booster (La Sota) at 36 days. Blood samples were collected (two birds per pen) at 13, 23, 31 and 37 days post-primary immunization.

At day 42, the chicks fed on 15 g/kg dried whey diet had significantly higher ($p < 0.05$) body weight (BW) than the control chicks. Probiotic had no significant effect on BW, feed intake or feed:gain ratio, except for feed intake in the growing period for the chicks fed on a diet supplemented by probiotic, where there was a significantly lowered feed intake than in control chicks. Dried whey and probiotic had no significant effect on antibody response. The results of this study show that including dried whey at up to 15 g/kg diet, especially in growing diets, can improve the BW of chicks.

POSTER *42*
Effect of feeding enzyme-supplemented rice bran on the performance of broiler chicks

M. Torki and M. Falahati

Animal and Poultry Science Department, Agricultural Faculty, Razi University, Imam Avenue, Kermanshah, Iran

Rice bran contains non-starch polysaccharides (NSPs), of which arabinose and xylose are predominant. Much of the 16–18 g phosphorus (P)/kg in rice bran occurs as phytic acid P. Not only is the availability of P very low in rice bran, but some other minerals can be complexes with phytate and this availability is further reduced. The influence of various dietary enzyme supplementations on the performance of 216 unsexed Ross broiler chicks fed on wheat/soybean meal/rice bran-based diets was investigated.

The experiment was conducted as a 4×3 factorial arrangement of treatments. Within the factorial, four iso-nitrogenous and iso-energetic (12.12 MJ/kg ME and 208 g crude protein) NRC-recommended diets containing four levels of rice bran (0, 7.5, 15.0 and 22.5%) were evaluated. Each diet was supplemented with either phytase (0.5 g Natuphos/kg of diet), NSPase (0.17 g of Grindazyme GP 15,000/kg of diet) or no enzyme as control.

There was a significant effect of enzyme on body weight gain during the period of 21–42 days of age ($p < 0.05$). There was a significant effect of rice bran on feed intake between 21–42 days of age ($p < 0.05$). Even though diet effects on feed conversion efficiency (FCE) during the periods of 7–21 and 21–42 days of age were significant ($p < 0.05$), there was no significant effect ($p > 0.05$) of dietary rice bran on FCE between 42–49 days of age. FCE was significantly influenced by the enzyme inclusion ($p < 0.01$).

POSTER 43
Gut health and poultry meat quality

T.I. Fotina

Veterinary Medicine Department, Sumy State University of Agriculture, 40021 Sumy, Ukraine

A healthy bird can live in equilibrium with *Escherichia coli* organisms in its lower gut, the acidic conditions of the higher gut preventing colonization of the gut lining and penetration into deeper tissues and bloodstream. Certain strains are more important than others and mediate their effects through the release of damaging toxins, which lead to tissue damage, toxaemia and death. *E. coli* is lying in wait, waiting for a change in the delicate balance between the bird's immune system and organisms in the intestines or environment. But moreover, when the balance in the gut is changed and *E. coli* starts to cause problems, the biochemical composition of meat could be changed also.

In our experiment we studied changes in meat quality during eschirichiosis. We established that the moisture content of meat increased by 1.77–2.27%, protein content by 1.02–3.0% while fat and calorie content decreased by 4.2% and 130.66–134.01 KJ, correspondingly. Essential amino acids concentration decreased by 1.08–1.27% and non-essential amino acid content increased by 2.58–2.69%. Mineral content (Ca, K, Al, Cu, Co), vitamin concentrations (vitamins A, E) and biological value of meat were all decreased in the meat from infected poultry. So we suggested that gut health is important to the health status of poultry and poultry product quality. Experimental work has also shown that water is a potent vector for spreading a variety of bacterial species, including *Salmonella*, *Campylobacter* and *Pasteurella* – as well as *E. coli* – from bird to bird in a house within a short space of time.

POSTER *44*
Changes in morphological properties of the small intestine of broiler chickens fed on diets containing higher levels of soluble, non-starch polysaccharides and enzymes

F. Shariatmadari,[1] S.D. Sharifi,[1] M. Teshfam[2] and A. Yaghobfar[3]

[1] *Faculty of Agriculture, Tarbiat Modarres University, Tehran, Iran;* [2] *Faculty of Veterinary Medicine, University of Tehran, Tehran, Iran;* [3] *Animal Science Research Institute, Karaj, Iran*

This experiment was carried out to investigate the effect of including hull-less barley (0, 10, 20 and 30%) and enzymes (0, 300 and 600g/ton) on broiler chicken performance. Nine hundred and sixty broiler chicks in a 4×3 factorial arrangement (each treatment consisted of four replicates with 20 birds in each replicate) were reared up to 49 days of age.

The results showed that an increase in concentration of soluble NSPs (barley) depressed the performances (body weight and feed conversion ratio) of broiler chickens ($p < 0.01$). Enzymes had beneficial effects only at lower barley inclusion level (10%) and not at the higher level (30%). The significant lower apparent digestibility of nutrients from diets containing higher barley inclusion levels is suggested to be the main factor for the depression effect on performances reported above.

Increasing NSP concentration increased the weight of small intestine ($p < 0.01$), decreased the percentage of tongue-shaped villi (but increased other types – leaf, finger, bridge convoluted) and reduced the heights and widths of villi significantly ($p < 0.01$). Enzyme supplementation of diets had little effect on the morphological properties of the small intestine. NSP concentration had no effect on *E. coli*, *Lactobacillus* or *Clostridium* count in caeca up to 21 days of age, but the *Lactobacillus* count increased while *E. coli* and *Clostridium* decreased significantly.

It is concluded that the most adverse effect seen by the presence of barley in the diet is due to morphological damages of the digestive tract and that enzymes could, to some extent, alter the negative effect of NSP presence at lower levels in later stages of growth.

POSTER *45*
Sequence analysis of the toll-like receptor 7 from gallinaceous and non-gallinaceous species

D.R. Kapczynski

Southeast Poultry Research Laboratory, USDA-ARS, 934 College Station Road, Athens, Georgia 30605, USA

The immune system can be divided into two functional components – the innate and adaptive – which differ in their mechanism of pathogen recognition. The innate immune response is responsible for detecting invading microorganisms during the initial stages of infection, which is a crucial determinant of disease resistance or susceptibility. The innate system uses germ-line encoded receptors, termed pattern-recognizing receptors (PRR), to recognize highly conserved, pathogen-associated molecular patterns (PAMP) from infectious agents. The Toll-like receptor (TLR) family members are membrane-bound PRR responsible for recognizing the presence of invading microorganisms through PAMP. Recently, several chicken TLR have been identified, including TLR7, which is reported to be involved in recognition of ribonucleic acid components characteristic of viral genomes (e.g. ssRNA).

In this study, we amplified a 416 bp product of TLR7 using RT-PCR with mRNA derived from 13 different avian species. The nucleotide and predicted amino acid sequences were compared to previously reported TLR 7 sequences from different sources. Preliminary results indicate 99% similarity between chicken and turkey species, and 64 or 70% similarity to human and murine species, respectively. The phylogenetic relationship of various TLR7 from avian and mammalian origin will be discussed.

POSTER *46*
Pathogenic characterization of infectious bursal disease virus isolated from broiler chickens in Bangladesh

M.A. Islam,[1] M.M. Khatun,[1] M.M. Rahman,[1] M.R. Islam[2] and M.T. Hossain[1]

[1]*Department of Microbiology and Hygiene, Bangladesh Agricultural University, Mymensingh-2202, Bangladesh;* [2]*Department of Pathology, Bangladesh Agricultural University, Mymensingh-2202, Bangladesh*

Infectious bursal disease (IBD) is an acute, contagious, viral disease of young chickens (3–6 weeks of age) characterised by diarrhoea, dehydration, voiding of blood and straining during defaecation and inflammation, followed by atrophy of the bursa of Fabricius. IBD is caused by a double-stranded RNA virus belonging to the family *Birnaviridae*. IBD is endemic in Bangladesh and causes serious economic losses in the poultry industry due to high morbidity and mortality.

Therefore, an attempt was undertaken to establish pathogenic characterization of infectious bursal disease virus (IBDV) from a field outbreak of IBD. Bursae of Fabricius (BF) were collected from broiler chickens during a field outbreak of IBD at the Bangladesh Agricultural University (BAU) poultry farm. Isolation of IBDV from bursal tissues was performed in chickens (5 weeks of age) by inoculation of 10% bursal homogenate via the intranasal, intraocular and intracloacal routes. Agar gel immunodiffusion (AGID) test was performed to detect viral antigen in the BF. Five IBDV isolates were obtained by chicken inoculation.

Clinical signs manifested by chickens included reduced feed and water intake, marked depression and diarrhoea. One hundred per cent morbidity and 10% mortality were recorded. Ulcerative enteritis, oedematous swelling of the BF with yellowish and gelatinous transudate covering the serosal surface were observed at day 3 post-hatch. Histopathological lesions of BF consisted of extensive lymphoid necrosis and complete lymphoid depletion. It may be concluded from the results that the field outbreak of IBD at the BAU poultry farm was caused by a very virulent pathotype of IBDV.

POSTER 47
Gut health and nutrition

T. Acamovic and J.D. Brooker

Avian Science Research Centre, Scottish Agricultural College, West Mains Road, Edinburgh, EH9 3JG, UK; (address for communication) SAC, Ayr Campus, Ayr, KA6 5HW, UK; e-mail: thomas.acamovic@sac.ac.uk

The provision of nutrients in a suitable form, balance and availability is essential for the normal health of animals. Similarly, since the gastrointestinal tract (GIT) and organs constitute about 25–30% of the body weight of poultry, the requirement for nutrients by these organs is variable, but substantial. The requirement of the GIT for energy amounts to 20–25% of the bird's total energy. If there is perturbation in the health of the gut and/or the supply of nutrients, those nutrient requirements and utilization will change substantially, thereby altering flock productivity and the financial viability of poultry production.

There are, therefore, inextricable interactions between the GIT, the nutrients present therein and anti-nutrients or toxins, as well as the microflora that will influence GIT health and, ultimately, bird growth. The physical nature of nutrients will also have an impact. The health of poultry flocks, especially in Europe, is now becoming influenced much more by management as well as variation in the quantity and quality of nutrients from different ingredients. Thus, the use of ingredients that, in the past, had little effect on nutrient utilization by poultry, may now have greater effects because of the absence of in-feed antibiotics.

Recently, there has been an increased focus on the interactions between ingredients, nutrients and GIT health, as well as nutrient utilization, microfloral changes and the factors that influence these. Aspects of these will be reviewed and presented in this poster.

INDEX